Heat, Energy, or Work Equivalents

(ft)(lb$_f$)	kWh	hp-hr	Btu	calorie*	Joule
0.7376	2.773×10^{-7}	3.725×10^{-7}	9.484×10^{-4}	0.2390	1
7.233	2.724×10^{-6}	3.653×10^{-6}	9.296×10^{-3}	2.3438	9.80665
1	3.766×10^{-7}	5.0505×10^{-7}	1.285×10^{-3}	0.3241	1.356
2.655×10^{6}	1	1.341	3.4128×10^{3}	8.6057×10^{5}	3.6×10^{6}
1.98×10^{6}	0.7455	1	2.545×10^{3}	6.4162×10^{5}	2.6845×10^{6}
74.73	2.815×10^{-5}	3.774×10^{-5}	9.604×10^{-2}	24.218	1.0133×10^{2}
3.086×10^{3}	1.162×10^{-3}	1.558×10^{-3}	3.9657	1×10^{3}	4.184×10^{3}
7.7816×10^{2}	2.930×10^{-4}	3.930×10^{-4}	1	2.52×10^{2}	1.055×10^{3}
3.086	1.162×10^{-6}	1.558×10^{-6}	3.97×10^{-3}	1	4.184

*The thermochemical calorie = 4.184 J; the IT calorie = 4.1867 J (see Sec. 4.1).

Pressure Equivalents

mm Hg	in. Hg	bar	atm	kPa
1	3.937×10^{-2}	1.333×10^{-3}	1.316×10^{-3}	0.1333
25.40	1	3.387×10^{1}	3.342×10^{-2}	3.387
750.06	29.53	1	0.9869	100.0
760.0	29.92	1.013	1	101.3
75.02	0.2954	1.000×10^{-2}	9.872×10^{-3}	1

Ideal Gas Constant R

1.987 cal/(g mol)(K)

1.987 Btu/(lb mol)(°R)

10.73 (psia)(ft^3)/(lb mol)(°R)

8.314 (kPa)(m^3)/(kg mol)(K) = 8.314 J/(g mol)(K)

82.06 (cm^3)(atm)/(g mol)(K)

0.08206 (l)(atm)/(g mol)(K)

21.9 (in Hg)(ft^3)/(lb mol)(°R)

0.7302 (ft^3)(atm)/(lb mol)(°R)

BASIC PRINCIPLES
AND CALCULATIONS
IN CHEMICAL ENGINEERING

PRENTICE-HALL INTERNATIONAL SERIES
IN THE PHYSICAL AND CHEMICAL ENGINEERING SCIENCES

NEAL R. AMUNDSON, EDITOR, *University of Houston*

ADVISORY EDITORS

ANDREAS ACRIVOS, *Stanford University*
JOHN DAHLER, *University of Minnesota*
THOMAS J. HANRATTY, *University of Illinois*
JOHN M. PRAUSNITZ, *University of California*
L. E. SCRIVEN, *University of Minnesota*

BASIC PRINCIPLES AND CALCULATIONS IN CHEMICAL ENGINEERING

fourth edition

DAVID M. HIMMELBLAU

University of Texas

PRENTICE-HALL, INC., Englewood Cliffs, New Jersey 07632

Library of Congress Cataloging in Publication Data

Himmelblau, David Mautner,
    Basic principles and calculations in chemical
engineering.

    (Prentice-Hall international series in the
physical and engineering sciences)
    Bibliography: p.
    Includes index.
    1. Chemical engineering—Tables. I. Title.
II. Series.
TP151.H5 1982   660.2   81-19175
ISBN 0-13-066498-7   AACR2

To Betty (again)

Editorial/production supervision and interior design by *Mary Carnis*
Manufacturing buyer: *Joyce Levatino*

Printed in the United States of America

10 9 8 7 6 5 4 3 2 1

ISBN 0-13-066498-7

Prentice-Hall International, Inc., *London*
Prentice-Hall of Australia Pty. Limited, *Sydney*
Prentice-Hall of Canada, Ltd., *Toronto*
Prentice-Hall of India Private Limited, *New Delhi*
Prentice-Hall of Japan, Inc., *Tokyo*
Prentice-Hall of Southeast Asia Pte. Ltd., *Singapore*
Whitehall Books Limited, *Wellington, New Zealand*

CONTENTS

2   Material Balances   94

3   Gases, Vapors, Liquids, and Solids   201

4   Energy Balances   313

5   Applications of Combined Material and Energy Balances   450

PREFACE

This text is intended to serve as an introduction to the principles and techniques used in the field of chemical, petroleum, and environmental engineering. The central themes of the book involve (1) learning how to formulate and solve (a) material balances, (b) energy balances, and (c) both simultaneously; (2) developing problem solving skills; and (3) becoming familiar with the use of units, physical properties, and the behavior of gases and liquids. At the front of each chapter is an information flow diagram that shows how the topics discussed in the chapter relate to the objective of being able to solve successfully problems involving material and energy balances. For the fourth edition I have added a number of new features described below that make both teaching and self-study easier.

A good introductory book to chemical engineering principles and calculations should (1) explain the fundamental concepts in not too stilted language together with generous use of appropriate equations and diagrams; (2) provide sufficient examples with detailed solutions to clearly illustrate (1); (3) present ideas in small packages that are easily identified as part of a larger framework; (4) include tests and answers that enable the reader to evaluate his or her accomplishments; and (5) provide the instructor with a wide selection of problems and questions to evaluate student competence. All of these features have been built into the fourth edition.

Piaget has argued that human intelligence proceeds in stages from the concrete to the abstract and that one of the big problems in teaching is that the teachers are formal reasoners (using abstraction) while many students are still concrete thinkers or at best in transition to formal operational thinking. I believe that this is true. Consequently, most topics are initiated with simple illustrations that illustrate the basic ideas. In this text the topics are presented in order of easy assimilation rather than in a strictly logical order. The organization is such that easy material is alternated with difficult material in order to give a "breather" after passing over each hump. For example, discussion of unsteady-state (lumped) balances has been deferred until the final chapter because experience has shown that most students lack the mathematical and engineering maturity to absorb these problems simultaneously with the steady-state balances.

A principle of educational psychology is to reinforce the learning experience by providing detailed guided practice following each new principle. We all have found from experience that there is a vast difference between having a student understand a principle and in establishing his or her ability to apply it. By the use of numerous detailed examples following each brief section of text, it is hoped that straightforward, orderly methods of procedure can be instilled along with some insight into the principles involved. Furthermore, the wide variety of problems at the end of each chapter, about one-fourth of which are accompanied by answers, offer practice in the application of the principles explained in the chapter.

The text does not describe the chemical process industries per se, although certain aspects of these are introduced by the use of problems. Also, the topics of vapor–liquid equilibria, chemical equilibria, and kinetics have been omitted because of a lack of space and because an appreciation of these topics requires more background information and maturity on the part of the student than has been assumed for the topics that are discussed. On the other hand, the topic of real gases has been included because far too much attention is devoted in scientific courses to the ideal gases, leaving a very misleading impression. Those who wish to skip this particular section can do so without interfering with any of the subsequent material. More topics have been included in the text than can be covered in one semester so that an instructor has some choice as to pace of instruction and topics to include. Many references are given at the end of each chapter in the hope that those who need supplementary information will pursue their interest further. With the assistance of a series of study guides and notes designed by the instructor, the text will fit in very neatly with self-paced instruction.

Let me now mention some of the new features of the fourth edition, features that were not present in earlier editions.

At the begining of each section is a list of objectives to be achieved by the reader stated in such a way that attainment can be readily measured. Whether the objectives should go at the beginning or end of the section is a matter of controversy; that the text should have objectives is not. But we often present our objectives by such broad, fuzzy statements that neither the student nor the teacher can ascertain whether or not students have achieved them. (Unfortunately, this situation does not seem to inhibit the testing of students.) Each set of objectives is quite concrete and has a corresponding set of self-assessment questions and problems at the end of the respective section.

Self-assessment tests have been included to provide readers with questions and answers that assist in them in appraising and developing their knowledge about a particular topic. Self-assessment is intended to be an educational experience for a student. The availability of answers to the self-assessment questions together with supplementary reading citations for further study is an inherent characteristic of self-assessment. To help the reader think about the concepts and decide whether to study further is one reason for having appraisal questions. Such an approach can be contrasted to a test in which the items are designed to discriminate between test takers and to allow the tester to draw conclusions about the competence of participants. Thus self-assessment is not focused on satisfying others about a student's knowledge.

How should you as a reader use the self-assessment problems? Some readers will start with the questions. Others will read the problems. Still others will refer to the answers first. These approaches and others are all acceptable if at the end of the procedure you can say: "Yes, this has been a worthwhile experience" or "I have learned something." The self-assessment materials are not a timed exercise so work at your pace. I suggest that you first go through the questions and problems, and write down the responses you think are most appropriate for the questions that are easy for you. Compare your responses with the answers in Appendix A. In those cases where you differ, read the section again, examine the examples, try some of the homework problems at the end of the chapter, or look up the references if necessary. In those cases in which you agree, but you felt uncomfortable with the subject matter, and the subject is significant to you, be sure to solve some of the problems at the end of the chapter for which answers are provided. After doing the easy problems, tackle the more difficult ones.

The solution of word problems has long been a discouraging task both for students and teachers. Why do students have so much trouble with word problems? Some educators say that students do not read a problem and assimilate the information therein adequately. But it seems to me that the main difficulty is that once the problem is assimilated, a student who has not developed a suitable strategy for problem solving does not know how to proceed. Structural difficulties in the problems, complexity, and so on make a new problem seem quite different than those encountered previously. Students do not know how to begin, where to focus their attention, or how to monitor their own progress toward a solution.

In this edition special attention has been devoted to presenting a sound strategy for solving material balance and energy balance problems, one that can be used again and again as a framework for solving word problems. All the examples showing how to solve material and energy balances have been reformulated according to this strategy (see Table 2.1). In teaching I ask students to memorize the strategy and apply it in all their homework problems and exams. I discourage the use of self-devised heuristic algorithms, or "cookbook" methods pointing out that they may be successful for one class of problems but fail quite dismally for others. By this means a student is guided into forming generalized patterns of attack in problem solving that can be used successfully in connection with unfamiliar types of problems. The text is designed to acquaint the student with a sufficient number of fundamental concepts so that he can (1) continue with his training, and (2) start finding solutions to new types of problems on his own. It offers practice in finding out what the problem is, defining it, collecting data, analyzing and breaking down information, assembling the basic ideas into patterns, and, in effect, doing everything but testing the solution experimentally.

Two perplexing problems remain for an author in preparing a new edition. One concerns the extent to which SI units should be used. It is evident that the transition to SI is well under way in a large number of chemical engineering departments in the United States, although some resistance to the system still exists. I believe that SI is an important system of measurment that chemical engineers must be able to deal with, but feel that chemical engineering students must be familiar with a variety of systems for some years to come. Consequently, it is premature to convert an undergraduate text entirely to the SI system of units even though most journals have done so. As a compromise, a little more than one-half of the text, examples, and problems and most of the tables employ SI units. For convenience, some of the crucial tables, such as the steam tables, are presented in both American engineering and SI units.

A second perplexing problem is to what extent and in what manner should problems involving the preparation of or use of computer codes be introduced into the text. If computer techniques are to be integrated into the classroom successfully, it is wise to start early in the game, but on the other hand the preparation (as opposed to use) of computer programs assumes that the student will have had some instruction in programming prior to or concurrently with, the use of this text. The selection of appropriate problems and the illustration of good computer habits, pointing out instances in which computer solutions are not appropriate as well as instances when they are, is important. Again, a compromise has been reached. At the end of each chapter you will find a few problems calling for the preparation of a complete computer code. Other problems that are more appropriately solved with the aid of a computer program, particularly a "canned" library code, than with a calculator, are designated by an asterisk (*) after the problem number. I have found that the use of canned computer programs reduces the tedium of many of the trial-and-error solution techniques and regenerates interest in the learning process.

I would indeed be ungrateful if I did not express many thanks to the hundreds of students who have participated in the preparation of both the original and the revised editions of this book. Drs. Robert S. Schechter and James B. Riggs have been kind enough to contribute a number of new problems, and far too many instructors using the text have contributed their corrections and suggestions for me to list them all by name. However, I do wish to express my appreciation for their kind assistance.

David M. Himmelblau
Austin, Texas

1

INTRODUCTION TO ENGINEERING CALCULATIONS

Environmental and economic concerns are more widespread today than ever. The tendency to dichotomize (split into two groups) environment and production, benefit and cost has an adverse effect on progress. In reality a wide spectrum of options exists in decision making, and for you to make wise decisions requires some preparation. For you to learn how to appreciate and treat the problems that will arise in our modern technology, especially in the technology of the future, it is necessary to learn certain basic principles and practice their application. This text describes the principles of making material and energy balances and illustrates their application in a wide variety of ways.

We begin in this chapter with a review of certain background information. You have already encountered most of the concepts to be presented in basic chemistry and physics courses. Why, then, the need for a review? First, from experience we have found it necessary to restate these familiar basic concepts in a somewhat more general and clearer fashion; second, you will need practice to develop your ability to analyze and work engineering problems. To read and understand the principles discussed in this chapter is relatively easy; to apply them to different unfamiliar situations is not. An engineer becomes competent in his or her profession by mastering the techniques developed by one's predecessors—thereafter comes the time to pioneer new ones.

The chapter begins with a discussion of units, dimensions, and conversion factors, and then goes on to review some terms you should already be acquainted with, such as:

(a) Mole and mole fraction

(b) Density and specific gravity

(c) Measures of concentration

(d) Temperature

(e) Pressure

It then provides some clues on "how to solve problems," which should be of material aid in all the remaining portions of your work. Finally, the principles of stoichiometry are reviewed, and the technique of handling incomplete reactions is illustrated. Figure 1.0 shows the relation of the topics to be discussed to each other and to the

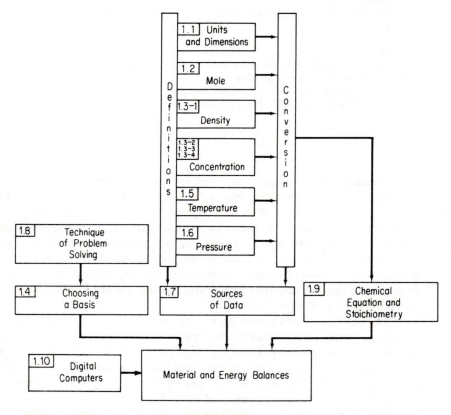

Fig. 1.0 Hierarchy of topics to be studied in this chapter (section numbers are in the upper left-hand corner of the boxes).

ultimate goal of being able to solve problems involving both material and energy balances.

Section 1.1   Units and Dimensions

Your objectives in studying this section are to be able to:

1. Add, subtract, multiply, and divide units associated with numbers.
2. Convert one set of units in a function or equation into another equivalent set for mass, length, area, volume, time, energy, and force.
3. Specify the basic and derived units in the SI and American engineering systems for mass, length, volume, density, and time, and their equivalences.
4. Explain the difference between weight and mass.
5. Define and know how to use the gravitational conversion factor g_c.
6. Apply the concepts of dimensional consistency to determine the units of any term in a function.

At some time in every student's life comes the exasperating sensation of frustration in problem solving. Somehow, the answers or the calculations do not come out as expected. Often this outcome arises because of inexperience in the handling of units. The use of units or dimensions along with the numbers in your calculations requires more attention than you probably have been giving to your computations in the past, but it will help avoid such annoying experiences. The proper use of dimensions in problem solving is not only sound from a logical viewpoint—it will also be helpful in guiding you along an appropriate path of analysis from what is at hand through what has to be done to the final solution.

Dimensions are our basic concepts of measurement such as *length*, *time*, *mass*, *temperature*, and so on; *units* are the means of expressing the dimensions, such as *feet* or *centimeters* for length, or *hours* or *seconds* for time. Units are associated with some quantities you may have previously considered to be dimensionless. A good example is *molecular weight*, which is really the mass of one substance per mole of that substance. This method of attaching units to all numbers which are not fundamentally dimensionless has the following very practical benefits:

(a) It diminishes the possibility of inadvertent inversion of any portion of the calculation.

(b) It reduces the calculation in many cases to simple ratios, which can be easily manipulated on a hand-held calculator.

(c) It reduces the intermediate calculations and eliminates considerable time in problem solving.

(d) It enables you to approach the problem logically rather than by remembering a formula and plugging numbers into the formula.

(e) It demonstrates the physical meaning of the numbers you use.

Every freshman knows that what you get from adding apples to oranges is fruit salad! The rule for handling units is essentially quite simple: treat the units as you would algebraic symbols. You can add, subtract, or equate like units such as pounds, watts, and so on, but not unlike units. Thus the operation

$$5 \text{ kilograms} + 3 \text{ calories}$$

is meaningless because the dimensions of the two terms are different. The numerical operation

$$10 \text{ pounds} + 5 \text{ grams}$$

can be performed (because the dimensions are the same, mass) *only* after the units are transformed to be the same, either pounds, or grams, or ounces, and so on. In multiplication and division, you can multiply or divide different units, such as (10 centimeters ÷ 4 seconds) = 2.5 centimeters/second, but you cannot cancel them. The units contain a significant amount of information content that cannot be ignored. They also serve as guides in efficient problem solving, as you will see shortly.

EXAMPLE 1.1 Dimensions and Units

Add the following:

(a)   1 foot + 3 seconds

(b)   1 horsepower + 300 watts

Solution

The operation indicated by

$$1 \text{ ft} + 3 \text{ sec}$$

has no meaning since the dimensions of the two terms are not the same. One foot has the dimensions of length, whereas 3 sec has the dimensions of time. In the case of

$$1 \text{ hp} + 300 \text{ watts}$$

the dimensions are the same (energy per unit time) but the units are different. You must transform the two quantities into like units, such as horsepower, watts, or something else, before the addition can be carried out. Since 1 hp = 746 watts,

$$746 \text{ watts} + 300 \text{ watts} = 1046 \text{ watts}$$

EXAMPLE 1.2 Conversion of Units

If a plane travels at twice the speed of sound (assume that the speed of sound is 1100 ft/sec), how fast is it going in miles per hour?

Solution

$$\frac{2 \mid 1100 \text{ ft} \mid 1 \text{ mi} \mid 60 \text{ sec} \mid 60 \text{ min}}{\mid \text{sec} \mid 5280 \text{ ft} \mid 1 \text{ min} \mid 1 \text{ hr}} = 1500 \frac{\text{mi}}{\text{hr}}$$

or

$$\frac{2 \mid 1100 \text{ ft} \mid 60 \frac{\text{mi}}{\text{hr}}}{\mid \text{sec} \mid 88 \frac{\text{ft}}{\text{sec}}} = 1500 \frac{\text{mi}}{\text{hr}}$$

You will note in Example 1.2 the use of what is called the *dimensional equation*. It contains both units and numbers. The initial speed, 2200 ft/sec, is multiplied by a number of ratios (termed *conversion factors*) of equivalent values of combinations of time, distance, and so on, to arrive at the final desired answer. The ratios used are simple well-known values and thus the conversion itself should present no great problem. Of course, it is possible to look up conversion ratios, which will enable the length of the calculation to be reduced; for instance, in Example 1.2 we could have used the conversion factor of 60 mi/hr equals 88 ft/sec. However, it usually takes less time to use values you know than to look up shortcut conversion factors in a handbook. Common conversion ratios are listed on the inside cover.

The dimensional equation has vertical lines set up to separate each ratio, and these lines retain the same meaning as an × or multiplication sign placed between each ratio. The dimensional equation will be retained in this form throughout most of this text to enable you to keep clearly in mind the significance of units in problem solving. It is recommended that units always be written down next to the associated numerical value (unless the calculation is very simple) until you become quite familiar with the use of units and dimensions and can carry them in your head.

At any point in the dimensional equation you can determine the consolidated net units and see what conversions are still required. This may be carried out formally, as shown below by drawing slanted lines below the dimensional equation and writing the consolidated units on these lines, or it may be done by eye, mentally canceling and accumulating the units.

$$\frac{2 \times 1100 \text{ ft} \mid 1 \text{ mi} \mid 60 \text{ sec} \mid 60 \text{ min}}{\text{sec} \mid 5280 \text{ ft} \mid 1 \text{ min} \mid 1 \text{ hr}}$$

$$\frac{\text{ft}}{\text{sec}} \qquad \frac{\text{mi}}{\text{sec}} \qquad \frac{\text{mi}}{\text{min}}$$

EXAMPLE 1.3 Use of Units

Change 400 in.³/day to cm³/min.

Solution

$$\frac{400 \text{ in.}^3}{\text{day}} \left| \frac{(2.54 \text{ cm})^3}{(1 \text{ in.})} \right| \frac{1 \text{ day}}{24 \text{ hr}} \left| \frac{1 \text{ hr}}{60 \text{ min}} \right. = 4.56 \frac{\text{cm}^3}{\text{min}}$$

In this example note that not only are the numbers raised to a power, but the units also are raised to the same power.

> There shall be one measure of wine throughout our kingdom, and one measure of ale, and one measure of grain . . . and one breadth of cloth. . . . And of weights it shall be as of measures.

So reads the standard measures clause of the *Magna Carta* (June 1215). The standards mentioned were not substantially revised until the nineteenth century. When the American colonies separated from England, they retained, among other things, the weights and measures then in use. It is probable that at that time these were the most firmly established and widely used weights and measures in the world.

No such uniformity of weights and measures existed on the European continent. Weights and measures differed not only from country to country but even from town to town and from one trade to another. This lack of uniformity led the National Assembly of France during the French Revolution to enact a decree (May 8, 1790) that called upon the French Academy of Sciences to act in concert with the Royal Society of London to "deduce an invariable standard for all of the measures and all weights." Having already an adequate system of weights and measures, the English did not participate in the French undertaking. The result of the French endeavor has evolved into what is known as the metric system.

Certainly in the American–British system there is ample room for confusion. The old question: "Which is heavier, a pound of feathers or a pound of gold?" indeed has a catch, one that is not obvious. The pound of feathers is heavier, because gold is measured in troy units; the appropriate pound comprises only 12 rather than 16 oz. But an ounce of feathers is lighter than an ounce of gold, because the troy ounce (or apothecary ounce) is heavier than the avoirdupois ounce! This example illustrates one of the problems of "customary" units: the use of one name for several different units.

The metric system became preferred by scientists of the nineteenth century, partly because it was intended to be an international system of measurement, partly because the units of measurement were theoretically supposed to be independently reproducible, and partly because of the simplicity of its decimal nature. These scientists proceeded to derive new units for newly observed physical phenomena, basing the new units on elementary laws of physics and relating them to the units of mass and length of the metric system. The Americans and the British adapted their system of measurements to the requirements of new technology as time went on for both business and commerce as well as scientific uses despite the fact that the other countries, one after another, turned toward the metric system.

Problems in the specification of the units for electricity and magnetism led to numerous international conferences to rectify inconsistencies and culminated in 1960 in the eleventh General Conference on Weights and Measures adopting the SI (Système International) system of units. As of this date the United States is the last large country not employing or engaged in transforming to some form of the SI units.

Tables 1.1 and 1.2 show the most common systems of units used by engineers in the last few decades. Note that the SI, the cgs, the fps (English absolute), and the British engineering systems all have three basically defined units and that the fourth unit is derived from these three defined units.

Many of the derived units in the SI system are given special names (see Table 1.2) honoring physicists and have corresponding symbols. The unit of force, for example, has the symbols $kg \cdot m \cdot s^{-2}$; for convenience, this combination has been given the name *newton* and the symbol N. Similarly, the unit of energy is the newton $\cdot$ meter,

$$N \cdot m \Leftrightarrow m^2 \cdot kg \cdot s^{-2}$$

with the short name *joule*, and symbol J. The unit of power is the *watt*, defined as one joule per second.

Only the American engineering system has four basically defined units. Consequently, in the American engineering system you have to use a conversion factor, g_c, a constant whose numerical value is not unity, to make the units come out properly. We can use Newton's law to see what the situation is with regard to conversion of units:

$$F = Cma \tag{1.1}$$

where[1] F = force
    m = mass
    a = acceleration
    C = a constant whose numerical value and units depend on those selected for F, m, and a

In the cgs system the unit of force is defined as the dyne; hence if C is selected to be $C = 1$ dyne/(g)(cm)/sec², then when 1 g is accelerated at 1 cm/sec²

$$F = \frac{1 \text{ dyne}}{\frac{(g)(cm)}{sec^2}} \left| \frac{1 \text{ g}}{} \right| \frac{1 \text{ cm}}{sec^2} = 1 \text{ dyne}$$

---

[1]A list of the nomenclature is included at the end of the book.

TABLE 1.1 COMMON SYSTEMS OF UNITS

	Length	Time	Mass	Force	Energy	Temperature	Remarks
Absolute (Dynamic) Systems							
Cgs	centi-meter	second	gram	dyne*	erg, joule, or calorie	°K, °C	Formerly common scientific
Fps (ft-lb-sec or English absolute)	foot	second	pound	poundal*	ft poundal	°R, °F	
SI	meter	second	kilogram	newton*	joule*	K, °C	Internationally adopted units for ordinary and scientific use
Gravitational Systems							
British engineering	foot	second	slug*	pound weight	Btu (ft)(lb) (ft)(lb$_f$)	°R, °F	
American engineering	foot	second, hour	pound mass (lb$_m$)	pound force (lb$_f$)	Btu or (hp)(hr)	°R, °F	Used by chemical and petroleum engineers in the United States

*Unit derived from basic units; all energy units are derived.

TABLE 1.2 SI UNITS ENCOUNTERED IN THIS TEXT

Basic SI Units

Physical Quantity	Name of Unit	Symbol for Unit*
Length	metre, meter	m
Mass	kilogramme, kilogram	kg
Time	second	s
Thermodynamic temperature	degree kelvin	K
Amount of substance	mole	mol

Derived SI Units

Physical Quantity	Name of Unit	Symbol for Unit*	Definition of Unit
Energy	joule	J	$kg \cdot m^2 \cdot s^{-2}$
Force	newton	N	$kg \cdot m \cdot s^{-2} \backsimeq J \cdot m^{-1}$
Power	watt	W	$kg \cdot m^2 \cdot s^{-3} \backsimeq J \cdot s^{-1}$
Frequency	hertz	Hz	cycle/s
Area	square meter		m^2
Volume	cubic meter		m^3
Density	kilogram per cubic meter		$kg \cdot m^{-3}$
Velocity	meter per second		$m \cdot s^{-1}$
Angular velocity	radian per second		$rad \cdot s^{-1}$
Acceleration	meter per second squared		$m \cdot s^{-2}$
Pressure	newton per square meter, pascal		$N \cdot m^{-2}$, Pa
Specific heat	joule per (kilogram·kelvin)		$J \cdot kg^{-1} \cdot K^{-1}$

Alternate Units

Physical Quantity	Allowable Unit	Symbol for Unit*
Time	minute	min
	hour	h
	day	d
	year	a
Temperature	degree Celsius	°C
Volume	litre, liter (dm³)	L
Mass	tonne, ton (Mg)	t
	gram	g
Pressure	bar (10⁵ Pa)	bar

*Symbols for units do not take a plural form, but plural forms are used for the unabbreviated names.

Similarly, in the SI system in which the unit of force is defined to be the newton (N), if $C = 1$ N/(kg)(m)/s², then when 1 kg is accelerated at 1 m/s²,

$$F = \frac{1 \text{ N}}{\frac{(\text{kg})(\text{m})}{\text{s}^2}} \left| \frac{1 \text{ kg}}{} \right| \frac{1 \text{ m}}{\text{s}^2} = 1 \text{ N}$$

However, in the American engineering system we ask that the numerical value of the force and the mass be essentially the same at the earth's surface. Hence, if a mass of 1 lb$_m$ is accelerated at g ft/sec², where g is the acceleration of gravity (about 32.2 ft/sec² depending on the location of the mass), we can make the force be 1 lb$_f$ by choosing the proper numerical value and units for C:

$$F = (C)\frac{1 \text{ lb}_m}{} \left| \frac{g \text{ ft}}{\text{sec}^2} = 1 \text{ lb}_f \right. \tag{1.2}$$

Observe that for Eq. (1.2) to hold, the units of C have to be

$$C \infty \frac{\text{lb}_f}{\text{lb}_m \left(\dfrac{\text{ft}}{\text{sec}^2} \right)}$$

A numerical value of 1/32.174 has been chosen for the constant because 32.174 is the numerical value of the average acceleration of gravity (g) at sea level at 45° latitude when g is expressed in ft/sec². The acceleration of gravity, you may recall, varies a few tenths of 1 % from place to place on the surface of the earth and changes considerably as you rise up from the surface as in a rocket. With this selection of units and with the number 32.174 employed in the denominator of the conversion factor, the inverse of C is given the special symbol g_c:

$$g_c = 32.174 \frac{(\text{ft})(\text{lb}_m)}{(\text{sec}^2)(\text{lb}_f)} \tag{1.3}$$

Division by g_c achieves exactly the same result as multiplication by C in Newton's law. You can see, therefore, that in the American engineering system we have the convenience that the numerical value of a pound mass is also that of a pound force if the numerical value of the ratio g/g_c is equal to 1, as it is approximately in most cases.

Similarly, a 1-lb mass also usually weighs 1 lb (weight is the force required to support the mass at rest). However, you should be aware that these two quantities g and g_c are *not* the same. Note also that the pound (mass) and the pound (force) are *not* the same units in the American engineering system even though we speak of *pounds* to express force, weight, or mass. Furthermore, if a satellite with a mass of 1 lb (mass) "weighing" 1 lb (force) on the earth's surface is shot up to a height of 50 mi, it will no longer "weigh" 1 lb (force), although its mass will still be 1 lb (mass).

Briefly, the source of the confusion over the word *weight* is as follows. In careful technical communication, weight means, strictly, *force of gravity*. Nearly all teachers and writers in physics, engineering, and related fields in technical communication are careful to use the term with only this meaning. On the other hand, in ordinary language most people, including scientists and engineers, omit the designation of "force" or "mass" associated with the pound or kilogram but pick up the meaning from the context of the statement. No one gets confused by the fact that a man is 6 feet tall but has only two feet. Similarly, you should interpret the statement that a bottle "weighs" 5 kg as meaning the bottle has a mass of 5 kg and is attracted to the earth's surface with a force equal to

$$(5 \text{ kg})(9.80 \text{ m/s}^2) = 49 \text{ (kg)(m)(s}^{-2}) \qquad \text{if } g = 9.80 \text{ m/s}^2$$

In a spring scale at equilibrium the upward force of the spring on a mass balances the downward force of gravity so that the "weight" of a mass can be easily measured.

Perhaps one of these days people will speak of "massing" instead of "weighing" and "mass-minding" instead of "weight-watching," but probably not soon.

Additional details concerning units and dimensions can be found in the articles by Silberg and McKetta,[2] and by Whitney,[3] in various books,[4-6] and in the references at the end of the chapter. For systems used in the USSR, consult *Meas. Tech. (USSR) (English Trans.)*, no. 10 (April 1964).

EXAMPLE 1.4 Use of g_c

One hundred pounds of water is flowing through a pipe at the rate of 10.0 ft/sec. What is the kinetic energy of this water in (ft)(lb$_f$)?

Solution

$$\text{kinetic energy} = K = \tfrac{1}{2}mv^2$$

Assume that the 100 lb of water means the mass of the water. (This would be numerically identical to the weight of the water if $g = g_c$.)

$$K = \frac{1}{2} \left| \frac{100 \text{ lb}_m}{} \right| \left(\frac{10.0 \text{ ft}}{\text{sec}} \right)^2 \left| \frac{}{32.174 \frac{\text{(ft)(lb}_m)}{\text{(sec}^2)\text{(lb}_f)}} \right. = 155 \text{ (ft)(lb}_f)$$

---

[2] I. H. Silberg and J. J. McKetta, Jr., *Petrol. Refiner*, v. 32, pp. 179–183 (April 1953); 147–150 (May 1953).

[3] H. Whitney, "The Mathematics of Physical Quantities," *Amer. Math. Monthly*, v. 75, pp. 115, 227 (1968).

[4] D. C. Ipsen, *Units, Dimensions, and Dimensionless Numbers*, McGraw-Hill, New York, 1960.

[5] S. J. Kline, *Similitude and Approximation Theory*, McGraw-Hill, New York, 1965.

[6] E. F. O'Day, *Physical Quantities and Units (A Self-Instructional Programmed Manual)*, Prentice-Hall, Englewood Cliffs, N.J., 1967.

EXAMPLE 1.5 Use of g_c

What is the potential energy in $(ft)(lb_f)$ of a 100-lb drum hanging 10 ft above the surface of the earth with reference to the surface of the earth?

Solution

$$\text{potential energy} = P = mgh$$

Assume that the 100 lb means 100 lb mass; g = acceleration of gravity = 32.2 ft/sec².

$$P = \frac{100\ lb_m}{} \left| \frac{32.2\ ft}{sec^2} \right| \frac{10\ ft}{} \left| \frac{}{32.174\ \frac{(ft)(lb_m)}{(sec^2)(lb_f)}} \right| = 1000\ (ft)(lb_f)$$

Notice that in the ratio of g/g_c, or 32.2 ft/sec² divided by 32.174 (ft/sec²)(lb_m/lb_f), the numerical values are almost equal. A good many people would solve the problem by saying that 100 lb × 10 ft = 1000 (ft)(lb) without realizing that in effect they are canceling out the numbers in the g/g_c ratio.

EXAMPLE 1.6 Weight

What is the difference in the weight in newtons of a 100-kg rocket at height of 10 km above the surface of the earth, where g = 9.76 m/s², as opposed to its weight on the surface of the earth, where g = 9.80 m/s²?

Solution

The weight in newtons can be computed in each case from Eq. (1.1) with $a = g$ if we ignore the tiny effect of centripetal acceleration resulting from the rotation of the earth (less than 0.3%):

$$\text{weight difference} = \frac{100\ kg}{} \left| \frac{(9.80 - 9.76)\ m}{s^2} \right| \frac{1\ N}{\frac{(kg)(m)}{s^2}} = 4.00\ N$$

Note that the concept of weight is not particularly useful in treating the dynamics of long-range ballistic missiles or of earth satellites because the earth is both round and rotating.

You should develop some facility in converting units from the SI system into the American engineering system and vice versa, since these are the two sets of units in this text. Certainly you are familiar with the common conversions in both the American engineering and the SI systems. If you have forgotten, Table 1.3 lists a short selection of essential conversion factors. Memorize them. Common abbreviations and symbols also appear in this table.

The distinction between uppercase and lowercase letters should be followed, even if the symbol appears in applications where the other lettering is in uppercase

TABLE 1.3 BASIC CONVERSION FACTORS*

Dimension	American Engineering	SI	Conversion: American Engineering to SI
Length	12 in. = 1 ft 3 ft = 1 yd 5280 ft = 1 mi	10 mm† = 1 cm† 100 cm† = 1 cm	1 in. = 2.54 cm 3.28 ft = 1 m
Volume	1 ft³ = 7.48 gal	1000 cm³† = 1 L†	35.31 ft³ = 1.00 m³
Density	1 ft³ H_2O = 62.4 lb$_m$	1 cm³† H_2O = 1 g 1 m³ H_2O = 1000 kg	—
Mass	1 ton$_m$ = 2000 lb$_m$	1000 g = 1 kg	1 lb = 0.454 kg
Time	1 min = 60 sec 1 hr = 60 min	1 min† = 60 s 1 h† = 60 min†	—

*Some conversion factors in this table are approximate but have sufficient precision for engineering calculations.

†An acceptable but not preferred unit in the SI system.

style. Unit abbreviations have the same form for both singular and plural, and they are not followed by a period (except in the case of inches).

Other useful conversion factors are discussed in subsequent sections of this book.

One of the best features of the SI system is that units and their multiples and submultiples are related by standard factors designated by the prefixes indicated in Table 1.4. Prefixes are not preferred for use in denominators (except for kg). Do not use double prefixes; that is, use nanometer, not millimicrometer. The strict use of these prefixes leads to some amusing combinations of noneuphonious sounds, such as nanonewton, nembujoule, and so forth. Also, some confusion is certain to arise because the prefix M can be confused with m as well as with M ∽ 1000 derived from the Roman numerical. When a compound unit is formed by multiplication of two or more other units, its symbol consists of the symbols for the separate units joined by a centered dot (e.g., N·m for newton meter). The dot may be omitted in the case of familiar units such as watthour (symbol Wh) if no confusion will result, or if the symbols are separated by exponents, as in N·m²kg⁻². Hyphens should not be used in symbols for compound units. Positive and negative exponents may be used with the symbols for units. If a compound unit is formed by division of one unit by another, its symbol consists of the symbols for the separate units either separated by a solidus or multiplied by using negative powers (e.g., m/s or m·s⁻¹ for meters per second). We do not use the center dot for multiplication in this text but use parentheses around the symbol instead, a procedure less subject to misinterpretation in handwritten material.

TABLE 1.4 SI PREFIXES

Factor	Prefix	Symbol	Factor	Prefix	Symbol
10^{18}	exa	E	10^{-1}	deci*	d
10^{15}	penta	P	10^{-2}	centi*	c
10^{12}	tera	T	10^{-3}	milli	m
10^{9}	giga	G	10^{-6}	micro	μ
10^{6}	mega	M	10^{-9}	nano	n
10^{3}	kilo	k	10^{-12}	pico	p
10^{2}	hecto*	h	10^{-15}	femto	f
10^{1}	deka*	da	10^{-18}	atto	a

*Avoid except for areas and volumes.

Note that in the SI system the symbols are *not* abbreviations. A symbol such as dm has to be taken as a whole. When prefixes are used with symbols raised to a power, the prefix is also raised to the same power. You should interpret cm³ as $(10^{-2} \cdot m)^3 = 10^{-6}(m^3)$, not as $10^{-2}(m^3)$.

Also watch out for ambiguities such as:

$$N \cdot m \text{ (newton} \times \text{meter) vs. mN (millinewton)}$$

$$m \cdot s \text{ (meter} \times \text{second) vs. ms (millisecond)}$$

and the use of symbols for units jointly with mathematical symbols leading to such unusual formulas as

$$N = 625 \text{ N} \quad \text{(the first } N \text{ is the normal force)}$$

$$F = m (32 \text{ m} \cdot s^{-2}) \quad \text{(the first } m \text{ stands for mass)}$$

EXAMPLE 1.7 Application of Dimensions

A simplified equation for heat transfer from a pipe to air is

$$h = \frac{0.026 G^{0.6}}{D^{0.4}}$$

where h = heat transfer coefficient, Btu/(hr)(ft²)(°F)
$\quad G$ = mass rate of flow, lb_m/(hr)(ft²)
$\quad D$ = outside diameter of the pipe, (ft)
If h is to be expressed in cal/(min)(cm²)(°C), what should the new constant in the equation be in place of 0.026?

Solution

To convert to cal/(min)(cm²)(°C), we set up the dimensional equation as follows:

$$h = \frac{0.026 G^{0.6} \text{Btu}}{D^{0.4}(\text{hr})(\text{ft}^2)(°\text{F})} \left| \frac{252 \text{ cal}}{1 \text{ Btu}} \right| \frac{1 \text{ hr}}{60 \text{ min}} \left| \left(\frac{1 \text{ ft}}{12 \text{ in.}}\right)^2 \right.$$

$$\left| \left(\frac{1 \text{ in.}}{2.54 \text{ cm}}\right)^2 \right| \frac{1.8°\text{F}}{1°\text{C}} = 2.11 \times 10^{-4} \frac{G^{0.6} \text{ cal}}{D^{0.4}(\text{min})(\text{cm}^2)(°\text{C})}$$

(*Note:* See Sec. 1.4 for a detailed discussion of temperature conversion factors.)

If G and D are to be used in units of

$$G': \quad \frac{\text{g}}{(\text{min})(\text{cm}^2)}$$

$$D': \quad \text{cm}$$

an additional conversion is required:

$$h' = \frac{2.11 \times 10^{-4} \left[\frac{G'(\text{g})}{(\text{min})(\text{cm}^2)} \left| \frac{1 \text{ lb}_m}{454 \text{ g}} \right| \frac{60 \text{ min}}{1 \text{ hr}} \left| \left(\frac{2.54 \text{ cm}}{1 \text{ in.}}\right)^2 \right| \left(\frac{12 \text{ in.}}{1 \text{ ft}}\right)^2 \right]^{0.6}}{\left[D'(\text{cm}) \left| \left(\frac{1 \text{ in.}}{2.54 \text{ cm}}\right) \right| \left(\frac{1 \text{ ft}}{12 \text{ in.}}\right) \right]^{0.4} \frac{1}{(\text{min})(\text{cm}^2)(°\text{C})}}$$

$$= 1.48 \times 10^{-2} \frac{(G')^{0.6} \text{ cal}}{(D')^{0.4}(\text{min})(\text{cm}^2)(°\text{C})}$$

Dimensional considerations can also be used to help identify the dimensions of terms or quantities in terms in an equation. Equations must be dimensionally consistent; that is, each term in an equation must have the same net dimensions and units as every other term to which it is added or subtracted. The use of dimensional consistency can be illustrated by an equation that represents gas behavior and is known as van der Waals' equation, an equation that is discussed in more detail in Chap. 3:

$$\left(p + \frac{a}{V^2} \right)(V - b) = RT$$

Inspection of the equation shows that the constant a must have the dimensions of [(pressure)(volume)^2] in order for the expression in the first set of parentheses to be consistent throughout. If the units of pressure are atm and those of volume are cm^3, a will have the units specifically of [(atm)(cm)^6]. Similarly, b must have the same units as V, or in this particular case the units of cm^3.

*Self-Assessment Test**

1. Convert 20 gal/hr to m^3/s.

2. On Phobos, the inner moon of Mars, the acceleration of gravity is 3.78 ft/sec^2. Suppose that an astronaut were walking around on Phobos, and this person plus his space suit and equipment had an earth weight of 252 lb$_f$.

*Answers to the self-assessment test problems will be found in Appendix A.

(*a*) What is the mass in lb$_m$ of the astronaut plus his suit and equipment?

(*b*) How much would the space-suited astronaut weigh in pounds on Phobos?

3. Prepare a table in which the rows are: length, area, volume, mass, and time. Make two columns, one for the SI and the other for the American engineering systems of units. Fill in each row with the name of the unit, and in a third column, show the numerical equivalency (i.e., 1 ft = 0.3048 m).

4. An orifice meter is used to measure flow rate in pipes. The flows rate is related to the pressure drop by an equation of the form

$$u = c\sqrt{\frac{\Delta P}{\rho}}$$

where u = fluid velocity; ΔP is the pressure drop
   ρ = density of the flowing fluid
   c = a constant of proportionality

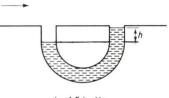

$h = 1.5$ in. Hg

What are the dimensions of c in the SI system of units?

5. What are the value and units of g_c?

6. In the SI system of units, the weight of a 180-lb man standing on the surface of the earth is approximately:
   (*a*) 801 N   (*b*) 81.7 kg
   (*c*) Neither of these   (*d*) Both of these

7. The thermal conductivity k of a liquid metal is predicted via the empirical equation $k = A \exp (B/T)$, where k is in J/(s)(m)(K) and A and B are constants. What are the units of A? Of B?

8. Fill in the following blanks:
   (*a*) _____ grams of prevention is worth _____ kilograms of cure.
   (*b*) A miss is as good as _____ kilometers.
   (*c*) Big sale on _____ liter hats!

Section 1.2   The Mole Unit

> ***Your objectives in studying this section are to be able to:***
>
> 1. Define a kilogram mole, pound mole, and a gram mole.
>
> 2. Convert from moles to mass and the reverse for any chemical compound given the molecular weight.
>
> 2. Calculate molecular weights from the molecular formula.

What is a mole? The best answer is that a *mole* is a certain number of molecules atoms, electrons, or other specified types of particles.[7] In particular, the 1969 International Committee on Weights and Measures approved the mole (symbol mol in the SI system) as being "the amount of a substance that contains as many elementary entities as there are atoms in 0.012 kg of carbon 12." Thus in the SI system the mole contains a different number of molecules than it does in the American engineering system. In the SI system a mole has about 6.023×10^{23} molecules; we shall call this a *gram mole* (symbol *g mol*) to avoid confusion even though in the SI system of units the official designation is simply mole (abbreviated mol). We can thereby hope to avoid the confusion that might well occur with 10^3 g mol (which we will call 1 kg mol) or with the American engineering system *pound mole* (abbreviated *lb mol*), which has $6.023 \times 10^{23} \times 454$ molecules. Thus a pound mole of a substance has more mass than does a gram mole of the substance.

Here is another way to look at the mole unit. To convert the number of moles to mass, we make use of the molecular weight:

$$\text{the g mol} = \frac{\text{mass in g}}{\text{molecular weight}} \tag{1.4}$$

$$\text{the lb mol} = \frac{\text{mass in lb}}{\text{molecular weight}} \tag{1.5}$$

or

$$\text{mass in g} = (\text{mol. wt.})(\text{g mol}) \tag{1.6}$$

$$\text{mass in lb} = (\text{mol. wt.})(\text{lb mol}) \tag{1.7}$$

Furthermore, there is no reason why you cannot carry out computations in terms of ton moles, kilogram moles, or any corresponding units if they are defined analogously to Eqs. (1.4) and (1.5). If you read about a unit such as a kilomole (kmol) without an associated mass specification, assume that it refers to the SI system (here 10^3 g mol or 1 kg mol).

Values of the molecular weights (relative molar masses) are built up from the tables of atomic weights based on an arbitrary scale of the relative masses of the elements. The terms atomic "weight" and molecular "weight" are universally used by chemists and engineers instead of the more accurate terms atomic "mass" or molecular "mass." Since weighing was the original method for determining the comparative atomic masses, as long as they were calculated in a common gravitational field, the relative values obtained for the atomic "weights" were identical with those of the atomic "masses."

Appendix B lists the atomic weights of the elements. On this scale of atomic weights, hydrogen is 1.008, carbon is 12.01, and so on. (In most of our calculations we shall round these off to 1 and 12, respectively, for convenience.)

[7]For a discussion of the mole concept, refer to the series of articles in *J. Chem. Educ.*, v. 38, pp. 549–556 (1961), and to M. L. McGlashan, *Phys. Educ.*, p. 276 (July 1977).

A compound is composed of more than one atom, and the molecular weight of the compound is nothing more than the sum of the weights of the atoms of which it is composed. Thus H_2O consists of 2 hydrogen atoms and 1 oxygen atom, and the molecular weight of water is $(2)(1.008) + 16.000 = 18.02$. These weights are all relative to the ^{12}C atom as 12.0000, and you can attach any unit of mass you desire to these weights; for example, H_2 can be 2.016 g/g mol, 2.016 lb/lb mol, 2.016 ton/ton mol, and so on. Gold (Au) is 196.97 oz/oz mol!

You can compute average molecular weights for mixtures of constant composition even though they are not chemically bonded if their compositions are known accurately. Later (Example 1.12) we show how to calculate the average molecular weight of air. Of course, for a material such as fuel oil or coal whose composition may not be exactly known, you cannot determine an exact molecular weight, although you might estimate an approximate average molecular weight good enough for engineering calculations.

EXAMPLE 1.8 Use of Molecular Weights

If a bucket holds 2.00 lb of NaOH (mol. wt. = 40.0), how many
(a) Pound moles of NaOH does it contain?
(b) Gram moles of NaOH does it contain?

Solution

Basis: 2.00 lb of NaOH

(a)
$$\frac{2.00 \text{ lb NaOH}}{} \left| \frac{1 \text{ lb mol NaOH}}{40.0 \text{ lb NaOH}} = 0.050 \text{ lb mol NaOH} \right.$$

(b$_1$)
$$\frac{2.00 \text{ lb NaOH}}{} \left| \frac{1 \text{ lb mol NaOH}}{40.0 \text{ lb NaOH}} \right| \frac{454 \text{ g mol}}{1 \text{ lb mol}} = 22.7 \text{ g mol}$$

or

(b$_2$)
$$\frac{2.00 \text{ lb NaOH}}{} \left| \frac{454 \text{ g}}{1 \text{ lb}} \right| \frac{1 \text{ g mol NaOH}}{40.0 \text{ g NaOH}} = 22.7 \text{ g mol}$$

EXAMPLE 1.9 Use of Molecular Weights

How many pounds of NaOH are in 7.50 g mol of NaOH?

Solution

Basis: 7.50 g mol of NaOH

$$\frac{7.50 \text{ g mol NaOH}}{} \left| \frac{1 \text{ lb mol}}{454 \text{ g mol}} \right| \frac{40 \text{ lb NaOH}}{1 \text{ lb mol NaOH}} = 0.66 \text{ lb NaOH}$$

Self-Assessment Test

1.   What is the molecular weight of acetic acid (CH_3COOH)?

2.   What is the difference between a kilogram mole and a pound mole?

3.   Convert 39.8 kg of NaCl per 100 kg of water to kilogram moles of NaCl per kilogram mole of water.

4.   How many pound moles of $NaNO_3$ are there in 100 lb?

5.   One pound mole of CH_4 per minute is fed to a heat exchanger. How many kilograms is this per second?

Section 1.3   Conventions in Methods of Analysis and Measurement

> *Your objectives in studying this section are to be able to:*
>
> 1. Define density and specific gravity.
> 2. Calculate the density of a liquid or solid given its specific gravity and the reverse.
> 3. Look up the density or specific gravity of a liquid or solid in reference tables.
> 4. Interpret the meaning of specific gravity data taken from reference tables.
> 5. Specify the common reference material(s) used to determine the specific gravity of liquids and solids.
> 6. Convert the composition of a mixture from mole fraction (or percent) to mass (weight) fraction (or percent) and the reverse.
> 7. Transform a material from one measure of concentration to another, including mass/volume, moles/volume, ppm, molarity, normality, and molality.
> 8. Calculate the mass or number of moles of each component in a mixture given the percent (or fraction) composition, and the reverse, and compute the average molecular weight.
> 9. Convert a composition given in mass (weight) percent to mole percent, and the reverse.

There are certain definitions and conventions which we mention at this point since they will be used constantly throughout the book. If you memorize them now, you will immediately have a clearer perspective and save considerable trouble later.

1.3-1 Density

Density is the ratio of mass per unit volume, as, for example, kg/m³ or lb/ft³. It has both a numerical value and units. To determine the density of a substance, you must find both its volume and its mass or weight. If the substance is a solid, a common method to determine its volume is to displace a measured quantity of inert liquid. For example, a known weight of a material can be placed into a container of liquid of known weight and volume, and the final weight and volume of the combination measured. The density (or specific gravity) of a liquid is commonly measured with a hydrometer (a known weight and volume is dropped into the liquid and the depth to which it penetrates into the liquid is noted) or a Westphal balance (the weight of a known slug is compared in the unknown liquid with that in water). Gas densities are quite difficult to measure; one device used is the Edwards balance, which compares the weight of a bulb filled with air to the same bulb when filled with the unknown gas.

In most of your work using liquids and solids, density will not change very much with pressure, but for precise measurements for common substances you can always look up in a handbook the variation of density with pressure. The change in density with temperature is illustrated in Fig. 1.1 for liquid water and liquid ammonia. Figure 1.2 illustrates how density also varies with composition. In the winter you may put antifreeze in your car radiator. The service station attendant checks the concentration of antifreeze by measuring the specific gravity and, in effect, the density of the radiator solution after it is mixed thoroughly. He has a little thermometer in his hydrometer kit in order to be able to read the density at the proper temperature.

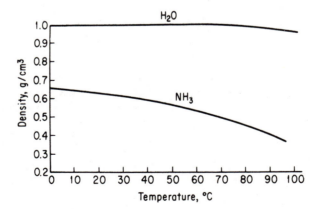

Fig. 1.1 Densities of liquid H_2O and NH_3 as a function of temperature.

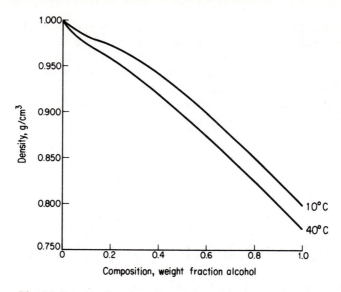

Fig. 1.2 Density of a mixture of ethyl alcohol and water as a function of composition.

1.3-2   Specific Gravity

Specific gravity is commonly thought of as a dimensionless ratio. Actually, it should be considered as the ratio of two densities—that of the substance of interest, A, to that of a reference substance. In symbols:

$$\text{sp gr} = \text{specific gravity} = \frac{(\text{lb/ft}^3)_A}{(\text{lb/ft}^3)_{\text{ref}}} = \frac{(\text{g/cm}^3)_A}{(\text{g/cm}^3)_{\text{ref}}} = \frac{(\text{kg/m}^3)_A}{(\text{kg/m}^3)_{\text{ref}}} \qquad (1.8)$$

The reference substance for liquids and solids is normally water. Thus the specific gravity is the ratio of the density of the substance in question to the density of water. The specific gravity of gases frequently is referred to air, but may be referred to other gases, as discussed in more detail in Chap. 3. Liquid density can be considered to be nearly independent of pressure for most common calculations, but, as indicated in Fig. 1.1 it varies somewhat with temperature; therefore, to be very precise when referring to specific gravity, state the temperature at which each density is chosen. Thus

$$\text{sp gr} = 0.73 \, \frac{20°}{4°}$$

can be interpreted as follows: the specific gravity when the solution is at 20°C and the reference substance (water) at 4°C is 0.73. Since the density of water at 4°C is very close to $1.0000\ \text{g/cm}^3$ in the SI system, the numerical values of the specific gravity and density in this system are essentially equal. Since densities in the American engineering system are expressed in lb/ft^3 and the density of water is about $62.4\ \text{lb/ft}^3$, it can be seen that the specific gravity and density values are not numerically equal in the American engineering system.

In the petroleum industry the specific gravity of petroleum products is usually reported in terms of a hydrometer scale called °API. The equation for the API scale is

$$°\text{API} = \frac{141.5}{\text{sp gr}\dfrac{60°}{60°}} - 131.5 \tag{1.9}$$

or

$$\text{sp gr}\,\frac{60°}{60°} = \frac{141.5}{°\text{API} + 131.5} \tag{1.10}$$

The volume and therefore the density of petroleum products vary with temperature, and the petroleum industry has established 60°F as the standard temperature for volume and API gravity. The °API is being phased out as SI units are accepted for densities.

There are many other systems of measuring density and specific gravity that are somewhat specialized, for example, the Baume (°Be) and the Twaddell (°Tw) systems. Relationships among the various systems of density may be found in standard reference books.

1.3-3   Specific Volume

The specific volume of any compound is the inverse of the density, that is, the volume per unit mass or unit amount of material. Units of specific volume might be ft^3/lb_m, $\text{ft}^3/\text{lb mole}$, cm^3/g, bbl/lb_m, m^3/kg, or similar ratios.

1.3-4   Mole Fraction and Weight Fraction

Mole fraction is simply the moles of a particular substance divided by the total number of moles present. This definition holds for gases, liquids, and solids. Similarly, the weight fraction is nothing more than the weight of the substance divided by the total weight of all substances present. Mathematically, these ideas can be expressed as

$$\text{mole fraction of } A = \frac{\text{moles of } A}{\text{total moles}} \tag{1.11}$$

$$\text{weight fraction of } A = \frac{\text{weight of } A}{\text{total weight}} \tag{1.12}$$

Mole percent and weight percent are the respective fractions times 100.

1.3-5 Analyses

The analyses of gases such as air, combustion products, and the like are usually on a dry basis—that is, water vapor is excluded from the analysis. Such an analysis is called an *Orsat* analysis. In practically all cases the analysis of the gas is on a volume basis, which for the ideal gas is the same as a mole basis. Consider a typical gas analysis, that of air, which is approximately

$$\begin{array}{l} 21\% \text{ oxygen} \\ \underline{79\% \text{ nitrogen}} \\ 100\% \text{ total} \end{array}$$

This means that any sample of air will contain 21% oxygen by volume and also 21 mole % oxygen. (Percent, you should remember, is nothing more than the fraction times 100; consequently, the mole fraction of oxygen is 0.21.)

Analyses of liquids and solids are usually given by weight percent, but occasionally by mole percent. In this text analyses of liquids and solids will always be assumed to be weight percent unless otherwise stated.

EXAMPLE 1.10 Mole Fraction and Weight Fraction

An industrial-strength drain cleaner contains 5.00 kg of water and 5.00 kg of NaOH. What are the weight fraction and mole fraction of each component in the bottle of cleaner?

Solution

Basis: 10.0 kg of total solution

component	kg	weight fraction	mol. wt.	kg mol	mole fraction
H_2O	5.00	$\frac{5.00}{10.0} = 0.500$	18.0	0.278	$\frac{0.278}{0.403} = 0.699$
NaOH	5.00	$\frac{5.00}{10.0} = 0.500$	40.0	0.125	$\frac{0.125}{0.403} = 0.311$
Total	10.00	1.000		0.403	1.00

The kilogram moles are calculated as follows:

$$\frac{5.00 \text{ kg } H_2O}{} \left| \frac{1 \text{ kg mol } H_2O}{18.0 \text{ kg } H_2O} \right. = 0.278 \text{ kg mol } H_2O$$

$$\frac{5.00 \text{ kg NaOH}}{} \left| \frac{1 \text{ kg mol NaOH}}{40.0 \text{ kg NaOH}} \right. = 0.125 \text{ kg mol NaOH}$$

Adding these quantities together gives the total kilogram moles.

EXAMPLE 1.11 Density and Specific Gravity

If dibromopentane has a specific gravity of 1.57, what is its density in (a) g/cm^3? (b) lb_m/ft^3? and (c) kg/m^3?

Solution

No temperatures are cited for the dibromopentane or the reference compound (presumed to be water); hence we assume that the temperatures are the same and that water has a density of 1.00×10^3 kg/m^3 (1.00 g/cm^3).

(a)
$$\frac{1.57 \dfrac{g\,DBP}{cm^3} \bigg| 1.00 \dfrac{g\,H_2O}{cm^3}}{1.00 \dfrac{g\,H_2O}{cm^3}} = 1.57 \frac{g\,DBP}{cm^3}$$

(b)
$$\frac{1.57 \dfrac{lb_m\,DBP}{ft^3} \bigg| 62.4 \dfrac{lb_m\,H_2O}{ft^3}}{1.00 \dfrac{lb_m\,H_2O}{ft^3}} = 97.9 \frac{lb_m\,DBP}{ft^3}$$

(c)
$$\frac{1.57\,g\,DBP}{cm^3} \bigg| \left(\frac{100\,cm}{1\,m}\right)^3 \bigg| \frac{1\,kg}{1000\,g} = 1.57 \times 10^3 \frac{kg\,DBP}{m^3}$$

Note how the units of specific gravity as used here clarify the calculation.

EXAMPLE 1.12 Average Molecular Weight of Air

Basis: 100 lb mol of air

component	moles = percent	mol. wt.	lb	weight %
O_2	21.0	32	672	23.17
N_2*	79.0	28.2	2228	76.83
Total	100		2900	100.00

*Includes Ar, CO_2, Kr, Ne, and Xe, and is called atmospheric nitrogen. The molecular weight is 28.2.

Table 1.5 lists the detailed composition of air.

The average molecular weight is 2900 lb/100 lb mol = 29.00.

Do **not** attempt to get an average specific gravity or average density for a mixture by multiplying the individual component specific gravities or densities by the respective mole fractions of the components in the mixture and summing the products.

TABLE 1.5 COMPOSITION OF CLEAN, DRY AIR NEAR SEA LEVEL

Component	Percent by Volume = Mole Percent
Nitrogen	78.084
Oxygen	20.9476
Argon	0.934
Carbon dioxide	0.0314
Neon	0.001818
Helium	0.000524
Methane	0.0002
Krypton	0.000114
Nitrous oxide	0.00005
Hydrogen	0.00005
Xenon	0.0000087
Ozone	
Summer	0–0.000007
Winter	0–0.000002
Ammonia	0–trace
Carbon monoxide	0–trace
Iodine	0–0.000001
Nitrogen dioxide	0–0.000002
Sulfur dioxide	0–0.0001

1.3-6   Concentrations

Concentration means the quantity of some solute per fixed amount of solvent or solution in a mixture of two or more components, for example:

(a) Weight per unit volume (lb_m of solute/ft³, g of solute/L, lb_m of solute/bbl, kg of solute/m³).

(b) Moles per unit volume (lb mol of solute/ft³, g mol of solute/liter, g mol of solute/cm³).

(c) Parts per million—a method of expressing the concentration of extremely dilute solutions. Ppm is equivalent to a weight fraction for solids and liquids because the total amount of material is of a much higher order of magnitude than the amount of solute; it is essentially a mole fraction for gases. Why?

(d) Other methods of expressing concentration with which you should be familiar are molarity (mol/liter), normality (equivalents/liter), and molality (mole/kg of solvent).

A typical example of the use of these concentration measures is the set of guidelines by which the Environmental Protection Agency defined the extreme levels at which the five most common air pollutants could harm individuals over periods of time.

(a) Sulfur dioxide: 365 μg/m^3 averaged over a 24-hr period

(b) Particulate matter: 260 μg/m^3 averaged over a 24-hr period

(c) Carbon monoxide: 10 mg/m^3 (9 ppm) when averaged over an 8-hr period; 40 mg/m^3 (35 ppm) when averaged over 1 hr

(d) Nitrogen dioxide: 100 μg/m^3 averaged over 1 year

It is important to remember that in an ideal solution, such as in gases or in a simple mixture of hydrocarbon liquids or compounds of like chemical nature, the volumes of the components may be added without great error to get the total volume of the mixture. For the so-called nonideal mixtures this rule does not hold, and the total volume of the mixture is bigger or smaller than the sum of the volumes of the pure components.

Because of its universal interest, a few remarks on the term *proof* as referred to an ethanol–water solution are appropriate. You may have read that the proof is two times the percent alcohol by volume. More precisely, the common proof scale used by the U.S. Internal Revenue Service defines the proof as twice the ratio of the volume of alcohol at 60°F divided by the volume of the alcohol–water solution at 60°F times 100. Since there is a contraction when water and ethanol are mixed, a 100-proof spirit consists of 50 parts by volume of ethanol (sp gr 0.7939 60°F/60°F) and 53.73 parts by volume of water that result in a volume of solution equivalent to 100 parts after mixing. Such a solution would be 42.5% alcohol by weight of specific gravity 0.943 60°F/60°F. Thus in terms of the quantity of alcohol being purchased, you are now prepared to establish that a fifth (i.e., one-fifth of a gallon = 25.6 oz) of 80-proof vodka at $4.69 is not a better buy than a quart of 100-proof vodka at $5.99!

Self-Assessment Test

1. Answer the following questions true (T) or false (F).
   (*a*) Density and specific gravity for mercury are the same.
   (*b*) The inverse of density is the specific volume.
   (*c*) Parts per million expresses a mole ratio.
   (*d*) Concentration of a component in a mixture does not depend on the amount of the mixture.

2. A cubic centimeter of mercury weighs 13.6 g at the earth's surface. What is the density of the mercury?

3. What is the approximate density of water at room temperature?

4. For liquid HCN, a handbook gives: sp gr 10°C/4°C = 1.2675. What does this mean?

5. For ethanol, a handbook gives: sp gr 60°F/60°F = 0.79389. What is the density of ethanol at 60°F?

6. Commercial sulfuric acid is 98% H_2SO_4 and 2% H_2O. What is the mole ratio of H_2SO_4 to H_2O?

7. A container holds 1.704 lb of HNO_3/lb of H_2O and has a specific gravity of 1.382 at 20°C. Compute the composition in the following ways:
(*a*) Weight percent HNO_3
(*b*) Pounds HNO_3 per cubic foot of solution at 20°C
(*c*) Molarity at 20°C

8. The specific gravity of steel is 7.9. What is the volume in cubic feet of a steel ingot weighing 4000 lb?

9. A solution contains 25% salt by weight in water. The solution has a density of 1.2 g/cm³. Express the composition as:
(*a*) Kilograms salt per kilogram of H_2O
(*b*) Pounds of salt per cubic foot of solution

10. A liquefied mixture of *n*-butane, *n*-pentane, and *n*-hexane has the following composition in percent:

n-C_4H_{10}	50
n-C_5H_{12}	30
n-C_6H_{14}	20

For this mixture, calculate:
(*a*) The weight fraction
(*b*) The mole fraction
(*c*) The mole percent of each component
(*d*) The average molecular weight

Section 1.4   Basis

Your objectives in studying this section are to be able to:

1. State the three questions useful in selecting a basis.

2. Apply the three questions to problems and select a suitable basis or sequences of bases.

Have you noted in previous examples that the word *basis* has appeared at the top of the computations? This concept of basis is vitally important both to your understanding of how to solve a problem and also to your solving it in the most expeditious manner. The basis is the reference chosen by you for the calculations you plan to make in any particular problem, and a proper choice of basis frequently makes the problem much easier to solve. The basis may be a period of time—for example, hours, or a given mass of material—such as 5 kg of CO_2 or some other convenient quantity. In selecting a sound basis (which in many problems is predetermined for you but in some problems is not so clear), you should ask yourself the following questions:

(a) What do I have to start with?
(b) What do I want to find out?
(c) What is the most convenient basis to use?

These questions and their answers will suggest suitable bases. Sometimes, when a number of bases seem appropriate, you may find it is best to use a unit basis of 1 or 100 of something, as, for example, kilograms, hours, moles, cubic feet. For liquids and solids when a weight analysis is used, a convenient basis is often 1 or 100 lb or kg; similarly, 1 or 100 moles is often a good choice if a mole analysis is used for a gas. The reason for these choices is that the fraction or percent automatically equals the number of pounds, kilograms, or moles, respectively, and one step in the calculations is saved.

EXAMPLE 1.13 Choosing a Basis

Aromatic hydrocarbons form 15 to 30% of the components of leaded fuels and as much as 40% of nonleaded gasoline. The carbon/hydrogen ratio helps to characterize the fuel components. If a fuel is 80% C and 20% H by weight, what is the C/H ratio in moles?

Solution

If a basis of 100 lb or kg of oil is selected, percent = pounds or kilograms.

Basis: 100 kg of oil (or 100 lb of oil)

component	kg = percent or lb = percent	mol. wt.	kg mol or lb mol
C	80	12.0	6.67
H	20	1.008	19.84
Total	100		

Consequently, the C/H ratio in moles is

$$C/H = 6.67/19.84 = 0.33$$

EXAMPLE 1.14 Choosing a Basis

Most processes for producing high-energy-content gas or gasoline from coal include some type of gasification step to make hydrogen or synthesis gas. Pressure gasification is preferred because of its greater yield of methane and higher rate of gasification.

Given that a 50.0-kg test run of gas averages 10.0% H_2, 40.0% CH_4, 30.0% CO, and 20.0% CO_2, what is the average molecular weight of the gas?

Solution

The obvious basis is 50.0 kg of gas ("what I have to start with"), but a little reflection will show that such a basis is of no use. You cannot multiply *mole percent* of this gas times kg and expect the answer to mean anything. Thus the next step is to choose a "convenient basis," which is 100 kg or lb mol of gas, and proceed as follows:

Basis: 100 kg or lb mol of gas

component	percent = kg or lb mol	mol. wt.	kg or lb
CO_2	20.0	44.0	880
CO	30.0	28.0	840
CH_4	40.0	16.04	642
H_2	10.0	2.02	20
Total	100.0		2382

$$\text{average molecular weight} = \frac{2382 \text{ kg}}{100 \text{ kg mol}} = 23.8 \text{ kg/kg mol}$$

It is important that your basis be indicated at the beginning of the problem so that you will keep clearly in mind the real nature of your calculations and so that anyone checking your problem will be able to understand on what basis they are performed. If you change bases in the middle of the problem, a new basis should be indicated at that time. Many of the problems that we shall encounter will be solved on one basis and then at the end will be shifted to another basis to give the desired answer. The significance of this type of manipulation will become considerably clearer as you accumulate more experience.

EXAMPLE 1.15 Changing Bases

A sample of medium-grade bituminous coal analysis is as follows:

component	percent
S	2
N	1
O	6
Ash	11
Water	3

The residuum is C and H in the mole ratio H/C = 9. Calculate the composition of the coal with the ash and the moisture omitted.

Solution

Take as a basis 100 kg of coal, for then percent = kilograms.

Basis: 100 kg of coal

The sum of the S + N + O + ash + water is

$$2 + 1 + 6 + 11 + 3 = 23 \text{ kg}$$

Hence the C + H must be $100 - 23 = 77$ kg.

To determine the kilograms of C and H, we have to select a new basis. Is 77 kg satisfactory? No. Why? Because the H/C ratio is in terms of moles, not weight (mass).

Basis: 100 kg mol of C + H

component	mole fraction	kg mol	mol. wt.	kg
H	$\dfrac{9}{1+9} = 0.90$	90	1.008	90.7
C	$\dfrac{1}{1+9} = 0.10$	10	12	120
Total	1.00	100		210.7

Finally, to return to the original basis, we have

$$\text{H:} \quad \frac{77 \text{ kg}}{} \left| \frac{90.7 \text{ kg H}}{210.7 \text{ kg total}} \right. = 33.15 \text{ kg H}$$

$$\text{C:} \quad \frac{77 \text{ kg}}{} \left| \frac{120 \text{ kg C}}{210.7 \text{ kg total}} \right. = 43.85 \text{ kg C}$$

and we can prepare a table summarizing the results.

component	kg	wt. fraction
C	43.85	0.51
H	33.15	0.39
S	2	0.02
N	1	0.01
O	6	0.07
Total	86.0	1.00

The ability to choose the basis that requires the fewest steps in solution can only come with practice. You can quickly accumulate the necessary experience if, as you look at each problem illustrated in this text, you determine first in your own mind what the basis should be and then compare your choice with the selected basis. By this procedure you will quickly obtain the knack of choosing a sound basis.

Self-Assessment Test

1.   What are the three questions you should ask yourself in selecting a basis?

2.   What would be good initial bases to select in solving Problems 1.13, 1.18, 1.30, and 1.47?

Section 1.5   Temperature

> *Your objectives in studying this section are to be able to:*
> 1. Define temperature.
> 2. Explain the difference between absolute temperature and relative temperature.
> 3. Convert a temperature in any of the four scales (°C, K, °F, °R) to any of the others.
> 4. Convert an expression involving units of temperature and temperature difference to other units of temperature and temperature difference.
> 5. Know the reference points for the four temperature scales.

Our concept of temperature probably originated with our physical sense of hot or cold. Attempts to be more specific and quantitative led to the idea of a temperature scale and the thermometer—a device to measure how hot or cold something is. We are all familiar with the thermometer used in laboratories which holds mercury sealed inside a glass tube or the alcohol thermometer used to measure outdoor temperatures.

Although we do not have the space to discuss in detail the many methods of measuring temperature, we can point out some of the other more common techniques with which you are probably already familiar:

(a) The voltage produced by a junction of two dissimilar conductors changes with temperature and is used as a measure of temperature (the *thermocouple*).

(b) The property of changing electrical resistance with temperature gives us a device known as the *thermistor*.

(c) Two thin strips of metal bonded together at one end expand at different rates with change of temperature. These strips assist in the control of the flow of water in the radiator of an automobile and in the operation of air conditioners and heating systems.

(d) High temperatures can be measured by devices called *pyrometers*, which note the radiant energy leaving a hot body.

Figure 1.3 illustrates the appropriate ranges for various temperature-measuring devices.

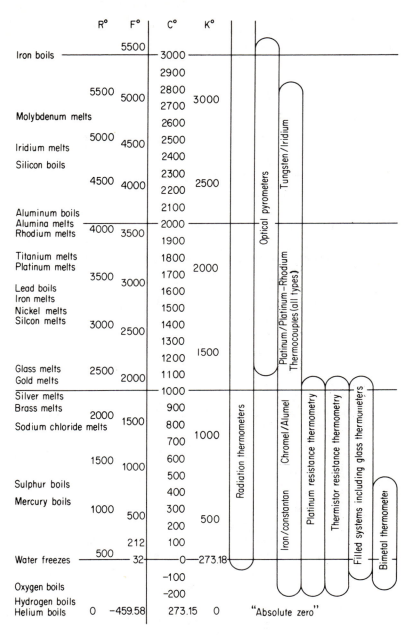

Fig. 1.3 Temperature measuring instruments span the range from near absolute zero to beyond 3000 K. The chart indicates the preferred methods of thermal instrumentation for various temperature regions.

   As you also know, temperature is a measure of the thermal energy of the random motion of the molecules of a substance at thermal equilibrium. Temperature is normally measured in degrees Fahrenheit or Celsius (centigrade). The common scientific scale is the *Celsius* scale,[8] where 0° is the ice point of water and 100° is the normal boiling point of water. In the early 1700s, Fahrenheit, a glassblower by trade, was able to build mercury thermometers that gave temperature measurements in reasonable agreement with each other. The *Fahrenheit* scale is the one commonly used in everyday life in the United States. Its reference points are of more mysterious origin, but it is reported that the fixed starting point, or 0° on Fahrenheit's scale, was that produced by surrounding the bulb of the thermometer with a mixture of snow or ice and sal ammoniac; the highest temperature was that at which mercury began to boil. The distance between these two points was divided into 600 parts or degrees. By trial Fahrenheit found that the mercury stood at 32 of these divisions when water just began to freeze, and 212 divisions when the thermometer was immersed in boiling water. In the SI system, temperature is measured in kelvins, a unit named after the famous Lord Kelvin. Note that in the SI system the degree symbol is suppressed (e.g., the boiling point of water is 373 K).

   The Fahrenheit and Celsius scales are *relative* scales; that is, their zero points were arbitrarily fixed by their inventors. Quite often it is necessary to use *absolute* temperatures instead of relative temperatures. Absolute temperature scales have their zero point at the lowest possible temperature which we believe can exist. As you may know, this lowest temperature is related both to the ideal gas laws and to the laws of thermodynamics. The absolute scale which is based on degree units the size of those in the Celsius (centigrade) scale is called the *kelvin* scale; the absolute scale which corresponds to the Fahrenheit degree scale is called the *Rankine* scale in honor of a Scottish engineer. The relations between relative temperature and absolute temperature are illustrated in Fig. 1.4. We shall round off absolute zero on the Rankine scale of $-459.58°$ to $-460°F$; similarly, $-273.15°C$ will be rounded off to $-273°C$.

   You should recognize that the unit degree (i.e., the unit temperature difference on the kelvin–Celsius scale) is not the same size as that on the Rankine–Fahrenheit scale. If we let $\Delta°F$ represent the unit temperature difference in the Fahrenheit scale, $\Delta°R$ be the unit temperature difference in the Rankine scale, and $\Delta°C$ and ΔK be the analogous units in the other two scales, you should be aware that

$$\Delta°F = \Delta°R \tag{1.13}$$

$$\Delta°C = \Delta K \tag{1.14}$$

Also, if you keep in mind that the $\Delta°C$ is larger than the $\Delta°F$,

$$\frac{\Delta°C}{\Delta°F} = 1.8 \quad \text{or} \quad \Delta°C = 1.8\,\Delta°F \tag{1.15}$$

---

[8] As originally devised by Celsius in 1742, the freezing point was designated as 100°. Officially, °C now stands for *degrees Celsius*.

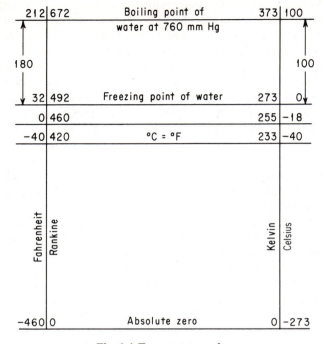

Fig. 1.4 Temperature scales.

$$\frac{\Delta K}{\Delta °R} = 1.8 \quad \text{or} \quad \Delta K = 1.8 \, \Delta °R \tag{1.16}$$

Unfortunately, the symbols $\Delta °C$, $\Delta °F$, ΔK, and $\Delta °R$ are not in standard usage, and consequently the proper meaning of the symbols $°C$, $°F$, K, and $°R$, as either the temperature or the unit temperature difference, *must be interpreted* from the context of the equation or sentence being examined.

You should learn how to convert one temperature to another with ease. The relations between $°R$ and $°F$ and between K and $°C$ are, respectively,

$$T_{°R} = T_{°F}\left(\frac{1 \, \Delta °R}{1 \, \Delta °F}\right) + 460 \tag{1.17}$$

$$T_K = T_{°C}\left(\frac{1 \, \Delta K}{1 \, \Delta °C}\right) + 273 \tag{1.18}$$

Because the relative temperature scales do not have a common zero at the same temperature, as can be seen from Fig. 1.4, the relation between $°F$ and $°C$ is

$$T_{°F} - 32 = T_{°C}\left(\frac{1.8 \, \Delta °F}{1 \, \Delta °C}\right) \tag{1.19}$$

After you have used Eqs. (1.17)–(1.19) a bit, they will become so familiar that temperature conversion will become an automatic reflex. During your "learning period," in case you forget them, just think of the appropriate scales side by side as in Fig. 1.4, and put down the values for the freezing and boiling points of water.

From Fig. 1.4 you will notice that $-40°$ is a common temperature on both the centigrade and Fahrenheit scales. Therefore, the following relationships hold:

$$T_{°F} = (T_{°C} + 40)\left(\frac{1.8\,\Delta°F}{1\,\Delta°C}\right) - 40 \tag{1.20}$$

$$T_{°C} = (T_{°F} + 40)\left(\frac{1\,\Delta°C}{1.8\,\Delta°F}\right) - 40 \tag{1.21}$$

In Fig. 1.4 all the values of the temperatures have been rounded off, but more significant figures can be used. $0°C$ and its equivalents are known as *standard conditions of temperature*.

EXAMPLE 1.16 Temperature Conversion

Convert $100°C$ to (a) K, (b) °F, and (c) °R.

Solution

(a)
$$(100 + 273)°C\frac{1\,\Delta K}{1\,\Delta°C} = 373\ K$$

or with suppression of the Δ symbol,

$$(100 + 273)°C\frac{1\ K}{1°C} = 373\ K$$

(b)
$$(100°C)\frac{1.8°F}{1°C} + 32°F = 212°F$$

(c)
$$(212 + 460)°F\frac{1°R}{1°F} = 672°R$$

or

$$(373K)\frac{1.8°R}{1\ K} = 672°R$$

The suppression of the Δ symbol perhaps makes the temperature relations more familiar looking. Just be careful that you do not cancel the symbols °C, and so on, in the numerator and denominator when one symbol represents a temperature and the other the temperature difference!

EXAMPLE 1.17 Temperature Conversion

The thermal conductivity of aluminum at $32°F$ is 117 Btu/(hr)(ft²)(°F/ft). Find the equivalent value at $0°C$ in terms of Btu/(hr)(ft²)(K/ft).

Solution

Since 32°F is identical to 0°C, the value is already at the proper temperature. The "°F" in the denominator of the thermal conductivity actually stands for $\Delta°F$, so that the equivalent value is

$$\frac{117 \ (Btu)(ft)}{(hr)(ft^2)(\Delta°F)} \ \left| \ \frac{1.8 \ \Delta°F}{1 \ \Delta°C} \ \right| \ \frac{1 \ \Delta°C}{1 \ \Delta K} = 211 \ (Btu)/(hr)(ft^2)(K/ft)$$

or with the Δ symbol suppressed,

$$\frac{117 \ (Btu)(ft)}{(hr)(ft^2)(°F)} \ \left| \ \frac{1.8°F}{1°C} \ \right| \ \frac{1°C}{1 \ K} = 211 \ (Btu)/(hr)(ft^2)(K/ft)$$

EXAMPLE 1.18 Temperature Conversion

The heat capacity of sulfuric acid has the units cal/(g mol)(°C) and is given by the relation

$$\text{heat capacity} = 33.25 + 3.727 \times 10^{-2}T$$

where T is expressed in °C. Modify the formula so that the resulting expression gives units of Btu/(lb mol)(°R) and T is in °R.

Solution

The units of °C in the denominator of the heat capacity are $\Delta°C$, whereas the units of T are °C. First, substitute the proper relation in the formula to convert T in °C to T in °R, and then convert the units in the resulting expression to those requested.

$$\text{heat capacity} = \left\{ 33.25 + 3.727 \times 10^{-2}\left[(T_{°R} - 460 - 32)\frac{1}{1.8} \right] \right\}$$

$$\frac{cal}{(g \ mol)(°C)} \ \left| \ \frac{1 \ Btu}{252 \ cal} \ \right| \ \frac{454 \ g \ mol}{1 \ lb \ mol} \ \left| \ \frac{1°C}{1.8°R} \right.$$

$$= 23.06 + 2.071 \times 10^{-2}T_{°R}$$

Self-Assessment Test

1. What are the reference points of (*a*) the Celsius and (*b*) Fahrenheit scales?

2. How do you convert a *temperature difference*, Δ, from Fahrenheit to Celsius?

3. Is the unit temperature difference $\Delta°C$ a larger interval than $\Delta°F$? Is 10°C higher than 10°F?

4. In Appendix E, the heat capacity of sulfur is $C_p = 3.63 + 0.640T$, where C_p is in cal/(g mol)(K) and T is in K. Convert so that C_p is in cal/(g mol)(°F) with T in °F.

5. Complete the following table with the proper equivalent temperatures:

°C	°F	K	°R
−40.0			
	77.0		
		698	
			69.8

6. Suppose that you are given a tube partly filled with an unknown liquid and are asked to calibrate a scale on the tube in °C. How would you proceed?

Section 1.6   Pressure

Your objectives in studying this section are to be able to:

1. Define pressure, atmospheric pressure, barometric pressure, standard pressure, and vacuum.
2. Explain the difference between absolute pressure and relative pressure (gauge pressure).
3. List four ways to measure pressure.
4. Convert from gauge pressure to absolute pressure and the reverse.
5. Convert a pressure measured in one set of units to another set, including kPa, mm Hg, in. H_2O, ft H_2O, atm, in. Hg, and psi using the standard atmosphere or density ratios of liquids.

Pressures, like temperatures, can be expressed by either absolute or relative scales. Pressure is defined as "force per unit area." Figure 1.5 shows a column of mercury

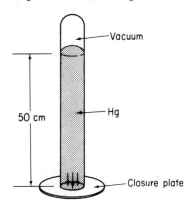

Figure 1.5 Pressure.

held in place by a sealing plate. Suppose that the column of mercury has an area of 1 cm² and is 50 cm high. From Table D.1 we can find that the sp gr at 20°C and hence the density, essentially, of the Hg is 13.55 g/cm³. Thus the force exerted on the 1-cm² section of that plate by the column of mercury is

$$F = \frac{13.55\ g}{cm^3} \left| \frac{50\ cm}{} \right| \frac{1\ cm^2}{} \left| \frac{980\ cm}{s^2} \right| \frac{1\ kg}{1000\ g} \left| \frac{1m}{100\ cm} \right| \frac{1\ N}{\frac{1\ (kg)(m)}{s^2}}$$

$$= 6.64\ N$$

The pressure on the section of the plate covered by the mercury is

$$p = \frac{6.64\ N}{1\ cm^2} \left| \left(\frac{100\ cm}{1\ m}\right)^2 = 6.64 \times 10^4 \frac{N}{m^2}$$

If we had started with units in the American engineering system, the pressure could be similarly computed as

$$p = \frac{846\ lb_m}{1\ ft^3} \left| \frac{50\ cm}{} \right| \frac{1\ in.}{2.54\ cm} \left| \frac{1\ ft}{12\ in.} \right| \frac{32.2\ ft}{sec^2} \left| \frac{32.174(ft)(lb_m)}{(sec)^2(lb_f)} \right|$$

$$= 1387 \frac{lb_f}{ft^2}$$

Whether relative or absolute pressure is measured in a pressure-measuring device depends on the nature of the instrument used to make the measurements. For example, an open-end manometer (Fig. 1.6) would measure a relative pressure, since the reference for the open end is the pressure of the atmosphere at the open end of the manometer. On the other hand, closing off the end of the manometer (Fig. 1.7) and creating a vacuum in the end results in a measurement against a complete vacuum, or against "no pressure." This measurement is called *absolute pressure*. Since absolute pressure is based on a complete vacuum, a fixed reference point which is unchanged regardless of location or temperature or weather or other factors,

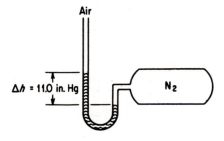

Fig. 1.6 Open-end manometer showing a pressure above atmospheric pressure.

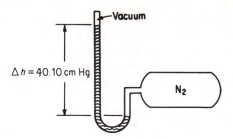

Fig. 1.7 Absolute pressure manometer.

absolute pressure then establishes a precise, invariable value which can be readily identified. Thus the zero point for an absolute pressure scale corresponds to a perfect vacuum, whereas the zero point for a relative pressure scale usually corresponds to the pressure of the air which surrounds us at all times, and as you know, varies slightly.

If the mercury reading is set up as in Fig. 1.8, with the dish open to the atmosphere, the device is called a *barometer* and the reading of atmospheric pressure is termed *barometric pressure*.

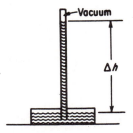

Fig. 1.8 A barometer.

An understanding of the principle upon which a manometer operates will aid you in recognizing the nature of the pressure measurement taken from it. As shown in Fig. 1.6 for an open-end U-tube manometer, if the pressure measured for the N_2 is greater than atmospheric, the liquid is forced downward in the leg to which the pressure source is connected and upward in the open leg. Eventually, a point of hydrostatic balance is reached in which the manometer fluid stabilizes, and the difference in the pressure exerted by the height of the fluid in the open leg relative to that in the leg attached to the pressure source is exactly equal to the difference between the atmospheric pressure and the applied pressure in the other leg. If vacuum were applied instead of pressure to the same leg of the manometer, the fluid column would rise on the vacuum side. Again, the difference in pressure between the pressure source in the tank and atmospheric pressure is measured by the difference in the height of the two columns of fluid. Water and mercury are commonly used indicating fluids for manometers; the readings thus can be expressed in "inches of water," "inches of mercury," "mm of fluid flowing," and so on. (In ordinary engineering calculations we ignore the vapor pressure of mercury and minor changes in the density of mercury due to temperature changes in making pressure measurements.)

Another type of common measuring device is the visual *Bourdon gauge* (Fig. 1.9), which normally (but not always) reads zero pressure when open to the atmosphere. The pressure-sensing device in the Bourdon gauge is a thin metal tube with an elliptical cross section closed at one end which has been bent into an arc. As the pressure increases at the open end of the tube, it tries to straighten out, and the movement of the tube is converted into a dial movement by gears and levers. Figure

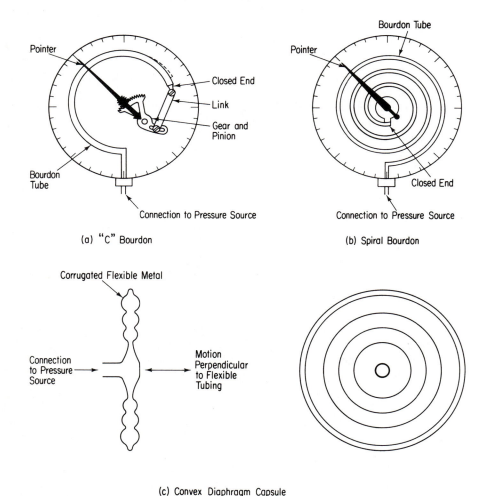

(a) "C" Bourdon

(b) Spiral Bourdon

(c) Convex Diaphragm Capsule

Fig. 1.9 Bourdon and diaphragm pressure-measuring devices.

1.9 also illustrates a diaphragm capsule gauge. Figure 1.10 indicates the pressure ranges for the various pressure-measuring devices.

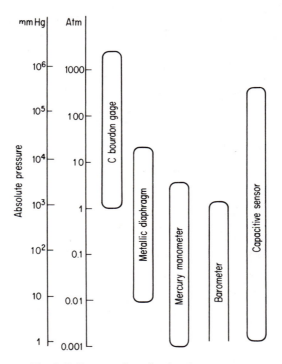

Fig. 1.10 Ranges of application for pressure-measuring devices.

Pressure scales may be temporarily somewhat more confusing than temperature scales since the reference point or zero point for the relative pressure scales is not constant, whereas in the temperature scales the boiling point or the freezing point of water is always a fixed value. However, you will become accustomed to this feature of the pressure scale with practice.

The relationship between relative and absolute pressure is illustrated in Figs. 1.11 and 1.12 and is also given by the following expression:

$$\text{gauge pressure} + \text{barometer pressure} = \text{absolute pressure} \qquad (1.22)$$

Equation (1.22) can be used only with consistent units. Note that you must add the atmospheric pressure (i.e., the barometric pressure) to the gauge, or relative pressure (or manometer reading if open on one end), in order to get the absolute pressure.

Another term applied in measuring pressure which is illustrated in Figs. 1.11 and 1.12 is *vacuum*. In effect, when you measure pressure as "inches of mercury vacuum," you reverse the usual direction of measurement and measure from the barometric pressure down to the vacuum, in which case a perfect vacuum would be the highest pressure that you could achieve. This procedure would be the same as evacuating the air at the top of a mercury barometer and watching the mercury rise up in the barometer as the air is removed. The vacuum system of measurement of pressure is commonly used in apparatus which operate at pressures less than atmos-

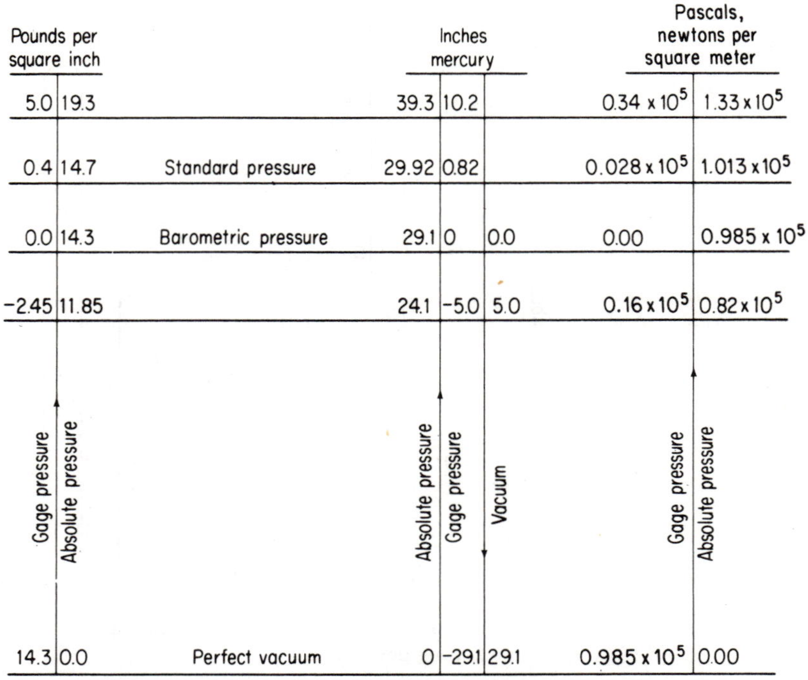

Pounds per square inch			Inches mercury			Pascals, newtons per square meter	
5.0	19.3		39.3	10.2		0.34×10^5	1.33×10^5
0.4	14.7	Standard pressure	29.92	0.82		0.028×10^5	1.013×10^5
0.0	14.3	Barometric pressure	29.1	0	0.0	0.00	0.985×10^5
−2.45	11.85		24.1	−5.0	5.0	0.16×10^5	0.82×10^5
Gage pressure	Absolute pressure		Absolute pressure	Gage pressure	Vacuum	Gage pressure	Absolute pressure
14.3	0.0	Perfect vacuum	0	−29.1	29.1	0.985×10^5	0.00

Fig. 1.11 Pressure comparisons when barometer reading is 29.1 in Hg.

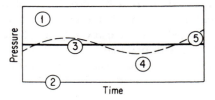

Fig. 1.12 Pressure terminology. The standard atmosphere is shown by the heavy horizontal line. The dashed line illustrates the atmospheric (barometric) pressure, which changes from time to time. Point ① in the figure is a pressure of 19.3 psi referred to a complete vacuum or 5 psi referred to the barometric pressure; ② is the complete vacuum, ③ represents the standard atmosphere, and ④ illustrates a negative relative pressure or a pressure less than atmospheric. This type of measurement is described in the text as a vacuum type of measurement. Point ⑤ also indicates a vacuum measurement, but one that is equivalent to an absolute pressure above the standard atmosphere.

pheric—such as a vacuum evaporator or vacuum filter. A pressure that is only slightly below barometric pressure may sometimes be expressed as a "draft" (which is identical to the vacuum system) in inches of water, as, for example, in the air supply to a furnace or a water cooling tower.

As to the units of pressure, we have shown in Fig. 1.11 three common systems: pounds per square inch (psi), inches of mercury (in. Hg), and pascals. Pounds per square inch absolute is normally abbreviated "psia," and "psig" stands for "pounds per square inch gauge." For other units, be certain to carefully specify whether they are gauge or absolute; for example, state "300 kPa absolute" or "12 cm Hg gauge." Figure 1.11 compares the relative and absolute pressure scales in terms of the three systems of measurement when the barometric pressure is 29.1 in. Hg (14.3 psia). Other systems of expressing pressure exist; in fact, you will discover that there are as many different units of pressure as there are means of measuring pressure. Some of the other most frequently used systems are

(a) Millimeters of mercury (mm Hg)

(b) Feet of water (ft H_2O)

(c) Atmospheres (atm)

(d) Bars (bar)

(e) Kilograms (force) per square centimeter (kg_f/cm^2)

To sum up our discussion of pressure and its measurement, you should now be acquainted with:

(a) Atmospheric pressure—the pressure of the air and the atmosphere surrounding us which changes from day to day

(b) Barometric pressure—the same as atmospheric pressure, called "barometric pressure" because a barometer is used to measure atmospheric pressure

(c) Absolute pressure—a measure of pressure referred to a complete vacuum, or zero pressure

(d) Gauge pressure—pressure expressed as a quantity measured from (above) atmospheric pressure (or some other reference pressure)

(e) Vacuum—a method of expressing pressure as a quantity below atmospheric pressure (or some other reference pressure)

You definitely must not confuse the standard atmosphere with atmospheric pressure. The *standard atmosphere* is defined as the pressure (in a standard gravitational field) equivalent to 1 atm or 760 mm Hg at 0°C or other equivalent value, whereas atmospheric pressure is a variable and must be obtained from a barometer each time you need it. The standard atmosphere may not actually exist in any part of the world except perhaps at sea level on certain days, but it is extremely useful in converting from one system of pressure measurement to another (as well as being useful in several other ways to be considered later).

Expressed in various units, the *standard atmosphere* is equal to

1.000	atmospheres (atm)
33.91	feet of water (ft H_2O)
14.7	(14.696, more exactly) pounds per square inch absolute (psia)
29.92	(29.921, more exactly) inches of mercury (in. Hg)
760.0	millimeters of mercury (mm Hg)
1.013×10^5	pascals (Pa) or newtons per square meter (N/m²)

To convert from one set of pressure units to another, it is convenient to use the relationships among the standard pressures as shown in the examples below.

If pressures are measured by means of the height of a column of liquid[9] other than mercury or water (for which the standard pressure is known), it is easy to convert from one liquid to another by means of the following expression. Examine Fig. 1.8. Assume that instead of a vacuum at the top of the column a pressure p_0 exists. Recall from physics that

$$p = \rho g h + p_0 \tag{1.23}$$

where p = pressure at the bottom of the column of the fluid
ρ = density of the fluid
g = acceleration of gravity
p_0 = pressure at the top of the column of fluid
h = height of the fluid column

By equating the same pressure recorded by the two columns of fluid with different densities, but for the same p_0, you will get the ratio between the heights of two columns of liquid. If one of these liquids is water, it is then easy to convert to any of the more commonly used liquids as follows:

$$\rho g h = \rho_{H_2O} g h_{H_2O} \qquad \text{or} \qquad \frac{h}{h_{H_2O}} = \frac{\rho_{H_2O}}{\rho}$$

How can you arrange to get the same p_0 for each column? Will leaving the upper end of each of the columns open to the atmosphere accomplish this objective?

EXAMPLE 1.19 Pressure Conversion

Convert 35 psia to inches of mercury.

Solution

It is desirable to use the ratio of 14.7 psia to 29.92 in. Hg, an identity, to carry out this conversion.

[9]Sometimes these liquid columns are referred to as "heads" of liquid.

Basis: 35 psia

$$\frac{35 \text{ psia} \mid 29.92 \text{ in. Hg}}{14.7 \text{ psia}} = 71.25 \text{ in. Hg}$$

an identity

EXAMPLE 1.20 Pressure Conversion

The density of the atmosphere decreases with increasing altitude. When the pressure is 340 mm Hg, how many inches of water is it? How many kilopascals?

Solution

Basis: 340 mm Hg

$$\frac{340 \text{ mm Hg} \mid 33.91 \text{ ft H}_2\text{O} \mid 12 \text{ in.}}{760 \text{ mm Hg} \mid 1 \text{ ft}} = 182 \text{ in. H}_2\text{O}$$

$$\frac{340 \text{ mm Hg} \mid 1.013 \times 10^5 \text{ N/m}^2 \mid 1 \text{ kN}}{760.0 \text{ mm Hg} \mid 1000 \text{ N}} = 45.4 \text{ kN/m}^2 = 45.4 \text{ kPa}$$

EXAMPLE 1.21 Pressure Conversion

The pressure gauge on a tank of CO_2 used to fill soda-water bottles reads 51.0 psi. At the same time the barometer reads 28.0 in. Hg. What is the absolute pressure in the tank in psia? See Fig. E1.21.

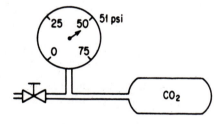

Fig. E1.21

Solution

The pressure gauge is reading psig, not psia.

From Eq. (1.22) the absolute pressure is the sum of the gauge pressure and the atmospheric (barometric) pressure expressed in the same units.

Basis: Barometric pressure = 28.0 in. Hg

$$\text{atmospheric pressure} = \frac{28.0 \text{ in. Hg} \mid 14.7 \text{ psia}}{29.92 \text{ in. Hg}} = 13.78 \text{ psia}$$

(*Note:* Atmospheric pressure does not equal 1 standard atm.) The absolute pressure in the tank is

$$51.0 + 13.78 = 64.78 \text{ psia}$$

EXAMPLE 1.22 Pressure Conversion

Air is flowing through a duct under a draft of 4.0 cm H_2O. The barometer indicates that the atmospheric pressure is 730 mm Hg. What is the absolute pressure of the gas in inches of mercury? See Fig. E1.22.

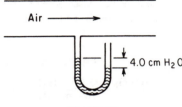

Fig. E1.22

Solution

Again we have to employ consistent units, and it appears in this case that the most convenient units are those of inches of mercury.

Basis: 730 mm Hg

$$\text{atmospheric pressure} = \frac{730 \text{ mm Hg}}{} \left| \frac{29.92 \text{ in. Hg}}{760 \text{ mm Hg}} \right. = 28.9 \text{ in. Hg}$$

Basis: 4.0 cm H_2O draft (under atmospheric)

$$\frac{4.0 \text{ cm } H_2O}{} \left| \frac{1 \text{ in}}{2.54 \text{ cm}} \right| \frac{1 \text{ ft}}{12 \text{ in.}} \left| \frac{29.92 \text{ in. Hg}}{33.91 \text{ ft } H_2O} \right. = 0.12 \text{ in. Hg}$$

In what other way could you make the conversion?

Since the reading is 4.0 cm H_2O draft (under atmospheric), the absolute reading in uniform units is

$$28.9 - 0.12 = 28.8 \text{ in. Hg absolute}$$

EXAMPLE 1.23 Vacuum Pressure Reading

Small animals such as mice can live at reduced air pressures down to 20 kPa (although not comfortably). In a test a mercury manometer attached to a tank as shown in Fig. E1.23 reads 64.5 cm Hg and the barometer reads 100 kPa. Will the mice survive?

Solution

Basis: 64.5 cm Hg below atmospheric

We ignore any temperature corrections to convert the mercury to 0°C. Then, since the vacuum reading on the tank is 64.5 cm Hg below atmospheric, the absolute pressure in the tank is

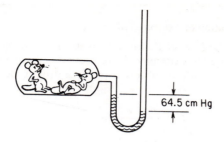

64.5 cm Hg

Fig. E1.23

$$100 \text{ kPa} - \frac{64.5 \text{ cm Hg}}{} \left| \frac{1.013 \text{ kPa}}{76.0 \text{ cm Hg}} \right. = 100 - 86 = 14 \text{ kPa absolute}$$

The mice probably will not survive.

In some instances the fluids in the legs of the manometer are not the same. Examine Fig. 1.23. When the columns of fluids are at equilibrium (it may take some

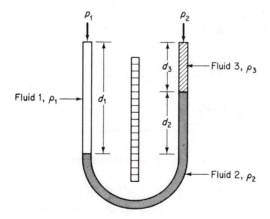

Fig. 1.13 Manometer with three fluids.

time!) the relation between p_1, p_2 and the heights of the various columns of fluid is

$$p_1 + \rho_1 d_1 g = p_2 + \rho_2 g d_2 + \rho_3 g d_3 \tag{1.24a}$$

Can you show for the case in which $\rho_1 = \rho_3 = \rho$ that the manometer expression reduces to

$$p_1 - p_2 = (\rho_2 - \rho)g d_2 \tag{1.24b}$$

Finally, suppose that fluids 1 and 3 are gases. Can you ignore the gas density ρ relative to the manometer fluid density? For what types of fluids?

EXAMPLE 1.24 Calculation of Pressure Difference

In measuring the flow of fluids in a pipeline, a differential manometer, as shown in Fig. E1.24, can be used to determine the pressure difference across an orifice plate. The flow rate can be calibrated with the observed pressure drop. Calculate the pressure drop $p_1 - p_2$ in pascals.

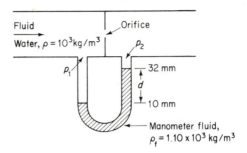

Fig. E1.24

Solution

Apply Eq. (1.24b), as the densities of the fluids above the manometer fluid are the same.

$$p_1 - p_2 = (\rho_f - \rho)gd_2$$

$$= \frac{(1.10 - 1.00)10^3 \text{ kg}}{\text{m}^3} \left| \frac{9.807 \text{ m}}{\text{s}^2} \right| \frac{(22)(10^{-3}) \text{ m}}{} \left| \frac{1 \text{ (N)(s}^2)}{\text{(kg)(m)}} \right| \frac{1 \text{ (Pa)(m}^2)}{1 \text{ (N)}}$$

$$= 21.6 \text{ Pa}$$

Self-Assessment Test

1. Write down the equation to convert gauge pressure to absolute pressure.

2. List the values and units of the standard atmosphere for six different methods of expressing pressure.

3. List the equation to convert vacuum pressure to absolute pressure.

4. Convert a pressure of 800 mm Hg to the following units:
(*a*) psia   (*b*) kPa
(*c*) atm   (*d*) ft H_2O

5. Your textbook lists five types of pressures: atmospheric pressure, barometric pressure, gauge pressure, absolute pressure, and vacuum pressure as shown on top of page 49.
(*a*) What kind of pressure is measured by A?
(*b*) What kind of pressure is measured by B?
(*c*) What would be the reading in the following setup, assuming that pressure and temperature inside and outside the helium tank are the same as in parts (a) and (b)?

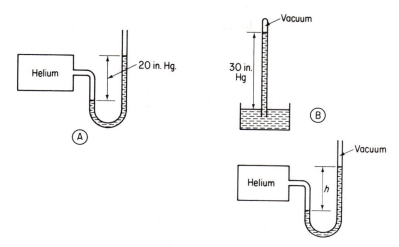

6. An evaporator shows a reading of 40 kPa vacuum. What is the absolute pressure in the evaporator in kilopascals?

Section 1.7   Physical and Chemical Properties of Compounds and Mixtures

Your objectives in studying this section are to be able to:

1. Cite five sources of physical property data.
2. Go to the library and locate the five sources of data.
3. Retrieve data from the five sources to use in your calculations, employing the indexes and tables of contents.

Publications of industrial associations provide a tremendous storehouse of data. For example, the American Petroleum Institute publishes both the *Manual on Disposal of Refinery Wastes*[10] and the *Technical Data Book—Petroleum Refining.*[11] Tables 1.6, 1.7 and 1.8 are typical of the tables that can be found in the former.

If you know the chemical formula for a pure compound, you can look up many of its physical properties in standard reference books, such as

(a) Perry's *Chemical Engineers' Handbook*[12]

[10]American Petroleum Institute, New York, 1969.

[11]Ibid., 2nd ed., 1970.

[12]J. H. Perry and C. H. Chilton, eds., *Chemical Engineers' Handbook*, 5th ed., McGraw-Hill, New York, 1973.

(b) *Handbook of Physics and Chemistry*[13]

(c) Lange's *Handbook*[14]

(d) *The Properties of Gases and Liquids*[15]

Also, specialized texts in your library frequently list properties of compounds. For example, *Fuel Flue Gases*[16] lists considerable data on gas mixtures which are primarily of concern in the gas utility field. Appendix D in this text tabulates the chemical formula, molecular weight, melting point, boiling point, and so on, for most, but not all of the compounds involved in the problems at the end of the chapters.

Many of the materials we talk about and use every day are not pure compounds, and it is considerably more trouble to obtain information about the properties of these materials. Fortunately, for materials such as coal, coke, petroleum products, and natural gas—which are the main sources of energy in this country—tables and formulas are available in reference books and handbooks which list some of the specific gross properties of these mixtures. Some typical examples of natural gas analyses and ultimate analyses of petroleum and petroleum products as shown in Tables 1.6 through 1.10 point out the variety of materials that can be found.

Since petroleum and petroleum products represent complex mixtures of hydrocarbons and other organic compounds of various types together with various impurities, if the individual components cannot be identified, then the mixture is treated as

TABLE 1.6 DISSOLVED SOLIDS IN FRESH WATERS OF THE UNITED STATES

	Concentration (mg/liter)		
Component	In 5% of the Waters Is Less Than	In 50% of the Waters Is Less Than	In 95% of the Waters Is Less Than
Total dissolved solids	72.0	169.0	400.0
Bicarbonate, HCO_3	40.0	90.0	180.0
Sulfate, SO_4	11.0	32.0	90.0
Chloride, Cl	3.0	9.0	170.0
Nitrate, NO_3	0.2	0.9	4.2
Calcium, Ca	15.0	28.0	52.0
Magnesium, Mg	3.5	7.0	14.0
Sodium, Na, and potassium, K	6.0	10.0	85.0
Iron, Fe	0.1	0.3	0.7

SOURCE: *Manual on Disposal of Refinery Wastes*, American Petroleum Institute, New York, 1969, pp. 2–4.

[13] *Handbook of Chemistry and Physics*, CRC Press, West Palm Beach, Fla., annually.

[14] N. A. Lange, *Handbook of Chemistry*, 12th ed., McGraw-Hill, New York, 1979.

[15] R. C. Reid and T. K. Sherwood, *The Properties of Gases and Liquids*, 3rd ed., McGraw-Hill, New York, 1977.

[16] *Fuel Flue Gases*, American Gas Association, New York, 1941.

a uniform compound. Usually, the components in natural gas can be identified, and thus their individual physical properties can be looked up in the reference books mentioned above or in Appendix D. As discussed in Chap. 3 under gases, the properties of a pure gas when mixed with another gas are often the sum of the properties of the pure components. On the other hand, liquid petroleum crude oil and petroleum

TABLE 1.7 CHARACTERISTICS OF WATERS THAT SUPPORT A VARIED AND PROFUSE FISH FAUNA—A CRITERION OF QUALITY

Characteristic	Value of Characteristic		
	In 5% of the Waters Is Less Than	In 50% of the Waters Is Less Than	In 95% of the Waters Is Less Than
pH value	6.7	7.6	8.3
		Concentration (mg/liter)	
Dissolved oxygen, O_2	5.0	6.8	9.8
Carbon dioxide, CO_2			
Free	0.1	1.5	5.0
Fixed	8.0	45.0	95.0
Ammonia, NH_3	0.5	1.5	2.5
		Mho (25°C)	
Specific conductivity	$(50)(10^{-6})$	$(270)(10^{-6})$	$(1100)(10^{-6})$

SOURCE: *Manual on Disposal of Refinery Wastes*, American Petroleum Institute, New York, 1969, pp. 2–4.

TABLE 1.8 REFINERY BIOLOGICAL TREATMENT UNIT FEED CHARACTERISTICS

	Ranges Reported*
Chlorides, mg/liter	200–960
COD, mg/liter	140–640
BOD_5, mg/liter	97–280
Suspended solids, mg/liter	80–450
Alkalinity, mg/liter as $CaCO_3$	77–210
Temperature, °F	69–100
Ammonia nitrogen, mg/liter	56–120
Oil, mg/liter	23–130
Phosphate, mg/liter	20–97
Phenolics, mg/liter	7.6–61
pH	7.1–9.5
Sulfides, mg/liter	1.3–38
Chromium, mg/liter	0.3–0.7

SOURCE: *Manual on Disposal of Refinery Wastes*, American Petroleum Institute, New York, 1969, pp. 2–4.

*Values are the averages of the minima and maxima reported by 12 refineries treating total effluent. Individual plants have reported data well outside many of these ranges.

fractions are such complicated mixtures that their physical properties are hard to estimate from the pure components (even if known) unless the mixture is very simple. As a result of the need for methods of predicting the behavior of petroleum stocks, empirical correlations have been developed in recent years for many of the physical properties we want to use. These correlations are based upon the °API, the Universal Oil Products characterization factor K, the boiling point, and the apparent molecular weight of the petroleum fraction. These parameters in turn are related to five or six relatively simple tests of the properties of oils. Some of the details of these tests, the empirical parameters, and the properties that can be predicted from these parameters will be found in the *API Technical Dada Book* and in Appendix K.

TABLE 1.9   TYPICAL DRY GAS ANALYSES

Type	Analysis (vol. %—excluding water vapor)									
	CO_2	O_2	N_2	CO	H_2	CH_4	C_2H_6	C_3H_8	C_4H_{10}	$C_5H_{12}^+$
Natural gas	6.5					77.5	16.0			
Natural gas, dry*	0.2		0.6			99.2				
Natural gas, wet*	1.1					87.0	4.1	2.6	2.0	3.4
Natural gas, sour†	(H_2S 6.4)					58.7	16.5	9.9	5.0	3.5
Butane							2.0	3.5	75.4	*n*-butane
									18.1	Isobutane
									Illuminants	
Reformed refinery oil	2.3	0.7	4.9	20.8	49.8	12.3	5.5	5.5	3.7	
Coal gas, by-product	2.1	0.4	4.4	13.5	51.9	24.3			3.4	
Producer gas	4.5	0.6	50.9	27.0	14.0	3.0				
Blast furnace gas	5.4	0.7	8.3	37.0	47.3	1.3				
Sewage gas	22.0		6.0		2.0	68.0				

SOURCE: *Fuel Flue Gases*, American Gas Association, New York, 1941, p. 20.
*Dry gas contains much less propane (C_3H_8) and higher hydrocarbons than wet gas does.
†*Sour* implies that the gas contains significant amounts of hydrogen sulfide.

TABLE 1.10   ULTIMATE ANALYSIS OF PETROLEUM CRUDE

Type	Sp Gr	At °C	Weight %				
			C	H	N	O	S
Pennsylvania	0.862	15	85.5	14.2			
Humbolt, Kan.	0.921		85.6	12.4			0.37
Beaumont, Tex.	0.91		85.7	11.0	2.61		0.70
Mexico	0.97	15	83.0	11.0	1.7		
Baku USSR	0.897		86.5	12.0		1.5	

SOURCE: Data from W. L. Nelson, *Petroleum Refinery Engineering*, 4th ed., McGraw-Hill, New York, 1958.

Self-Assessment Test

1. What are five sources of data on physical properties?

2. In what reference book might you find data on:
   (a) Boiling point of inorganic liquids?
   (b) Gas compositions for refinery gases?
   (c) Vapor pressures of organic liquids?
   (d) Chemical formula and properties of protocatechuic acid $(3-, 4-)$?

3. List the page numbers on which the items described in question 2 can be found in the reference you select.

Section 1.8 Technique of Problem Solving

> *Your objectives in studying this section are to be able to:*
>
> 1. Memorize the eight components of effective problem solving.
> 2. Apply the components to all types of problems.

Weiler's Law: Nothing is impossible for the person who doesn't have to do it.

Howe's Law: Every person has a scheme which will not work.

The 90/90 Law: The first 10% of the task takes 90% of the time. The remaining 90% takes the remaining 10%.

Gordon's Law: If a project is not worth doing, it's not worth doing well.

Slack's Law: The least you will settle for is the most you can expect to get.

O'Toole's Commentary: Murphy was an optimist.[17]

If you can form good habits of problem solving early in your career, you will save considerable time and avoid many frustrations in all aspects of your work, in and out of school. Polya[18] recommends the use of four steps for solving problems and puzzles: define, plan, carry out the plan, and look back. The key features of this strategy are the interaction among the steps and the interplay between critical and creative thinking. A number of other helpful references are listed at the end of this chapter if you need assistance. Figure 1.14 sketches the general concept involved in problem solving. First, you must identify what result you are to achieve, that is, what the problem is. Then you must define the system, perhaps with the aid of a diagram. Various physical constraints will apply as well as the time available for you to work on the solution. In almost all cases you will have to look up data and make

[17]Arthur Bloch, *Murphy's Law and Other Reasons Why Things Go Wrong*! Price/Stern/Sloan, New York, 1977.

[18]G. Polya, *How to Solve It*, 2nd ed. Doubleday Anchor, Garden City, N.Y., 1957.

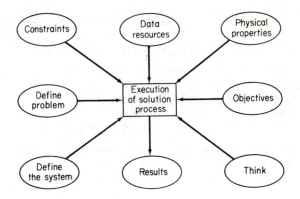

Fig. 1.14 Elements of problem solving.

use of general laws. Finally, the results will have to be presented properly so that you can communicate them to someone else.

Novice prospective engineers usually have the most difficulty with:

(a) Translating the physical problem into a mathematical description
(b) Relating the physical problem to theory
(c) Simplifying a complex problem
(d) Learning how to ask the right questions

In solving material and energy balance problems, you should:

(a) Read the available information through thoroughly and understand what is required for an answer. Sometimes, as in life, the major obstacle is to find out what the problem really is.
(b) Determine what additional data are needed, if any, and obtain this information.
(c) Draw a simplified picture of what is taking place and write down the available data. You may use boxes to indicate processes or equipment, and lines for the flow streams.
(d) Pick a basis on which to start the problem, as discussed in Sec. 1.3–6.
(e) If a chemical equation is involved, write it down and make sure that it is balanced.

By this time you should have firmly in mind what the problem is and a reasonably clear idea of what you are going to do about it; however, if you have not seen exactly how to proceed from what is available to what is wanted, then you should:

(f) Decide what formulas or principles are governing in this specific case and what types of calculations and intermediate answers you will need to get the final answer. If alternative procedures are available, try to decide which are the most expedient. If an unknown cannot be found directly, give it a letter symbol, and proceed as if you knew it.

(g) Make the necessary calculations in good form, being careful to check the arithmetic and units as you proceed.

(h) Determine whether the answer seems reasonable in view of your experience with these types of calculations.

You can work backward as well as forward in solving problems if the forward sequence of steps to take is not initially clear. Problems that are long and involved should be divided into parts and attacked systematically piece by piece.

If you can assimilate this procedure and make it a part of yourself—so that you do not have to think about it step by step—you will find that you will be able to materially improve your speed, performance, and accuracy in problem solving.

> The major difference between problems solved in the classroom and in the plant lies in the quality of the data available for the solution. Plant data may be of poor quality, inconclusive, inadequate, or actually conflicting, depending on the accuracy of sampling, the type of analytical procedures employed, the skill of the technicians in the operation of analytical apparatus, and many other factors. The ability of an engineer to use the stoichiometric principles for the calculation of problems of material balance is only partly exercised in the solution of problems, even of great complexity, if solved from adequate and appropriate data. The remainder of the test lies in his ability to recognize poor data, to request and obtain usable data, and, if necessary, to make accurate estimates in lieu of incorrect or insufficient data.[19]

Self-Assessment Test

1. Prepare an information flow diagram showing the sequence (serial and parallel) of eight steps to be used in effective problem solving.

2. Take any example in Sec. 1.3, 1.5, or 1.6 and prepare an information flow diagram of your thought process in solving the example. Make a tree showing how the following classes of information are connected (put the solution as the last stage of the tree at the bottom):
(*a*) Information stated in the problem
(*b*) Information implied or inferred from the problem statement
(*c*) Information from your memory (internal data bank!)
(*d*) Information from an external data bank (reference source)

[19]B. E. Lauer, *The Material Balance*, Work Book Edition, Dept. Chem. Eng., Univ. of Colorado, Boulder, Colorado, 1954, p. 89.

(e) Information determined by reasoning or calculations
Label each class with a different-type box (circle, square, diamond, etc.) and let arrows connect the boxes to show the sequence of information flow for your procedure.

3. What should you do if you experience the following difficulties in solving problems?
   (a) No interest in the material and no clear reason to remember
   (b) Cannot understand after reading the material
   (c) Read to learn "later"
   (d) Rapidly forget what you have read
   (e) Form of study is inappropriate

Section 1.9   The Chemical Equation and Stoichiometry

Your objectives in studying this section are to be able to:

1. Write and balance chemical reaction equations.
2. Calculate the stoichiometric quantities of reactants and products given the chemical equation.
3. Define excess reactant, limiting reactant, conversion, degree of completion, and yield in a reaction.
4. Identify the limiting and excess reactants and calculate the percent excess reactant(s), the percent conversion, the percent completion, and yield for a chemical reaction with the reactants being in nonstoichiometric proportions.
5. Know the products of common reactions given the reactants.

As you already know, the chemical equation provides a variety of qualitative and quantitative information essential for the calculation of the combining weights of materials involved in a chemical process. Take, for example, the combustion of heptane as shown below. What can we learn from this equation?

$$C_7H_{16} + 11O_2 \longrightarrow 7CO_2 + 8H_2O \tag{1.25}$$

First, make sure that the equation is balanced. Then we can see that 1 mole (*not* lb_m or kg) of heptane will react with 11 moles of oxygen to give 7 moles of carbon dioxide plus 8 moles of water. These may be lb mol, g mol, kg mol, or any other type of mole, as shown in Fig. 1.15. One mole of CO_2 is formed from each $\frac{1}{7}$ mole of C_7H_{16}. Also, 1 mole of H_2O is formed with each $\frac{7}{8}$ mole of CO_2. Thus the equation tells us in terms of moles (*not* mass) the ratios among reactants and products.

Stoichiometry (stoi-ki-om-e-tri)[20] deals with the combining weights of elements and compounds. The ratios obtained from the numerical coefficients in the chemical equation are the stoichiometric ratios that permit you to calculate the moles of one substance as related to the moles of another substance in the chemical equation. If the basis selected is to be mass (lb_m, kg) rather than moles, you should use the follow-

[20]From the Greek *stoicheion*, basic constituent, and *metrein*, to measure.

C_7H_{16}	$+$	$11O_2$	$\longrightarrow$	$7CO_2$	$+$	$8H_2O$

Qualitative information

heptane	reacts with	oxygen	to give	carbon dioxide	and	water

Quantitative information

1 molecule of heptane	reacts with	11 molecules of oxygen	to give	7 molecules of carbon dioxide	and	8 molecules of water
6.023×10^{23} molecules of C_7H_{16}	$+$	$11(6.023 \times 10^{23})$ molecules of O_2	$\longrightarrow$	$7(6.023 \times 10^{23})$ molecules of CO_2	$+$	$8(6.023 \times 10^{23})$ molecules of H_2O
1 g mole of C_7H_{16}	$+$	11 g moles of O_2	$\longrightarrow$	7 g moles of CO_2	$+$	8 g moles of H_2O
1 kg mole of C_7H_{16}	$+$	11 kg moles of O_2	$\longrightarrow$	7 kg moles of CO_2	$+$	8 kg moles of H_2O
1 lb mole of C_7H_{16}	$+$	11 lb moles of O_2	$\longrightarrow$	7 lb moles of CO_2	$+$	8 lb moles of H_2O
1 ton mole of C_7H_{16}	$+$	11 ton moles of O_2	$\longrightarrow$	7 ton moles of CO_2	$+$	8 ton moles of H_2O
$1(100)$ g of C_7H_{16}	$+$	$11(32)$ g of O_2	$=$	$7(44)$ g of CO_2	$+$	$8(18)$ g of H_2O
100 g		352 g		308 g		144 g

452 g	$=$	452 g
452 kg	$=$	452 kg
452 ton	$=$	452 ton
452 lb	$=$	452 lb

Fig. 1.15 The chemical equation.

ing method in solving problems involving chemical equations: (1) Use the molecular weight to calculate the number of moles of the substance equivalent to the basis; (2) change this number of moles into the corresponding number of moles of the desired product or reactant by multiplying by the proper stoichiometric ratio, as determined by the chemical equation; and (3) then change the moles of product or reactant to a weight basis. These steps are indicated in Fig. 1.16 for the reaction above. You can combine these steps in a single dimensional equation, as shown in the examples below, for ease of calculations.

An assumption implicit in the above is that the reaction takes place exactly as written in the equation and proceeds to 100% completion. When reactants, products, or degree of completion of the actual reaction differ from the assumptions of the equation, additional data must be made available to indicate the actual status or situation.

Basis : 10.0 kg C_7H_{16}

Component	Mol. wt
C_7H_{16}	100.1
O_2	32.0
CO_2	44.0
H_2O	18.0

$$1 \text{ kg mole} \qquad\qquad 7 \text{ kg mole}$$

$$C_7H_{16} \quad + \quad 11O_2 \quad \longrightarrow \quad 7CO_2 \quad + \quad 8H_2O$$

$$\frac{10.0 \text{ kg } C_7H_{16}}{\dfrac{100.1 \text{ kg } C_7H_{16}}{\text{kg mole } C_7H_{16}}} = 0.100 \text{ kg mole } C_7H_{16} \longrightarrow \frac{0.700 \text{ kg mole } CO_2}{\dfrac{1 \text{ kg mole } CO_2}{44.0 \quad CO_2}} = 30.8 \text{ kg } CO_2$$

$$10.0 \text{ kg } C_7H_{16} \text{ yields } 30.8 \text{ kg } CO_2$$

Fig. 1.16 Stoichiometry.

EXAMPLE 1.25 Stoichiometry

In the combustion of heptane, CO_2 is produced. Assume that you want to produce 500 kg of dry ice per hour and that 50% of the CO_2 can be converted into dry ice, how many kilograms of heptane must be burned per hour?

Solution

Basis: 500 kg of dry ice (or 1 hr)

Mol. wt. heptane $= 100$.

Chemical equation as in Fig. 1.15

$$\frac{500 \text{ kg dry ice}}{} \left| \frac{1 \text{ kg } CO_2}{0.5 \text{ kg dry ice}} \right| \frac{1 \text{ kg mole } CO_2}{44 \text{ kg } CO_2} \left| \frac{1 \text{ kg mole } C_7H_{16}}{7 \text{ kg mole } CO_2} \right.$$

$$\left| \frac{100 \text{ kg } C_7H_{16}}{1 \text{ kg mole } C_7H_{16}} = 325 \text{ kg } C_7H_{16} \right.$$

Since the basis of 500 kg of dry ice is identical to 1 hr, 325 kg of C_7H_{16} must be burned per hour. Note that kilograms are first converted to moles, then the chemical equation is applied, and finally moles are converted to kilograms again for the final answer.

EXAMPLE 1.26 Stoichiometry

Corrosion of pipes in boilers by oxygen can be alleviated through the use of sodium sulfite. Sodium sulfite removes oxygen from boiler feedwater by the following reaction:

$$2Na_2SO_3 + O_2 \longrightarrow 2Na_2SO_4$$

How many pounds of sodium sulfite are theoretically required to remove the oxygen from

8,330,000 lb of water (10^6 gal) containing 10.0 parts per million (ppm) of dissolved oxygen and at the same time maintain a 35% excess of sodium sulfite? See Fig. E1.26.

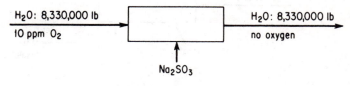

H₂O: 8,330,000 lb
10 ppm O₂   H₂O: 8,330,000 lb
                                             no oxygen

Na₂SO₃

Fig. E1.26

Solution

Additional data: mol. wt. of Na_2SO_3 is 126. Chemical equation: $2Na_2SO_3 + O_2 \longrightarrow 2Na_2SO_4$.

Basis: 8,330,000 lb of H_2O with 10 ppm O_2 or 83.3 lb of O_2

$$\frac{8{,}330{,}000 \text{ lb } H_2O}{} \left| \frac{10 \text{ lb } O_2}{\underbrace{(1{,}000{,}000 - 10 \text{ lb } O_2) \text{ lb } H_2O}_{\textbf{effectively same as 1,000,000}}} \right. = 83.3 \text{ lb } O_2$$

$$\frac{8{,}330{,}000 \text{ lb } H_2O}{} \left| \frac{10 \text{ lb } O_2}{10^6 \text{ lb } H_2O} \right| \frac{1 \text{ lb mol } O_2}{32 \text{ lb } O_2} \left| \frac{2 \text{ lb mol } Na_2SO_3}{1 \text{ lb mol } O_2} \right.$$

$$\left| \frac{126 \text{ lb } Na_2SO_3}{1 \text{ lb mol } Na_2SO_3} \right| \frac{1.35}{1} = 885 \text{ lb } Na_2SO_3$$

EXAMPLE 1.27 Stoichiometry

A limestone analyzes

$CaCO_3$	92.89%
$MgCO_3$	5.41%
Insoluble	1.70%

(a) How many pounds of calcium oxide can be made from 5 tons of this limestone?
(b) How many pounds of CO_2 can be recovered per pound of limestone?
(c) How many pounds of limestone are needed to make 1 ton of lime?

Solution

Read the problem carefully to fix in mind exactly what is required. Lime will include all the impurities present in the limestone which remain after the CO_2 has been driven off. Next, draw a picture of what is going on in this process. See Fig. E1.27.

To complete the preliminary analysis you need the following chemical equations:

$$CaCO_3 \longrightarrow CaO + CO_2$$
$$MgCO_3 \longrightarrow MgO + CO_2$$

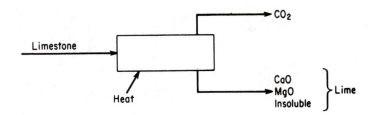

Fig. E1.27

Additional data:

	CaCO₃	MaCO₃	CaO	MgO	CO₂
Mol. wt.:	100	84.3	56.0	40.3	44

Basis: 100 lb of limestone

This basis was selected because pounds = percent.

component	lb = percent	lb mol ⟶	lime	lb	CO₂ (lb)
CaCO₂	92.89	0.9289	CaO	52.0	40.9
MgCO₂	5.41	0.0641	MgO	2.59	2.82
Insoluble	1.70		Insoluble	1.70	
Total	100.00	0.9930	Total	56.3	43.7

Note that the total pounds of products equal the 100 lb of entering limestone. Now to calculate the quantities originally asked for:

(a) $$\text{CaO produced} = \frac{52.0 \text{ lb CaO}}{100 \text{ lb stone}} \Big| \frac{2000 \text{ lb}}{1 \text{ ton}} \Big| 5 \text{ ton} = 5200 \text{ lb CaO}$$

(b) $$\text{CO}_2 \text{ recovered} = \frac{43.7 \text{ lb CO}_2}{100 \text{ lb stone}} = 0.437 \text{ lb, or}$$

$$\frac{0.9930 \text{ lb mol CaCO}_3 + \text{MgCO}_3}{100 \text{ lb stone}} \Big| \frac{1 \text{ lb mole CO}_2}{1 \text{ lb mol CaCO}_3 + \text{MgCO}_3}$$

$$\Big| \frac{44 \text{ lb CO}_2}{1 \text{ lb mol CO}_2} = 0.437 \text{ lb}$$

(c) $$\text{Limestone required} = \frac{100 \text{ lb stone}}{56.3 \text{ lb lime}} \Big| \frac{2000 \text{ lb}}{1 \text{ ton}} = 3560 \text{ lb stone}$$

In industrial reactions you will rarely find exact stoichiometric amounts of materials used. To make a desired reaction take place or to use up a costly reactant, excess reactants are nearly always used. This excess material comes out together with, or perhaps separately from, the product—and sometimes can be used again. Even if stoichiometric quantities of reactants are used, but if the reaction is not complete or there are side reactions, the products will be accompanied by unused reactants as well as side products. In these circumstances some new definitions must be understood:

(a) *Limiting reactant* is the reactant that is present in the smallest stoichiometric amount.

(b) *Excess reactant* is a reactant in excess of the limiting reactant. The *percent excess* of a reactant is based on the amount of any excess reactant above the amount required to react with the limiting reactant according to the chemical equation, or

$$\% \text{ excess} = \frac{\text{moles in excess}}{\text{moles required to react with limiting reactant}} (100)$$

where the moles in excess frequently can be calculated as the total available moles of a reactant less the moles required to react with the limiting reactant. A common term, *excess air*, is used in combustion reactions; it means the amount of air available to react that is in excess of the air theoretically required to *completely* burn the combustible material. The *required* amount of a reactant is established by the limiting reactant and is for all other reactants the corresponding stoichiometric amount. **Even if only part of the limiting reactant actually reacts, the required and excess quantities are based on the entire amount of the limiting reactant.**

Air requirements for combustion vary with the need to ensure full utilization of the fuel's heating value but not generate excessive air pollutants. The excess air required in practice depends on the type of fuel, the furnace, and the burner. Fuel oil, for instance, requires 5 to 20% excess air depending on burner design. Excess air is recognized as a routine measure of heater performance.

Three other terms that are used in connection with chemical reactions have less clear-cut definitions: conversion, selectivity, and yield. No universally agreed upon definitions exist for these terms—in fact, quite the contrary. Rather than cite all the possible usages of these terms, many of which conflict, we shall define them as follows:

(c) *Conversion* is the fraction of the feed or some material in the feed that is converted into products. What the basis in the feed is for the calculations and into what products the basis is being converted must be clearly specified or endless confusion results. Conversion is related to the *degree of completion* of a reaction, which is usually the percentage or fraction of the limiting reactant converted into products.

(d) *Selectivity* is the ratio of the moles of a particular (usually the desired) product produced to the moles of another (usually undesired) product produced in a set of reactions.

(e) *Yield*, for a single reactant and product, is the weight or moles of final product divided by the weight or moles of initial reactant (P lb of product

A per *R* lb of reactant *B*). If more than one product and more than one reactant are involved, the reactant upon which the yield is to be based must be clearly stated.

Suppose that we have a reaction sequence as follows:

$$A \longrightarrow B \longrightarrow C$$
$$\searrow C$$

with *B* the desired product and *C* the undesired one. The yield of *B* is the mass (moles) of *B* produced/mass (moles) of *A* consumed, and the selectivity of *B* is the mass (moles) of *B*/mass (moles) of *C* produced.

The terms "yield" and "selectivity" are terms that measure the degree to which a desired reaction proceeds relative to competing alternative (undesirable) reactions. As a designer of equipment you want to maximize production of the desired product and minimize production of the unwanted products. Do you want high or low selectivity? Yield?

The employment of these concepts can best be illustrated by examples.

EXAMPLE 1.28 Limiting Reactant and Incomplete Reaction

Antimony is obtained by heating pulverized stibnite with scrap iron and drawing off the molten antimony from the bottom of the reaction vessel:

$$Sb_2S_3 + 3Fe \longrightarrow 2Sb + 3FeS$$

Suppose 0.600 kg of stibnite and 0.250 kg of iron turnings are heated together to give 0.200 kg of Sb metal.

Calculate:
- (a) The limiting reactant
- (b) The percentage of excess reactant
- (c) The degree of completion (fraction)
- (d) The percent conversion
- (e) The yield

Solution

The molecular weights needed to solve the problem and the gram moles forming the basis are:

component	kg	mol. wt.	g mol
Sb_2S_3	0.600	339.7	1.77
Fe	0.250	55.8	4.48
Sb	0.200	121.8	1.64
FeS		87.9	

(a) To find the limiting reactant, we examine the chemical reaction equation and note that if 1.77 g mol of Sb_2S_3 reacts, it requires $3(1.77) = 5.31$ g mol of Fe, whereas if 4.48 g mol of Fe reacts, it requires $(4.48/3) = 1.49$ g mol of Sb_2S_3 to be available. Thus Fe is present in the smallest stoichiometric amount and is the limiting reactant; Sb_2S_3 is the excess reactant.

(b) The percentage of excess reactant is

$$\% \text{ excess} = \frac{1.77 - 1.49}{1.49}(100) = 18.8\% \text{ excess } Sb_2S_3$$

(c) Although Fe is the limiting reactant, not all the limiting reactant reacts. We can compute from the 1.64 g mol of Sb how much Fe actually does react:

$$\frac{1.64 \text{ g mol Sb}}{} \left| \frac{3 \text{ g mol Fe}}{2 \text{ g mol Sb}} = 2.46 \text{ g mol Fe}\right.$$

If by the fractional degree of completion is meant the fraction conversion of Fe to FeS, then

$$\text{fractional degree of completion} = \frac{2.46}{4.48} = 0.55$$

(d) The percent conversion can be arbitrarily based on the Sb_2S_3 if our interest is mainly in the stibnite:

$$\frac{1.64 \text{ g mol Sb}}{} \left| \frac{1 \text{ g mol } Sb_2S_3}{2 \text{ g mol Sb}} = 0.82 \text{ g mol } Sb_2S_3\right.$$

$$\% \text{ conversion of } Sb_2S_3 \text{ to Sb} = \frac{0.82}{1.77}(100) = 46.3\%$$

(e) The yield will be stated as kilograms of Sb formed per kilogram of Sb_2S_3 that was fed to the reaction:

$$\text{yield} = \frac{0.200 \text{ kg Sb}}{0.600 \text{ kg } Sb_2S_3} = \frac{1}{3} \frac{\text{kg Sb}}{\text{kg } Sb_2S_3}$$

EXAMPLE 1.29 Limiting Reactant and Incomplete Reactions

Aluminum sulfate can be made by reacting crushed bauxite ore with sulfuric acid, according to the following equation:

$$Al_2O_3 + 3H_2SO_4 \longrightarrow Al_2(SO_4)_3 + 3H_2O$$

The bauxite ore contains 55.4% by weight of aluminum oxide, the remainder being impurities. The sulfuric acid solution contains 77.7% H_2SO_4, the rest being water.

To produce crude aluminum sulfate containing 1798 lb of pure aluminum sulfate, 1080 lb of bauxite ore and 2510 lb of sulfuric acid solution are used.

(a) Identify the excess reactant.

(b) What percentage of the excess reactant was consumed?

(c) What was the degree of completion of the reaction?

Solution

The pound moles of substances forming the basis of the problem can be computed as follows:

$$\frac{1798 \text{ lb } Al_2(SO_4)_3}{} \left| \frac{1 \text{ lb mol } Al_2(SO_4)_3}{342.1 \text{ lb } Al_2(SO_4)_3} = 5.26 \text{ lb mol}\right.$$

$$\frac{1080 \text{ lb bauxite}}{} \left| \frac{0.554 \text{ lb } Al_2O_3}{1 \text{ lb bauxite}} \right| \frac{1 \text{ lb mol } Al_2O_3}{101.9 \text{ lb } Al_2O_3} = 5.87 \text{ lb mol}$$

$$\frac{2510 \text{ lb acid}}{} \left| \frac{0.777 \text{ lb } H_2SO_4}{1 \text{ lb acid}} \right| \frac{1 \text{ lb mol } H_2SO_4}{98.1 \text{ lb } H_2SO_4} = 19.88 \text{ mol}$$

(a) Assume that Al_2O_3 is the limiting reactant. Then $5.87 \times 3 = 17.61$ lb mol of H_2SO_4 would be required, which is present. Hence H_2SO_4 is the excess reactant.

(b) The $Al_2(SO_4)_3$ actually formed indicates that

$$\frac{5.26 \text{ lb mol } Al_2(SO_4)_3}{} \left| \frac{3 \text{ lb mol } H_2SO_4}{1 \text{ lb mol } Al_2(SO_4)_3} = 15.78 \text{ lb mol } H_2SO_4 \text{ was consumed}\right.$$

$$\frac{15.78}{19.88}(100) = 79.4\%$$

(c) The fractional degree of completion was

$$\frac{5.26}{5.87} = 0.90$$

EXAMPLE 1.30 The Meaning of Selectivity and Yield

Two well-known reactions take place in the dehydrogenation of ethane:

$$C_2H_6 \longrightarrow C_2H_4 + H_2 \tag{a}$$

$$C_2H_6 + H_2 \longrightarrow 2CH_4 \tag{b}$$

Given the following product distribution (in the gas-phase reaction of C_2H_6 in the presence of H_2)

component	percent
C_2H_6	35
C_2H_4	30
H_2	28
CH_4	7
Total	100

what is (a) the selectivity of C_2H_4 relative to CH_4 and (b) the yield of C_2H_4 in kilogram moles of C_2H_4 per kilogram mole of C_2H_6?

Solution

Basis: 100 kg mol of products

(a) The selectivity (as defined) is

$$\frac{30 \text{ kg mol } C_2H_4}{7 \text{ kg mol } CH_4} = 4.29 \frac{\text{mol } C_2H_4}{\text{mol } CH_4}$$

(b) The moles of C_2H_6 entering into the reaction can be determined from the C_2H_4 and the CH_4 formed.

$$\frac{30 \text{ kg mol } C_2H_4}{} \left| \frac{1 \text{ kg mol } C_2H_6}{1 \text{ kg mol } C_2H_4} \right. = 30 \text{ kg mol } C_2H_6$$

$$\frac{7 \text{ kg mol } CH_4}{} \left| \frac{1 \text{ kg mol } C_2H_6}{2 \text{ kg mol } CH_4} \right. = \underline{3.5 \text{ kg mol } C_2H_6}$$

$$33.5 \text{ kg mol } C_2H_6$$

$$\frac{30 \text{ kg mol } C_2H_4}{33.5 \text{ kg mol } C_2H_6} = 0.90 \frac{\text{kg mol } C_2H_4}{\text{kg mol } C_2H_6}$$

You should remember that the chemical equation does not indicate the true mechanism of the reaction or how fast or to what extent the reaction will take place. For example, a lump of coal in air will sit unaffected at room temperature, but at higher temperatures it will readily burn. All the chemical equation indicates is the stoichiometric amounts required for the reaction and obtained from the reaction if it proceeds in the manner in which it is written. Also remember to make sure that the chemical equation is balanced before using it.

Self-Assessment Test

1. Write balanced reaction equations for the following reactions:
(**a**) C_9H_{18} and oxygen to form carbon dioxide and water
(**b**) FeS_2 and oxygen to form Fe_2O_3 and sulfur dioxide

2. If 1 kg of benzene (C_6H_6) is oxidized with oxygen, how many kilograms of O_2 are needed to convert all the benzene to CO_2 and H_2O?

3. The electrolytic manufacture of chlorine gas from a sodium chloride solution is carried out by the following reaction:

$$2NaCl + 2H_2O \longrightarrow 2NaOH + H_2 + Cl_2$$

How many kilograms of Cl_2 can one produce from 10 m³ of a brine solution containing 5% by weight of sodium chloride? The specific gravity of the solution relative to water at 4°C is 1.07.

4. Calcium oxide (CaO) is formed by decomposing limestone (pure $CaCO_3$). In one kiln the reaction goes to 70% completion.
(**a**) What is the composition of the solid product withdrawn from the kiln?
(**b**) What is the yield in terms of pounds of CO_2 produced per pound of limestone charged?

5. In problem 3, suppose that 50.0 kg of NaCl reacts with 10.0 kg of H_2O.

  (*a*) What is the limiting reaction?

  (*b*) What is the excess reactant?

  (*c*) What components will the product solution contain if the reaction is 60% complete?

Section 1.10   Digital Computers in Problem Solving

Your objectives in studying this section are to be able to:

1. Cite one computer code that you have available for use in solving material and energy balance problems.

2. Cite three difficulties in using such codes.

3. Cite two advantages of using such codes.

Digital computers, both large and small, have made a significant impact on problem solving in chemical engineering. At the same time the somewhat arbitrary distinction between calculators and computers is fading as a result of the greatly enhanced storage capacity and speed of hand-held devices. You must become skilled in the enployment of both types of devices even though the bulk of the problems in this text are formulated toward solution by hand-held devices. (Some of the problems, those denoted by asterisks and the terminal problems in each problem set at the ends of chapters, are more appropriate for computer solutions.)

Many routes of approach can be used in engineering education to familiarize students with the role of the computer in solving engineering problems. On the one hand, library computer programs ("canned" programs) can be employed to solve specific types of problems, absolving you of most of the programming chores. On the other hand, you can prepare the complete computer code to solve a problem from scratch. Wide differences in computers and the proliferation of programming languages have made the inclusion of detailed computer codes in this text inappropriate.

You should not attempt to justify the use of a computer as a time-saving device per se when working with a single problem. It is much better to look on the computer and the program as a tool that will enable you, after some experience, to solve complex problems expediently and that may save time and trouble for a simple problem, but may not. Modern high-speed computers can eliminate repetitive, time-consuming routine calculations. Also, computers can essentially eliminate numerical errors once the programs are verified. Typical problems encountered in this text that are most susceptible to computer solution are (a) the solution of simultaneous equations, (b) the solution of one or more nonlinear equations, (c) statistical data fitting, and (d) various types of iterative calculations. Typical industrial problems routinely solved are:

  (a) Heat and mass transfer coefficient correlations

  (b) Correlation of chemical structure with physical properties

(c) Analysis of gaseous hydrocarbon mixtures

(d) Particle size analysis

(e) Granular fertilizer formulation

(f) Evaluation of infrared analysis

(g) Optimum operating conditions

(h) Comprehensive economic evaluations

(i) Chemical–biological coordination and correlation

(j) Vapor–liquid equilibrium calculations

(k) Mass spectrometer matrix inversion

(l) Equipment design

One of the most significant changes in the use of computers has been in the meaning of the word *solution*. It is widely recognized that a mathematical closed-form solution, although unquestionably pleasing aesthetically and relatively universal in application, may not necessarily be interpreted easily. Curves and graphs may constitute a more desirable format for a solution, but a computer and a program essentially consist of a procedure whereby a numerical answer can be obtained to a specific problem. Repetitive solutions must be obtained to prepare tables, graphs, and so on, but these, of course, can be prepared automatically with an appropriate computer code.

The use of a canned computer program saves you considerable time and effort relative to programming your own code. Every major computer center has an extensive subroutine library, often stored on-line, for effecting extensive numerical calculations and/or simulation. Of course, such programs may still be fallible. When an unexpected answer appears, the user of the program must know enough about the problem to be able to account for the discrepancy. Often the trouble lies in the interpretation of what the program is intended to do, or it is caused by improper entry of parameters or data. If you can avoid these difficulties, you can concentrate on stating the problem correctly and estimating the expected answer; the mechanics of obtaining the answer would be left to the computer. Considerably more complex and realistic problems can be solved, furnishing a deeper understanding of the principles involved, if the latter are not obscured by concern with the problem-solving technique.

When you use the computer you have to be more precise than you have been accustomed to being in the past. Because of the nature of the computer languages (i.e., the necessity for very precise grammar and punctuation) it is unusual to solve completely correctly an engineering problem on the first approach. Experience has shown that, even in the most carefully prepared programs, there will be a certain percentage of errors, which may include programming logic, actual coding, data preparation, or data entry. For this reason, processing of a problem requires prior program verification in which a computer run of a sample problem is executed for which the results are known, can be estimated, or can be calculated by manual methods. To facilitate such a checkout as well as to guard against undetected and possible future data errors, a good program should include self-checking features strategically located in the program. If physically impossible conditions are created, the calculation should be stopped and the accumulated results printed out. You no doubt will require many tries before achieving success. The turnaround time (i.e., the time elapsed between submission of a program to the computer and its return for checking and possible resubmission in the case of error) must be fairly short if problems are to be solved in a reasonable amount of time. The use of remote time-sharing terminals can assist in the speedy resolution of programming errors. You will also be forced to think more logically and in greater detail than you have been accustomed to doing, if you do your own programming, about the problem and how to solve it. The computer is a rigid task master that requires precision in the statement of the problem and the flow of information needed to effect a solution. No ambiguity is permitted.

Self-Assessment Test

1.   List the items asked for in the objectives.

2.   Use a computer code or prepare one to solve one or more of the computer-oriented problems at the end of the chapter.

SUPPLEMENTARY REFERENCES

General

1. Benson, S. W., *Chemical Calculations*, 2nd ed., Wiley, New York, 1963.
2. Considine, D. M., and S. D. Ross, eds., *Handbook of Applied Instrumentation*, McGraw-Hill, New York, 1964.
3. Felder, R. M., and R. W. Rousseau, *Elementary Principles of Chemical Processes*, Wiley, New York, 1978.
4. Myers, A. L., and W. D. Seider, *Introduction to Chemical Engineering and Computer Calculations*, Prentice-Hall, Englewood Cliffs, N.J., 1976.
5. Ranz, W. E., *Describing Chemical Engineering Systems*, McGraw-Hill, New York, 1970.
6. Russell, T. W. F., and M. M. Denn, *Introduction to Chemical Engineering Analysis*, Wiley, New York, 1972.
7. Whitwell, J. C., and R. K. Toner, *Conservation of Mass and Energy*, Ginn/Blaisdell, Waltham, Mass., 1969.
8. Williams, E. T., and R. C. Johnson, *Stoichiometry for Chemical Engineers*, McGraw-Hill, New York, 1958.

Units and Dimensions

1. American National Metric Council, *Metric Guide for Educational Materials*, 1977.
2. American Petroleum Institute, *API Metric Manual*, Publication 2564, 1977.

3. Davies, W. G., and J. W. Moore, "Adopting SI Units in Introductory Chemistry," *J. Chem. Educ.*, 57, pp. 303–306 (1980).

4. Jones, E. A., and R. B. James, "Metrication in the Process Industries," *Chem. Eng. Progr.*, pp. 76–78 (May 1979).

5. Klinkenberg, A., "The American Engineering System of Units and Its Dimensional Constant g_c," *Indus. Eng. Chem.*, v. 61, no. 4, p. 53 (1960).

6. O'Dal, E. G., *Physical Quantities and Units*, Prentice-Hall, Englewood Cliffs, N.J., 1967.

7. Page, C. H., and P. Vigoureux, *The International System of Units (SI)*, NBS Spec. Publ. 330, January 1971. (For sale by the Superintendent of Documents, U.S. Government Printing Office, Washington, D.C. 20402, SD Catalog No. 13: 10: 330, price 50 cents.)

Problem Solving

1. Chorneyko, D. M., et al., "What Is Problem Solving?" *Chem. Eng. Educ.*, p. 132 (Summer 1979).

2. Leibold, B. G., et al., "Problem Solving," *Eng. Educ.*, p. 172 (November 1976).

3. Moore, R. F., et al., "Developing Style in Solving Problems," *Eng. Educ.*, p. 713 (April 1979).

4. Woods, D. R., "On Teaching Problem Solving," Parts I, II, *Chem. Eng. Educ.*, p. 86 (Spring 1977); p. 140 (Summer 1977).

PROBLEMS[21]

Section 1.1

1.1. An engineer went to the liquor supermarket—which advertised the lowest prices in town—and bought what he thought was the usual 1/2-gallon jug for $7.77, but the jug turned out to be in reality a 1.75-liter bottle. Was this a bargain in comparison with two "fifths" (of a gallon) at $7.99?

1.2. J. Charlton of Brighton, Ontario, claims that "the controversy about the metric system vs. the U.S. and English fps systems is being created by those persons who have to use the fps system the least. . . ." To shake the confidence of this group in "their favorite system," he says, "I have devised a little test."

(1) How many kinds of miles are there?

(2) How many gallons in a barrel?

(3) How many gallons in a drum?

(4) If you import a barrel of wine from France, how many gallons of wine do you get?

(5) Which is heavier, 11 oz of gold or 12 oz of lead?

(6) You are an electroplater. Your supplier sells you a solution containing 1 oz of gold chloride per gallon. What are you getting?

(7) Upon completing the foregoing quiz, you have an opportunity to buy

---

[21]An asterisk designates problems appropriate for computer solution. Refer also to the computer problems at the end of the chapter.

90-proof Canadian rye or 90-proof American rye at the same unit price. Which is the better deal?[22]

1.3. Fill in the values in the blanks in the following:

(a) Consider a tank that is to take a pressure of 10 psi. This is the equivalent of (approximately) _____ newtons/square meter (N/m^2), or _____ kilonewtons/square meter (kN/m^2), or _____ kilopascals (kPa).

(b) It may help to remember that 14.7 psi (_____ atm) = _____ kN/m^2 or approximately _____ MN/m^2.

(c) If your automobile tires take 24 psi of air, you will have to set the air-pump pressure to _____ kN/m^2; if you need 26 psi, the pressure will be _____ kN/m^2. If you are used to asking just for 30 lb of air in your car tires, you would in the future have to ask for _____ kilograms.

(d) An absolute pressure of 10 micrometers of mercury is almost exactly equivalent to _____ N/m^2, or _____ Pa.

(e) If you are designing a corrosion-resistant tank of rolled (annealed) Hastelloy alloy C, the tensile strength needed for your calculations is 896 MN/m^2, which is the equivalent of _____ psi. Perhaps it would be better to round it off to _____ GN/m^2.

1.4. The following appeared as a letter to the editor. Is the author correct? Explain.

> TO THE EDITOR: As a long-time proponent of a programmed switch to the metric system, I wish to applaud your policy of giving measurements in both metric and English units in many of your articles.
>
> The accuracy of some of your conversions, however, does not always merit such applause. On page 63 of your June issue under *Do You Know That*, the area of Guam is given as 212 mi^2 (340 km^2) and the area of the Alabama sinkhole area is given as 10 mi^2 (16 km^2). These are not correct conversions.

1.5. The liquor industry in making the change over to metric measurements used a 46.8-ml bottle for the single serving size (as used on airplanes and trains). Can this be a reasonable metric size? Who benefits from the new size, the drinker or the bottling company?

Data: A "jigger" is $1\frac{1}{2}$ fluid ounces. The old single serving size was $\frac{1}{10}$ pint.

1.6. The following test will measure your SIQ. List the correct answer.

(a) Which is the correct symbol?

(1) nm   (2) °K   (3) sec   (4) N/mm

(b) Which is the wrong symbol?

(1) MN/m^2 (2) GHz/s   (3) $kJ/(s \cdot m^3)$ (4) °C/M/s

(c) Atmospheric pressure is about:

(1) 100 Pa   (2) 100 kPa   (3) 10 MPa   (4) 1 GPa

(d) The temperature 0°C is defined as:

(1) 273.15°K   (2) Absolute zero

(3) 273.15 K   (4) The freezing point of water

---

[22]*Chem. Eng. News*, p. 60 (June 4, 1979).

(e) Which height and mass are those of a petite woman?

(1) 1.50 m, 45 kg   (2) 2.00 m, 95 kg   (3) 1.50 m, 75 kg   (4) 1.80 m, 60 kg

(f) Which is a recommended room temperature in winter?

(1) 15°C   (2) 20°C   (3) 28°C   (4) 45°C

(g) The watt is:

(1) One joule per second   (2) Equal to 1 kg·m²/s³

(3) The unit for all types of power   (4) All of the above

(h) What force may be needed to lift a heavy suitcase?

(1) 24 N   (2) 250 N   (3) 25 kN   (4) 250 kN

1.7. Your boss announced that the speed of the company Boeing 727 is to be cut from 525 mi/hr to 475 mi/hr to "conserve fuel," thus cutting consumption from 2200 gal/hr to 2000 gal/hr. How many gallons are saved in a 1000-mile trip?

1.8. It can be determined that the rate of energy loss from a person's body will normally vary by approximately 20 Btu/hr for each 1-°F change in skin temperature. Therefore, in going from a comfortably cool skin temperature of 90°F to a comfortably warm skin temperature of 93°F, the heat rejection rate is increased by only 60 Btu/hr. How much are the two rates, respectively, in watts?

1.9. The following slogan was used in Great Britain as a memory aid in conversion to the metric system:

> A meter measures three foot three.
> It's longer than a yard, you see.

and

> Two and a quarter pounds of jam
> Weigh about a kilogram.

Is the slogan reasonably correct in the conversion rates?

1.10. An overhaul of the state's tax collection policy is in view when the almost inevitable change to the metric system of measurement occurs. "It is clear that we can no longer collect 5 cents on a gallon of gasoline; we will need $0.013208602 per liter," said the director of the tax department. Is this the correct rate?

The real issue is: "Do we want to tax gasoline at 1.3208602 cents per liter, or 1.4, or 1.32; or is the appropriate rate at 1.5"? Decisions such as this, and there are many others, will have to be faced by the legislature."

1.11. The Mexican oil blowout of 1979–1980 was reported to discharge 30,000 bbl of oil per day, and to be equivalent to an oil slick 2 mi long, $\frac{1}{2}$ mi wide, and 1 in. thick. Is this report technically sound?

1.12. A technical publication describes a new model 20-hp Stirling (air cycle) engine that drives a 68-kW generator. Is this possible?

1.13. A freeze-dried coffee is advertised as "97% caffeine-free." Is it possible for this coffee to contain only 0.14% caffeine? Explain.

1.14. The following article is being submitted for publication to a journal that requires both SI and American engineering units to be displayed, the latter in parentheses. Fill in the parentheses with the values of units indicated.

Experimental Procedures

An artificial landfill was constructed based on the model of Qasim and Burchinal, differing from the prototype only in the degree of compaction. Three 1.0-m (-ft) sections of 10-cm (-in.) plastic tubing were bolted together and mounted vertically. Approximately 25 kg (lb) of a representative sample of solid wastes from which most of the glass, metal, and plastics had been removed was packed into the column in three layers, each separated by 15 cm (in.) of soil. The column was flooded with water and maintained at 25°C (°F). Leachate samples were removed at intervals through a tap at the bottom of the column. Two leachate samples of about 4 liters (gal) each were collected, the first after the freshly prepared column had been flooded for a month and the second after the landfill had stood for a year.

1.15. Rising gasoline prices have sparked a search for gasoline pumps that can cope with prices of more than 99.9 cents per gallon. Some California state officials have suggested that converting gasoline pumps to the metric system is the cheapest alternative. What would be the price shown per liter equivalent to $1.25 per gallon? Will it cost more to fill your gas tank with gas priced in $/liter than $/gal?

1.16. Draw an information flow diagram in which there are four nodes: length, time, mass, temperature. Link the following quantities to the nodes by means of lines (arcs): area, volume, speed, acceleration, force, energy, pressure, density, power, specific heat (heat capacity).

1.17. An elevator that weighs 10,000 lb is pulled up 10 ft between the first and second floors of a building 100 ft high. The greatest velocity the elevator attains is 3 ft/sec. How much kinetic energy does the elevator have in (ft)(lb$_f$) at this velocity?

1.18. Find the kinetic energy of a ton of water moving at 60 mi/hr expressed as
(a) (ft)(lb$_f$)
(b) joules
(c) (hp)(s)
(d) (watt)(s)
(e) (liter) (atm)

1.19. Calculate the kinetic and potential energy of a missile moving at 12,000 mi/hr above the earth where the acceleration due to gravity is 30 ft/s².

1.20. The following equation represents the change in concentration that occurs in a well-mixed tank containing an initial concentration of ethanol of C_0 to which pure water is added at a constant rate:

$$C = K \exp\left(\frac{-QT}{V}\right)$$

where C = concentration in mass per unit volume
   Q = rate at which water is charged to the tank
   V = volume of the tank
   T = time
   K = a constant of integration to be determined from the initial conditions

(a) What are the units in which K will be expressed?

(b) What are the correct units to represent the rate of water flow Q?

1.21. An experimental investigation of the rate of mass transfer of SO_2 from an airstream into water indicated that the mass transfer coefficient could be correlated by an equation of the form

$$k_x = Ku^{0.487}$$

where k_x is the mass transfer coefficient in mol/(cm²)(s) and u is the velocity in cm/s. Does the constant K have dimensions? What are they? If the velocity is to be expressed in ft/s, and we want to retain the same form of the relationship, what would the units of K' have to be if k_x is still mol/(cm²)(s), where K' is the new coefficient in the formula.

1.22. The equation for pressure drop due to friction for fluids flowing in a pipe is

$$\Delta p = \frac{2fL\rho v^2}{D}$$

where Δp = pressure drop
   v = velocity
   L = length of the pipe
   ρ = density of the fluid
   D = diameter of the pipe

Is the equation dimensionally consistent? What are the units of the *friction factor* f? Use units of the ft–lb–sec system for Δp, L, ρ, v, and D; is the equation then consistent?

1.23. The equation for the flow of water through a nozzle is as follows:

$$q = C\sqrt{\frac{2g}{1 - (d_1/d_2)^4}}\left(A\sqrt{\frac{\Delta p}{\rho}}\right)$$

where q = volume flowing per unit time
   g = local gravitational acceleration
   d_1 = smaller nozzle diameter
   d_2 = larger nozzle diameter
   A = area of nozzle outlet
   Δp = pressure drop across nozzle
   ρ = density of fluid flowing
   C = dimensionless constant

State whether this equation is dimensionally consistent. Show how you arrived at your conclusion.

1.24. The rate of mass transfer between a gas and a liquid in countercurrent flow is given by the equation

$$\frac{dm}{dt} = k_x A\,\Delta c$$

where k_x = mass transfer coefficient, cm/s
    A = area available for transfer, cm²
    Δc = concentration difference between the material in the gas phase and the concentration in the liquid phase, g mol/cm³
    t = time, s

What are the units of m?
    If the equation above is replaced by

$$\frac{dm}{dt} = k'_x A \, \Delta p$$

where Δp = partial pressure difference and has the units of kilopascals, what are the units of k'_x?

1.25. A useful dimensionless number called the *Reynolds number* is

$$\frac{DU\rho}{\mu}$$

where D = diameter or length
    U = some characteristic velocity
    ρ = fluid density
    μ = fluid viscosity

Calculate the Reynolds number for the following cases:

	1	2	3	4
D	2 in.	20 ft	1 ft	2 mm
U	10 ft/s	10 mi/hr	1 m/s	3 cm/s
ρ	62.4 lb/ft³	1 lb/ft³	12.5 kg/m³	25 lb/ft³
μ	0.3 lb$_m$/(hr)(ft)	0.14 × 10⁻⁴ lb$_m$/(s)(ft)	2 × 10⁻⁶ centipoise (cP)	1 × 10⁻⁶ centipoise

1.26. The Colburn equation for heat transfer is

$$\left(\frac{h}{CG}\right)\left(\frac{C\mu}{k}\right)^{2/3} = \frac{0.023}{(DG/\mu)^{0.2}}$$

where C = heat capacity, Btu/(lb fluid)(°F)
    μ = viscosity, lb/(hr)(ft)
    k = thermal conductivity, Btu/(hr)(ft²)(°F)/ft
    D = pipe diameter, ft
    G = mass velocity, lb/(hr)(ft²) cross section
What are the units of the heat transfer coefficient h?

Section 1.2

1.27. Write the formula and calculate the molecular weight of:
    (a) Silver sulfate
    (b) Anhydrous barium chloride
    (c) Propyl benzoate
    (d) Thyroxin (thyro-oxyindol)

1.28. Write the formula and calculate the molecular weight of:
    (a) Nickel bromide
    (b) Anhydrous chromic sulfate
    (c) Benzyl alcohol
    (d) Cortisone

1.29. How many pounds of compound are contained in each of the following?
    (a) 64 g of $BaMnO_4$
    (b) 200 g mol of ferric ammonium oxalate
    (c) 40 lb mol of dimethyl amine
    (d) 11 kg mol of nicotine

1.30. Convert the following:
    (a) 105 g mol of potassium oxide to g
    (b) 2 lb mol of aluminum acetate to lb
    (c) 114 g mol of oxalic acid to lb
    (d) 3 lb of quinine to g mol
    (e) 40 g of trichlorobenzene to lb mol

1.31. Comment on the statement: "A mole is an amount of substance containing the same number of atoms as 12 g of pure carbon 12."

1.32. Convert the following:
    (a) 120 g mol of NaCl to g
    (b) 120 g of NaCl to g mol
    (c) 120 lb moles of NaCl to lb
    (d) 120 lb of NaCl to lb mol
    (e) 120 lb mol of NaCl to g
    (f) 120 g mol of NaCl to lb
    (g) 120 lb of NaCl to g mol
    (h) 120 g of NaCl to lb mol
    (i) 120 g mol of NaCl to lb mol
    (j) 120 g of NaCl to lb

1.33. The structural formulas on the next page are for vitamins:
    (a) How many pounds of compound are contained in each of the following (do for each vitamin):
        (1) 2.00 g mol
        (2) 16 g
    (b) How many grams of compound are contained in each of the following (do for each vitamin):
        (1) 1.00 lb mol
        (2) 12 lb

Vitamin	Structural formula	Dietary sources	Deficiency symptoms
Vitamin A	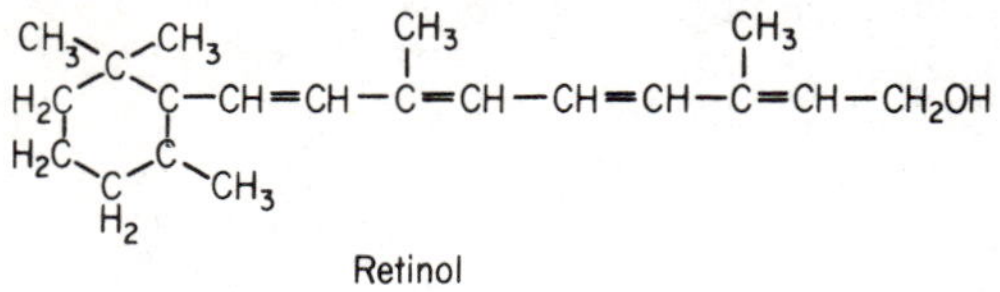 Retinol	Fish liver oils, liver, eggs, fish, butter, cheese, milk; a precursor, β-carotene, is present in green vegetables, carrots, tomatoes, squash	Night blindness, eye inflammation
Ascorbic acid (vitamin C)	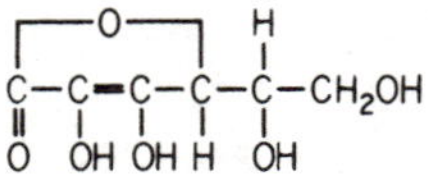	Citrus fruit, tomatoes, green peppers, strawberries, potatoes	Scurvy
Vitamin D	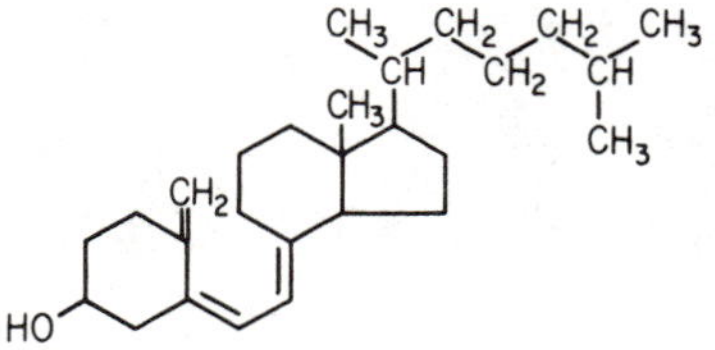 Vitamin D_3	Fish liver oils, butter, vitamin-fortified milk, sardines, salmon; the body also obtains this compound when ultraviolet light converts 7-dehydrocholesterol in the skin to vitamin D	Rickets, osteomalacia, hypoparathyroidism

Fig. P1.33

Section 1.3

1.34. A proposed New Mexico coal gasification project is predicted to produce about 573,000 lb/hr of wet ash from one unit. It would form a pile 231 ft at the base, 100 ft high, and almost 7 mi long in 1 year! Based on these estimates, what was the density used for the ash in pounds per cubic foot?

1.35. The specific gravity of diethanolamine (DEA) at 60°F/60°F is 1.096. On a day when the temperature is 60.0°F, a 1347-gal volume of DEA is carefully metered into a storage tank. This volume corresponds to how many pounds of DEA?

1.36. The density of a certain solution is 8.80 lb/gal at 80°F. How many cubic feet will be occupied by 10,010 lb of this solution at 80°F?

1.37. The density of benzene at 60°F is 0.879 g/cm³. What is the specific gravity of benzene at 60°F/60°F?

1.38. A hydrometer is constructed so that the reference temperature is 25°C and the measurement is made at 25°C. In a handbook the specific gravity of 10% acetic acid in water is given as $1.0125 \frac{25°C}{4°C}$. Compute what the value of the reading should be in the hydrometer.

1.39. Given a water solution that contains 1.704 kg of HNO_3/kgH_2O and has a specific gravity of 1.382 at 20°C, express the composition in the following ways:
(a) Weight percent HNO_3
(b) Pounds HNO_3 per cubic foot of solution at 20°C
(c) Molarity (gram moles of HNO_3 per liter of solution at 20°C)

1.40. Two immiscible liquids are allowed to separate in a vessel. One liquid has a specific gravity of 0.936; the second liquid weighs 9.63 lb/gal. A block that is 9 in. by 9 in. by 9 in. and weighs 25.8 lb is dropped into the vessel. Will the block float at the top, stop at the interface where the two liquids are separated, or sink? What fraction of the volume of the block is in one or both of the liquids?

1.41. A sulfuric acid solution containing 70% pure H_2SO_4 and 30% water is found to have a specific gravity, compared to water, of 1.75.

(a) What would be the weight (in pounds) of 1000 gal of this acid solution?

(b) What would be the weight (in pounds) of pure H_2SO_4 in the 1000 gal of part (a)?

(c) How many pounds of pure NaOH would be required to neutralize exactly the 1000 gal of acid solution in part (a)?

(d) If the sodium hydroxide [in part (c)] were present in the form of a 55% solution of NaOH in water, how many pounds of this caustic solution would be required for the neutralization of the acid?

(e) How many pounds of the caustic solution would you use if 20% excess of NaOH were to be employed?

(f) What would be the density of the caustic solution (in pounds per gallon) if the solution had a specific gravity of 1.23?

(g) How many gallons of caustic solution would be the quantity to be used in part (e)?

1.42. In a handbook you find that the conversion between °API and density is 0.800 density = 45.28 °API. Is this a misprint?

1.43. Five thousand barrels of 28°API gas oil is blended with 20,000 bbl of 15°API fuel oil. What is the °API (API gravity) of the mixture? What is the density in lb/gal and lb/ft³?

1.44. One barrel each of gasoline (55°API), kerosene (40°API), gas oil (31°API), and isopentane (96°API) are mixed. What is the composition of the mixture expressed in weight percent and volume percent? What is the API gravity and the density of the mixture in g/cm³ and lb/gal?

1.45. NIOSH sets standards for CCl_4 in air at 12.6 mg/m³ of air (a time weighted average over 40 hr). The CCl_4 found in a sample is 4800 ppb (parts per billion; billion = 10^9). Does the sample exceed the NIOSH standard? Be careful!

1.46. Considerable controversy has existed for years over the uses of water in the Colorado River basin. Nature's price for the conversion of the Colorado into a string of sapphire lakes is that a huge amount of water evaporates uselessly into the air. This concentrates the salinity in the water left behind, and agricultural runoff adds much more salt. The result: About half of the 11 million tons of dissolved salts carried through the river yearly is the result of human activities.

Salinity grows with the southward flow. Farmers near Yuma, Arizona, and across the border in California, where a cluster of irrigation districts take a last mighty gulp through the All-American and Gila canals before passing the remnant trickle to Mexico, have a rule of thumb: Put a foot of water on an acre of land and you apply a ton of salt along with it.

What is this concentration expressed as:

(a) lb/ft³?

    (b) mg/liter?

    (c) ppm?

    (d) percent?

1.47. If 100 cm³ of ethyl alcohol is added to 100 cm³ of water at 20°C, the total volume of the mixture is only 193 cm³. In other words, volumes are not additive for this system as they are for hydrocarbon mixtures.

    (a) If the proof of alcohol–water mixtures is defined as twice the volume percent alcohol, and volume percent is defined as the volumes of pure alcohol obtainable from 100 volumes of mixture, determine accurately the weight and mole percent alcohol in a 100-proof mixture. The specific gravity at 20°/4°C of the mixture is 0.9303, of pure alcohol 0.7893, and of pure water 0.9982, all at 20°C.

    (b) Why is it unscientific to express compositions as *volume percent* in this system? Why is it done commercially?

1.48. Alcoholic content does vary from one brand to another, says Jim Carg of the Alcoholic Beverage Commission. He defines beer as a malt beverage containing one-half of 1 % or more of alcohol by volume and not more than 4 % of alcohol by weight. Anything over 4 % is considered malt liquor.

    Old Snowshoes brand has 3.85 % alcohol by volume. Is it classified as a malt liquor? See Problem 1.47 for definitions.

1.49. The following table shows the annual inputs of phosphorus to Lake Erie.

	Short tons/yr
Source	
Lake Huron	2,240
Land drainage	6,740
Municipal waste	19,090
Industrial waste	2,030
	30,100
Outflow	4,500
Retained	25,600

SOURCE: *23rd Report to Committee on Government Operations*, U.S. Government Printing Office, Washington, D.C., 1970.

    (a) Convert the retained phosphorus to concentration in micrograms per liter assuming that Lake Erie contains 1.2×10^{14} gal of water and that average phosphorus retention time is 2.60 yr.

    (b) What percentage of the input comes from municipal waste?

    (c) What percentage of the input comes from detergents, assuming they represent 70 % of the municipal waste?

    (d) If 10 ppb of phosphorus triggers nuisance algal blooms, as has been reported in some documents, would removing 30% of the phosphorus in the municipal waste and all the phosphorus in the industrial waste be effective in reducing the eutrophication (i.e., the unwanted algal blooms) in Lake Erie?

(e) Would removing all the phosphate in detergents help?

Section 1.4

1.50. You have 100 lb of gas of the following composition:

CH_4	30%
H_2	10%
N_2	60%

What is the average molecular weight of this gas?

1.51. Calculate the average molecular weight of the following gas mixture (volume percent is given in the column to the right):

CO_2	2.0
CO	10.0
O_2	8.0
N_2	75.0
H_2O	5.0
	100.0

1.52. A fuel gas is reported to analyze, on a mole basis, 20% methane, 5% ethane, and the remainder CO_2. Calculate the analysis of the fuel gas on a mass percentage basis.

1.53. The table lists the typical composition of northwestern Colorado power plant ash:

Mineral	Concentration (%)
P_2O_5	1.0
SiO_2	40.5
Fe_2O_3	21.5
Al_2O_3	22.0
TiO_2	0.5
CaO	7.5
MgO	1.5
SO_3	4.5
K_2O	0.5
Na_2O	0.5
Total	100.0

What is the composition in mole percent?

1.54. Typical municipal refuse contains the following materials:

SAMPLE MUNICIPAL REFUSE COMPOSITION—U.S. EAST COAST

Physical (wt %)		Rough Chemical (wt %)	
Cardboard	7	Moisture	28.0
Newspaper	14	Carbon	25.0
Miscellaneous paper	25	Hydrogen	3.3
Plastic film	2	Oxygen	21.1
Leather, molded	2	Nitrogen	0.5
plastics, rubber		Sulfur	0.1
Garbage	12	Glass, ceramics, etc.	9.3
Grass and dirt	10	Metals	7.2
Textiles	3	Ash, other inerts	5.5
Wood	7	Total	100.0
Glass, ceramics, stones	10		
Metallics	8		
Total	100		

SOURCE: E. R., Kaiser, "Refuse Reduction Processes," in *Proceedings*, *The Surgeon General's Conference on Solid Waste Management for Metropolitan Washington*, U.S. Public Health Service Publication No. 1729, Government Printing Office, Washington, D.C., July 1967, p. 93.

(a) What is the rough chemical analysis on a water-free basis?

(b) Give a rough estimate of the volume reduction of the municipal waste if it is completely incinerated. Clearly state any assumptions you make.

1.55. A liquid mixture of compounds A, B, and C containing 10 kg of A analyzes 25% B and contains 1.5 moles of C per mole of B. The respective molecular weights of A, B, and C are 56, 58, and 72; and the specific gravities are 0.58, 0.60, and 0.67. Calculate the analysis of the mixture in mole percent, the molecular weight of the mixture, the volume percent A on a B-free basis (ignore volume change on mixing), the density of the mixture, and the total number of moles of the mixture.

Section 1.5

1.56. In a report on the record low temperatures in Antarctica, *Chemical and Engineering News* said at one point that "the mercury dropped to $-76°C$." In what sense is that possible? Mercury freezes at $-39°C$.

1.57. "Further, the degree Celsius is exactly the same as a kelvin. The only difference is that zero degree Celsius is 273.15 kelvin. Use of Celsius temperature gives us one less digit in most cases" from [*Eng. Educ.*, p. 678 (April 1977)]. Comment on the above quotation. Is it correct? If not, in what way or sense is it wrong?

1.58. (a) Convert 70°F to °C, K, °R.

   (b) Convert 210 K to °C, °F, °R.

1.59.   Calculate all temperatures from the one value given:

	(a)	(b)	(c)	(d)	(e)	(f)	(g)	(h)
°F	140				1000			
°R			500			1000		
K		298					1000	
°C				−40				1000

1.60.   The emissive power of a blackbody depends on the fourth power of the temperature and is given by

$$W = AT^4$$

where W = emissive power, Btu/(ft²)(hr)
  A = Stefan–Boltzmann constant, 0.171×10^{-8} Btu/(ft²)(hr)(°R)⁴
  T = temperature, °R
What is the value of A in the units J/(m²)(s)(K⁴)?

1.61.   Mercury boils at 630 K. What is its boiling temperature expressed in °C? In °F? In °R?

1.62.   The inhabitants of Betelgeuse, in the constellation Orion, have contacted us by radio, and in communicating with them we find that the zero of their temperature scale is based on the freezing point of methane (−182.5°C) and that the kindling point of wood (451°F) is 100 on their scale. Derive an equation relating °B as used by the Betelgeusians to °F. What is 122°C in °B?

1.63.   Show that:
(a) $T_{°F} = 2[T_{°C} + \frac{1}{10}(160 - T_{°C})]$
(b) $T_{°F} = [(T_{°C} \times 2) - \frac{1}{10}(T_{°C} \times 2)] + 32$

1.64.* The temperature scale we use at present is linear, forming an ordered sequence of numbers starting at 0°K,

$$+0\,K \cdots 273.16\,K \cdots + \infty\,K$$

Kelvin proposed the relation

$$T = e^{\psi}$$

where ψ depends on the thermal properties of the system and T is the absolute temperature (K). It is possible to construct another temperature scale, the ψ scale, or the logarithmic scale. The logarithmic nature of the ψ scale assures that the zero value on the linear absolute scale (T scale) will be reached asymptotically (i.e., $T = 0$ K) when $\psi = -\infty$. Comparing both these scales, the inaccessibility of the absolute zero follows generically, not being imaginable from the linear T scale alone. Plot the ψ vs. T scales, and calculate the value of ψ for $T = 1$ K and 273 K.

Section 1.6

1.65.   The amplitude of sound at any given point is expressed as sound-pressure level (SPL). Its physical unit is the decibel, which is given as

$$SPL = 20 \ln (p/p_0) \qquad \text{in dB}$$

where p is the sound pressure being measured and p_0 is a reference pressure, usually 20 micronewtons per square meter (μN/m²). The reference pressure of 20 μN/m² is approximately equal to the lowest pressure which a young person with normal hearing can barely detect at a frequency of 1000 hertz. Other measures of sound pressure may be encountered in the literature, such as dynes per square centimeter, microbars, and pounds per square inch.

Common examples of representative SPL (in dB) include:

Business office	50
Speech at 3 ft	65
Subway at 20 ft	95
Jet aircraft at 35 ft	130
On gantry during Saturn V launch	172

Convert the dB values of 0, 20, and those in the list above to:

(a) N/m²

(b) μbar

(c) psi

1.66.   The accompanying table shows the effect of pressure differentials during ascent and descent. During ascent, the pressure in the external ear canal and nose drops, creating a positive differential within the middle ear. The eardrum bulges outward and air leaves the middle ear. During descent, a negative pressure differential exists, the eardrum is pushed inward, but air cannot reenter the middle ear without voluntary effort. Once high differentials exist, it is difficult or impossible to force air into the middle ear voluntarily; avoidance of such differentials requires frequent attention during rapid descents. Descents of less than 500 ft/min in the lower atmosphere (0.25 psia/min) are usually tolerated by inexperienced air passengers without difficulty, although modern pressurization controllers are usually operated at perhaps half this rate.

TYPE OF EAR COMPLAINTS ENCOUNTERED BECAUSE OF PRESSURE DIFFERENCES

Complaint	Range (mm Hg)	
	Ascent	Descent
No sensation; hearing is normal (level flight)	0	0
Feeling of fullness in ears	+3 to 5	−3 to −5
More fullness, lessened sound intensity	+10 to 15	−10 to −15
Fullness, discomfort, tinnitus in ears; ears usually "pop" as air leaves middle ear—desire to clear ears; if this is done, symptoms stop	+15 to 30	−15 to −30
Increasing pain, tinnitus, and dizziness	+30	−30 to −60
Severe and radiating pain, dizziness, and nausea		−60 to −80
Voluntary clearing becomes difficult or impossible		−100
Eardrum ruptures		−200 or more

Convert the data in the table to the following units:
- **(a)** psi
- **(b)** kPa
- **(c)** atm
- **(d)** bar
- **(e)** in. Hg
- **(f)** ft H_2O
- **(g)** kg_f/cm^2

1.67. Suppose that a submarine inadvertently sinks to the bottom of the ocean at a depth of 1000 m. It is proposed to lower a diving bell to the submarine and attempt to enter the conning tower. What must the minimum air pressure be in the diving bell at the level of the submarine to prevent water from entering into the bell when the opening valve at the bottom is cracked open slightly? Give your answer in absolute kilopascals. Assume that seawater has a constant density of 1.024 g/cm^3.

1.68. What is the gauge pressure (in $lb_f/in.^2$) at a depth of 4.50 mi below the surface of the sea if the water temperature averages 60°F? The specific gravity of seawater at 60°F/60°F is 1.042 and is assumed to be independent of pressure.

1.69. Air in a scuba diver's tank shows a pressure of 300 kPa absolute. What is the pressure in:
- **(a)** atm?
- **(b)** psia?
- **(c)** in. Hg?

1.70. A pressure gauge on a welder's tank gives a reading of 22.4 psig. The barometric pressure is 28.6 in. Hg. Calculate the absolute pressure in the tank in:
- **(a)** lb_f/ft^2
- **(b)** in. Hg
- **(c)** newtons/(meter)²
- **(d)** feet of water

1.71. Air cannot be used for diving at depths of greater than 150 ft because of nitrogen's narcotic effects. Divers cite the "martini law": Every 50 ft of depth is equivalent to drinking one martini. A depth of 1000 ft is equivalent to how many:
- **(a)** martinis?
- **(b)** lb_f/ft^2?
- **(c)** newtons/(meter)²?

Assume that the density of sea water is constant at 63.9 lb/ft^3.

1.72. A water tank at the top of a 10-story building is vented to the atmosphere. What would be the pressure recorded in kilopascals at the bottom floor, 20 m below the water level, on a Bourdon gauge, when flow does not occur? What is it in psi?

1.73. A centrifugal pump is to be used to pump water from a lake to a storage tank that is 148 ft above the surface of the lake. The pumping rate is to be 25.0 gal/min, and the water temperature is 60°F.

   The pump on hand can develop a pressure of 50.0 psig when it is pumping at a rate of 25.0 gal/min. (Neglect pipe friction kinetic energy effects, or factors involving pump efficiency.)
- **(a)** How high (in feet) can the pump raise the water at this flow rate and temperature?
- **(b)** Is this pump suitable for the intended service?

1.74. A manometer uses kerosene, sp gr 0.82, as the fluid. A reading of 5 in. on the mano-meter is equivalent to how many millimeters of mercury?

1.75. The pressure gauge on the steam condenser for a turbine indicates 26.2 in. Hg of vacuum. The barometer reading is 30.4 in. Hg. What is the pressure in the condenser in psia?

1.76. A pressure gauge on a process tower indicates a vacuum of 3.53 in. Hg. The barom-eter reads 29.31 in. Hg. What is the absolute pressure in the tower in millimeters of mercury?

1.77. John Long says he calculated from a formula that the pressure at the top of Pikes Peak is 9.75 psia. John Green says that it is 504 mm Hg because he looked it up in a table. Which John is right?

1.78. In the configuration shown in Fig. P1.78, calculate the absolute pressure at point A. (The valve is closed.)

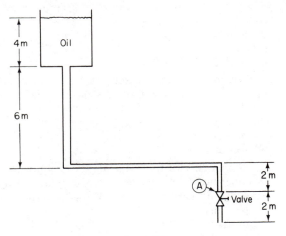

Fig. P1.78

barometric pressure = 98.5 kPa

specific gravity of oil = 0.85

density of water = 10^3 kg/m³

1.79. The pressure reading on the gauge shown in Fig. P1.79 is 2.4 psi. If the liquid level in the glass pipe connecting the two tanks is 10 ft below the bottom of the closed tank, is the pressure gauge working properly?

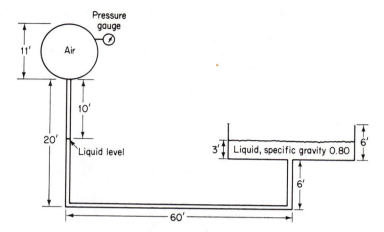

Fig. P1.79

1.80. The indicating liquid in the manometer shown in Fig. P1.80 is water, and the other liquid is benzene. These two liquids are essentially insoluble in each other. If the manometer reading is $\Delta Z = 36.3$ cm water, what is the pressure difference in kPa? The temperature is 25°C.

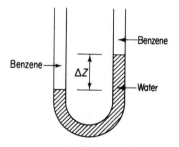

Fig. P1.80

1.81. Water at 60.0°F is flowing through a pipeline, and the line contains on orifice meter for flow measurement. An orifice meter is a circular plate with a hole in the center that is placed in the pipe. The ends of a U-tube manometer 48.0 in. long are connected across the orifice meter. See Fig. P1.81. If the maximum expected pressure

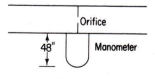

Fig. P1.81

drop across the meter is 3.55 psi, what is the minimum specific gravity at 60°F/60°F that the manometer liquid can have when used with this equipment if the liquid is to stay in the manometer?

1.82. Examine Fig. P1.82. The barometric pressure is 720 mm Hg. The density of the oil is 0.80 g/cm³. The Bourdon gauge reads 33.1 psig. What is the pressure in kPa of the gas?

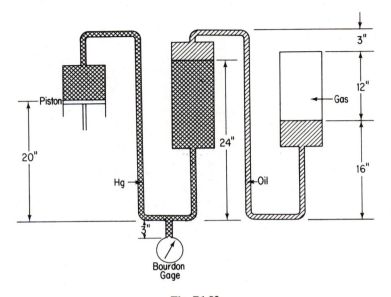

Fig. P1.82

Section 1.7

1.83. Go to your library and look at three reference books listing information about the properties of industrial chemicals. Prepare an abstract of the contents of each volume in 200 to 300 words.

1.84. Does Perry's *Chemical Engineers' Handbook* contain information on densities of alcohol–water mixtures? osmotic pressure of sodium chloride solutions? corrosion properties of metals?

1.85. List four journals that publish physical property data and the libraries in which they are located.

Section 1.8

1.86. Prepare an information flow diagram linking the following concepts:
(a) Solution process
(b) Given knowledge
(c) Results
(d) Objectives
(e) Data resources that are used in problem solving

1.87. Based on your experience to date, what are three general categories of laws or principles used in problem solving?

1.88. How would you estimate the rate at which a spherical water-filled capsule in outer space cools. List the types of information needed to solve such a problem, the assumptions that must be made, and what physical principles that you have studied in physics might be used to help solve the problem. Draw a picture of the process and indicate what the independent and dependent variables involved in the process might be.

1.89. Read the introductory portions of the book by Moshe F. Rubenstein, *Problem Solving* (Prentice-Hall, Englewood Cliffs, N.J., 1975), and compare the techniques outlined there with those listed in Sec. 1.8.

Section 1.9

1.90. $BaCl_2 + Na_2SO_4 \longrightarrow BaSO_4 + 2NaCl$
   (a) How many grams of barium chloride will be required to react with 5.00 g of sodium sulfate?
   (b) How many grams of barium chloride are required for the precipitation of 5.00 g of barium sulfate?
   (c) How many grams of barium chloride are needed to produce 5.00 g of sodium chloride?
   (d) How many grams of sodium sulfate are necessary for the precipitation of the barium of 5.00 g of barium chloride?
   (e) How many grams of sodium sulfate have been added to barium chloride if 5.00 g of barium sulfate is precipitated?
   (f) How many pounds of sodium sulfate are equivalent to 5.00 lb of sodium chloride?
   (g) How many pounds of barium sulfate are precipitated by 5.00 lb of barium chloride?
   (h) How many pounds of barium sulfate are precipitated by 5.00 lb of sodium sulfate?
   (i) How many pounds of barium sulfate are equivalent to 5.00 lb of sodium chloride?

1.91. $AgNO_3 + NaCl \longrightarrow AgCl + NaNO_3$
   (a) How many grams of silver nitrate will be required to react with 5.00 g of sodium chloride?
   (b) How many grams of silver nitrate are required for the precipitation of 5.00 g of silver chloride?
   (c) How many grams of silver nitrate are equivalent to 5.00 g of sodium nitrate?
   (d) How many grams of sodium chloride are necessary for the precipitation of the silver of 5.00 g of silver nitrate?
   (e) How many grams of sodium chloride have been added to silver nitrate if 5.00 g of silver chloride is precipitated?
   (f) How many pounds of sodium chloride are equivalent to 5.00 lb of sodium nitrate?
   (g) How many pounds of silver chloride are precipitated by 5.00 lb of silver nitrate?
   (h) How many pounds of silver chloride are precipitated by 5.00 lb of sodium chloride?
   (i) How many pounds of silver chloride are equivalent to 5.00 lb of silver nitrate?

1.92. When benzene (C_6H_6) is burned with oxygen, how many pounds of oxygen are required to burn 10 lb of benzene to carbon dioxide and water vapor?

1.93. A plant makes liquid CO_2 by treating dolomitic limestone with commercial sulfuric acid. The dolomite analyzes 68.0% $CaCO_3$, 30.0% $MgCO_3$, and 2.0% SiO_2; the acid is 94% H_2SO_4 and 6% H_2O. Calculate:
(a) Pounds of CO_2 produced per ton dolomite treated
(b) Pounds of commercial acid required per ton of dolomite treated
Data mol. wt.: $CaCO_3$, 100; $MgCO_3$, 84.3; SiO_2, 60; H_2SO_4, 98.

1.94. Chlorine dioxide is widely used as a bleaching agent for pulp and occasionally used as a disinfectant for drinking water. The first plants for pulp bleaching with chlorine dioxide were put in operation in Sweden and Canada. The first plant to disinfect drinking water with chlorine dioxide was in Zurich, Switzerland. Since ClO_2 does not enter into aqueous chlorination reactions, as chlorine does, its use as a disinfectant could obviate the presence of chloroorganics in drinking water.

The manufacture of chlorine dioxide by the Munch Process is based on the reaction of HCl with $NaClO_3$ solution. As a typical feature, it yields a residual solution with a relatively high content of $NaClO_3$. This solution is sent back to the chlorate electrolysis plant in which the sodium chlorate reduced to NaCl is reoxidized to $NaClO_3$.

In the reaction tower, ClO_2 and Cl_2 are obtained by the reaction of $NaClO_3$ and HCl as reflected by the following equations:

$$NaClO_3 + 2HCl \longrightarrow ClO_2 + \tfrac{1}{2}Cl_2 + NaCl + H_2O \qquad \text{(a)}$$

$$NaClO_3 + 6HCl \longrightarrow 3Cl_2 + NaCl + 3H_2O \qquad \text{(b)}$$

Experience has indicated that, on the average, the minimum molar ratio of the gas is $m = 1.25$, where the molar ratio of the gas is designated by

$$\frac{\text{mol } ClO_2}{\text{mol } Cl_2} = m$$

For 1 kg of ClO_2, what are the corresponding quantities (in kilograms) in reaction (a)?

1.95. A limestone analyzes

$CaCO_3$	92.89%
$MgCO_3$	5.41%
insoluble	1.70%

How many pounds of calcium oxide can be produced from 5 tons of this limestone?

1.96. Some states and cities are placing a ban on laundry detergents containing more than 20% phosphates. In reading the detailed provisions of one law, it prohibits the sale of any detergent with a phosphorus pentoxide content of over 20% by weight. How much is this when expressed as weight percent sodium triphosphate, the compound found in the detergent?

1.97. Suppose that the following reaction is carried out:

$$2NaOH + H_2SO_4 \longrightarrow Na_2SO_4 + 2H_2O$$

How many kilogram moles of H_2O will result from the complete reaction of 156 kg of NaOH with the H_2SO_4?

1.98. When sodium nitrate is treated with sulfuric acid, the two most important reactions that occur are:

$$2NaNO_3 + H_2SO_4 \longrightarrow Na_2SO_4 + 2HNO_3 \qquad (1)$$

$$NaNO_3 + H_2SO_4 \longrightarrow NaHSO_4 + HNO_3 \qquad (2)$$

For a given weight of sulfuric acid, which reaction yields the most nitric acid? Which reaction would be used commercially? Explain, with reference to your source of information.

1.99. To produce HCN, what is the difference in cost per pound of using KCN (98%) at 50 cents per pound vs. a mixture of KCN (65%) and NaCN (25%) at 60 cents per pound (for the mixture)?

1.100. In submarines, spacecraft, and other self-contained systems, various bases are used to react with and remove CO_2 from the atmosphere. A typical reaction is

$$2LiOH + CO_2 \longrightarrow Li_2CO_3 + H_2O$$

Water is evolved as a vapor. What is the theoretical maximum removal (complete reaction) of CO_2 expressed as kilograms of CO_2 per kilogram of reactant for the following reactants?

Substance	Formula
Lithium oxide	Li_2O
Lithium hydroxide	$LiOH$
Sodium oxide	Na_2O
Sodium hydroxide	$NaOH$

1.101. Sulfuric acid production plants add to the already growing problem of atmospheric pollution with sulfur dioxide and sulfur trioxide released from the processing. However, sulfuric acid production is one way of ameliorating the problem of what to do with the sulfur dioxide released in the winning of metals from their ores. For example, sulfuric acid can be produced as a by-product in the manufacture of zinc from sulfide ores. The ore analyzes 65.0% ZnS and 35.0% inert impurities by weight. The ore is burned in a furnace, and the resulting SO_2 is converted to SO_3 in a catalytic reactor. The SO_3 is then absorbed by water to give sulfuric acid, the final product being 2.0% water and 98.0% H_2SO_4 by weight. A total of 99.0% of the sulfur in the ore is recovered in the acid. The chemical reactions are:

$$ZnS + 1.5O_2 \longrightarrow ZnO + SO_2$$
$$SO_2 + 0.5O_2 \longrightarrow SO_3$$
$$SO_3 + H_2O \longrightarrow H_2SO_4$$

(a) Calculate the weight of product sulfuric acid (98.0% H_2SO_4 by weight) produced in a zinc plant that processes 186 metric tons of ore per day.

(b) How much water is required per day?

1.102. On a car for maximum economy the air/fuel ratio is known to be 15.1 : 1. Note that although gases are involved, this ratio is actually the mass of air per mass of gasoline. Suppose that the gasoline is C_8H_{18}. Write the chemical reaction equation with proper number of moles in front of each compound. Do not forget to include the nitrogen. The actual air/fuel ratio used in cars ranges from 12.5 (at full throttle) to 14 : 1 (a richer mix) because the power from the engine peaks in this range.

If incomplete complete combustion occurs because of limited O_2, so that some CO is found instead of CO_2, rewrite the combustion balance to include the CO. Let x be the amount of CO formed.

1.103. Removal of CO_2 from a manned spacecraft has been accomplished by absorption with lithium hydroxide according to the following reaction:

$$2LiOH(s) + CO_2(g) \longrightarrow Li_2CO_3(s) + H_2O(l)$$

(a) If 1.00 kg of CO_2 is released per day per person, how many kilograms of LiOH are required per day per person?

(b) What is the penalty (i.e., the percentage increase) in weight if the cheaper NaOH is substituted for LiOH?

1.104. In the manufacture of chlorine gas, a solution of sodium chloride in water is electrolyzed in specially designed cells. The overall reaction is

$$2NaCl + 2H_2O \longrightarrow 2NaOH + H_2 + Cl_2$$

At the end of one batch run, the sodium hydroxide in a particular cell was 10.4% by weight, and the cell contained 20,000 lb of solution (to the nearest 100 lb). How many pounds of NaCl were required to produce 1850 lb of Cl_2?

1.105. Oxygen can be produced by heating potassium chlorate, which costs 60 cents per pound, or potassium nitrate, which costs 40 cents per pound. Which one produces oxygen at the lower cost?

$$2KClO_3 \longrightarrow 2KCl + 3O_2$$
$$2KNO_3 \longrightarrow 2KNO_2 + O_2$$

1.106. Sulfuric acid can be manufactured by the contact process according to the following reactions:

$$S + O_2 \longrightarrow SO_2 \qquad\qquad (1)$$
$$2SO_2 + O_2 \longrightarrow 2SO_3 \qquad\qquad (2)$$
$$SO_3 + H_2O \longrightarrow H_2SO_4 \qquad\qquad (3)$$

You are asked as part of the preliminary design of a sulfuric acid plant with a design capacity of 2000 tons/day of 66°Be (Baumé) (93.2% H_2SO_4 by weight) to calculate the following:

(a) How many tons of pure sulfur are required per day to run this plant?

(b) How many tons of oxygen are required per day?

(c) How many tons of water are required per day for reaction (3)? Do not include the water used to dilute the acid to 66°Be.

(d) How many tons of water are required per day to dilute the H_2SO_4 to 66°Be (water beyond that calculated in (c))?

1.107. A reactor is designed to carry out the saponification of methyl acetetate:

$$CH_3COOCH_3 + H_2O \longrightarrow CH_3COOH + CH_3OH$$

MW: 74   18   60   32

It is found that 3 tons of methanol (CH_3OH) is produced per 1000 tons of feed to the reactor. The feedstream contains 0.2 g mol/liter of methyl acetate (CH_3COOCH_3) in water. What is the percent excess of water in the feed? What is the percent conversion of the reaction?

1.108. Fuel oil (assumed to be $C_{12}H_{26}$) is injected into a furnace and burned with exactly 1.50 times the theoretically required air for complete combustion. What is the exit composition of the stack gases? What is the composition on a dry basis?

1.109. The gas obtained by burning pure carbon in excess oxygen analyses:

CO_2	75 mole %
CO	14 mole %
O_2	11 mole %

What was the percent excess oxygen used? What was the yield of carbon dioxide in kilograms of CO_2 per kilogram of C burned?

1.110. One can view the blast furnace from a simple viewpoint as a process in which the principal reaction is

$$Fe_2O_3 + 3C \longrightarrow 2Fe + 3CO$$

but some other undesired side reactions occur, mainly

$$Fe_2O_3 + C \longrightarrow 2FeO + CO$$

After mixing 600.0 lb of carbon (coke) with 1.00 ton of pure iron oxide, Fe_2O_3, the process produces 1200.0 lb of pure iron, 183 lb of FeO, and 85.0 lb of Fe_2O_3. Calculate the following items:

(a) The percentage of excess carbon furnished, based on the principal reaction

(b) The percentage conversion of Fe_2O_3 to Fe

(c) The pounds of carbon used up and the pounds of CO produced per ton of Fe_2O_3 charged

1.111. Aluminum sulfate is used in water treatment and in many chemical processes. It can be made by reacting crushed bauxite with 77.7% sulfuric acid by weight. The bauxite ore contains 55.4% aluminum oxide by weight, the remainder being impurities. To produce crude aluminum sulfate containing 2000 lb of pure aluminum sulfate, 1080 lb of bauxite and 2510 lb of sulfuric acid solution (77.7% acid) are used.
(a) Identify the limiting reactant.
(b) What was the fraction conversion of Al_2O_3 to $Al_2(SO_4)_3$?
(c) What was the degree of completion of the reaction?
Molecular weights:

$$Al_2O_3 = 101.9$$

$$H_2SO_4 = 98.1$$

$$Al_2(SO_4)_3 = 342.1$$

1.112. The burning of limestone, $CaCO_3 \longrightarrow CaO + CO_2$, goes only 70% to completion in a certain kiln. What is the composition (weight percent) of the solid withdrawn from such a furnace? How many kilograms of CO_2 are produced per kilogram of limestone charged? Assume that the limestone is pure $CaCO_3$.

1.113. Phosgene gas is probably most famous for being the first toxic gas used offensively in World War I, but it is also used extensively in the chemical processing of a wide variety of materials. Phosgene can be made by the catalytic reaction between CO and chlorine gas in the presence of a carbon catalyst. The chemical reaction is

$$CO + Cl_2 \longrightarrow COCl_2$$

Suppose that you have measured the reaction products from a given reactor and found that they contained 3.00 kg mol of chlorine, 10.00 kg mol of phosgene, and 7.00 kg mol CO. Calculate the following:
(a) The percentage of excess reactant
(b) The percentage conversion of the limiting reactant
(c) The kilogram moles of phosgene formed per kilogram mole of total reactants fed to the reactor

PROBLEMS TO PROGRAM ON THE COMPUTER

1.1. Write a program to convert °C into K, °F, and °R at 1°C intervals. Have the program arranged so that you can do the following:
(a) Read in a starting value of °C.
(b) Read in a stopping value of °C.
(c) Use an IF loop to convert °C to the other temperature scales at 1°C intervals between the starting and stopping values, inclusive.
(d) Print out the four temperatures for each conversion with appropriate column headings.
(e) Use comment lines where appropriate to indicate your procedure in conversion.

1.2. Write a computer program to convert millimeters of mercury into inches of mercury, feet of H_2O, and psia at 2 mm Hg (even-numbered) increments below 760 mm Hg. Have the computer arranged so that you can do the following:

(a) Read in a starting value for the millimeters of mercury.

(b) Stop at 760 mm Hg (the last value to be converted).

(c) Read in the necessary conversion factors, such as CPSIA in

$$PSIA = CPSIA * HGMM \text{ for psia} = 0.01933 \text{ mm Hg}$$

(d) Use one-dimensional arrays to store values.

(e) Use a DO loop to compute all values. You must compute the number of iterations required with your program.

(f) Print out the results after all the conversions have been made.

 (1) Print your name at the top of the first page.

 (2) Print appropriate column headings at the top of each new page.

 (3) Print only 50 lines of conversion results per page.

1.3. Prepare a computer program to give the specific gravity of a fluid as a function of °Be (Baumé) and °API over the interval 0°–30° in 1° increments. Equation (1.10) gives the relation between specific gravity and °API (for liquids less dense than water), while the relationships for °Be are

liquids more dense than water

$$\text{sp gr} = \frac{145.0}{145.0 - °\text{Be}}$$

liquids less dense than water

$$\text{sp gr} = \frac{140.0}{130.0 + °\text{Be}}$$

Read the °API or °Be into the program as data.

INDUSTRIAL CHEMICAL DATA

HOW MADE	MAJOR END USES	MAJOR DERIVATIVES	ANNUAL PRODUCTION, PRICES (USA 1981)
	Ammonia	NH₃	
Catalytic reaction of nitrogen from air and hydrogen from natural gas	Fertilizer 80%, plastics and fibers 10%, explosives 5%	Nitric acid 20%, urea 20%, ammonium phosphates 15%	18.7×10^6 tons $135–$145/ton
	Lime	CaO	
Calcining limestone (mainly CaCO₃) to make quicklime, CaO; then adding water to make hydrated lime, Ca(OH)₂	Metallurgy (mainly steel flux) 45%, chemical manufacture 10%, potable water treatment 10%, sewage and pollution control 5%, pulp and paper manufacture 5%		19.4×10^6 tons $35–$50/ton

Adapted from Chemical and Engineering News by permission of the American Chemical Society.

2

MATERIAL
BALANCES

Conservation laws occupy a special place in science and engineering. Common statements of these laws take the form of "mass (energy) is neither created nor destroyed," "the mass (energy) of the universe is constant," "the mass (energy) of any isolated system is constant," or equivalent statements. To refute a conservation law, it would be sufficient to find just one example of a violation.

But what degree of accuracy is needed to prove a violation? In an industrial environment, in spite of considerable effort, it is not possible to make a 99.9% closure of a mass balance. The results of errors in measurement in determining the amount of material in tanks, bins, drums, bags, and bottles; in obtaining representative samples; in analyzing the samples; and in determining the fate of raw materials fed into the system and products removed from it preclude such accuracy.

Our belief in the validity of the conservation laws rests on the experiences of Lavoisier and many of the scientists following in his path who studied chemical changes quantitatively and found invariably that the sum of the weights of the substances entering into a reaction equaled the sum of the weights of the products of the reaction. Thus our collective experience has been summed up and generalized as the law of the conservation of matter. Of course, we must exclude processes involving nuclear transformations, or else extend our law to include the conservation of both energy and matter. (Insofar as we shall be concerned here, the word *process* will be taken to mean a series of physical operations on or physical or chemical changes in some specified material.)

Why study mass balances as a separate topic? You will find that mass balance calculations are almost invariably a prerequisite to all other calculations in the solution of both simple and complex chemical engineering problems. Furthermore, skills that you develop in analyzing mass balances are easily transferred to other types of balances and other types of problems.

In this chapter we discuss the principle of the conservation of matter and how it can be applied to engineering calculations, making use of the background information discussed in Chap. 1. Figure 2.0 shows the relations between the topics

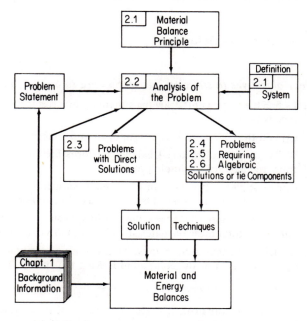

Fig. 2.0 Hierarchy of topics to be studied in this chapter (section numbers are in the upper left-hand corner of the boxes).

discussed in this chapter and the general objective of making material and energy balances. In approaching the solution of material balance problems, we first consider how to analyze them in order to clarify the method and the procedure of solution. The aim will be to help you acquire a generalized approach to problem solving so that you may avoid looking upon each new problem, unit operation, or process as entirely new and unrelated to anything you have seen before. As you scrutinize the examples used to illustrate the principles involved in each section, explore the method of analysis, but avoid memorizing each example by rote, because, after all, they are only samples of the myriad problems that exist or could be devised on the subject of material balances. Most of the principles we consider are of about the same degree of complexity as the law of compensation devised by some unknown, self-made philosopher who said: "Things are generally made even somewhere or some place. Rain always is followed by a dry spell, and dry weather follows rain. I have found it an invariable rule that when a man has one short leg, the other is always longer!"

In working these problems you will find it necessary to employ some engineering judgment. You think of mathematics as an exact science. For instance, suppose that it takes 1 man 10 days to build a brick wall; then 10 men can finish it in 1 day. Therefore, 240 men can finish the wall in 1 hr, 14,400 can do the job in 1 min, and with 864,000 men the wall will be up before a single brick is in place! Your password to success is the famous IBM motto: THINK.

Section 2.1   The Material Balance

Your objectives in studying this section are to be able to:

1. Define the system and draw the system boundaries for which the material balance is to be made.

2. Explain the difference between an open and a closed system.

3. Write the general material balance is words including all terms. Be able to apply the balance to simple problems.

4. Cite examples of processes in which no accumulation takes place; no generation or consumption takes place; no mass flow in and out takes place.

5. Apply the material balance equation for the simplified case of input = output to the total mass of material and to an individual species.

6. Explain the circumstances in which mass of a compound entering the system equals the mass of the compound leaving the system. Repeat for moles.

To take into account the flow of material in and out of a system, the generalized law of the conservation of mass is expressed as a material balance. A *material balance* is nothing more than an accounting for mass flows and changes in inventory of mass for a system. Examine Fig. 2.1. Equation (2.1) describes in words the principle of the material balance applicable to processes both with and without chemical reaction:

$$\left\{\begin{array}{c}\text{mass}\\\text{accumulation}\\\text{within}\\\text{the}\\\text{system}\end{array}\right\} = \left\{\begin{array}{c}\text{mass}\\\text{input}\\\text{through}\\\text{system}\\\text{boundaries}\end{array}\right\} - \left\{\begin{array}{c}\text{mass}\\\text{output}\\\text{through}\\\text{system}\\\text{boundaries}\end{array}\right\} + \left\{\begin{array}{c}\text{mass}\\\text{generation}\\\text{within}\\\text{the}\\\text{system}\end{array}\right\} - \left\{\begin{array}{c}\text{mass}\\\text{consumption}\\\text{within}\\\text{the}\\\text{system}\end{array}\right\} \qquad (2.1)$$

In Eq. (2.1) the generation and consumption terms in this text refer to gain or loss

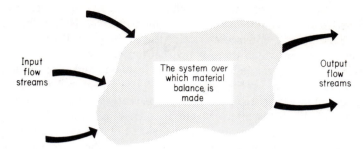

Fig. 2.1 Any enclosed volume or system over which material balances are to be made. We are not concerned with the internal details, only with the passage of material across the volume boundaries.

by chemical reaction. The accumulation may be positive or negative. You should understand in reading the words in Eq. (2.1) that the equation refers to a time interval of any desired length, including a year, hour, or second, or a differential time.

Equation (2.1) reduces to Eq. (2.2) for cases in which there is no generation (or usage) of material within the system,

$$\text{accumulation} = \text{input} - \text{output} \tag{2.2}$$

and reduces further to Eq. (2.3) when there is, in addition, no accumulation within the system,

$$\text{input} = \text{output} \tag{2.3}$$

If there is no flow in and out of the system, Eq. (2.1) reduces to the basic concept of the conservation of one species of matter within an enclosed isolated system:

$$\text{accumulation} = \text{generation} - \text{consumption} \tag{2.3a}$$

Inherent in the formulation of each of the balances above is the concept of a system for which the balance is made. By *system* we mean any arbitrary portion or whole of a process as set out specifically by the engineer for analysis. Figure 2.2 shows a system in which flow and reaction take place; note particularly that the system boundary is formally circumscribed about the process itself to call attention

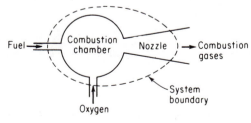

Fig. 2.2 Flow (open) system with combustion.

to the importance of carefully delineating the system in each problem you work. An *open* (or flow) *system* is one in which material is transferred across the system boundary, that is, enters the system, leaves the system, or both. A *closed* (or batch) *system* is one in which there is no such transfer *during the time interval of interest.* Obviously, if you charge a reactor with reactants and take out the products, and the reactor is designated as the system, material is transferred across the system boundary. But you might ignore the transfer, and focus attention solely on the process of reaction that takes place only after charging is completed and before the products are withdrawn. Such a process would occur within a closed system.

In nearly all the problems in this chapter the mass accumulation term will be zero; that is, primarily *steady-state* problems will be considered. (See Chap. 6 for cases in which the mass accumulation is not zero.) For a steady-state process we can say, "What goes in must come out." Illustrations of this principle can be found below.

Material balances can be made for a wide variety of materials, at many scales of size for the system and in various degrees of complication. To obtain a true perspective as to the scope of material balances, examine Figs. 2.3 and 2.4. Figure

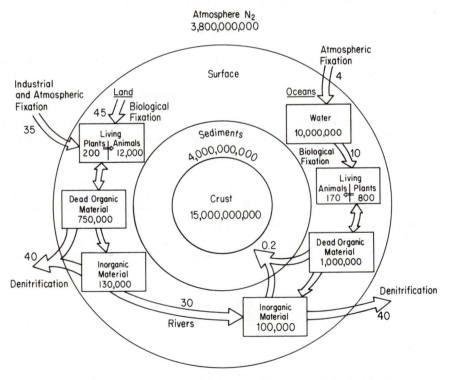

Fig. 2.3 Distribution and annual rates of transfer of nitrogen in the biosphere (in millions of metric tons).

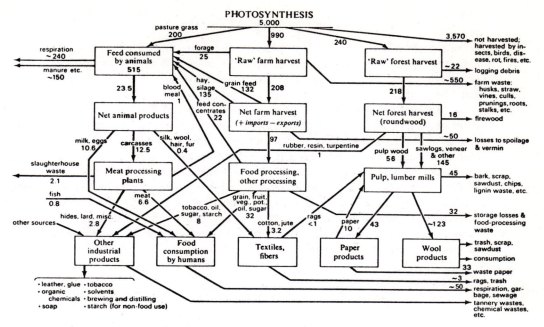

All figures refer to millions of tons of dry organic matter per year.

Fig. 2.4 Production and disposal of products of photosynthesis. (Taken from A. V. Kneese, R. U. Ayres, and R. C. d'Arge, *Economics and the Environment*, Resources for the Future, Inc., Washington, D.C., 1970, p. 32.)

2.3 delineates the distribution of nitrogen in the biosphere as well as the annual transfer rates, both in millions of metric tons. The two quantities known with high confidence are the amount of nitrogen in the atmosphere and the rate of industrial fixation. The apparent precision of the other figures reflects primarily an effort to preserve indicated or probable ratios among different inventories. Because of the extensive use of industrially fixed nitrogen, the amount of nitrogen available to land plants may significantly exceed the nitrogen returned to the atmosphere by denitrifying bacteria in the soil. Figure 2.4 resulted from an analysis of the organic wastes arising from the processing of food and forest products. Changes in any of the streams can pinpoint sources of improvement or deterioration in environmental quality.

We should also note in passing that balances can be made on many other quantities in addition to mass. Balances on dollars are common (your bank statement, for example) as are balances on the number of entities, as in traffic counts, population balances, and social services.

In the process industries, material balances assist in the planning for process design, in the economic evaluation of proposed and existing processes, in process control, and in process optimization. For example, in the extraction of soybean oil from soybeans, you could calculate the amount of solvent required per ton of soybeans or the time needed to fill up the filter press, and use this information in the design of equipment or in the evaluation of the economics of the process. All sorts of raw materials can be used to produce the same end product, and quite a few different types of processing can achieve the same end result, so that case studies (simulations) of the processes can assist materially in the financial decisions that must be made.

Material balances are also used in the hourly and daily operating decisions of plant managers. If there are one or more points in a process where it is impossible or uneconomical to collect data, then if sufficient other data are available, by making a material balance on the process it is possible to get the information you need about the quantities and compositions at the inaccessible location. In most plants a mass of data is accumulated on the quantities and compositions of raw materials, intermediates, wastes, products, and by-products that is used by the production and accounting departments and that can be integrated into a revealing picture of company operations. A schematic outline of the material balance control in the manufacture of phenol is shown in Fig. 2.5.

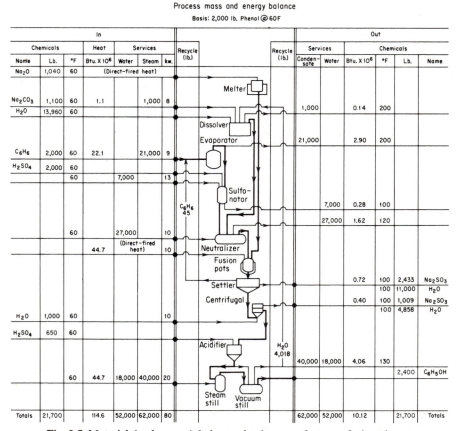

Fig. 2.5 Material (and energy) balances in the manufacture of phenol presented in the form of a ledger sheet. [Taken from *Chem. Eng.*, p. 117 (April 1961), by permission.]

EXAMPLE 2.1 Water Balance for a River Basin

Water balances on river basins for a season or for a year can be used to check predicted groundwater infiltration, evaporation, or precipitation in the basin. Prepare a water balance, in symbols, for a large river basin, including the physical processes indicated in Fig. E2.1 (all symbols are for 1 year and S = storage or inventory).

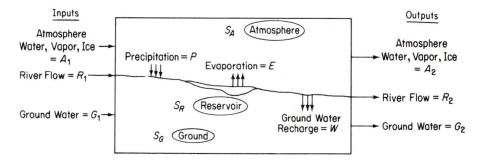

Fig. E2.1

Solution

Equation (2.2) applies in as much as there is no reaction in the system. Each term in Eq. (2.2) can be represented by symbols defined in Fig. E2.1. Let the subscript t_2 designate the end of the year and t_1 designate the beginning of the year. If the system is chosen to be the atmosphere plus the river plus the ground, then the accumulation is

$$(S_{At_2} - S_{At_1}) + (S_{Rt_2} - S_{Rt_1}) + (S_{Gt_2} - S_{Gt_1})$$

The inputs are $A_1 + R_1 + G_1$ and the outputs are $A_2 + R_2 + G_2$. Consequently, the material balance is

(a)
$$(S_{At_2} - S_{At_1}) + (S_{Rt_2} - S_{Rt_1}) + (S_{Gt_2} - S_{Gt_1})$$
$$= (A_1 - A_2) + (R_1 - R_2) + (G_1 - G_2)$$

If the system is chosen to be just the river (including the reservoir), the material balance would be

(b)
$$\underset{\text{accumulation}}{S_{Rt_2} - S_{Rt_1}} = \underset{\text{input}}{(R_1 + P)} - \underset{\text{output}}{(R_2 + E + W)}$$

and analogous balances could be made for the water in the atmosphere and in the ground. Can you make them?

One important point to always keep in mind is that the basic material balance is a balance on mass, not on volume or moles. Thus, in a process in which a chemical reaction takes place so that compounds are generated and/or used up, you will have to employ the principles of stoichiometry discussed in Chap. 1 in making a material balance.

Let us analyze first some very simple examples in the use of the material balance.

EXAMPLE 2.2 Material Balance

Hydrogenation of coal to give hydrocarbon gases is one method of obtaining gaseous fuels with sufficient energy content for the future. Figure E2.2 shows how a free-fall fluidized-bed reactor can be set up to give a product gas of high methane content.

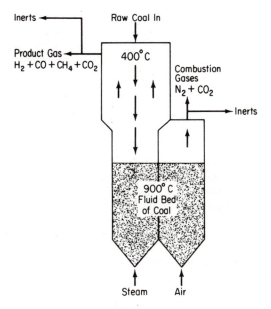

Fig. E2.2

Suppose, first, that the gasification unit is operated without steam at room temperature (25°C) to check the air flow rates and that cyclones separate the solids from the gases effectively at the top of the unit and that no accumulation (buildup) of coal, or ash, or gases, occurs in the unit. If 1200 kg of coal per hour (assume that the coal is 80% C, 10% H, and 10% inert material) is dropped through the top of the reactor:

(a) How many kilograms of coal leave the reactor per hour?
(b) If 15,000 kg of air per hour is blown into the reactor, how many kilograms of air per hour leave the reactor?
(c) If the reactor operates at the temperatures shown in Fig. E2.2 and with the addition of 2000 kg of steam (H_2O vapor) per hour, how many kilograms per hour of combustion and product gases leave the reactor per hour, assuming complete combustion of the coal?

Solution

Basis: 1 hr

This will be a steady-state process.
(a) Since 1200 kg/hr of coal enters the reactor, and none remains inside, 1200 kg/hr must leave the reactor.
(b) Similarly, 15,000 kg/hr of air must leave the reactor.
(c) All the material except the inert portion of the coal leaves as a gas. Consequently, we can add up the total mass of material entering the unit, subtract the inert material, and obtain the mass of combustion gases by difference:

$$\frac{1200 \text{ kg coal}}{} \left| \frac{10 \text{ kg inert}}{100 \text{ kg coal}} \right. = 120 \text{ kg inert}$$

entering material	kg
Coal	1,200
Air	15,000
Steam	2,000
Total	18,200 − 120 = 18,080 kg/hr of gases

EXAMPLE 2.3 Material Balance

(a) If 300 lb of air and 24.0 lb of carbon are placed in a reactor (see Fig. E2.3) at 600°F and after complete combustion no material remains in the reactor, how many pounds of carbon will have been removed? How many pounds of oxygen? How many pounds total?
(b) How many moles of carbon and oxygen enter? How many leave the reactor?
(c) How many total moles enter the reactor and how many leave the reactor?

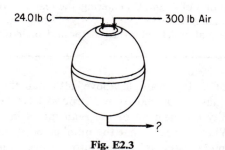

24.0 lb C ——————   —————— 300 lb Air

?

Fig. E2.3

Solution

This is a problem with no accumulation.
(a) The total material leaving the reactor would be 324 lb. Since air has 21.0% O_2 and 79.0% N_2,

Basis: 300 lb of air

$$\frac{300 \text{ lb air}}{} \left| \frac{1 \text{ lb mol air}}{29.0 \text{ lb air}} \right| \frac{21.0 \text{ lb mol O}_2}{100 \text{ lb mol air}} = 2.18 \text{ lb mol O}_2$$

$$\frac{24 \text{ lb C}}{} \left| \frac{1 \text{ lb mol C}}{12.0 \text{ lb C}} \right. = 2.00 \text{ lb mol C}$$

The oxygen required to completely burn 24.0 lb of C is 2.00 lb mol; consequently,

$$2.18 - 2.00 = 0.18$$

is the number of moles of unused oxygen, and this is equivalent to $(0.18)(32) = 5.76$ lb of O_2 as such leaving the reactor. In addition, 88 lb of CO_2 leave and

$$\frac{2.18 \text{ lb mol O}_2}{} \left| \frac{79.0 \text{ lb mol N}_2}{21.0 \text{ lb mol O}_2} \right| \frac{28.2 \text{ lb N}_2}{1 \text{ lb mol N}_2} = 230 \text{ lb N}_2$$

The moles of N_2 are

$$\frac{230 \text{ lb N}_2}{} \left| \frac{1 \text{ lb mol N}_2}{28.2 \text{ lb N}_2} \right. = 8.20 \text{ lb mol N}_2$$

Summing up all input and exit streams, we have

in	lb	lb mol	out	lb	lb mol
O_2	70	2.18	O_2	5.76	0.18
N_2	230	8.20	N_2	230	8.20
C	24	2.00	CO_2	88	2.00
Total	324	12.38		324	10.38

There will be no carbon leaving as carbon because all the carbon burns to CO_2; the carbon leaves in a combined form with some of the oxygen.

(b) The moles of C and O_2 entering the reactor are shown in part (a); no moles of C, as such, leave.

(c) 12.38 total moles enter the reactor and 10.38 total moles leave.

Note that in the example above, although the total mass put into a process and the total mass recovered from a process have been shown to be equal, there is no such equality on the part of the *total* moles in and out, *if* a chemical reaction takes place. What is true is that the number of atoms of an element (such as C, O, or even oxygen expressed as O_2) put into a process must equal the atoms of the same element leaving the process. In Example 2.3 the total moles in and out are shown to be unequal, but the atoms of O in (expressed as moles of O_2) equal the atoms of O (similarly expressed as O_2) leaving:

Component Moles in		Component Moles out	
O_2	2.18	O_2	0.18
		O_2 in CO_2	2.00
Total $= O_2$	2.18	Total $= O_2$	2.18
N_2	8.20	N_2	8.20
C	2.00	C in CO_2	2.00

Keeping in mind the remarks above for processes involving chemical reactions, we can summarize the circumstances under which the input equals the output for *steady-state processes* (no accumulation) as follows:

	Type of Balance	Equality Required for Input and Output of Steady-State Process	
		Without Chemical Reaction	With Chemical Reaction
Total	Total mass	Yes	Yes
balances	Total moles	Yes	No*
	Mass of a pure compound	Yes	No*
Component	Moles of a pure compound	Yes	No*
balances	Mass of an atomic species	Yes	Yes
	Moles of an atomic species	Yes	Yes

*Equality may occur by chance.

How many independent material balance equations can be written? Without a reaction taking place, an independent equation can be written to balance each chemical compound because the various chemical species introduced do not change their identities. Let us say that the chemical component balances comprise in total C equations where $C =$ the number of components. In addition, you can write a total material balance, for a grand total of $(C + 1)$ equations. However, only C of them are independent equations. Do you understand why? In cases in which a chemical reaction occurs the component material balances have to be based on the elements themselves (H, O, S) or their equivalents (H_2, O_2, S_4) and not the compounds. Do you understand why?

We now turn to more detailed consideration of problems *in which the accumulation term is zero*. The basic task is to turn the problem, expressed in words, into a quantitative form, expressed in mathematical symbols and numbers, and then solve the mathematical statements.

Self-Assessment Test

1. Draw a sketch of the following processes and place a dashed line around the system:
   (a) Tea kettle
   (b) Fireplace
   (c) Swimming pool

2. Show the materials entering and leaving the systems in problem 1. Designate the time interval of reference and classify the system as open or closed.

3. Write down the general energy balance in words. Simplify it, stating the assumptions made in each simplification.

4. Classify the following processes as (1) batch, (2) flow, (3) neither, or (4) both on a time scale of one day:
   (a) Oil storage tank at a refinery
   (b) Flush tank on a toilet
   (c) Catalytic converter on an automobile
   (d) Gas furnace in a home

5. What is a steady-state process?

6. Do the input and output of material in Fig. 2.4 agree? Why not? Repeat for Fig. 2.5 for the chemicals.

7. Define a material balance.

8. Answer the following true (T) or false (F).
   (a) If a chemical reaction occurs, the total masses entering and leaving the system for a steady-state process are equal.
   (b) In combustion, all the moles of C that enter a steady-state process exit from the process.
   (c) If you have C components entering a system, you can write $C + 1$ independent material balances.

9. List the circumstances for a steady-state process in which the number of moles entering the system equals the number of moles leaving the system.

Section 2.2 Program of Analysis of Material Balance Problems

> *Your objectives in studying this section are to be able to:*
>
> 1. Define what the term "solution of a material balance problem" means.
> 2. Ascertain that a unique solution exists for a problem using the given data, or ascertain that additional information is needed (and get it).
> 3. Decide which equations to use if you have redundant equations.
> 4. Solve a set of n independent equations containing n variables whose values are unknown.

5. Retain in memory and recall as needed the implicit constraints in a problem.
6. Prepare material flow diagrams from word problems.
7. Translate word problems and the associated diagrams into material balances with properly defined symbols for the unknown variables and consistent units for steady-state processes with and without chemical reaction.
8. State the maximum number of independent equations that can be generated in a specific problem.
9. Recite the 10 steps used to analyze material balance problems so that you have an organized strategy for solving material balance problems.

Much of the remaining portion of this chapter demonstrates the techniques of analyzing and solving problems involving material balances. By *solving* we mean obtaining a *unique* solution. Later portions of the text consider the case of combined material and energy balances. Since these material balance problems all involve the same principle, although the details of the applications of the principle may differ slightly, we shall examine in this section a generalized method of analyzing such problems which can be applied to the solution of any type of material balance problem.

We are going to discuss a method of analysis of material balance problems that will enable you to understand, first, how similar these problems are, and second, how to solve them in the most expeditious manner. For some types of problems the method of approach is relatively simple and for others it is more complicated, but the important point is to regard problems in distillation, crystallization, evaporation, combustion, mixing, gas absorption, or drying not as being different from each other but as being related from the viewpoint of how to proceed to solve them.

An orderly method of analyzing problems and presenting their solutions represents training in logical thinking that is of considerably greater value than mere knowledge of how to solve a particular type of problem. Understanding how to approach these problems from a logical viewpoint will help you to develop those fundamentals of thinking that will assist you in your work as an engineer long after you have read this material.

Consider the two basic types of information you have to work with in analyzing material balance problems:

(a) The total mass (weight) in each stream entering and leaving the system

(b) The composition of each stream entering and leaving the system

Of course, if a chemical reaction takes place inside the system, the equations for the reaction and/or the extent of the reaction are important additional pieces of information.

As previously mentioned, we shall assume that the process we are analyzing is taking place in the steady state (i.e., that there is no accumulation or depletion of material). Even if the process is of the batch type in which there is no flow in and out, let us pretend that the initial material is pushed into the vessel and that the final material is removed from the vessel. By this hypothesis we can imagine a batch process converted into a fictitious flow process, and talk about the "streams" entering and leaving even though in the real process nothing of the sort happens (except over the entire period of time under consideration). After you do this three or four times, you will see that it is a very convenient technique, although at the beginning it may seem a bit illogical.

There is no point in beginning to solve a set of material balances until you can be certain that the equations have a unique solution. Let us look at some problems to ascertain what types and how much information is needed about the two categories listed above so that you can be sure that a unique solution to a problem is possible. You should recall from your experience in mathematics that each equation consists of parameters (coefficients and perhaps variables) of known values, and one or more variables whose value(s) is (are) to be determined by solving the equation. In the analysis of a material balance problem you should always check first to make sure that the number of independent equations and the number of variables whose values are to be determined by solving the equations are equal. What happens if they are not equal? You either have a case with redundant information (fewer unknown variables than independent equations) or an infinite number of solutions (more unknown variables than independent equations).

Let us start the analysis of material balance problems by examining a simple problem without any chemical reaction being involved. Look at Fig. 2.6(a). An independent material balance equation equivalent to Eq. (2.3) can be written for each species involved in the process defined by the system boundary. We will use the symbol ω with an appropriate subscript to denote the mass fraction of a component in the streams F, W, and P, respectively. Each mass balance will have the form

$$\omega_{i,F}F = \omega_{i,P} + \omega_{i,W}W \tag{2.4}$$

but we will, of course, insert the known values of the stream flow rates and mass fractions instead of the symbols where possible. Here are the three balances:

	in	=	*out*	
EtOH:	$(0.50)(100)$	=	$(0.80)(60) + \omega_{\text{EtOH},W}(W)$	(2.4a)
H_2O:	$(0.40)(100)$	=	$(0.05)(60) + \omega_{H_2O,W}(W)$	(2.4b)
MeOH:	$(0.10)(100)$	=	$(0.15)(60) + \omega_{\text{MeOH},W}(W)$	(2.4c)

In addition, we can write a balance for the total material in and out:

	in		*out*	
Total:	$(1.00)\ 100$	=	$(1.00)(60) + (\omega_{\text{EtOH},W}$	
			$+ \omega_{H_2O,W} + \omega_{\text{MeOH},W})(W)$	(2.4d)

Remember that implicit constraints (equations) exist in the problem formulation because of the definition of mass fraction: namely, that the sum of the mass fractions in each stream must be unity:

$$\omega_{\mathrm{EtOH}} + \omega_{\mathrm{H_2O}} + \omega_{\mathrm{MeOH}} = 1 \tag{2.5}$$

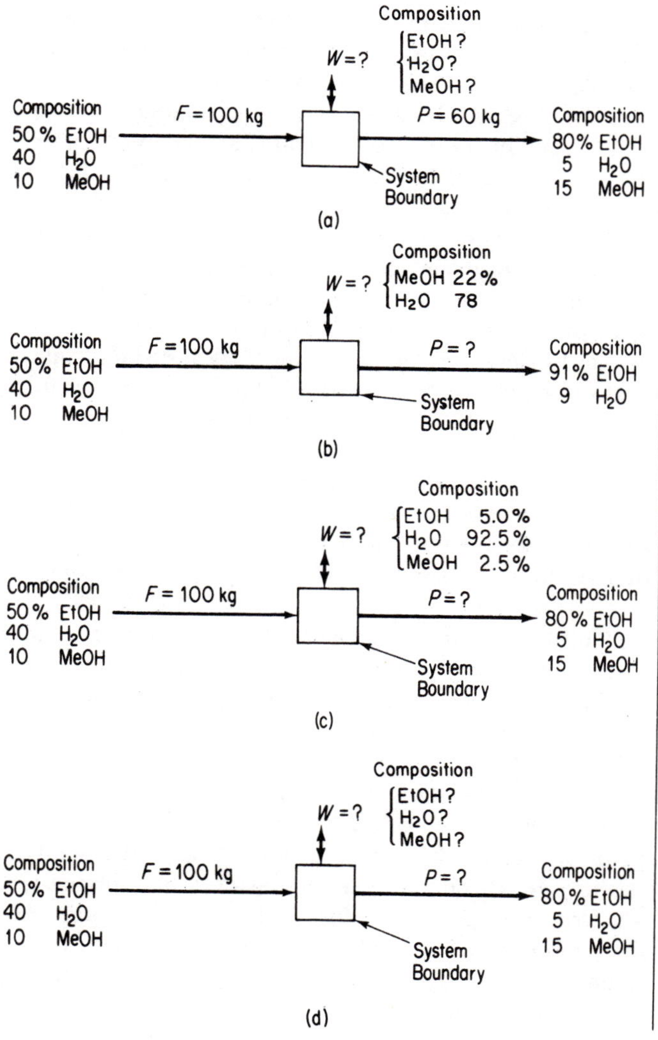

Parameters			
Known		Unknown	
Mass	Stream	Mass	Stream
2	6	1	3
1	9†	2	0
1	9	2	0
1	6	2	3

† A stream with only two components has 0% for the third component.

Fig. 2.6 Typical material balance problems (no chemical reaction involved).

First, let us count the number of independent equations. You should recognize that not all of the four mass balances, Eqs. (2.4a)–(2.4d), are independent equations. Note how the sum of the three component balances (2.4a), (2.4b), and (2.4c) equals the total mass balance. *The number of independent mass balance equations is equal to the number of components.* In any problem you can substitute the total material balance for any one of the component material balances if you plan to solve two or more equations simultaneously. Appendix L describes how to ascertain in a formal way whether the equations in a set of linear equations are independent.

Normally, we would have a statement equivalent to Eq. (2.5) for each stream, but in the problem specified in Fig. 2.6(a), only one such equation need be written because the mass fractions in the F and P streams are known a priori.

Thus we have four independent equations to solve, 2.4a–2.4c plus 2.5. How many variables can have unknown values if a unique solution is to be obtained? Four! Let us count the number of variables whose values are unknown: W, $\omega_{EtOH,W}$, $\omega_{H_2O,W}$, and $\omega_{MeOH,W}$. The problem specifications worked out quite well—we have the same number of independent equations as unknown values of variables, and can solve the four equations to get a unique solution. To solve the equations as simply as possible, in this particular case you should note by inspection that the equations are not badly coupled together by the unknowns. Furthermore, you intuitively would introduce Eq. (2.5) into Eq. (2.4d), solve the latter for W, and then solve Eqs. (2.4a)–(2.4c).

What would you do if the count of independent equations and unknown values of variables did not match up? The best procedure is to review your analysis of the problem to make sure you have not ignored some equation(s) or variable(s), double counted, forgotten to look up some missing data, or made some error in your assumptions.

Look at Fig. 2.6(b) and 2.6(c). Make a count of the known and unknown stream and composition values. Compare your count with the numbers listed in the parameter count on the right-hand side of Fig. 2.6. Do you agree with the count? How many independent equations will be needed for each case to solve uniquely for the unknown variables? Can you write them down?

Look at Fig. 2.6(d). Count the known and unknown parameters. Do you agree with the count shown? Note how for each of the first three figures in Fig. 2.6 the number of independent material balances was not less than the number of unknown quantities. How many independent material balances and implicit mass fraction equations can you write for the case of Fig. 2.6(d)? To solve uniquely for the five unknown quantities present in Fig. 2.6(d), one additional piece of information or equation must be obtained, or no unique solution is possible. (In some problems no basis is cited, and the number of unknowns is one greater than the number of independent equations. In such a case, you can select an arbitrary basis of 1 or 100 kg, lb, mol, and so on, to provide the essential extra piece of information needed to obtain a unique solution.)

To sum up the analysis of Fig. 2.6, in Fig. 2.6(a) there are four unknowns and four independent equations; in both Fig. 2.6(b) and 2.6(c) there are two unknowns and three equations, so that redundant information exists (use the most accurate information in solving the equations); and in Fig. 2.6(d) there are five unknowns and only four equations, so that the problem has no unique solution as it stands and hence is indeterminate.

Now let us look at material balances in which chemical reactions are involved. How does the procedure for analysis differ from the previous analysis for the case without chemical reaction? As mentioned in Sec. 2.1, because moles of a species and total moles are not conserved when a chemical reaction takes place, you must make the balances on total mass and mass or moles of each atomic species or multiple thereof (i.e., hydrogen expressed as H_2). For example, in Fig. 2.7(a) you can write a total mass (not mole) balance and a carbon, a hydrogen, a nitrogen, and an oxygen balance. The carbon balance might be in terms of C, and the hydrogen, nitrogen, and oxygen balances in terms of H_2, N_2, and O_2, respectively. Depending

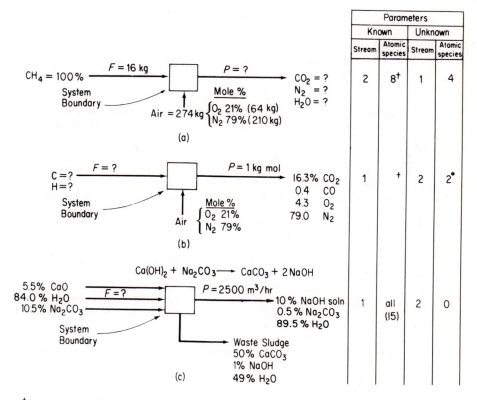

†Components not listed are present in 0%. and thus are known

Fig. 2.7 Typical material balance problems (with chemical reaction).

on the number of unknown values of variables in the problem, you would not necessarily have to use each of the balances. Also keep in mind that in using balances on atomic species, both mass and moles are conserved.

Figure 2.7(a) and 2.7(b) illustrate cases in which information about the chemical reaction is not given. In such cases you are expected to know the reaction equation and conversion, or be able to look it up in a chemistry text. Count the number of known and unknown parameters. Do you agree with the count shown in Fig. 2.7? In Fig. 2.7(a) each stream can include up to four species: C, H, O, N. In the CH_4 and air streams all the mole fractions of the compounds are known (some are zero), and thus it is easy to calculate the mole composition of the entering streams in terms of, say, C and H_2 for CH_4 directly from the molecular formula ($\frac{1}{3}C$ and $\frac{2}{3}H_2$), and O_2 and N_2 for air from known information.

In counting up the number of independent equations[1] do not forget the implicit equations that the sum of the mole fractions of the components in a stream is unity. Thus in Fig. 2.7(a), you can make four independent material balances and one summation of mole fractions equation (for P) so that the problem presented has a unique solution. Can you make the same statement about Figs. 2.7(b) and (c)? Hint: Where did the hydrogen go in Fig. 2.7b?

Let us now formally list the prescription for success in analyzing material balance problems. If you use the steps in Table 2.1 as a mental checklist each time you

TABLE 2.1 STRATEGY FOR ANALYZING MATERIAL BALANCE PROBLEMS
("A problem recognized is a problem half-solved." Ann Landers)

---

1. Draw a sketch of the process; define the system by a boundary.
2. Label the flow of each stream and the associated compositions with symbols.
3. Put all the known values of compositions and stream flows on the figure by each stream; calculate additional compositions from the given data as necessary. Or, at least initially identify the known parameters in some fashion.
4. List by symbols each of the unknown values of the stream flows and compositions, or at least mark them distinctly in some fashion.
5. List the number of independent balances that can be written; ascertain that a unique solution is possible. If not, look for more information or check your assumptions.
6. Select a basis.
7. Select an appropriate set of balances to solve; write the balances down with type of balance listed by each one. Do not forget the implicit balances for mass or mole fractions.
8. State whether the problem is to be solved by direct addition or subtraction, by the tie component method, or by an algebraic method.
9. Solve the equations. Each calculation must be made on a consistent basis.
10. Check your answers by introducing them, or some of them, into the material balances. Are the equations satisfied?

---

[1] In a few special cases in which reactions take place, the number of independent equations is less than the number of atomic species. Refer to Appendix L.1 for a discussion of this point.

start to work on a problem, you will have achieved the major objective of this chapter and substantially added to your professional skills. These steps do not have to be carried out in exactly the order prescribed and you may repeat steps as the problem specification becomes clearer. But they are all essential.

Now we turn to the analysis of simple combinations of units. Suppose that the system as defined is comprised of three subsystems as indicated in Fig. 2.8. The analysis of the problem presented in Fig. 2.8 follows the guidelines listed in Table

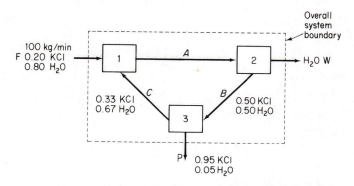

Fig. 2.8 Flow diagram of a system comprised of three subsystems. No reaction takes place.

2.1. You can make material balances for the subsystems and overall system—just make sure that the balances selected for solution are independent! We can use the same symbols as before. How many values of the variables are unknown? There will be seven in all: $W, P, A, B, C, \omega_{KCl,A}$ and $\omega_{H_2O,A}$. How many independent equations must be written to obtain a unique solution? Seven. What are the names of such a set of equations? One set might be:

Unit 1, total: $100 + C = A$

Unit 1, KCl: $(0.20)(100) + (0.33)(C) = (\omega_{KCl})(A)$

Unit 2, total: $A = W + B$

Unit 2, KCl: $(\omega_{KCl,A})(A) = (0.50)B$

Overall, total: $F = W + P$

Overall, KCl: $(0.20)F = (0.95)P$

$\sum \omega_i = 1$: $\omega_{KCl,A} + \omega_{H_2O,A} = 1$

Other sets are possible. Write down a different set. Did you note that the set we have used has been chosen so as to include as few of the unknown variables as possible in a given equation; that is, we made component balances on KCl and not on H_2O.

EXAMPLE 2.4 Independent Material Balances

Examine Fig. E2.4. The composition of each stream is as follows:

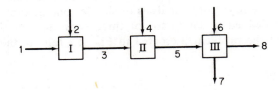

Fig. E2.4

(1) Pure *A*

(2) Pure *B*

(3) *A* and *B*, split known

(4) Pure *C*

(5) *A*, *B*, and *C*, split known

(6) Pure *D*

(7) *A* and *D*, split known

(8) *B* and *C*, split known

How many independent material balances could be generated to solve this problem?

Solution

	number of balances
At Unit I, two components are involved	2
At Unit II, three components are involved	3
At Unit III, four components are involved	4
Total	9

Can you show that one or more component balances around systems I and II, or II and III in Example 2.4, or the entire set of three units, will add *no* additional independent balances to the set of component material balances made on each individual unit? Can you show that a total balance on each unit, or on units I + II, or II + III, or around the entire system of three units, will add *no* additional independent balances? Can you substitute one of the indicated alternate material balances for a component balance? Yes (as long as the precision of the balance is about the same).

Past experience has indicated that the two major difficulties you will experience in solving material balance problems have nothing to do with the mechanics of solving the problem (once they are properly formulated!). Instead, you will find that the two stumbling blocks are:

(a) After reading the problem you do not really understand what the process in the problem is all about. Help yourself achieve understanding by sketching the system—just a box with arrows as in Figs. 2.6 and 2.7 will do.

(b) After a first reading of the problem you are confused as to which values of compositions and flows are known and unknown. Help reduce the confusion by putting the known values down on your diagram as you discover them or calculate them, and by putting question marks or alphabetic symbols for the unknown quantities. Try to record the compositions as weight, or mole, fractions (or percents) as you proceed so as to assist in clarifying the choice of basis and the material balances that can be employed. Because you will have to set up and solve a material balance for each unknown you enter on the diagram, try to keep the number of unknown labels to as few as possible.

Problems in which the mass (weight) of one stream and the composition of one stream are unknown can be solved without difficulty by direct addition or direct subtraction. Problems in which all the compositions are known and two or more of the weights are unknown require some slightly more detailed calculations. If a tie component exists which makes it possible to establish the relationship between the unknown weights and the known weights, the problem solution may be simplified. (The tie component is discussed in detail in Sec. 2.5.) When there is no direct or indirect tie component available, algebra must be used to relate the unknown masses to the known masses.

Self-Assessment Test

1. What does the concept "solution of a material balance problem" mean?

2. (a) How many values of unknown variables can you compute from one independent material balance?
    (b) From three?
    (c) From four material balances, three of which are independent?

3. A water solution containing 10% acetic acid is added to 20 kg/min of a water solution containing 30% acetic acid. The product P of the combination leaves at the rate of 100 kg/min. What is the composition of P? For this process,
    (a) Determine how many independent material balances can be written.
    (b) List the names of the balances.
    (c) Determine how many unknown variables can be solved for.
    (d) List their names and symbols.
    (e) Determine the composition of P.
    In your solution set out the 10 steps specifically and label them.

4. Can you solve these three material balances for F, D, and P?

$$0.1F + 0.3D = 0.2P$$
$$0.9F + 0.7D = 0.8P$$
$$F + \quad D = \quad P$$

5. Cite two ways to solve a set of linear equations.

6. If you want to solve a set of independent equations that contain fewer unknown variables than equations (the overspecified problem), how should you proceed with the solution?

7. What is the major category of implicit constraints (equations) you encounter in material balance problems?

8. If you want to solve a set of independent equations that contain more unknown variables than equations, what must you do to proceed with the solution?

9. How many values of the set variables in the process shown in the figure are unknown? List them. The streams contain two components, 1 and 2.

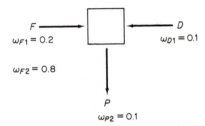

10. How many material balances are needed to solve problem 9? Is the number the same as the number of unknown variables? Explain.

Section 2.3   Material Balance Problems with Direct Solution

> **Your objectives in studying this section are to be able to:**
>
> 1. Define flue gas, stack gas, Orsat analysis, dry basis, wet basis, theoretical air (oxygen), required air (oxygen), and excess air (oxygen).
> 2. Given two of the three factors: entering air (oxygen), excess air (oxygen), and required air (oxygen), compute the third factor.
> 3. Apply the 10-step strategy to solve problems (with or without chemical reaction) having a direct solution (i.e., problems in which the equations are decoupled so that algebra is not needed).

Problems in which the mass balances are decoupled can be solved without the use of algebra. An example would be a problem in which one mass (weight) and one composition are unknown. It can be solved by direct addition or subtraction, as shown in the examples below. You can save time and effort by bypassing the use of formal algebra. You may find it necessary to make some brief preliminary calculations in order to decide whether or not all the information about the compositions and weights that you would like to have is available. Of course, in a stream containing just one component, the composition is known, because that component is 100% of the stream.

In dealing with problems involving combustion, you should become acquainted with a few special terms:

(a) *Flue* or *stack gas*—all the gases resulting from a combustion process including the water vapor, sometimes known as *wet basis*.

(b) *Orsat analysis or dry basis*—all the gases resulting from the combustion process not including the water vapor. (Orsat analysis refers to a type of gas analysis apparatus in which the volumes of the respective gases are measured over and in equilibrium with water; hence each component is saturated with water vapor. The net result of the analysis is to eliminate water as a component being measured.)

Pictorially, we can express this classification for a given gas as in Fig. 2.9. To convert from one analysis to another, you have to ratio the percentages for the components as shown in Example 2.10.

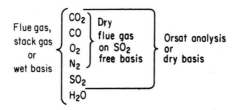

Fig. 2.9 Comparison of gas analyses on different bases.

(c) *Theoretical air* (or *theoretical oxygen*)—the amount of air (or oxygen) required to be brought into the process for *complete* combustion. Sometimes this quantity is called the *required* air (or oxygen).

(d) *Excess air* (or *excess oxygen*)—in line with the definition of excess reactant given in Chap. 1, excess air (or oxygen) would be the amount of air (or oxygen) **in excess of that required for complete combustion** as computed in (c).

Even if only partial combustion takes place, as, for example, C burning to both CO and CO_2, the excess air (or oxygen) is computed as if the process of combustion produced only CO_2. The percent excess air is identical to the percent excess O_2 (a more convenient method of calculation):

$$\% \text{ excess air} = 100\frac{\text{excess air}}{\text{required air}} = 100\frac{\text{excess } O_2/0.21}{\text{required } O_2/0.21} \qquad (2.6)$$

Note that the ratio 1/0.21 of air to O_2 cancels out in Eq. (2.6). Percent excess air may also be computed as

$$\% \text{ excess air} = 100\frac{O_2 \text{ entering process} - O_2 \text{ required}}{O_2 \text{ required}} \qquad (2.7)$$

or

$$\% \text{ excess air} = 100\frac{\text{excess } O_2}{O_2 \text{ entering} - \text{excess } O_2}$$

since

$$O_2 \text{ entering process}$$
$$= O_2 \text{ required for complete combustion} + \text{excess } O_2 \qquad (2.8)$$

The precision of these different relations for calculating the percent excess air may not be the same. If the percent excess air and the chemical equation are given in a problem, you know how much air enters with the fuel, and hence the number of unknowns is reduced by one.

In the burning of coal, you may wonder how to treat the oxygen found in most coals in some combined form. Just assume that the oxygen is already combined with some of the hydrogen in the coal in the proper proportions to make water. Consequently, the O_2 does not enter into the combustion process, and the hydrogen equivalent to this oxygen does not enter the combustion reaction either. No corresponding oxygen is counted toward the required oxygen need in the air. Only the remaining or "net" hydrogen requires oxygen from the air to form water vapor on burning.

EXAMPLE 2.5 Excess Air

Fuels for motor vehicles other than gasoline are being eyed because they generate lower levels of pollutants than does gasoline. Compressed ethylene has been suggested as a source of economic power for vehicles. Suppose that in a test 20 lb of C_2H_4 is burned with 400 lb of air to produce 44 lb of CO_2 and 12 lb of CO. What was the percent excess air?

Solution

$$C_2H_4 + 3O_2 \longrightarrow 2CO_2 + 2H_2O$$
$$\text{Basis: 20 lb of } C_2H_4$$

Since the percentage of excess air is based on the complete combustion of C_2H_4 to CO_2 and H_2O, the fact that combustion is not complete has no influence on the definition of "excess air." The required O_2 is

$$\frac{20 \text{ lb } C_2H_4}{} \left| \frac{1 \text{ lb mol } C_2H_4}{28 \text{ lb } C_2H_4} \right| \frac{3 \text{ lb mol } O_2}{1 \text{ lb mol } C_2H_4} = 2.14 \text{ lb mol } O_2$$

The entering O_2 is

$$\frac{400 \text{ lb air}}{} \left| \frac{1 \text{ lb mol air}}{29 \text{ lb air}} \right| \frac{21 \text{ lb mol } O_2}{100 \text{ lb mol air}} = 2.90 \text{ lb mol } O_2$$

The percent excess air is

$$100 \times \frac{\text{excess } O_2}{\text{required } O_2} = 100 \frac{\text{entering } O_2 - \text{required } O_2}{\text{required } O_2}$$

$$\% \text{ excess air} = \frac{2.90 \text{ lb mol } O_2 - 2.14 \text{ lb mol } O_2}{2.14 \text{ lb mol } O_2} \left| \frac{100}{} = 35.5\% \right.$$

EXAMPLE 2.6 Excess Air

A salesperson comes to the door selling a service designed to check "chimney rot." He explains that if the CO_2 content of the gases leaving the chimney rises above 15%, it is dangerous to your health, is against the city code, and causes your chimney to rot. On checking the flue gas from the furnace he finds it is 30% CO_2. Suppose that you are burning natural gas which is about 100% CH_4 and that the air supply is adjusted to provide 130% excess air. Do you need his service?

Solution

Let us calculate the actual percentage of CO_2 in the gases from the furnace assuming that complete combustion takes place. See Fig. E2.6a. The 130% excess air means 130% of the air required for complete combustion of CH_4. The chemical reaction is

$$CH_4 + 2O_2 \longrightarrow CO_2 + 2H_2O$$

Step 1 (see Table 2.1): The system is defined by Fig. E2.6a; note that a chemical reaction occurs.

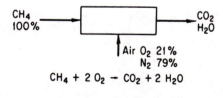

Fig. E2.6a

Steps 2, 3, and 4: For clarity in labeling, Fig. E2.6a is repeated as Fig. E2.6b with all the species listed by each related stream, and designated either by a check ($\checkmark$) as known (or can easily be calculated) or by a question mark (?) as unknown initially.

Step 5: We can only make four material balances, C, H_2, O_2, and N_2, and one summation of mole fractions; hence more information is needed.

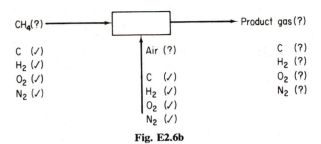

CH$_4$(?) ──────→ [ ] ──────→ Product gas(?)

C (✓)   Air (?)   C (?)
H$_2$ (✓)   H$_2$ (?)
O$_2$ (✓)   C (✓)   O$_2$ (?)
N$_2$ (✓)   H$_2$ (✓)   N$_2$ (?)
                         O$_2$ (✓)
                         N$_2$ (✓)

Fig. E2.6b

Step 6: By picking a basis of 1 mole of CH$_4$, we can reduce the unknown quantities by one, leaving six.

Basis: 1 mole of CH$_4$

Step 3 Repeated: The quantity of air is ascertained by the specification of the excess air and is computed as follows: On the basis of 1 mole of CH$_4$, 2 moles of O$_2$ are required for complete combustion, or

$$\frac{2 \text{ mol O}_2 \mid 1.00 \text{ mol air}}{0.21 \text{ mol O}_2} = 9.52 \text{ mol air required}$$

composed of 2 moles of O$_2$ and 7.52 moles of N$_2$. The entering air is 9.52 (1.30) = 12.4 moles of air, composed of 2.60 moles of O$_2$ and 9.80 moles of N$_2$. Now we have left five quantities with unknown values. We can in principle make four independent material balances (C, H$_2$, O$_2$, N$_2$), and use one sum of the mole fraction relations for the product gas, so that the problem can be solved for the quantity of CO$_2$.

Steps 7, 8, and 9: Summing up our calculations so far we have entering the furnace together with 1 mole of C and 2 moles of H$_2$:

	air	O$_2$	N$_2$
Required	9.52	2.00	7.54
Excess	12.4	2.60	9.80
Total	21.9	4.60	17.3

Since all the entering C exits as CO$_2$ and all the H exits as H$_2$O, based on the chemical equation the composition of the gases from the furnace should be as follows:

$$\text{C balance:} \quad \frac{1 \text{ mol C in} \mid 1 \text{ mol CO}_2 \text{ out}}{1 \text{ mol CH}_4 \text{ in} \mid 1 \text{ mol C out}} = 1 \text{ mol CO}_2 \text{ out}$$

$$\text{H}_2 \text{ balance:} \quad \frac{2 \text{ mol H}_2 \text{ in} \mid 1 \text{ mol H}_2\text{O out}}{1 \text{ mol CH}_4 \text{ in} \mid 1 \text{ mol H}_2 \text{ out}} = 2 \text{ mol H}_2\text{O out}$$

component	moles	percent
CH_4	0	0
CO_2	1.00	4.4
H_2O	2.00	8.7
O_2	2.60	11.3
N_2	{ 7.54 / 9.80	75.6
	22.9	100.0

The salesman's line seems to be rot all the way through.

Step 10: Check the answer.

EXAMPLE 2.7 Drying

A wet paper pulp is found to contain 71% water. After drying it is found that 60% of the original water has been removed. Calculate the following:
    (a) The composition of the dried pulp
    (b) The mass of water removed per kilogram of wet pulp

Solution

Steps 1, 2, 3, and 4:

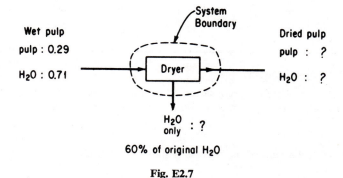

Fig. E2.7

Step 5: Only two independent material balances (pulp and H_2O) and one sum of mass fractions (for the dried pulp) can be written so that more information is needed to reduce the unknown quantities to three (or less).

Step 6: Pick a convenient basis.

Basis: 1 kg of wet pulp

Steps 7, 8, and 9: From the statement about the mass of water removed, we can calculate that

$$H_2O \text{ removed} = 0.60(0.71) = 0.426 \text{ kg}$$

Steps 5 (Repeated), 8, and 9: A unique solution is now possible. By direct subtraction the water that exits in the dried pulp is

$$in - out = \textbf{remainder}$$

$$H_2O \text{ balance:} \quad H_2O \text{ in pulp} = 0.71 - 0.426 = 0.284 \text{ kg}$$

or alternatively

$$0.40(0.71) = 0.284 \text{ kg } H_2O$$

All of the pulp exits in the dried pulp:

$$in = \textbf{remainder}$$

$$\text{dry pulp balance:} \quad 0.29 = 0.29$$

The composition of the dried pulp is

component	kg	percent
Pulp (dry)	0.29	50.5
H_2O	0.284	49.5
Total	0.574	100.0

Step 10: We check to make sure that total kg in = total kg out

$$\textbf{wet pulp} = \textbf{water} + \textbf{dried pulp}$$

$$1.00 = 0.426 + 0.574$$

EXAMPLE 2.8 Crystallization

A tank holds 10,000 kg of a saturated solution of $NaHCO_3$ at 60°C. You want to crystallize 500 kg of $NaHCO_3$ from this solution. To what temperature must the solution be cooled?

Solution

Step 1: A diagram of the process is shown in Fig. E2.8a.

Step 3: Additional data are needed on the solubility of $NaHCO_3$ as a function of temperature. From any handbook you can find

temp. (°C)	solubility g $NaHCO_3$/100 g H_2O
60	16.4
50	14.45
40	12.7
30	11.1
20	9.6
10	8.15

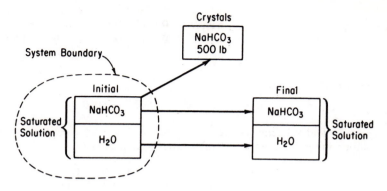

Fig. E2.8a

Because the initial solution is saturated at 60°C and the final solution is also saturated at some temperature to be determined, we can only calculate the composition of the initial solution at this stage:

$$\frac{16.4 \text{ g NaHCO}_3}{16.4 \text{ g NaHCO}_3 + 100 \text{ g H}_2\text{O}} = 0.141 \quad \text{or} \quad 14.1\% \text{ NaHCO}_3$$

The remainder of the solution is water, or 85.9%.

Step 6: Take a basis.

Basis: 10,000 kg of saturated solution at 60°C

Steps 1, 2, and 3 Repeated:

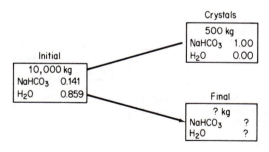

Fig. E2.8b

Steps 4 and 5: We are missing the mass of the final solution and its composition, but can make two material balances and one summation of mass fractions balance, so a unique solution is possible.

Steps 7, 8, and 9: A direct solution is possible by subtraction.

$$
\begin{array}{llll}
 & \textbf{initial} & -\textbf{ crystals} & = \textbf{final} \\
\text{NaHCO}_3 \text{ balance:} & 0.141(10,000) - & 500 & = \ 910 \\
\text{H}_2\text{O balance:} & 0.859(10,000) - & 0 & = 8590 \\
\hline
 & & & 9500
\end{array}
$$

Step 10: Check on total:

$$9500 + 500 = 10,000$$

To find the temperature of the final solution, calculate the composition of the final solution in terms of grams of $NaHCO_3/100$ gram of H_2O.

$$\frac{910 \text{ g NaHCO}_3}{8590 \text{ g H}_2\text{O}} = \frac{10.6 \text{ g NaHCO}_3}{100 \text{ g H}_2\text{O}}$$

Thus the temperature to which the solution must be cooled is (using linear interpolation)

$$30°C - \frac{11.1 - 10.6}{11.1 - 9.6}(10.0°C) = 27°C$$

EXAMPLE 2.9 Distillation

A novice manufacturer of alcohol for gasohol is having a bit of difficulty with her still. The operation is shown in Fig. E2.9. She finds that she is losing too much alcohol in the bottoms (waste). Calculate the composition of the bottoms for her and the weight of alcohol lost in the bottoms.

Solution

Steps 1, 2, and 3:

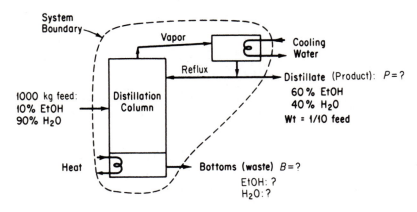

Fig. E2.9

Step 4: The unknown quantities are the bottoms and its composition plus the distillate.

Step 6: Select as the basis the given feed

Basis: 1000 kg of feed

Step 3 Continued: We are given that P is $\frac{1}{10}$ of F, so that

$$P = 0.1(1000) = 100 \text{ kg}$$

Steps 5, 7, 8, and 9: We can make two material balances and one sum of mass fractions balance for the system so that the problem has a unique solution. The solution can be computed directly by subtraction

	kg feed in	−	kg distillate out	=	kg bottoms out	*percent*
EtOH balance:	0.10(1000)	−	0.60(100)	=	40	4.4
H₂O balance:	0.90(1000)	−	0.40(100)	=	860	95.6
					900	100.0

Step 10: 900 kg B + 100 kg P = 1000 kg F.

EXAMPLE 2.10 Combustion

Ethane is initially mixed with oxygen to obtain a gas containing 80% C_2H_6 and 20% O_2 that is then burned with 200% excess air. Eighty percent of the ethane goes to CO_2, 10% goes to CO, and 10% remains unburned. Calculate the composition of the exhaust gas on a wet basis.

In this problem to save space we do not explicitly outline the steps in the analysis and solution of the problem. Use your mental checklist, nevertheless, to ascertain that each step is indeed taken into account.

Solution

We know the composition of the air and fuel gas; if a weight of fuel gas is chosen as the basis, the weight of air can easily be calculated. However, it is wasted effort to convert to a weight basis for this type of problem. Since the total moles entering and leaving the boiler are not equal, if we look at any one component and employ the stoichiometric principles discussed in Chap. 1, together with Eq. (2.1), we can easily obtain the composition of the stack gas. The net generation term in Eq. (2.1) can be evaluated from the stoichiometric equations listed below. The problem can be worked in the simplest fashion by choosing a basis of 100 moles of entering gas. See Fig. E2.10.

Basis: 100 lb mol of fuel

$$C_2H_6 + \tfrac{7}{2}O_2 \longrightarrow 2CO_2 + 3H_2O$$
$$C_2H_6 + \tfrac{5}{2}O_2 \longrightarrow 2CO + 3H_2O$$

.   The total O_2 entering is 3.00 times the required O_2 (100% required plus 200% excess). Let us calculate the required oxygen:

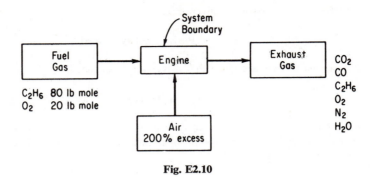

Fig. E2.10

O_2 (for complete combustion):

$$\frac{80 \text{ lb mol } C_2H_6 \mid 3.5 \text{ lb mol } O_2}{1 \text{ lb mol } C_2H_6} = 280 \text{ lb mol } O_2$$

Required O_2:

$$280 - 20 = 260 \text{ lb mol } O_2$$

(*Note:* The oxygen used to completely burn the fuel is reduced by the oxygen already present in the fuel to obtain the oxygen required in the entering air.)

Next we calculate the input of O_2 and N_2 to the system:

O_2 entering with air:

$$3(260 \text{ lb mol } O_2) = 780 \text{ lb mol } O_2$$

N_2 entering with air:

$$\frac{780 \text{ mol } O_2 \mid 79 \text{ lb mol } N_2}{21 \text{ lb mol } O_2} = 2930 \text{ lb mol } N_2$$

Now we apply our stoichiometric relations to find the components generated within the system:

$$\frac{80 \text{ lb mol } C_2H_6 \mid 2 \text{ lb mol } CO_2 \mid 0.8}{1 \text{ lb mol } C_2H_6} = 128 \text{ lb mol } CO_2$$

$$\frac{80 \text{ lb mol } C_2H_6 \mid 3 \text{ lb mol } H_2O \mid 0.8}{1 \text{ lb mol } C_2H_6} = 192 \text{ lb mol } H_2O$$

$$\frac{80 \text{ lb mol } C_2H_6 \mid 2 \text{ lb mol } CO \mid 0.1}{1 \text{ lb mol } C_2H_6} = 16 \text{ lb mol } CO$$

$$\frac{80 \text{ lb mol } C_2H_6 \mid 3 \text{ lb mol } H_2O \mid 0.1}{1 \text{ lb mol } C_2H_6} = 24 \text{ lb mol } H_2O$$

To determine the O_2 remaining in the exhaust gas, we have to find how much of the available (800 lb mol) O_2 combines with the C and H.

$$\frac{80 \text{ lb mol } C_2H_6 \mid 3.5 \text{ lb mol } O_2 \mid 0.8}{\mid 1 \text{ lb mol } C_2H_6 \mid} = 224 \text{ lb mol } O_2 \text{ to burn to } CO_2 \text{ and } H_2O$$

$$\frac{80 \text{ lb mol } C_2H_6 \mid 2.5 \text{ lb mol } O_2 \mid 0.1}{\mid 1 \text{ lb mol } C_2H_6 \mid} = \underline{20} \text{ lb mol } O_2 \text{ to burn to } CO \text{ and } H_2O$$

$$244 \text{ lb mol } O_2 \text{ total "used up" by reaction}$$

By an oxygen (O_2) balance we get

$$O_2 \text{ out} = 780 \text{ lb mol} + 20 \text{ lb mol} - 244 \text{ lb mol} = 556 \text{ lb mol } O_2$$

From a water balance we find

$$H_2O \text{ out} = 192 \text{ lb mol} + 24 \text{ lb mol} = 216 \text{ lb mol } H_2O$$

The balances on the other compounds—C_2H_6, CO_2, CO, N_2—are too simple to be formally listed here.

Summarizing these calculations, we have:

component	lb mol			percent in exhaust gas
	fuel	air	exhaust gas	
C_2H_6	80	—	8	0.21
O_2	20	780	556	14.42
N_2	—	2930	2930	76.01
CO_2	—	—	128	3.32
CO	—	—	16	0.42
H_2O	—	—	216	5.62
Total	100	3710	3854	100.00

On a dry basis we would have (the water is omitted):

component	lb mol	percent
C_2H_6	8	0.23
O_2	556	15.29
N_2	2930	80.52
CO_2	128	3.52
CO	16	0.44
Total	3638	100.00

Once you have become familiar with these types of problems, you will find the tabular form of solution illustrated in the continuation of Example 2.10, on the following page, a very convenient form to use.

EXAMPLE 2.10 (cont.) Alternative Manner of Presentation

Initial Component	lb mol = percent	lb mol O_2 Required	Actual lb mol O_2 Used	lb mol Formed of CO_2	H_2O	CO	C_2H_6 Left	Reactions
C_2H_6	64 $\Big\}$ 80	$\frac{7}{2}(80) = 280$	$\frac{7}{2}(64) = 224$	128	192			$C_2H_6 + \frac{7}{2}O_2 \rightarrow 2CO_2 + 3H_2O$
C_2H_6	8		$\frac{5}{2}(8) = 20$		24	16		$C_2H_6 + \frac{5}{2}O_2 \rightarrow 2CO + 3H_2O$
C_2H_6	8						8	$C_2H_6 \rightarrow C_2H_6$
O_2	20							
	$\overline{100}$	$\overline{260}$	$\overline{244}$	$\overline{128}$	$\overline{216}$	$\overline{16}$	$\overline{8}$	

(lb mol O_2 Required column also shows: $= 280$, -20, $\overline{260}$)

O_2 entering with air: $3(260) = 780$

N_2 entering with air: $780\frac{79}{21} = 2930$

O_2 in exit gas: $780 + 20 - 244 = 556$

air   fuel used

summary of material balances

Total mol in $\neq$ total mol out

Total lb in = total lb out: (each compound has to be multiplied by its molecular weight)

balances on atomic species
$$\begin{cases} \text{lb mol C in} = \text{lb mol C out}: & 2(80) = 128 + 16 + 2(8) \\ \text{lb mol } H_2 \text{ in} = \text{lb mol } H_2 \text{ out}: & 3(80) = 216 + 3(8) \\ \text{lb mol } O_2 \text{ in} = \text{lb mol } O_2 \text{ out}: & 20 + 780 = 128 + \frac{1}{2}(216) + \frac{1}{2}(16) + 556 \\ \text{lb mol } N_2 \text{ in} = \text{lb mol } N_2 \text{ out}: & 2930 = 2930 \end{cases}$$

balances on compounds
$$\begin{cases} \text{lb mol } C_2H_6 \text{ in} = \text{lb mol } C_2H_6 \text{ out} + \text{lb mol } C_2H_6 \text{ consumed}: & 80 = 8 + 72 + 0 \\ \text{lb mol } CO_2 \text{ in} = \text{lb mol } CO_2 \text{ out} - \text{lb mol } CO_2 \text{ generated}: & 0 = 128 - 128 \\ \text{lb mol CO in} = \text{lb mol CO out} - \text{lb mol CO generated}: & 0 = 16 - 16 \\ \text{lb mol } H_2O \text{ in} = \text{lb mol } H_2O \text{ out} - \text{lb mol } H_2O \text{ generated}: & 0 = 216 - 216 \\ \text{lb mol } N_2 \text{ in} = \text{lb mol } N_2 \text{ out}: & 2930 = 2930 + 0 \\ \text{lb mol } O_2 \text{ in} = \text{lb mol } O_2 \text{ out} + \text{lb mol } O_2 \text{ consumed}: & 800 = 556 + 244 \end{cases}$$

Self-Assessment Test

1. Explain the difference between flue gas analysis and Orsat analysis; wet basis and dry basis for a gas.

2. What does "SO_2 free basis" mean?

3. Write down the equation relating percent excess air to required air and entering air.

4. Will the percent excess air always be the same as the percent excess oxygen in combustion (by oxygen)?

5. In a combustion process in which a specified percentage of excess air is used, and in which CO is one of the products of combustion, will the analysis of the resulting exit gases contain more or less oxygen than if all the carbon had gone to CO_2?

In solving the following problems, be sure to employ the 10 steps listed in Table 2.1.

6. Pure carbon is burned in oxygen. The flue-gas analysis is:

$$
\begin{array}{ll}
CO_2 & 75 \text{ mole } \% \\
CO & 14 \text{ mole } \% \\
O_2 & 11 \text{ mole } \%
\end{array}
$$

What was the percent excess oxygen used?

7. Toluene, C_7H_8, is burned with 30% excess air. A bad burner causes 15% of the carbon to form soot (pure C) deposited on the walls of the furnace. What is the Orsat analysis of the gases leaving the furnace?

8. A cereal product containing 55% water is made at the rate of 500 kg/hr. You need to dry the product so that it contains only 30% water. How much water has to be evaporated per hour?

Section 2.4   Material Balances Using Algebraic Techniques

Your objectives in studying this section are to be able to:

1. Write a set of independent material balance equations for a process.
2. Solve one or two simultaneous nonlinear equations.
3. Solve a set of linear equations.
4. Apply the 10-step strategy to solve steady-state problems (with or without chemical reaction) requiring the use of algebra.

As illustrated in Figs. 2.6 and 2.7, problems can be posed or formulated in different ways depending on the type of information available on the process streams and their respective compositions. The problems treated in Sec. 2.3 were quite easy to solve, once the problem had been converted from words into numbers, because the missing information pertained to a single stream. Only simple addition or subtraction was required to find the unknown quantities. Other types of material balance problems can be solved by writing the balance formally, and assigning letters or symbols to represent the unknown quantities. Each unknown stream, or component, is assigned a letter to replace the unknown value in the total mass balance or the component mass balance, as the case may be. Keep in mind that for each unknown so introduced you will have to write one independent material balance if the set of equations you form is to have a unique solution.

If more than one piece of equipment or more than one junction point is involved in the problem to be solved, you can write material balances for each piece of equipment and a balance around the whole process. However, since the overall balance is nothing more than the sum of the balances about each piece of equipment, not all the balances will be independent. Appendix L discusses how you can determine whether linear equations are independent or not. Under some circumstances, particularly if you split a big problem into smaller parts to make the calculations easier, you may want to make a material balance about a *mixing point*. As illustrated in Fig. 2.10, a mixing point is nothing more than a junction of three or more streams and can be designated as a system in exactly the same fashion as any other piece of equipment.

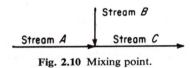

Fig. 2.10 Mixing point.

Examine Fig. 2.11. Which streams have the same composition? Is the composition of stream 5 the same as the average composition inside the unit? Streams 5, 6, and 7 must have the same composition, and presumably if no reaction takes place, the output composition is the properly weighted average of the input compositions 3 and 4.

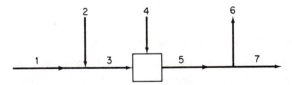

Fig. 2.11 Multiple mixing points.

Some brief comments are now appropriate as to how to solve sets of coupled simultaneous equations. If only two or three linear material balances are written, the unknown variables can be solved for by substitution. If the material balances consist of large sets of linear equations, you will find suggestions for solving them in Appendix L. If the material balance is a nonlinear equation, it can be plotted by hand or by using a computer routine, and the root(s), that is, the crossing(s) of the horizontal axis, located. See Fig. 2.12. If two nonlinear material balances have to be solved, the equations can be plotted and their intersection(s) located. With many simultaneous nonlinear equations to be solved, the use of computer routines is essential; for linear equations, computer routines are quick and prevent human error. By making a balance for each component for each defined system, in most instances a set of independent equations can be obtained whether linear or nonlinear. Total mass balances may be substituted for one of the component mass balances. Similarly, an overall balance around the entire system may be substituted for one of the subsystem balances. By following these rules, you should encounter no difficulty in generating sets of independent material balances for any process.

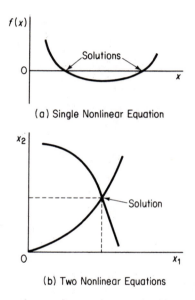

Fig. 2.12 Solution of nonlinear material balances by graphical techniques.

Illustrations of the use of algebraic techniques to solve material balance problems can be found below in this section and in the next section (on tie components).

Example 2.11 Mixing

Dilute sulfuric acid has to be added to dry charged batteries at service stations to activate a battery. You are asked to prepare a batch of new acid as follows. A tank of old weak battery acid (H_2SO_4) solution contains 12.43% H_2SO_4 (the remainder is pure water). If 200 kg of 77.7% H_2SO_4 is added to the tank, and the final solution is 18.63% H_2SO_4, how many kilograms of battery acid have been made? See Fig. E2.11.

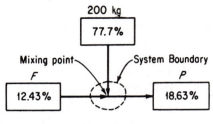

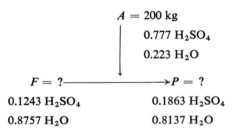

Fig. E2.11

Solution

Steps 1, 2, and 3:

$$A = 200 \text{ kg}$$
$$0.777 \text{ H}_2\text{SO}_4$$
$$0.223 \text{ H}_2\text{O}$$

$$F = \text{?} \longrightarrow P = \text{?}$$

0.1243 H$_2$SO$_4$ 0.1863 H$_2$SO$_4$

0.8757 H$_2$O   0.8137 H$_2$O

Step 4: The unknown stream flows are F and P.

Step 5: Two independent material balances can be written; hence a unique solution exists.

Step 6: Select a basis of what is given.

Basis: 200 kg of 77.7% H$_2$SO$_4$ solution

Steps 7 and 8:

type of balance	in	=	out	
Total	$F + 200$	=	P	(a)
component				
H$_2$SO$_4$	$F(0.1243) + 200(0.777) = P(0.1863)$			(b)
H$_2$O	$F(0.8757) + 200(0.223) = P(0.8137)$			(c)

Step 9: An algebraic solution is required. Use of the total mass balance and one of the others is the easiest way to find P:

$$(P - 200)(0.1243) + 200(0.777) = P(0.1863)$$
$$P = 2110 \text{ kg acid}$$
$$F = 1910 \text{ kg acid}$$

Step 10: Check the answer using the H$_2$O balance.

EXAMPLE 2.12 Distillation

A typical distillation column is shown in Fig. E2.12 together with the known information for each stream. Calculate the kilograms of distillate per kilogram of feed and per kilogram of waste.

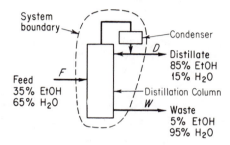

Fig. E2.12

Solution

Steps 1, 2, and 3: Figure E2.12 already has inscribed on it the preliminary information.

Step 4: The unknown quantities are F, D, and W.

Step 5: Only two independent material balances can be written. More information is needed.

Step 6: Select a basis.

Basis: 1.00 kg of feed

Step 5 Continued: Now only D and W are unknown values so that a unique solution is possible.

Step 7:

type of balance	in	=	out
Total	1.00	=	$D + W$
EtOH	1.00(0.35)	=	$D(0.85) + W(0.05)$
H_2O	1.00(0.65)	=	$D(0.15) + W(0.95)$

Step 8: An algebraic solution is necessary.

Step 9: Solve for D with $W = (1.00 - D)$,

$$1.00(0.35) = D(0.85) + (1.00 - D)(0.05)$$
$$D = 0.375 \text{ kg/kg feed}$$

Since $W = 1 - 0.375 = 0.625$ lb,

$$\frac{D}{W} = \frac{0.375}{0.625} = \frac{0.60 \text{ kg}}{\text{kg}}$$

Step 10: Check your solution by using the H_2O balance.

EXAMPLE 2.13

Examine Fig. E2.13a, which represents a simple flow sheet. Only the value of D is known. What is the minimum number other measurements that must be made to determine all the other stream and composition values?

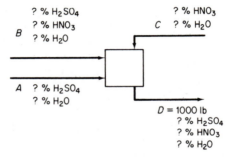

Fig. E2.13a

Solution

Here is the count of the number of unknown values taking into account that in any stream, one of the composition values can be determined by difference from 100. Do you remember why?

stream	degrees of freedom
A	2
B	3
C	2
D	2
Total	9

As example of the count, in stream A specification of A plus one composition makes it possible to calculate the other composition. In total only three independent material balance equations can be written (do you remember why?), leaving $9 - 3 = 6$ compositions and stream values that have to be specified, and three values to be solved for from the material balances.

Will any six values do? No. Only values can be specified that will leave a number of independent material balances equal to the number of unknown parameters. As an example of a satisfactory set of measurements, choose one composition in stream A, two in B, one in C, and two in D leaving the flows A, B, and C as unknowns. How about the selection of the following set of the measurements: A, B, C, two compositions in D, and one composition in B? Draw a diagram of the information, as in Fig. E2.13b ($\bullet$ = known quantity). Write down the three material balance equations. Are the three equations independent? You will find that they are not independent. Remember that the sum of the mass fractions is unity for each stream.

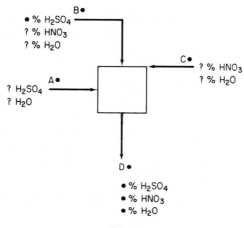

B$\bullet$

$\bullet$ % H_2SO_4
? % HNO_3
? % H_2O

C$\bullet$? % HNO_3
? % H_2O

A$\bullet$
? H_2SO_4
? H_2O

D$\bullet$

$\bullet$ % H_2SO_4
$\bullet$ % HNO_3
$\bullet$ % H_2O

Fig. E2.13b

EXAMPLE 2.14 Multiple Equipment

In your first assignment as a new company engineer you have been asked to set up an acetone recovery system to remove acetone from gas that was formerly vented but can no longer be handled that way, and to compute the cost of recovering acetone. Your boss has provided you with the design of the proposed system, as shown in Fig. E2.14. From the information given, list the flow rates (in kilograms per hour) of all the streams so that the size of the equipment can be determined, and calculate the composition of the feed to the still.

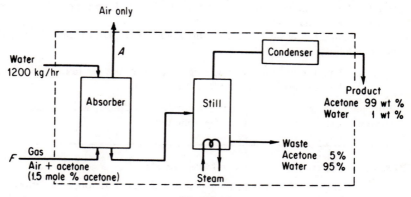

Air only

Water
1200 kg/hr

A

Absorber

Condenser

Still

Product
Acetone 99 wt %
Water   1 wt %

F Gas
Air + acetone
(1.5 mole % acetone)

Steam

Waste
Acetone 5%
Water   95%

Fig. E2.14

Solution

Steps 1, 2, and 3: Define the system to include all three units. Add a dashed line around the boxes in Fig. E2.14.

Step 3 Continued: Convert mole percent of the air–acetone feed to weight percent because all the other compositions are given in weight percent:

$$\text{mol. wt. acetone} = 58, \quad \text{mol. wt. air} = 29$$

Basis: 100 kg mol of gas feed

composition	kg mol	mol. wt.	kg	wt %
Acetone	1.5	58	87	2.95
Air	98.5	29	2860	97.05
Total	100.0		2947	100.00

Step 4: The unknown stream flows are F, A, B, P, and W. The steam is not involved in the material balance problem because it does not mix with the constituents of the still. All the compositions are known except the two in stream B. We do not need to use the (known) composition of the air in this problem.

Step 5: Here are the names of the material balances that can be made for each unit.

absorber	*still*
Water	Acetone
Air	Water
Acetone	

Add one summation of mass fractions equation for stream B. We do not need a specific balance on the condenser as all the material that enters exits unchanged (except as to phase). Because there are seven unknown quantities and only six independent equations, more information is needed to solve the complete problem.

Is it possible to solve the material balances for the overall system ignoring the internal details? The unknown parameters are F, A, P, and W and the balances are water, air, and acetone; hence the answer is no.

One additional bit of information would be enough to solve the problem as originally posed, such as

(a) The weight of entering gas per hour, product per hour, or waste per hour.
(b) The composition of the feed into the still.

With an extra piece of pertinent information available, a complete solution is possible; without it only partial results can be obtained.

EXAMPLE 2.15 Countercurrent Stagewise Mass Transfer

In many commercial processes such as distillation, extraction, absorption of gases in liquids, and the like, the entering and leaving streams represent two different phases that flow in opposite directions to each other, as shown in Fig. E2.15a. (The figure could just as well be laid on its side.)

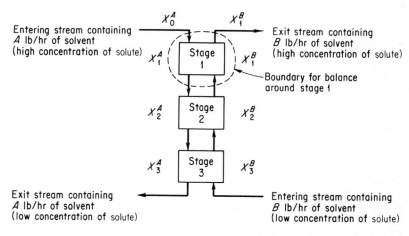

Entering stream containing
A lb/hr of solvent
(high concentration of solute)

X_0^A

X_1^B

Exit stream containing
B lb/hr of solvent
(high concentration of solute)

X_1^A Stage 1 X_1^B

Boundary for balance around stage 1

X_2^A Stage 2 X_2^B

X_3^A Stage 3 X_3^B

Exit stream containing
A lb/hr of solvent
(low concentration of solute)

Entering stream containing
B lb/hr of solvent
(low concentration of solute)

A, B = lb of stream less solute in the stream
= lb of solvent

Fig. E2.15a

This type of operation is known as *countercurrent* operation. If equilibrium is attained between each stream at each stage in the apparatus, calculations can be carried out to relate the flow rates and concentration of products to the size and other design features of the apparatus. We shall illustrate how a material balance can be made for such type of equipment. The letter X stands for the weight concentration of solute in pounds of solute per pound of stream, solute-free. The streams are assumed immiscible as in a liquid–liquid extraction process.

Around stage 1 the solute material balance is

in   out

$$\frac{A \text{ lb}}{\text{hr}} \left| \frac{X_0^A \text{ lb}}{\text{lb } A} + \frac{B \text{ lb}}{\text{hr}} \right| \frac{X_2^B \text{ lb}}{\text{lb } B} = \frac{A \text{ lb}}{\text{hr}} \left| \frac{X_1^A \text{ lb}}{\text{lb } A} + \frac{B \text{ lb}}{\text{hr}} \right| \frac{X_1^B \text{ lb}}{\text{lb } B}$$

or

$$A(X_0^A - X_1^A) = B(X_1^B - X_2^B) \tag{a}$$

Around stage 2 the solute material balance is

in   out

$$\frac{A \text{ lb}}{\text{hr}} \left| \frac{X_1^A \text{ lb}}{\text{lb } A} + \frac{B \text{ lb}}{\text{hr}} \right| \frac{X_3^B \text{ lb}}{\text{lb } B} = \frac{A \text{ lb}}{\text{hr}} \left| \frac{X_2^A \text{ lb}}{\text{lb } A} + \frac{B \text{ lb}}{\text{hr}} \right| \frac{X_2^B \text{ lb}}{\text{lb } B}$$

or

$$A(X_1^A - X_2^A) = B(X_2^B - X_3^B)$$

We could generalize that for any stage number n

$$A(X_{n-1}^A - X_n^A) = B(X_n^B - X_{n+1}^B) \tag{b}$$

Also, a solute material balance could be written about the sum of stage 1 and stage 2 as follows:

$$\underset{\text{in}}{A(X_0^A) + B(X_3^B)} = \underset{\text{out}}{A(X_2^A) + B(X_1^B)} \tag{c}$$

or to generalize for an overall solute material balance between the top end and the nth stage,

$$A(X_0^A - X_n^A) = B(X_1^B - X_{n+1}^B)$$

or for the $(n-1)$th stage,

$$A(X_0^A - X_{n-1}^A) = B(X_1^B - X_n^B)$$

Changing signs, we obtain

$$A(X_{n-1}^A - X_0^A) = B(X_n^B - X_1^B) \tag{d}$$

If we rearrange Eq. (d) assuming that A, B, X_0^A, and X_1^B are constants (i.e., steady-state operation) and X^A and X^B are the variables as we go from stage to stage, we can write

$$X_{n-1}^A = \frac{B}{A} X_n^B + \left(X_0^A - \frac{B}{A} X_1^B\right) \tag{e}$$

Equations (a)–(e) represent an unusual type of equation, one that gives the relationship between discrete points rather than continuous variables; it is called a *difference* equation. In Eq. (e) the locus of these points will fall upon a line with the slope B/A and an intercept $[X_0^A - (B/A)X_1^B]$, as shown in Fig. E2.15b.

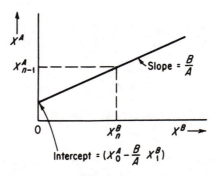

Fig. E2.15b

Self-Assessment Test

1. Write a set of independent equations for the system shown. The streams are A, B, C, D, and E. The components are 1, 2, and 3. The table lists the known mass fractions.

	A	B	C	D
1:	0.5	0.3	0.6	0.8
2:	0.3	0.2	0.4	0.2
3:	0.2	0.5		

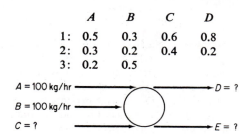

$A = 100$ kg/hr $\longrightarrow$

$B = 100$ kg/hr $\longrightarrow$

$C = ?$

$\longrightarrow D = ?$

$\longrightarrow E = ?$

2. In Fig. 2.10 the compositions are:

	A	B	C
NH_3:	0.10	0.05	0.15
Na_2CO_3:	0.20	0.10	0.25
H_2O:	0.70	0.85	0.60

Write down the material balances and solve for A and B given that $C = 100$ kg/hr.

3. Solve this equation for positive-real roots:

$$X^4 - X^3 - 10X^2 - X + 1 = 0$$

4. Solve these equations for a set of (x, y) that are real:

$$4.20x^2 + 8.80y^2 = 1.42$$
$$(x - 1.2)^2 + (y - 0.6)^2 = 1$$

In solving the following problems, be sure to employ the 10 steps listed in Table 2.1.

5. A coal analyzing 65.4% C, 5.3% H, 0.6% S, 1.1% N, 18.5% O, and 9.1% ash is burned so that all combustible is burnt out of the ash. The flue gas analyzes 13.00% CO_2, 0.76% CO, 6.17% O_2, 0.87% H_2, and 79.20% N_2. All the sulfur burns to SO_2, which is included in the CO_2 figure in the gas analysis (i.e., $CO_2 + SO_2 = 13.00\%$). The nitrogen from the coal may be ignored. Calculate:
(a) Pounds of coal fired per 100 lb mol of dry flue gas as analyzed
(b) Ratio of moles total combustion gases to moles of dry air supplied
(c) Total moles of water vapor in the stack gas per 100 lb of coal if the air is dry
(d) Percent excess air

6. A cellulose solution contains 5.2% cellulose by weight in water. How many kilograms of 1.2% solution are required to dilute 100 kg of 5.2% solution to 4.2%?

7. Hydrofluoric acid (HF) can be manufactured by treating calcium fluoride (CaF_2) with sulfuric acid (H_2SO_4). The other product of the reaction is calcium sulfate ($CaSO_4$). A sample of fluorospar (the raw material) contains 75% by weight CaF_2 and 25% inert (nonreacting) materials. The pure sulfuric acid used in the process is in 30% excess of that theoretically required. Most of the manufactured HF leaves the reaction chamber as a gas, but a solid cake is also removed from the reaction chamber that contains 5% of all the HF formed, plus $CaSO_4$, inerts, and unreacted sulfuric acid. How many kilograms of *cake* are produced per 100 kg of fluorospar charged to the process?

8. A solution contains 60% $Na_2S_2O_2$ and 1% soluble impurity in water. Upon cooling to 10°C, $Na_2S_2O_2 \cdot 5H_2O$ crystallizes out. The solubility of this hydrate is 1.4 lb $Na_2S_2O_2 \cdot 5H_2O$/lb free water. The crystals removed carry as adhering solution 0.06 lb solution/lb crystals. These are dried to remove the remaining water (but not the water of hydration). The final dry $Na_2S_2O_2 \cdot 5H_2O$ crystals must not contain more than 0.1% by weight of impurity. In order to meet this specification, the original solution, calculate:
   (a) The amount of water added before cooling
   (b) The percentage recovery of the $Na_2S_2O_2$ in the crystals

Section 2.5   Problems Involving Tie Components

> ### *Your objectives in studying this section are to be able to:*
>
> 1. Recognize and use a tie component in solving material balance problems.
> 2. Apply the 10-step strategy to solve steady-state problems (with or without chemical reaction) by the tie component technique.

A *tie component* (*element*) is material that goes from one stream into another without changing in any respect or having like material added to it or lost from it. If a tie component exists in a problem, in effect you can write a material balance that involves *only two* streams. Furthermore, the material balance has a particularly simple form, that of a ratio. For example, in Fig. 2.6(b) all the MeOH in stream F goes into stream W. Consequently, streams F and W can be related by a MeOH material balance:

$$\omega_{\text{MeOH},F}F = \omega_{\text{MeOH},W}W \qquad \text{or} \qquad \frac{\omega_{\text{MeOH},W}}{\omega_{\text{MeOH},F}} = \frac{F}{W}$$

$$0.10(100) = 0.22W \qquad \text{or} \qquad W = \frac{0.10}{0.22}(100)$$

To select a tie element, ask yourself the question: What component passes from one stream to another unchanged with constant weight (mass)? The answer is the tie component. Frequently, several components pass through a process with continuity so that there are many choices of tie elements. Sometimes the amounts of these components can be added together to give an overall tie component that will result in a smaller percentage error in your calculations than if an individual tie element had been used. It may happen that a minor constituent passes through with continuity, but if the percentage error for the analysis of this component is large, you should not use it as a tie component. Sometimes you cannot find a tie component by direct examination of the problem, but you still may be able to develop a hypothetical tie component or a man-made tie component which will be equally effective as a tie component for a single material. A tie component is useful even if you do not know all the compositions and weights in any given problem because the tie component enables you to put two streams on the same basis. Thus a *partial solution* can be effected even if the entire problem is not resolved.

Examine Fig. 2.7. Note how CH_4 is not a tie component in Fig. 2.7(a) but that C is a tie component. Is O_2 a tie component? Is N_2? State all of the tie components you observe in Fig. 2.7(b).

In solutions of problems involving tie components it is not always advantageous to select as the basis "what you have." Frequently, it turns out that the composition of one stream is given as a percentage composition, and then it becomes convenient to select as the basis 100 kg (or other mass) of the material because then kilograms equal percentage and the same numbers can be used to represent both. After you have carried out the computations on the basis of 100 kg (or another basis), the answer can be transformed to the basis of the amount of material actually given by using the proper conversion factor. For example, if you had 32.5 kg of material and you knew its composition on the basis of percentage, you could then take as your basis 100 kg of material, and at the end of the calculations convert your answer to the basis of 32.5 kg.

Tie component problems can, of course, be worked by algebraic means, but in combustion problems and in some of the more complicated types of industrial chemical calculations so many unknowns and equations are involved that the use of a tie component greatly simplifies the calculations. Use of a tie component also clarifies the internal workings of a process. Consequently, the examples below illustrate the use of both tie-component and algebraic solutions. Become proficient in both techniques, as you will have to draw upon both of them in your career as an engineer.

EXAMPLE 2.16 Drying

Fish caught by human beings can be turned into fish meal, and the fish meal can be used as feed to produce meat for human beings or used directly as food. The direct use of fish meal significantly increases the efficiency of the food chain. However, fish-protein concentrate, primarily for aesthetic reasons, is used as a supplementary protein food. As such, it competes with soy and other oilseed proteins.

In the processing of the fish, after the oil is extracted, the fish cake is dried in rotary drum dryers, finely ground, and packed. The resulting product contains 65% protein. In a given batch of fish cake that contains 80% water (the remainder is dry cake), 100 kg of water is removed, and it is found that the fish cake is then 40% water. Calculate the weight of the fish cake originally put into the dryer.

Solution

Steps 1, 2, and 3:

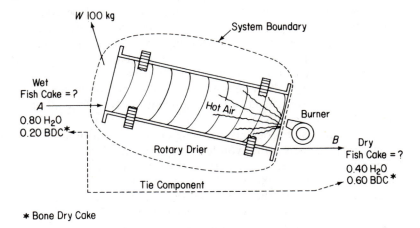

* Bone Dry Cake

Fig. E2.16

Step 4: The unknown stream flows are two: A and B. All the compositions are known.

Step 5: Two independent balances can be written so that a unique solution exists.

Step 6: Take a basis of what is given.

Basis: 100 kg of water evaporated

Steps 7 and 8: To solve the problem by use of algebra requires the use of two independent material balance equations.

$$in = out$$

Total balance: $A = B + W = B + 100$

BDC balance: $0.20A = 0.60B$

⎫ mass balances

Step 9: Notice that because the bone dry cake balance involves only two streams, you can set up a direct ratio of A to B, which is the essence of the tie component:

$$B = \frac{0.20\,A}{0.60} = \frac{1}{3}\,A$$

Introduction of this ratio into the total balance gives

$$A = \tfrac{1}{3}A + 100$$

$$\tfrac{2}{3}A = 100$$

$$A = 150 \text{ kg initial cake}$$

Steps 7, 8, and 9 Repeated: To solve the same problem by use of the tie component, take as a basis 100 lb of initial material (the 100 lb of H_2O will not serve as a useful basis).

Basis: 100 kg initial fish cake

initial − loss = final

component	*percent = kg*	*kg*	*kg*	*percent*
H_2O	80	100	?	40
BDC	20 ⟵	tie ⟶	20	60
Total	100		?	100

To make either a total or a water material balance for this problem, all streams must be placed on the same basis (100 kg of initial cake) by use of the tie component.

Tie final cake to initial cake by a tie component:

$$\frac{20 \text{ kg } \cancel{BDC}}{100 \text{ kg initial cake}} \left| \frac{100 \text{ kg final cake}}{60 \text{ kg } \cancel{BDC}} \right. = 33.3 \text{ kg final cake/100 kg initial cake}$$

Notice that the BDC is the same (the tie component) in both the initial and final stream and that the units of BDC can be canceled. We have now used one independent material balance equation, the BDC balance. Our second balance will be total balance:

$$100 \text{ kg initial} - 33.3 \text{ kg final} = 66.7 \text{ kg } H_2O \text{ evap./100 kg initial cake}$$

An alternative way to obtain this same value is to use a water balance.

$$\frac{20 \text{ kg } \cancel{BDC}}{100 \text{ kg initial cake}} \left| \frac{40 \text{ kg } H_2O \text{ in dry cake}}{60 \text{ kg } \cancel{BDC} \text{ in dry cake}} \right. = 13.3 \text{ kg } H_2O \text{ left/100 kg initial cake}$$

$$80 \text{ kg initial } H_2O - 13.3 \text{ kg final } H_2O = 66.7 \text{ kg } H_2O \text{ evap./100 kg initial cake}$$

Next a shift is made to the basis of 100 kg of H_2O evaporated as required by the problem statement:

$$\frac{100 \text{ kg initial cake}}{66.7 \text{ kg } H_2O \text{ evap.}} \left| \frac{100 \text{ kg } H_2O \text{ evap.}}{} \right. = 150 \text{ kg initial cake}$$

Another suitable starting basis would be to use 1 kg of bone dry cake as the basis.

Basis: 1 kg of BDC

	initial −	**loss =**	**final**
component	*fraction*	*kg*	*fraction*
H_2O	0.80	100	0.40
BDC	0.20		0.60
Total	1.00		1.00

initial H_2O − final H_2O = H_2O lost   water balance

$$\frac{0.80 \text{ kg } H_2O}{0.20 \text{ kg BDC}} - \frac{0.40 \text{ kg } H_2O}{0.60 \text{ kg BDC}} =$$

$$4.0 \quad - \quad 0.67 \quad = 3.33 \text{ kg } H_2O/\text{kg BDC}$$

Again we have to shift bases to get back to the required basis of 100 kg of H_2O evaporated:

$$\frac{1 \text{ kg } \cancel{BDC}}{3.33 \text{ kg } H_2O} \left| \frac{100 \text{ kg } H_2O \text{ evap.}}{} \right| \frac{1.0 \text{ kg initial cake}}{0.2 \text{ kg } \cancel{BDC}} = 150 \text{ kg initial cake}$$

You can see that the use of the tie component makes it quite easy to shift from one basis to another. If you now want to compute the pounds of final cake, all that is necessary is to write

Step 10: Check your answer.

$$\begin{array}{r} 150 \text{ kg initial cake} \\ -100 \text{ kg } H_2O \text{ evaporated} \\ \hline 50 \text{ kg final cake} \end{array}$$

Now that we have examined some of the correct methods of handling tie components, let us look at another method frequently attempted (not for long!) by some students before they become completely familiar with the use of the tie component. They try to say that

$$0.80 - 0.40 = 0.40 \text{ kg } H_2O \text{ evaporated}$$

$$\frac{100 \text{ kg}}{0.40 \text{ kg}} = 250 \text{ kg cake}$$

On the surface this looks like an easy way to solve the problem—what is wrong with the method? More careful inspection of the calculation shows that two different bases were used in subtracting $(0.80 - 0.40)$. The 0.80 is on the basis of 1.00 kg of initial cake, while the 0.40 is on the basis of 1.00 kg of final cake. You would not say $0.80/A - 0.40/B = 0.40/A$, would you? From our early discussion about units you can see that this subtraction is not a legitimate operation.

EXAMPLE 2.17 Dilution

Trace components can be used to measure flow rates in pipelines, process equipment, and rivers that might otherwise be quite difficult to measure. Suppose that the water analysis in a flowing creek shows 180 ppm Na_2SO_4. If 10 lb of Na_2SO_4 is added to the stream uniformly over a 1-hr period and the analysis downstream where mixing is complete indicates 3300 ppm Na_2SO_4, how many gallons of water are flowing per hour? See Fig. E2.17a.

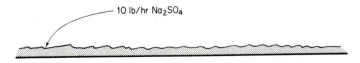

10 lb/hr Na_2SO_4

Fig. E2.17a

Solution

Steps 1, 2, and 3:

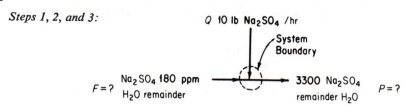

Fig. E2.17b

Steps 4 and 5: We have two unknown variables and two independent material balances, so a unique solution is possible.

Step 8: A tie component, the water, exists.

Step 6: Take a basis of 100 lb of F for convenience.

Basis: 100 lb of initial solution (water + Na_2SO_4 initially present)

After choosing a basis of 100 lb of initial solution, we know F and not P and Q.

component	*initial* lb = percent		*final* percent
H_2O	99.982	tie component	99.670
Na_2SO_4	0.018		0.330
Total	100.000		100.000

Steps 7, 8, 9:

$$\frac{0.33 \text{ lb } Na_2SO_4}{99.67 \text{ lb } H_2O} \left| \frac{99.982 \text{ lb } H_2O}{100 \text{ lb initial solution}} = 0.331 \text{ lb } NaS_2O_4/100 \text{ lb initial solution} \right.$$

The tie component of water comprises our first material balance. Now that the exit Na_2SO_4 concentration is on the same basis as the entering Na_2SO_4 concentration, by an Na_2SO_4 balance (our second balance) we find that the Na_2SO_4 added on the same basis is

$$0.331 - 0.0180 = 0.313 \text{ lb}/100 \text{ lb initial solution}$$

Basis: 1 hr

$$\frac{100 \text{ lb initial solution}}{0.313 \text{ lb } Na_2SO_4 \text{ added}} \left| \frac{10 \text{ lb added}}{1 \text{ hr}} \right| \frac{1 \text{ gal}}{8.35 \text{ lb}} = 383 \text{ gal/hr}$$

Algebraic Solution

$$\begin{array}{ll} \text{Total balance:} & F + 10 = P \\ \text{Water balance:} & 0.99982F = 0.99670P \end{array} \Bigg\} \text{ mass balances}$$

$$F + 10 = \frac{0.99982F}{0.99670}$$

The water balance involves only two streams and provides a direct ratio between F and P. One of the difficulties in the solution of this problem using a water balance is that the quantity

$$\frac{0.99982}{0.99670} - 1 = 0.00313$$

has to be determined quite accurately. It might be more convenient to use an Na_2SO_4 balance

$$0.00018F + 10 = 0.00330P$$

Solving the total balance and the Na_2SO_4 balance simultaneously,

$$F = 3205 \text{ lb/hr}$$

$$\frac{3205 \text{ lb}}{\text{hr}} \left| \frac{1 \text{ gal}}{8.35 \text{ lb}} \right. = 384 \text{ gal/hr}$$

EXAMPLE 2.18 Crystallization

The solubility of barium nitrate at 100°C is 34 g/100 g of H_2O and at 0°C is 5.0 g/100 g of H_2O. If you start with 100 g of $Ba(NO_3)_2$ and make a saturated solution in water at 100°C, how much water is required? If this solution is cooled to 0°C, how much $Ba(NO_3)_2$ is precipitated out of solution?

Tie-Component Solution

Steps 1, 2, 3, and 6:

$$\text{Basis: 100 g } Ba(NO_3)_2$$

The maximum solubility of $Ba(NO_3)_2$ in H_2O at 100°C is a saturated solution, or 34 g/100 g of H_2O. Thus the amount of water required at 100°C is

$$\frac{100 \text{ g } H_2O}{34 \text{ g } Ba(NO_3)_2} \left| \frac{100 \text{ g } Ba(NO_3)_2}{} \right. = 295 \text{ g } H_2O$$

If the 100°C solution is cooled to 0°C, the $Ba(NO_3)_2$ solution will still be saturated, and the problem now can be pictured as in Fig. E2.18.

Steps 4 and 5: F and C are unknown values, two independent material balances can be written [$Ba(NO_3)_2$ and H_2O], hence a unique solution is possible.

Steps 7, 8, and 9: Water is a tie.

$$\frac{295 \text{ g } H_2O}{} \left| \frac{5 \text{ g } Ba(NO_3)_2}{100 \text{ g } H_2O} \right. = 14.7 \text{ g } Ba(NO_3)_2 \text{ in final solution}$$

$$\begin{array}{ccccc}
\textbf{original} & - & \textbf{final} & = & \textbf{crystals} \\
100 \text{ g } Ba(NO_3)_2 & - & 14.7 \text{ g } Ba(NO_3)_2 & = & 85.3 \text{ g } Ba(NO_3)_2 \text{ precipitated}
\end{array}$$

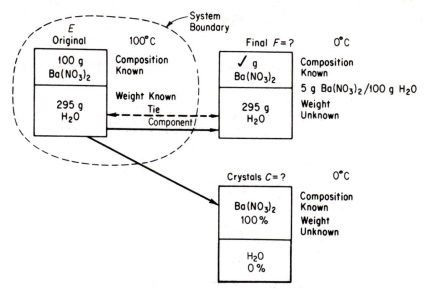

Fig. E2.18

Algebraic Solution

$$\text{H}_2\text{O balance:} \quad 295 = F\left(\frac{100}{100 + 5}\right)\frac{\text{g H}_2\text{O}}{\text{g final solution}}$$

$$\text{Total balance:} \quad 395 = C + F$$

$$F = 295\left(\frac{105}{100}\right) = 310$$

$$C = 395 - F = 85 \text{ g Ba(NO}_3)_2 \text{ crystals}$$

EXAMPLE 2.19 Combustion

The main advantage of catalytic incineration of odorous gases or other obnoxious substances over direct combustion is the lower cost. Catalytic incinerators operate at lower temperatures—500 to 900°C compared with 1100 to 1500°C for thermal incinerators—and use substantially less fuel. Because of the lower operating temperatures, materials of construction do not need to be as heat resistant, reducing installation and construction costs.

In a test run, a liquid having the composition 88% C and 12% H_2 is vaporized and burned to a flue gas (fg) of the following composition:

CO_2	13.4%
O_2	3.6%
N_2	83.0%
	100.0%

To compute the volume of the combustion device, determine how many pound moles of dry fg are produced per 100 kg of liquid feed. What was the percentage of excess air used? See Fig. E2.19.

Solution

Steps 1, 2, and 3: The necessary data are placed in Fig. E2.19. Do not forget the water vapor! And that a reaction occurs!

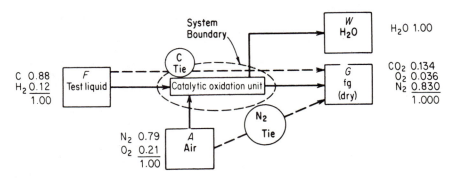

Fig. E2.19

Steps 4 and 5: F, A, W, and G are all unknown values; all the compositions are known. Four species material balances can be made; hence the problem has a unique solution.

Step 6: Select 100 moles of dry fg as a convenient basis.

Basis: 100 moles of dry fg

Steps 7 and 8: Because we do not have to answer any questions about the water in the exit flue-gas stream, two tie elements to relate the test fluid to the dry flue gas and the air to the dry flue gas would be sufficient to solve the problem. In examining the data to determine whether a tie element exists, we see that the carbon goes directly from the test fluid to the dry flue gas, and nowhere else, so carbon will serve as one tie component. The N_2 in the air all shows up in the dry flue gas, so N_2 can be used as another tie component. As explained before, there is no reason to convert all the available data to a weight basis, even though the moles of material entering and leaving the oxidation unit are not the same, because we can balance the moles of each species present. We can say, for example, that

kg C in = kg C out

kg mol C in = kg mol C out

We do have to determine the mole composition of the test liquid.

Basis: 100 kg of test fluid

$$\frac{88 \text{ kgC}}{} \left| \frac{1 \text{ kg mol C}}{12 \text{ kg C}} \right. = 7.33 \text{ kg mol C}$$

$$\frac{12 \text{ kg } H_2}{} \left| \frac{1 \text{ kg mol } H_2}{2 \text{ kg } H_2} \right. = 6.00 \text{ kg mol } H_2$$

(a) By using kilogram moles of C as the tie component, we obtain

$$\frac{100 \text{ kg mol dry fg}}{13.4 \text{ kg mol C}} \left| \frac{7.33 \text{ kg mol C}}{100 \text{ kg test fluid}} \right. = \frac{54.6 \text{ kg mol dry fg}}{100 \text{ kg test fluid}}$$

(b) The N_2 serves as the tie component to tie the air to the dry flue gas, and since we know the excess O_2 in the dry flue gas, the percentage of excess air can be computed.

Basis: 100 kg mol of dry fg

$$\frac{83.0 \text{ kg mol } N_2}{} \left| \frac{1.00 \text{ kg mol air}}{0.79 \text{ kg mol } N_2} \right. = \frac{105.0 \text{ kg mol air}}{100 \text{ kg mol dry fg}}$$

$$(105.0)(0.21) = O_2 \text{ entering} = 22.1 \text{ kg mol } O_2$$

Or, saving one step,

$$\frac{83.0 \text{ kg mol } N_2}{} \left| \frac{0.21 \text{ kg mol } O_2 \text{ in air}}{0.79 \text{ kg mol } N_2 \text{ in air}} \right. = \frac{22.1 \text{ kg mol } O_2 \text{ entering}}{100 \text{ kg mol dry fg}}$$

$$\% \text{ excess air} = 100 \frac{\text{excess } O_2}{O_2 \text{ entering} - \text{excess } O_2} = \frac{(100)3.6}{22.1 - 3.6} = 19.4\%$$

Step 10: To check this answer, let us work part (b) of the problem using as a basis

Basis: 100 kg of test fluid

The required oxygen for complete combustion is

component	kg	kg mol	kg mol O_2 required
C	88	7.33	7.33
H_2	12	6	3
Total			10.33

The excess oxygen is

$$\frac{3.6 \text{ kg mol } O_2}{100 \text{ kg mol fg}} \left| \frac{54.6 \text{ kg mol dry fg}}{100 \text{ kg liquid}} \right. = 1.97 \text{ kg mol}$$

$$\% \text{ excess air} = \frac{\text{excess } O_2}{\text{required } O_2} 100 = \frac{1.97(100)}{10.33} = 19.1\%$$

The answers computed on the two different bases agree reasonably well here, but in many combustion problems slight errors in the data will cause large differences in the calculated percentage of excess air. Assuming that no mathematical mistakes have been made (it is wise to check), the better solution is the one involving the use of the most precise data. Frequently, this turns out to be the method used in the check here—the one with a basis of 100 kg of test fluid.

If the dry flue gas analysis had shown some CO, as in the following hypothetica analysis:

$$
\begin{array}{ll}
CO_2 & 11.9\% \\
CO & 1.6\% \\
O_2 & 4.1\% \\
N_2 & 82.4\%
\end{array}
$$

then, on the basis of 100 moles of dry flue gas, we would calculate the excess air as follows:
O_2 entering with air:

$$
\frac{82.4 \text{ kg mol } N_2}{} \left| \frac{0.21 \text{ kg mol } O_2}{0.79 \text{ kg mol } N_2} \right. = 21.9 \text{ kg mol } O_2
$$

Excess O_2:

$$
4.1 - \frac{1.6}{2} = 3.3 \text{ kg mol}
$$

$$
\% \text{ excess air} = 100 \frac{3.3}{21.9 - 3.3} = 17.7\%
$$

Note that to get the true excess oxygen, the apparent excess oxygen in dry flue gas, 4.1 kg moles, has to be reduced by the amount of the theoretical oxygen not combining with the CO. According to the reactions

$$
C + O_2 \longrightarrow CO_2 \qquad C + \tfrac{1}{2}O_2 \longrightarrow CO
$$

for each mole of CO in the dry flue gas, $\tfrac{1}{2}$ mole of O_2 which should have combined with the carbon to form CO_2 did not do so. This $\tfrac{1}{2}$ mole carried over into the flue gas and inflated the value of the true excess oxygen expected to be in the flue gas. In this example, 1.6 kg mol of CO are in the flue gas, so that (1.6/2) kg mol of theoretical oxygen are found in the flue gas in addition to the true excess oxygen.

Algebraic Solution

A summary of an algebraic solution might be as follows. Let

$$
\begin{aligned}
F &= \text{kg test fluid} \\
G &= \text{kg mol dry fg} \\
A &= \text{kg mol air} \\
W &= \text{kg mol } H_2O \text{ associated with the flue gas}
\end{aligned}
$$

One of these unknowns can be replaced by the basis, say $F = 100$ kg of test fluid. Since three unknowns will remain, it will be necessary to use in the solution three independent material balances from the following set if W is to be calculated, or the first two balances if W can be ignored

Basis: 100 kg of test liquid

$$
\text{C balance:} \qquad \frac{100(0.88)}{12} = \frac{G \text{ kg mol dry fg}}{} \left| \frac{0.134 \text{ kg mol C}}{1 \text{ kg mol dry fg}} \right.
$$

$$
\text{N}_2 \text{ balance:} \quad 0.79A = 0.83G
$$

The other two component balances are:

$$O_2: \quad 0.21A = \frac{W \text{ kg mol } H_2O}{} \bigg| \frac{0.5 \text{ kg mol } O_2}{1 \text{ kg mol } H_2O}$$

$$+ \frac{G \text{ kg mol dry fg}}{} \bigg| \frac{(0.036 + 0.134) \text{ kg mol } O_2}{1 \text{ kg mol dry fg}}$$

$$H_2: \quad \frac{100(0.12)}{2} = \frac{W \text{ kg mol } H_2O}{} \bigg| \frac{1 \text{ kg mol } H_2}{1 \text{ kg mol } H_2O}$$

Note that the balances for the tie components C and N_2 (as well as H_2) involve only two streams, whereas the O_2 (and total) balance involves three streams. We shall use the C and N_2 balances.

$$G = \frac{88}{(0.134)(12)} = 54.7 \text{ kg mol dry fg}$$

$$A = \frac{0.83}{0.79}G \quad = 57.5 \text{ kg mol air}$$

$$O_2 \text{ entering} = 57.5(0.21) = 12.1 \text{ kg mol}$$

The remainder of the calculations follow those described above. Can you demonstrate that the dry flue gas analysis is slightly inaccurate?

EXAMPLE 2.20 Distillation

A continuous still is to be used to separate acetic acid, water, and benzene from each other. On a trial run the calculated data were as shown in Fig. E2.20. Data recording the benzene composition of the feed were not taken because of an instrument defect. Calculate the benzene flow in the feed per hour.

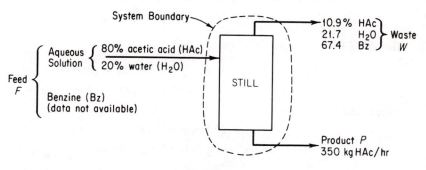

Fig. E2.20

Solution

Steps 1, 2, and 3: Examine Fig. E2.20. No reaction takes place.

Steps 4 and 5: Values of two streams, W and F, are not known, nor is the concentration of benzene in F. Three component material balances can be made so that the problem has a unique solution.

Steps 7 and 8: Two tie components (benzene and H_2O) exist to tie F to W. P could then be obtained by difference.

Step 6: Take a convenient basis of 100 kg of waste (100 kg of feed would involve more work since the benzene composition is unknown, but 100 kg of aqueous solution—80 percent HAc, 20% H_2O—would be a sound basis). Let $x =$ kg of benzene in the feed/100 kg of feed. Then we have the following:

Basis: 100 kg of waste

composition	*feed kg*	*waste* *percent = kg*	*product* *percent*
HAc	$(100 - x)0.80$	10.9	100.0
H_2O	$(100 - x)0.20$	21.7	0.0
Benzene	x	67.4	0.0
Total	100.0	100.0	100.0

Steps 7, 8, and 9: To start with, let us calculate the quantity of feed per 100 kg of waste. We have the water in the feed (21.7 kg) and the benzene in the feed (67.4 kg); these two materials appear only in the waste and not in the product and can act as tie components. All that is needed is the HAc in the feed. We can use water as the tie component to find this quantity.

$$\frac{21.7 \text{ kg } H_2O}{100 \text{ kg waste}} \left| \frac{0.80 \text{ kg HAc}}{0.20 \text{ kg } H_2O} \right. = 86.8 \text{ kg HAc/100 kg waste}$$

Do you get the same value if you use benzene as the tie component? The product now is

HAc in feed $-$ HAc in waste $=$ HAc in product   *HAc balance*

86.8 $-$ 10.9 $=$ 75.9

Step 10: Check by means of a total balance.

feed   = waste + product

$86.8 + 21.7 + 67.4 =$ 100 + 75.9

Take a new basis:

Basis: 350 kg of HAc product $\equiv$ 1 hr

$$\frac{67.4 \text{ kg benzene in feed}}{75.9 \text{ kg HAc in product}} \left| \frac{350 \text{ kg HAc product}}{\text{hr}} \right. = 311 \text{ kg benzene/hr}$$

Algebraic Solution

To employ algebra in the solution of this problem, we need an independent material balance for each unknown. The unknowns are F, W, and x:

Basis: 1 hr $\equiv$ 350 kg of HAc product

The equations are:

Total balance $\qquad\qquad\qquad\qquad F = W + 350$

H$_2$O balance $\qquad F\left[\dfrac{0.20(100 - x)}{100}\right] = W(0.217)$

HAc balance $\qquad F\left[\dfrac{0.80(100 - x)}{100}\right] = W(0.109) + 350$

Benzene balance $\qquad\qquad F\left[\dfrac{x}{100}\right] = W(0.674)$

Any three will be independent equations, but you can see the simultaneous solution of three equations involves more work than the use of tie components.

EXAMPLE 2.21 Economics

A chemical company buys coal at a contract price based on a specified maximum amount of moisture and ash. Your first job assignment is as purchasing agent. The salesperson for the Higrade Coal Co. offers you a contract for 10 carloads per month of coal with a maximum moisture content of 3.2% and a maximum ash content of 5.3% at $4.85 per ton (weighed at delivery). You accept this contract price. In the first delivery the moisture content of the coal as reported by your laboratory is 4.5% and the ash content is 5.6%. The billed price for this coal is $4.85 per ton—as weighed by the railroad in the switchyards outside your plant. The accounting department wants to know if this billing price is correct. Calculate the correct price.

Solution

What you really pay for is 91.5% combustible material with a maximum allowable content of ash and water. Based on this assumption, you can find the cost of the combustible material in the coal actually delivered.

Basis: 1 ton of coal as delivered

composition	delivered	contract coal
Combustible	0.899	0.915
Moisture plus ash	0.101	0.085
Total	1.000	1.000

We omit the detailed steps in the analysis and the figure, and just summarize the calculation here using the combustible as the tie element.

The contract calls for

$$\frac{\$4.85}{\text{ton contract}}\left|\frac{1 \text{ ton contract}}{0.915 \text{ ton comb.}}\right|\frac{0.899 \text{ ton comb.}}{1 \text{ ton del.}} = \$4.76/\text{ton del.}$$

The billed price is wrong.

Go through each of the usual steps in the analysis. Do you agree with the result?

If the ash content had been below 5.3 % but the water content above 3.2 %, presumably an adjustment might have to be made in the billing for excess moisture even though the ash was low.

Self-Assessment Test

1. Salt in crude oil must be removed before the oil undergoes processing in a refinery. The crude oil is fed to a washing unit where freshwater feed to the unit mixes with the oil and dissolves a portion of the salt contained in the oil. The oil (containing some salt but no water), being less dense than the water, can be removed from the top of the washer. The "spent" wash water (containing salt but no oil) is removed at the bottom of the washer. If the "spent" wash water contains 15 % salt and the crude oil contains 5 % salt, determine the concentration of salt in the "washed" oil product if the ratio of crude oil (with salt) to water used is 4:1.

2. A hydrocarbon fuel is burnt with excess air. The Orsat analysis of the flue gas shows 10.2 % CO_2, 1.0 % CO, 8.4 % O_2, and 80.4 % N_2. What is the atomic ratio of H to C in the fuel?

Section 2.6   Recycle, Bypass, and Purge Calculations

> *Your objectives in studying this section are to be able to:*
> 1. Draw a flow diagram for problems involving recycle, bypass, and purge.
> 2. Apply the 10-step strategy to solve steady-state problems (with and without chemical reaction) involving recycle, and/or bypass, and/or purge streams.
> 3. Solve problems in which a modest number of interconnected units are involved by making appropriate balances.
> 4. Use the overall conversion and single-pass (once-through) conversion concepts to solve recycle problems involving reactors.

Processes involving *feedback* or recycle of part of the product are frequently encountered in the chemical and petroleum industry as well as in nature. Figure 2.13 shows by a block diagram the character of the recycle stream. Figures 2.4 and 2.5 indicate some typical recycle streams in the world at large. As another example, in planning long space missions, all the food and water will have to be provided from stores on board the spacecraft. Figure 2.14 shows the recycle of O_2 and water by source.

Many industrial processes employ recycle streams. For example, in some drying operations, the humidity in the air is controlled by recirculating part of the wet air that leaves the dryer. In chemical reactions, to reduce the cost of materials the unreacted material may be separated from the product and recycled, as in the

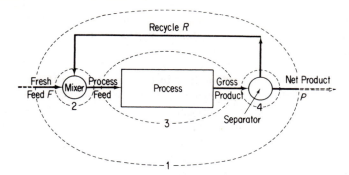

Fig. 2.13 Process with recycle (the numbers designate possible system boundaries for the material balances—see the text).

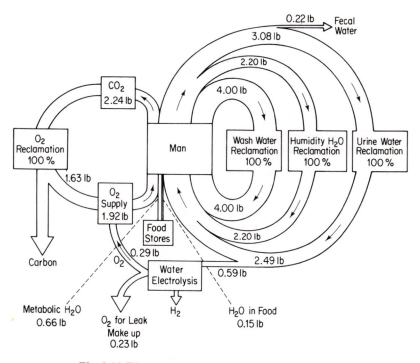

Fig. 2.14 Water and oxygen recycle in a space vehicle.

synthesis of ammonia. Another example of the use of recycling operations is in fractionating columns where part of the distillate is refluxed through the column to maintain the quantity of liquid within the column.

Do not let recycle streams confuse you. The steps in the analysis and solution of material balance problems involving recycle are the same as described in Table 2.1. With a little practice in solving problems involving recycle, you should experience little difficulty. The first point you should grasp concerning recycle calculations is that the process shown in Fig. 2.13 is in the *steady state*—no buildup or depletion of material takes place inside the process or in the recycle stream. The values of F, P, and R are *constant*. The second point is to note the distinction between *fresh feed* and *feed to the process*. The feed to the process itself is made up of two streams, the fresh feed and the recycle material. The gross product leaving the process is separated into two streams, the net product and the material to be recycled. In some cases the recycle stream may have the same composition as the gross product stream, while in other instances the composition may be entirely different depending on how the separation takes place and what happens in the process.

The third point to keep in mind is if a reaction takes place within the system, you must be given (or look up) information about the reaction stoichiometry and extent of reaction. Or, perhaps the question will be about the extent of conversion. In any case, the fraction of process feed converted to products is always an essential additional piece of information in material balance problems involving reaction and recycle.

In recycle problems you can make a total material balance as well as one for each component. Depending on the information available concerning the amount and composition of each stream, you can determine the amount and composition of the unknowns. If tie components are available, they simplify the calculations. If they are not available, algebraic methods should be used.

Material balances can be written for several different systems, four of which are shown by dashed lines in Fig. 2.13:

(a) About the entire process including the recycle stream, as indicated by the dashed lines marked 1 in Fig. 2.13

(b) About the junction point at which the fresh feed is combined with the recycle stream (marked 2 in Fig. 2.13)

(c) About the process only (marked 3 in Fig. 2.13)

(d) About the junction point at which the gross product is separated into recycle and net product (marked 4 in Fig. 2.13)

Only three of the four balances are independent for one component. However, balance 1 will not include the recycle stream, so that the balance will not be directly useful in calculating a value for the recycle R. Balances 2 and 4 do include R. It would also be possible to write a material balance for the combination of 2 and 3 or 3 and 4 and include the recycle stream.

EXAMPLE 2.22

Examine the flow sheet in Fig. E2.22, a flow sheet that contains recycle streams. What is the maximum number of independent material balances that can be written for the system if each stream contains three components, ethanol, acetone, and methanol?

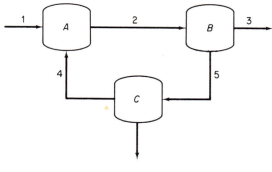

Fig. E2.22

Solution

Three independent material balances (corresponding to three components) can be written each for units *A, B,* and *C,* for a total of 9. Any other material balance, such as one for the combined subsystem *A* and *B* or *A* and *C,* can be obtained by appropriate combination of the nine component material balances. This is another example of the general observation that the maximum number of independent material balance equations equals the number of components in each unit (subsystem) added together unit by unit.

Can you obtain a unique solution for all the compositions and flows for any set of nine balances? A little reflection should lead you to the conclusion that the nine balances selected must be independent (with or without recycle) to obtain a unique solution.

EXAMPLE 2.23 Recycle without Chemical Reaction

A distillation column separates 10,000 kg/hr of a 50% benzene–50% toluene mixture. The product recovered from the condenser at the top of the column contains 95% benzene, and the bottoms from the column contain 96% toluene. The vapor stream entering the condenser from the top of the column is 8000 kg/hr. A portion of the product is returned to the column as reflux, and the rest is withdrawn for use elsewhere. Assume that the compositions of the streams at the top of the column (V), the product withdrawn (D), and the reflux (R) are identical. Find the ratio of the amount refluxed to the product withdrawn.

Solution

See Fig. E2.23. All the compositions are known and two weights are unknown. No tie components are evident in this problem; thus an algebraic solution is mandatory.

Steps 1, 2, and 3: See Fig. E2.23.

Steps 4 and 5: The unknown stream values are three: *W, R,* and *D.* Is the composition of *R* known? How? Is the composition of *V* known? Two independent material balances can be made for the still and two for the condenser; hence the problem has a unique solution. A balance around either unit would involve the stream *R.* A balance around either unit would involve the stream *R.*

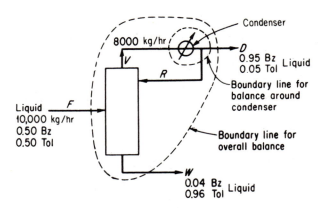

Fig. E2.23

Step 6: Select a basis, say 1 hr (or 10,000 kg of F).

Basis: 1 hr

Steps 7, 8, and 9: What balances to select to solve for R is somewhat arbitrary. We will choose to use overall balances first to get D (and W), and then use a balance on the condenser to get R. From the diagram it is apparent that no tie element exists, and consequently an algebraic solution is needed for D. Once D is obtained, R can be obtained by subtraction.

Overall Material Balances:

Total material:

$$F = D + W$$
$$10,000 = D + W$$

(a)

Component (benzene):

$$F\omega_F = D\omega_D + W\omega_W$$
$$10,000(0.50) = D(0.95) + W(0.04)$$

(b)

Solving (a) and (b) together, we obtain

$$5000 = (0.95)(10,000 - W) + 0.04W$$
$$W = 4950 \text{ kg/hr}$$
$$D = 5050 \text{ kg/hr}$$

Balance around the Condenser:

Total material:

$$V = R + D$$
$$8000 = R + 5050$$
$$R = 2950 \text{ kg/hr}$$
$$\frac{R}{D} = \frac{2950}{5050} = 0.584$$

EXAMPLE 2.24 Recycle without Chemical Reaction

Data are presented in Fig. E2.24a for an evaporator. What is the recycle stream in kilograms per hour?

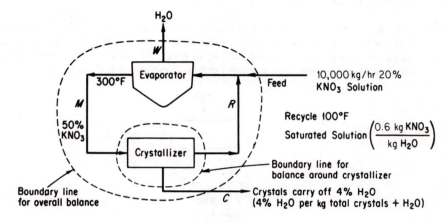

Fig. E2.24a

Solution

Steps 1, 2, 3, and 4: Figure E2.24a should be simplified with all the flows and compositions placed on it. Examine Fig. E2.24b. The composition of C can be computed from the data given in Fig. E2.24a, as can the composition of R.

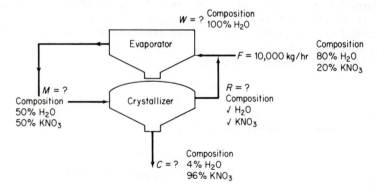

Fig. E2.24b

Step 5: We have four unknown values of variables, W, M, C, and R, and can make two component material balances each on two units of equipment; hence the problem has a unique solution.

Step 3 Continued: Compute the weight fraction compositions of R. On the basis of 1 kg of water, the saturated recycle stream contains (1.0 kg of H_2O + 0.6 kg of KNO_3) = 1.6 kg total. The recycle stream composition is

$$\frac{0.6 \text{ kg KNO}_3}{1 \text{ kg H}_2\text{O}} \left| \frac{1 \text{ kg H}_2\text{O}}{1.6 \text{ kg solution}} = 0.375 \text{ kg KNO}_3/\text{kg solution}\right.$$

or 37.5% KNO_3 and 62.5% H_2O.

Steps 6, 7, and 8: With all compositions now calculated, a search for a tie component shows that one exists for an overall balance—the KNO_3. Temporarily ignoring the interior streams, we can draw a picture as in Fig. E2.24c, in which the system is the overall process. Now we can calculate the amount of crystals (and H_2O evaporated, if wanted).

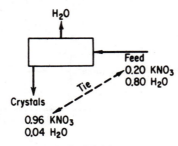

Fig. E2.24c

Basis: 1 hr ≡ 10,000 kg of feed

$$\frac{10,000 \text{ kg } F}{1 \text{ hr}} \left| \frac{0.20 \text{ kg KNO}_3}{1 \text{ kg } F} \right| \frac{1 \text{ kg crystals}}{0.96 \text{ kg KNO}_3} = 2083 \text{ kg/hr crystals}$$

Steps 7, 8, and 9 Continued: To determine the recycle stream R, we need to make a balance that involves the stream R. Either (a) a balance around the evaporator or (b) a balance around the crystallizer will do. The latter is easier since only three rather than four (unknown) streams are involved.

Total balance on crystallizer:

$$M = C + R$$
$$M = 2083 + R$$

Component (KNO_3) balance on crystallizer:

$$M\omega_M = C\omega_c + R\omega_R$$
$$0.5M = 0.96C + R(0.375)$$

Solving these two equations, we obtain

$$0.5(2083 + R) = 0.375R + 2000$$
$$R = 7670 \text{ kg/hr}$$

Step 10: Check the value of R using a material balance around the evaporator.

Now let us turn to recycle problems in which a chemical reaction occurs.

If a chemical reaction occurs in the process, the stoichiometric equation, the limiting reactant, and the fraction conversion all must be considered before beginning your calculations. Do you remember the concept of conversion as discussed in Chap. 1? With respect to Fig. 2.13, two types of conversion are used in describing the process if the process is a reactor:

(a) *Overall fraction conversion:*

$$\frac{\text{mass (moles) of reactant in fresh feed} - \text{mass (moles) of reactant in output of the overall process}}{\text{mass (moles) of reactant in fresh feed}}$$

(b) *Single-pass ("once-through") fraction conversion:*

$$\frac{\text{mass (moles) of reactant into the reactor} - \text{mass (moles) of reactant exiting the reactor}}{\text{mass (moles) of reactant into the reactor}}$$

Data about the fraction single-pass conversion provide the crucial information you need to calculate the amount of recycle in many problems. For example, examine Fig. 2.15, in which a simple recycle reactor converts $A \rightarrow B$. Suppose that you are

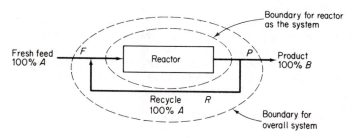

Fig. 2.15 Recycle problem.

given the data that 30% of the A is converted to B on a single pass through the reactor, and are asked to calculate the value of R, the recycle on the basis of 100 moles of fresh feed, F. We will make a single-pass balance for A with the reactor as the system.

First note that

$$(1.0F + 1.0R) = \text{moles of } A \text{ entering the reactor}$$

Then

$$(0.70)(1.0F + 1.0R) = \text{moles of } A \text{ unconverted}$$
$$= \text{moles of } A \text{ leaving the reactor} = R$$

Note that in Fig. 2.15 all of the A was recycled for simplicity, but such may not be the case in general. The single-pass mass (mole) balance on A provides the crucial information to evaluate R. Will an A balance or total balance on the overall process enable you to solve for R? Try one and see why not. When the fresh feed consists of more than one material, the conversion (or yield) must be stated for a single component, usually the limiting reactant, the most expensive reactant, or some similar compound.

EXAMPLE 2.25 Recycle with a Chemical Reaction

In a proposed process for the preparation of methyl iodide, 2000 kg/day of hydroiodic acid is added to an excess of methanol:

$$HI + CH_3OH \longrightarrow CH_3I + H_2O$$

If the produce contains 81.6 wt % CH_3I along with the unreacted methanol, and the waste contains 82.6 wt % hydroiodic acid and 17.4% H_2O, calculate, assuming that the reaction is 40% complete in the process vessel:

(a) The weight of methanol added per day
(b) The amount of HI recycled

Solution

Steps 1, 2, and 3: Sketch the process and place all the known and unknown flows and compositions on the diagram. Note that a chemical reaction is involved. In this problem all the compositions are known (in weight percent) as well as one weight—the HI input stream. The unknown quantities are the weights of the CH_3OH (M), the product (P), the waste (W), and the recycle stream. An overall balance about the whole process is all that is needed to calculate the M stream, but it will not include the recycle stream—see the dashed line in Fig. E2.25.

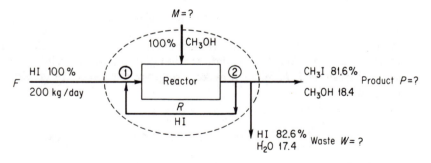

Fig. E2.25

Steps 4 and 5: Let us look at this problem from an overall viewpoint first (see the system defined by the dashed line in Fig. E2.25). All the compositions are known (in weight percent) as well as one flow rate—the HI input stream. The unknown quantities are the weights of the CH_3OH (M), the product (P), and the waste (W) streams. We can make species balances for H, I, O, and C so that a unique solution is possible for the overall problem.

How can the value of R be calculated? We need to make a material balance(s) about one of the mixing points, or about the reactor itself, so that stream R is included in the balance(s).

Steps 6, 7, 8, and 9: We first need to convert the product and waste streams to mole compositions in order to make use of the chemical equation and/or the elemental balances. A basis of 100 kg of waste is convenient, although 100 kg of product would also be a satisfactory basis.

Basis: 100 kg of waste

component	kg = percent	MW	kg mol
HI	82.6	128	0.646
H_2O	17.4	18	0.968
Total	100.0		

Basis: 100 kg of product

component	kg = percent	MW	kg mol	equivalent kg mol of CH_3OH
CH_3I	81.6	142	0.575	0.575
CH_3OH	18.4	32	0.575	0.575
Total				1.150

(a) No obvious species tie elements exist, but we could use a made-up CH_3 tie component to relate M to P. Also, according to the chemical equation for each mole of CH_3I produced there is associated 1 mole of H_2O, and 1 mole of HI is used up. Hence we can relate W to P, and F to W, and thus M to F. An alternate approach would be to use elemental balances and algebra.

Basis: 100 kg of waste

$$\frac{0.968 \text{ kg mol } H_2O}{100 \text{ kg W}} \left| \frac{1 \text{ kg mol HI}}{1 \text{ kg mol } H_2O} \right| \frac{128 \text{ kg HI}}{1 \text{ kg mol HI}} = \frac{124 \text{ kg HI}}{100 \text{ kg W}}$$

total HI entering $= F = 124 + 82.6 = 206.6$ kg HI/100 kg W

On the same basis, the CH_3OH entering is

$$\frac{0.968 \text{ kg mol } H_2O}{100 \text{ kg W}} \left| \frac{1 \text{ kg mol } CH_3 \text{ reacted}}{1 \text{ kg mol } H_2O} \right| \frac{1.150 \text{ kg mol } CH_3OH \text{ reacted plus unreacted}}{0.575 \text{ kg mol } CH_3 \text{ reacted}}$$

$$\left| \frac{32 \text{ kg } CH_3OH}{1 \text{ kg mol } CH_3OH} = \frac{61.9 \text{ kg } CH_3OH}{100 \text{ kg waste}} \right.$$

Basis: 1 day

$$\frac{2000 \text{ kg HI}}{1 \text{ day}} \left| \frac{100 \text{ kg waste}}{206.6 \text{ kg HI}} \right| \frac{61.9 \text{ kg } CH_3OH}{100 \text{ kg waste}} = 599 \text{ kg } CH_3OH/\text{day}$$

Also,

$$\frac{2000 \text{ kg HI}}{1 \text{ day}} \left| \frac{100 \text{ kg waste}}{206.6 \text{ kg HI}} = 968 \text{ kg waste/day} \right.$$

(b) To obtain the value of the recycle stream R, a balance must be made about a junction point or about just the reactor itself so as to include the stream R. A total balance about junction 1 indicates that (all subsequent bases are 1 day)

$$2000 + R = \text{reactor feed}$$

but this equation is not of much help since we do not know the rate of feed to the reactor. Similarly, for a balance on junction 2,

$$\text{gross product} = R + P + W$$

we can compute the value of the product and waste but not that of the gross product.

Will species balances about junction 1 or 2 help resolve the difficulty? Try them and see.

The only additional information available to us that has not yet been used in the fact that the reaction is 40% complete on one pass through the reactor. Consequently, 60% of the HI in the gross feed passes through the reactor unchanged. We can make an HI balance, then, about the reactor:

$$(2000 + R)(0.60) = R + 968(0.826)$$
$$400 = 0.40R$$
$$R = 1000 \text{ kg/day}$$

Our material balance around the reactor for the HI is known as a *once-through balance* and says that 60% of the HI that enters the reactor (as fresh feed plus recycle) leaves the reactor unchanged in the gross product (recycle plus product plus waste).

Step 10: As a check, solve the problem again using algebraic techniques by making I, C, and O balances for part (a)—the overall solution.

Two additional commonly encountered types of process streams are shown in Fig. 2.16:

(a) A bypass stream—one that skips one or more stages of the process and goes directly to another stage.

(b) A purge stream—a stream bled off to remove an accumulation of inerts or unwanted material that might otherwise build up in the recycle stream.

As an example of the use of a purge stream, consider the production of NH_3. Steam reforming, with feedstock natural gas, LPG, or naphtha, is the most widely accepted process for ammonia manufacture. The route includes four major chemical steps:

$$\text{Reforming:} \quad CH_4 + H_2O \longrightarrow CO + 3H_2$$
$$\text{Shift:} \quad CO + H_2O \longrightarrow CO_2 + H_2$$
$$\text{Methanation:} \quad CO + 3H_2 \longrightarrow H_2O + CH_4$$
$$\text{Synthesis:} \quad 3H_2 + N_2 \longrightarrow 2NH_3$$

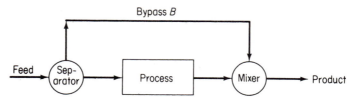

Fig. 2.16 Recycle, Bypass, and purge streams.

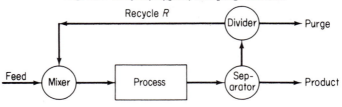

Fig. 2.17 Recycle stream with purge.

In the final stage, for the fourth reaction, the synthesis gas stream is approximately a 3:1 mixture of hydrogen to nitrogen, with the remainder about 0.9% methane and 0.3% argon.

Compressors step up the gas pressure from atmospheric to about 3000 psi—the high pressure that is needed to favor the synthesis equilibrium. Once pressurized and mixed with recycle gas, the stream enters the synthesis converter, where ammonia is catalytically formed at 400 to 500°C. The NH_3 is recovered as liquid via refrigeration, and the unreacted syngas is recycled.

In the synthesis step, however, some of the gas stream must be purged to prevent buildup of argon and methane. But purging causes a significant loss of hydrogen that could be used for additional ammonia manufacture, a loss that process designers seek to minimize.

Do you understand why the recycle process without a purge stream will cause an impurity to build up even though the recycle rate is constant? The purge rate is adjusted so that the amount of purged material remains below an acceptable specified level or so that the

$$\begin{Bmatrix} \text{rate of} \\ \text{accumulation} \end{Bmatrix} = 0 = \begin{Bmatrix} \text{rate of entering material} \\ \text{and/or production} \end{Bmatrix} - \begin{Bmatrix} \text{rate of purge} \\ \text{and/or loss} \end{Bmatrix}$$

Calculations for bypass and purge streams introduce no new principles or techniques beyond those presented so far. Two examples will make this clear.

EXAMPLE 2.26 Bypass Calculations

In the feedstock preparation section of a plant manufacturing natural gasoline, isopentane is removed from butane-free gasoline. Assume for purposes of simplification that the process and components are as shown in Fig. E2.26. What fraction of the butane-free gasoline is passed through the isopentane tower?

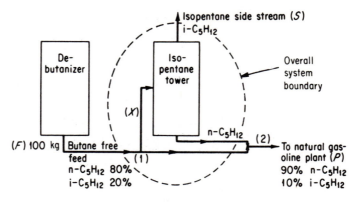

Fig. E2.26

Detailed steps will not be listed in the analysis and solution of this problem.

Solution

By examining the flow diagram you can see that part of the butane-free gasoline bypasses the isopentane tower and proceeds to the next stage in the natural gasoline plant. All the compositions are known. What kind of balances can we write for this process? We can write the following (each stream is designated by the letter F, S, or P):

Basis: 100 kg feed

(a) *Overall balances* (all compositions are known; a tie component exists):
Total material balance:

$$\textbf{in} \ = \ \textbf{out}$$
$$100 = S + P \tag{1}$$

Component balance (n-C_5):

$$\textbf{in} \quad = \quad \textbf{out}$$
$$100(0.80) = S(0) + P(0.90) \tag{2}$$

Using Eq. (2),

$$P = 100\left(\frac{0.80}{0.90}\right) = 89 \text{ kg}$$
$$S = 100 - 89 = 11 \text{ kg}$$

The overall balances will not tell us the fraction of the feed going to the isopentane tower; for this we need another balance.

(b) *Balance around isopentane tower:* Let $x =$ lb of butane-free gas going to isopentane tower.

Total material:

$$\begin{array}{cc} \textbf{in} = & \textbf{out} \\ x = 11 + n\text{-}C_5H_{12} \text{ stream (another unknown, } y) \end{array} \qquad (3)$$

Component (n-C_5):

$$x(0.80) = y \qquad \text{(a tie component)} \qquad (4)$$

Consequently, combining (3) and (4),

$$x = 11 + 0.8x$$
$$x = 55 \text{ kg} \qquad \text{or the desired fraction is } 0.55$$

Another approach to this problem is to make a balance at mixing points (1) or (2). Although there are no pieces of equipment at those points, you can see that streams enter and leave the junction.

(c) *Balance around mixing point* (2):

$$\text{material into junction} = \text{material out}$$

Total material:
$$(100 - x) + y = 89 \qquad (5)$$

Component (iso-C_5):

$$(100 - x)(0.20) + 0 = 89(0.10) \qquad (6)$$

Equation (6) avoids the use of y. Solving yields

$$20 - 0.2x = 8.9$$
$$x = 55 \text{ kg} \qquad \text{as before}$$

EXAMPLE 2.27 Purge

Ethylene oxide, a commercial reaction production of C_2H_4 and air, is formed according to the equation

$$C_2H_4 + \tfrac{1}{2}O_2 \longrightarrow CH_2CH_2O$$

The fresh feed into the reactor is comprised of 10 moles of air to 1 mole of ethylene. On a single pass through the reactor 25% of the C_2H_4 is converted to CH_2CH_2O. Makeup O_2 is fed to the recycle stream (see Fig. E2.27) and a purge stream maintains the rare gas concentration below 1% in the reactor feedstream. Calculate how many moles of purge there must be per mole of ethylene oxide produced. The known compositions of the streams are in Fig. E2.27. No CH_2CH_2O is in the purge stream. The ratio of O_2 to C_2H_4 entering the reactor is 2:1.

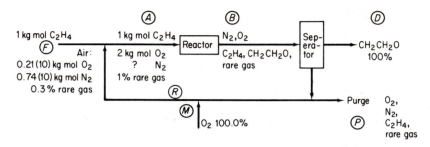

Fig. E2.27

Solution

Steps 1, 2, 3, and 4: We need to calculate the compositions of as many additional streams as possible. Select suitable bases to calculate the stream compositions. Start with the reactor feed and work forward through the process.

		F basis:		*A* basis:	*B* basis:
		1 kg mol C$_2$H$_4$		1 kg mol C$_2$H$_4$ entering the reactor	100 kg mol of *A*
component	kg mol	percent	kg mol	percent	kg mol
C$_2$H$_4$	1	9.1	1	$99/(3 + x)$	$0.75(99)/(3 + x)$
O$_2$	2.1	19.0	2	$2(99)/(3 + x)$	$1.88(99)/(3 + x)$
N$_2$	7.9	71.6	x	$x(99)/(3 + x)$	$x(99)/(3 + x)$
Total	11.0	99.7	$3 + x$	99.0	
Rare gas		0.3		1.0	1.0
C$_2$H$_4$O		—		—	$(0.25)(99)/(3 + x)$
Total		100.0		100.0	$[(2.88 + x)(99)/(3 + x)] + 1$

In the calculations we have let the mole fraction N$_2$ in stream *A* be x.

The purge stream *P* composition can be calculated by deleting the C$_2$H$_4$O from stream *B*.

	P basis:
	100 kg mol of *A*
component	kg mol
C$_2$H$_4$	$(0.75)(99)/(3 + x)$
O$_2$	$(1.88)(99)/(3 + x)$
N$_2$	$(x)(99)/(3 + x)$
Rare gas	1.0
Total	$[(2.63 + x)(99)/(3 + x)] + 1$

Steps 5 and 6: If *F* is selected as the basis, or 1 mole of C$_2$H$_4$ in *F*, then on an overall basis only *D*, *P*, and *M* are unknown values. We can make C (or C$_2$H$_4$), O$_2$, N$_2$, and rare gas balances so that the problem has a unique solution.

Steps 7, 8, and 9:   Make a rare gas and a N_2 balance on the entire system.

Basis: 1 kg mol of C_2H_4 in the feed F

$$\text{Rare gas:} \quad (0.003)(1) = \left[\frac{1.0}{\dfrac{(2.63 + x)(99)}{3 + x} + 1} \right] P$$

$$N_2: \quad (0.79)(10) = \left[\frac{99x/(3 + x)}{\dfrac{(2.63 + x)(99)}{3 + x} + 1} \right] P$$

The solution of these two equations reduces to the solution of a linear equation in x. Can you carry out the manipulations? Solve the equation to get $x = 26.6$ kg mol. Note how we had to use two tie components here to first eliminate P. Then from the N_2 balance, $P = 0.30$.

Step 10: Check P by the rare gas balance. Finally, the C_2H_4 in P is small

$$\left[\frac{(0.75)(99)/(3 + 26.6)}{\dfrac{(2.63 + 26.6)(99)}{3 + 26.6} + 1} \right] (0.30) = 0.0076 \text{ kg mol}$$

hence by a C_2H_4 balance,

$$D = 1 - 0.0076 \simeq 1.0$$

$$\frac{P}{D} = \frac{0.30}{1.0} = 0.30 \text{ kg mol } P/\text{kg mol } D$$

After a little practice you can size up a problem and visualize the simplest types of balances to make.

Up to now we have discussed material balances of a rather simple order of complexity. If you try to visualize all the operations that might be involved in even a moderate-sized plant, as illustrated in Fig. 2.5, the stepwise or simultaneous solution of material balances for each phase of the entire plant is truly a staggering task, but a task that can be eased considerably by the use of computer codes. Keep in mind that a plant can be described by a number of individual, interlocking material balances each of which, however tedious they are to set up and solve, can be set down according to the principles and techniques discussed in this chapter. In a practical case there is always the problem of collecting suitable information and evaluating its accuracy, but this matter calls for detailed familiarity with any specific process and is not a suitable topic for discussion here. We can merely remark that some of the problems you will encounter have such conflicting data or so little useful data that the ability to perceive what kind of data are needed is the most important attribute you can bring to bear in their solution.

In working with complex flow sheets with many interconnected units, you should

(a) Try to solve for the values of the recycle streams first by using known information. Start at one location and work your way around the loop.

(b) If insufficient information is provided to solve for the value of one or more recycle streams, do not be dismayed. Assign the value of the unknown stream a letter or name. Then proceed around the loop making balances to get an equation that can be solved for the unknown. Try to assign symbols to as few unknowns as possible because too many unknowns make the problem complex.

(c) Check the outcome of your recycle calculations to make sure they are correct when put in with the other values on the flowsheet of the process.

Self-Assessment Test

1. Explain what recycle and bypassing involve by means of words and also by a diagram.

2. Repeat for the term "purge."

3. If the feed to a process is stoichiometric and the subsequent separation process is complete so that all the unreacted reactants are recycled, what is the ratio of reactants in the recycle stream?

4. A material containing 75% water and 25% solid is fed to a granulator at a rate of 4000 kg/hr. The feed is premixed in the granulator with recycled product from a dryer which follows the granulator (to reduce the water concentration of the overall material in the granulator to 50% water, 50% solid). The product that leaves the dryer is 16.7% water. In the dryer, air is passed over the solid being dried. The air entering the dryer contains 3% water by weight (mass), and the air leaving the dryer contains 6% water by weight (mass).
   (*a*) What is the recycle rate to the granulator?
   (*b*) What is the rate of air flow to the dryer on a dry basis?

5. In the famous Haber process to manufacture ammonia, the reaction is carried out at pressures of 800 to 1000 atm and at 500 to 600°C using a suitable catalyst. Only a small fraction of the material entering the reactor reacts on one pass, so recycle is needed. Also, because the nitrogen is obtained from the air, it contains almost 1% rare gases (chiefly argon) that do not react. The rare gases would continue to build up in the recycle until their effect on the reaction equilibrium becomes adverse so that a small purge stream is used.

   The fresh feed composed of 75.16% H_2, 24.57% N_2, and 0.27% Ar is mixed with the recycled gas and enters the reactor with a composition of 79.52% H_2. The gas leaving the ammonia separator contains 80.01% H_2 and no ammonia. The product ammonia contains no dissolved gases. Per 100 moles of fresh feed:
   (*a*) How many moles are recycled and purged?
   (*b*) What is the percent conversion of hydrogen per pass?

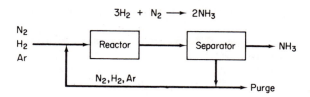

$$3H_2 + N_2 \longrightarrow 2NH_3$$

6. Ethyl ether is made by the dehydration of ethyl alcohol in the presence of sulfuric acid at 140°C:

$$2C_2H_5OH \longrightarrow C_2H_5OC_2H_5 + H_2O$$

A simplified process diagram is shown below:

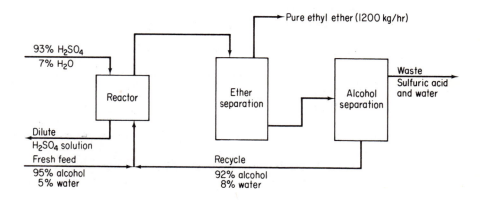

If 87% conversion of the alcohol fed to the reactor occurs per pass in the reactor, calculate:
(*a*) Kilograms per hour of fresh feed
(*b*) Kilograms per hour of recycle

SUPPLEMENTARY REFERENCES

1. Benson, S. W., *Chemical Calculations*, 2nd ed., Wiley, New York, 1963.
2. Felder, R. M., and R. W. Rousseau, *Elementary Principles of Chemical Processes*, Wiley, New York, 1978.
3. Myers, A. L., and W. D. Seider, *Introduction to Chemical Engineering and Computer Calculations*, Prentice-Hall, Englewood Cliffs, N.J., 1976.
4. Nash, L., *Stoichiometry*, Addison-Wesley, Reading, Mass., 1966.
5. Ranz, W. E., *Describing Chemical Engineering Systems*, McGraw-Hill, New York, 1970.
6. Russell, T. W. F., and M. M. Denn, *Introduction to Chemical Engineering Analysis*, Wiley, New York, 1972.

7. Schmidt, A. X., and H. L. List, *Material and Energy Balances*, Prentice-Hall, Englewood Cliffs, N.J., 1962.

8. Whitwell, J. C., and R. K. Toner, *Conservation of Mass and Energy*, Ginn/Blaisdell, Waltham, Mass., 1969.

9. Williams, E. T., and R. C. Johnson, *Stoichiometry for Chemical Engineers*, McGraw-Hill, New York, 1958.

PROBLEMS[2]

Section 2.1

2.1. Figure P2.1 shows how ethylene can be made by the M-B-E process from Kuwait kerosine. All flows are in pounds per hour. Does the flow sheet represent a satisfactory material balance? (see *Hydrocarbon Proc.*, p. 127 (July 1977), for a description of the M-B-E process.)

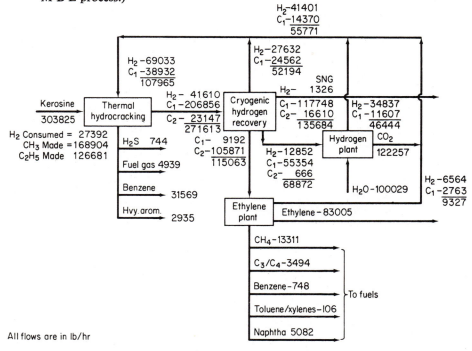

Fig. P2.1 Ethylene by M-B-E system from Kuwait kerosine.

2.2. Examine Fig. P2.2. Is the material balance satisfactory?

2.3. Does the plant shown in Fig. P2.3 yield a material balance that balances? Explain a possible reason for the outcome of your calculations.

[2]An asterisk designates problems appropriate for computer solution. Refer also to the computer problem at the end of the chapter.

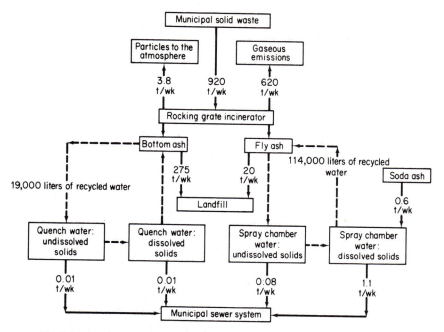

Fig. P2.2 Flow diagram of a municipal incinerator. Flow rates of solids are given in tons/week on a dry-weight basis.

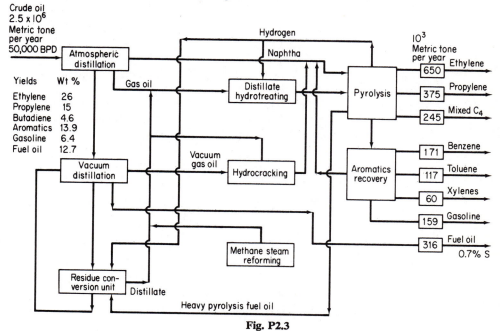

Fig. P2.3

2.4. The molten salt coal gasification project of Atomics International is shown in Fig. P2.4. Does this flow sheet represent satisfactory material balances on all units?

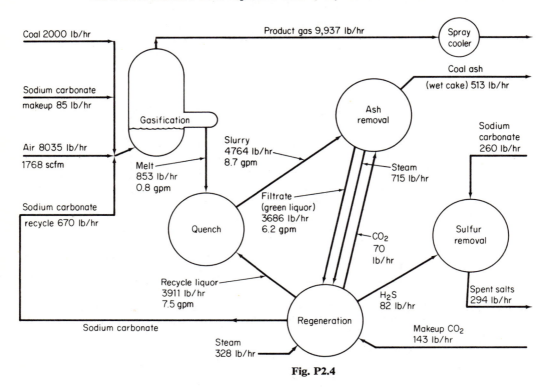

Fig. P2.4

2.5. Examine the processes in Fig. P2.5. Each box represents a system. For each, state whether:

(a) The process is in the (**1**) steady state, (**2**) unsteady state, or (**3**) unknown condition

(b) The system is (**1**) closed, (**2**) open, (**3**) neither, or (**4**) both

The wavy line represents the initial fluid level when the flows begin. In case (c), the tank stays full.

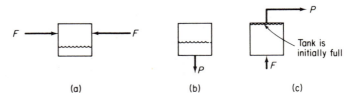

Fig. P2.5

2.6. The table shows the human daily material balance. The values given are the nominal amounts of substances consumed and excreted by a moderately active man of average size. From the data in the table, prepare a related water balance, and compute the amount in kilograms of metabolic water produced in one day by the average man.

HUMAN DAILY BALANCE

Commodity	Material Balance (kg)
Output	
Urine (95% H_2O)	1.429
Feces (75.8% H_2O)	0.132
Transpired H_2O	0.998
CO_2 (1.63 lb of O_2)	1.016
Other losses	0.064
Total	3.639
Input	
Food (dry weight)	0.680
O_2	0.832
H_2O	2.127
Total	3.639

Section 2.2

2.7. Make a set of all the independent material balances you can for the steady-state operation of the cooling tower in Fig. P2.7. Use F for the mass flow rate and ω for the mass fraction, with the appropriate subscripts. Suppose that F was the molar flow rate. How would your equations change?

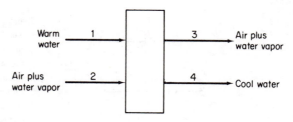

Fig. P2.7

2.8. See Fig. P2.8 on the next page.

$$F_1 + F_2 - F_3 = 0 \tag{1}$$

$$F_3 - F_4 - F_5 = 0 \tag{2}$$

$$F_1 + F_2 - F_4 - F_5 = 0 \tag{3}$$

For Fig. P2.8 how many independent equations are obtained from the overall balance around the entire system, Eq. (3), plus the balances on units A, Eq. (1), and B, Eq. (2)?

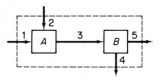

Fig. P2.8

2.9. What is the maximum number of independent material balances that can be written for the process in Fig. P2.9?

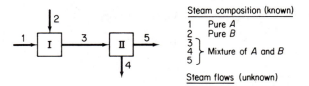

Steam composition (known)
1   Pure A
2   Pure B
3
4 } Mixture of A and B
5

Steam flows (unknown)

Fig. P2.9

2.10. What is the maximum number of independent material balances that can be written for the process in Fig. P2.10? The stream flows are unknown.

Suppose you find out that A and B are always combined in each of the streams in the same ratio. How many independent equations could you write?

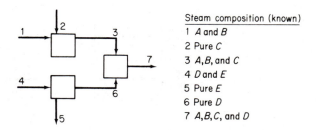

Steam composition (known)
1 A and B
2 Pure C
3 A,B, and C
4 D and E
5 Pure E
6 Pure D
7 A,B,C, and D

Fig. P2.10

2.11. The diagram in Fig. P2.11 represents a typical but simplified distillation column. Streams 3 and 6 consist of steam and water, and do not come in contact with the two components in the column. Write the overall material balances for the three sections of the column. How many independent equations would these balances represent?

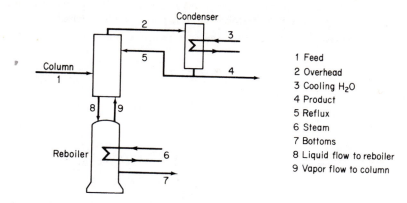

Fig. P2.11

1 Feed
2 Overhead
3 Cooling H_2O
4 Product
5 Reflux
6 Steam
7 Bottoms
8 Liquid flow to reboiler
9 Vapor flow to column

2.12. For the process shown in Fig. P2.12, what is the maximum number of material balance equations that can be written? Write them. How many independent material balance equations are there in the set?

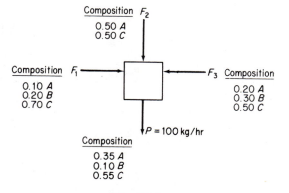

Composition F_2
0.50 A
0.50 C

Composition F_1
0.10 A
0.20 B
0.70 C

F_3 Composition
0.20 A
0.30 B
0.50 C

$P = 100$ kg/hr

Composition
0.35 A
0.10 B
0.55 C

Fig. P2.12

2.13. Examine the process in Fig. P2.13. No chemical reaction takes place, and x stands for

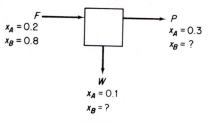

F
$x_A = 0.2$
$x_B = 0.8$

P
$x_A = 0.3$
$x_B = ?$

W
$x_A = 0.1$
$x_B = ?$

Fig. P2.13

mole fraction. How many stream values are unknown? How many compositions? Can this problem be solved uniquely for the unknowns?

2.14. Repeat Problem 2.13 for the process in Fig. P2.14.

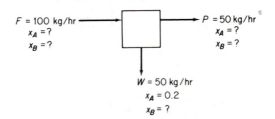

Fig. P2.14

2.15. You have been asked to check out the process shown in Fig. P2.15. What will be the minimum number of measurements to make in order to compute the value of each of the stream flow rates and stream compositions? Explain your answer. Can any arbitrary set of five be used; that is, can you measure just the three flow rates and two compositions? Can you measure just three compositions in stream F and two compositions in stream W? No chemical reaction takes place and x is the mole fraction of component A, B, or C.

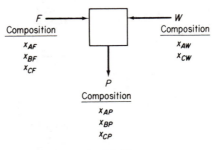

Fig. P2.15

Section 2.3

2.16. A synthesis gas analyzing 6.4% CO_2, 0.2% O_2, 40.0% CO, and 50.8% H_2 (the balance is N_2) is burned with 40% dry excess air. What is the composition of the flue gas?

2.17. Hydrogen-free carbon in the form of coke is burned:
(a) With complete combustion using theoretical air
(b) With complete combustion using 50% excess air
(c) Using 50% excess air but with 10% of the carbon burning to CO only
In each case calculate the gas analysis that will be found by testing the flue gases with an Orsat apparatus.

2.18. In the potential hydrogen-based fuel of the future, an engine burns a gas that is 96% hydrogen and 4% water vapor by volume. Complete combustion takes place because the excess air employed is 32%. Calculate the composition of the combustion product gas on a dry basis.

2.19. In the Deacon process for the manufacture of Cl_2, a dry mixture of HCl and air is passed over a heated catalyst that promotes the oxidation of HCl to Cl_2. The Deacon process can be reversed and HCl formed from Cl_2 and steam by removing the O_2 formed with hot coke as follows:

$$2Cl_2 + 2H_2O + C \longrightarrow 4HCl + CO_2$$

If chlorine cell gas analyzing 90% Cl_2 and 10% air is mixed with 5% excess steam and the mixture is passed through a hot coke bed at 900°C, the conversion of Cl_2 will be 80% complete, but all the O_2 in the air will react. Calculate the composition of the exit gases from the converter, assuming no CO formation.

2.20. The utilization of coal to produce gaseous fuels for heating and generating power as a supplement to the dwindling supplies of natural gas is a major research and development activity in the United States at the present time. Barriers to synthetic fuel production are more economic than technological, but there are still many technical problems to be resolved. In a test of pilot-plant coal gasifier with a production rate of 25.8 m³/hr, the output gas with a composition of 4.5% CO_2, 26% CO, 13% H_2, 0.5% CH_2, and 56% N_2 is burned with 10% excess air. Calculate the Orsat analysis of the combustion products.

2.21. In the analysis of stack gases from incinerators it is desirable to have a convenient means for measuring the amount of air in excess of the air needed to burn the fuel completely (i.e., stoichiometric combustion). The Orsat analysis provides information that can be used to estimate the percentage of excess air. It is proposed that the ratio of O_2 to N_2 would provide a good means of estimating the percentage of excess air where the fuel is cellulose ($C_6H_{10}O_5$) and the incinerator temperature is not excessively hot, causing the N_2 in the air to oxidize. Determine what the O_2/N_2 ratio is in the stack gas of an incinerator for the stoichiometric burning of cellulose and for 20%, 50%, and 100% excess air.

2.22. With increasing costs of energy, it has been suggested that municipal waste treatment plants convert the biomass in such plants to combustible gases instead of completely oxidizing the material to CO_2 and H_2O, or landfilling the resulting sludge. One process that forms a product about 65% CH_4 and 35% CO_2 (on a dry basis) is the anaerobic conversion of biomass to gases by microorganisms. Assume that H_2S is not present, or has been removed by processing.

If this particular gas is burned with 15% excess air, what is the final gas composition, and how many kilograms of combustion gas products (including H_2O) are produced per kilogram of gas burned?

2.23. Aviation gasoline is isooctane, C_8H_{18}. If it is burned with 20% excess air and 30% of the carbon forms carbon monoxide, what is the flue-gas analysis?

2.24. In the United States the five most common primary air pollutants (in tons emitted annually) are carbon monoxide, sulfur oxides, hydrocarbons, nitrogen oxides, and particulate matter. If a gas that analyzes 80% CH_4, 10% H_2, and 10% N_2 is burned with 40% excess air and 10% of the carbon forms CO, compute the Orsat (dry basis) analysis of the resulting flue gas. Will the nitrogen be oxidized? Recommend a method of eliminating the CO formed.

2.25. Synthetic fibers are made in a wet spinning process by pumping a polymer solution (density ρ in g/cm³, solids fraction ω_s) at a volumetric rate of Q (cm³/min) through a spinnerette with n holes. See Fig. P2.25. The filaments emerge into a bath containing both solvent and nonsolvent. Part of the solvent is lost from the filaments and an equivalent weight of nonsolvent is picked up; this solidifies the filaments. Next, the filaments are washed free of solvent as they are stretched by a factor S between two sets of rolls. Finally, the filaments are dried in a hot roll rotating at V_f feet per minute. Calculate the denier per filament (d.p.f.) (weight in grams of a filament 9000 m long) in terms of Q, n, ρ, ω_s, S (a constant), and V_f.

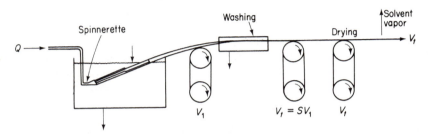

Fig. P2.25

2.26. The U.B.T. Development Company is selling a fuel cell that generates electrical energy by direct conversion of coal into electrical energy. The cell is quite simple—it works on the same principle as a storage battery and produces energy free from the Carnot cycle limitation. To make clear the outstanding advantages of the fuel cell, we can consider the reaction

$$\text{fuel} + \text{oxygen} = \text{oxidation products} \tag{1}$$

Reaction (1) is intended to apply to any fuel and always to 1 mole of the fuel. The fuel cell is remarkable in that it can convert chemical energy directly into work and bypass the wasteful intermediate conversion into heat. A typical fuel consists of

C	65%
H	5%
O	10%
S	4%
Ash	16%

Assume that no carbon is left in the ash and that the S is oxidized to SO_2. If 20% excess air is used for oxidation of the fuel, calculate the compositions of all the oxidation products.

2.27. Solvents emitted from industrial operations can become significant pollutants if not disposed of properly. A chromatographic study of the waste exhaust gas from a synthetic fiber plant has the following analysis in mole percent:

$$
\begin{array}{ll}
\text{CS}_2 & 40 \\
\text{SO}_2 & 10 \\
\text{H}_2\text{O} & 50
\end{array}
$$

It has been suggested that the gas be disposed of by burning with an excess of air. The gaseous combustion products are then emitted to the air through a smokestack. The local air pollution regulations say that no stack gas is to analyze more than 2% SO_2 (dry basis) analysis averaged over a 24-hr period. Calculate the minimum percent excess air that must be used to stay within this regulation.

Section 2.4

2.28. In a distillation train a liquid hydrocarbon containing 20 mole % ethane, 40 mole % propane, and 40 mole % butane is to be fractionated into essentially pure components as shown in Fig. P2.28. On the basis of $F = 100$ moles, what is P (in moles) and the composition of stream A?

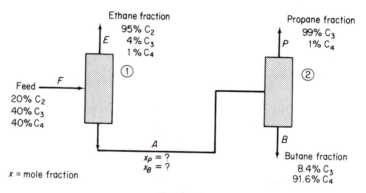

Fig. P2.28

2.29. To prepare a solution of 50.0% sulfuric acid, a dilute waste acid containing 28.0% H_2SO_4 is fortified with a purchased acid containing 96.0% H_2SO_4. How many kilograms of the purchased acid must be bought for each 100 kg of dilute acid?

2.30. A natural gas analyzes CH_4, 80.0% and N_2, 20.0%. It is burned under a boiler and most of the CO_2 is scrubbed out of the flue gas for the production of dry ice. The exit gas from the scrubber analyzes CO_2, 1.2%; O_2, 4.9%; and N_2, 93.9%. Calculate:
(a) The percentage of the CO_2 absorbed
(b) The percent excess air used

2.31. A gas stream from a refinery, containing 30 mole % ethane and 70% methane, enters an absorber where the two gases are almost completely separated (see Fig. P2.31). The methane-rich stream is to be sold to a customer. Calculate the moles of sales gas produced per mole of feed to the absorber.

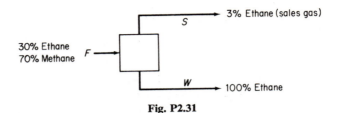

Fig. P2.31

2.32. Because of changes in demand, the customer wants to receive 10% ethane in the sales gas of Problem 2.31 instead of 3%. You are asked to draw a sketch labeling all streams carefully as to composition and amount, indicating how to accomplish the desired objective still using the same absorber. Note that the absorber itself is designed so that it only operates with the concentration conditions specified in Problem 2.31.

2.33. A fuel composed of ethane (C_2H_6) and methane (CH_4) in unknown proportions is burned in a furnace with oxygen-enriched air (50.0 mole % O_2). Your Orsat analysis is: 25% CO_2, 60% N_2, and 15% O_2. Find:
  (a) The composition of the fuel, that is, the mole percent methane in the methane–ethane mixture
  (b) The moles of oxygen-enriched air used per mole of fuel

2.34. In a gas-separation plant, the feed-to-butane splitter has the following constitutents:

Component	Mole %
C_3	1.9
i-C_4	51.6
n-C_4	46.0
C_5^+	0.6
Total	100.0

The flow rate is 5804 kg mol/day. If the overhead and bottoms streams from the butane splitter have the following compositions, what are the flow rates of the overhead and bottoms streams in kg mol/day?

	Mole %	
Component	Overhead	Bottoms
C_3	3.4	—
i-C_4	95.7	1.1
n-C_4	0.9	97.6
C_5^+	—	1.3
Total	100.0	100.0

Use an algebraic method of solution. Check your answer at least one way by using tie components. Do the results have the same accuracy if the tie is present in a small quantity?

2.35. The solubility of anhydrous magnesium sulfate at 20°C is 35.5 g/100 g of H_2O. How much $MgSO_4 \cdot 7H_2O$ must be dissolved in 100 kg of H_2O to form a saturated solution?

2.36. The solubility of anhydrous manganous sulfate at 20°C is 62.9 g/100 g of H_2O. How much $MnSO_4 \cdot 5H_2O$ must be dissolved in 100 lb of water to give a saturated solution?

2.37. A water solution contains 60% $Na_2S_2O_2$ together with 1% soluble impurity. Upon cooling to 10°C, $Na_2S_2O_2 \cdot 5H_2O$ crystallizes out. The solubility of this hydrate is 1.4 lb $Na_2S_2O_2 \cdot 5H_2O$/lb free water. The crystals removed carry as adhering solution 0.06 lb solution/lb crystals. These are dried to remove the remaining water (but not the water of hydration). The final dry $Na_2S_2O_2 \cdot 5H_2O$ crystals must not contain more than 0.1% impurity. To meet this specification, the original solution, before cooling, is further diluted with water. On the basis of 100 lb of the original solution, calculate:
(a) The amount of water added before cooling
(b) The percentage recovery of the $Na_2S_2O_2$ in the dried hydrated crystals

2.38. Twelve hundred pounds of $Ba(NO_3)_2$ is dissolved in sufficient water to form a saturated solution at 90°C, at which temperature the solubility is 30.6 g/100 g water. The solution is then cooled to 20°C, at which temperature the solubility is 8.6 g/100 g of water.
(a) How many pounds of water are required for solution at 90°C, and what weight of crystals is obtained at 20°C?
(b) How many pounds of water are required for solution at 90°C, and what weight of crystals is obtained at 20°C, assuming that 10 percent more water is to be used than necessary for a saturated solution at 90°C?
(c) How many pounds of water are required for solution at 90°C, and what weight of crystals is obtained at 20°C, assuming that the solution is to be made up 90% saturated at 90°C?
(d) How many pounds of water are required for solution at 90°C, and what weight of crystals is obtained at 20°C, assuming that 5% of the water evaporates on cooling and that the crystals hold saturated solution mechanically in the amount of 5% of their dry weight?

2.39. If 100 g of anhydrous Na_2SO_4 is dissolved in 200 g of H_2O and the solution is cooled until 100 g of $Na_2SO_4 \cdot 10H_2O$ crystallizes out, find:
(a) The composition of the remaining solution (*mother liquor*)
(b) The grams of crystals recovered per 100 g of initial solution

2.40. Exactly 1.000 kg of commercial $MnSO_4 \cdot 4H_2O$ is added to 0.083 kg of water. Afterward, the saturated solution is filtered to obtain dry crystals of $MnSO_4 \cdot H_2O$. The solubility of $MnSO_4$ in water is at 100°C is 0.361 kg of $MnSO_4$ per kilogram of water. How many kilograms of dry crystals will be obtained at 100°C? (Mol. wt. $MnSO_4 = 151$.)

2.41. A polymer blend is to be formed from the three compounds whose compositions and approximate formulas are listed in the table on the next page. Determine the percentages of each compound A, B, and C to be introduced into the mixture to achieve the desired composition.

| | Compound (%) | | | |
Composition	A	B	C	Desired Mixture
$(CH_4)_x$	25	35	55	30
$(C_2H_6)_x$	35	20	40	30
$(C_3H_8)_x$	40	45	5	40
Total	100	100	100	100

How would you decide to blend compounds A, B, C, and D [$(CH_4)_x = 10\%$, $(C_2H_6)_x = 30\%$, $(C_3H_8)_x = 60\%$] to achieve the desired mixture?

2.42. The feed to a distillation column is separated into net overhead product containing nothing with a boiling point higher than isobutane and bottoms containing nothing with a boiling point below that of propane. See Fig. P2.42. The composition of the feed is:

Component	Mole %
Ethylene	2.0
Ethane	3.0
Propylene	5.0
Propane	15.0
Isobutane	25.0
n-Butane	35.0
n-Pentane	15.0
Total	100.0

The concentration of isobutane in the overhead is 5.0 mole %, and the concentration of propane in the bottoms is 0.8 mole %. Calculate the composition of the overhead and bottoms streams per 100 moles of feed.

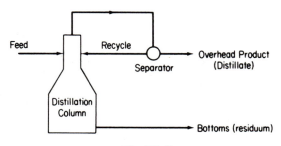

Fig. P2.42

2.43. A mixture of gaseous fuels is burned in an experimental furnace by conventional techniques. The combustion gas is analyzed in a mercury Orsat, and there is no reason to question the dependability of the results. If the combustion products are solely CO_2, H_2O, and N_2, what is the composition of the fuel for the case in which:

(a) The fuel is a mixture of methane and ammonia and the results of the Orsat are 6.2% CO_2, 4.6% O_2, and 89.2% N_2?

(b) The fuel is a mixture of methane and HCN and the results of the Orsat are 10.8% CO_2, 4.1% O_2, and 85.1% N_2?

2.44. A power company operates one of its boilers on natural gas and another on oil (for peak period operation). The analysis of the fuels is as follows:

Natural Gas	Oil
96% CH_4	$(CH_{1.8})_n$
4% CO_2	

When both boilers are on the line, the flue gas shows (Orsat analysis) 10.0% CO_2, 4.5% O_2, and the remainder N_2. What percentage of the total carbon burned comes from the oil? (*Hint:* Do not forget the H_2O in the stack gas.)

2.45. A power company operates one of its boilers on natural gas and another on oil. The analyses of the fuels show 96% CH_4, 2% C_2H_2, and 2% CO_2 for the natural gas and $C_nH_{1.8n}$ for the oil. The flue gases from both groups enter the same stack, and an Orsat analysis of this combined flue gas shows 10.0% CO_2, 0.63% CO, and 4.55% O_2. What percentage of the total carbon burned comes from the oil?

2.46. An automobile engine burning a fuel consisting of a mixture of hydrocarbons is found to give an exhaust gas analyzing 10.0% CO_2 by the Orsat method. It is known that the exhaust gas contains no oxygen or hydrogen. Careful metering of the air entering the engine and the fuel used shows that 12.4 lb of dry air enter the engine for every pound of fuel used.

(a) Calculate the complete Orsat gas analysis.

(b) What is the weight ratio of hydrogen to carbon in the fuel?

2.47. A fuel oil and a sludge are burned together in a furnace with dry air. Assume that the fuel oil contains only C and H.

	Sludge (%)		
Fuel Oil (%)	Wet	Dry	Flue Gas (%)
C = ?	water = 50	S = 32	SO_2 = 1.52
	solids = 50	C = 40	CO_2 = 10.14
H = ?		H_2 = 4	CO = 2.02
		O_2 = 24	O_2 = 4.65
			N_2 = 81.67

(a) Determine the weight percent composition of the fuel oil.

(b) Determine the ratio of kilograms of sludge to kilograms of fuel oil.

2.48. A low-grade pyrites containing 32% S is mixed with 10 lb of pure sulfur per 100 lb of pyrites so the mixture will burn readily, forming a burner gas that analyzes (Orsat) 13.4% SO, 2.7% O_2, and 83.9% N_2. No sulfur is left in the cinder. Calculate the percentage of the sulfur fired that burned to SO_3. (The SO_3 is not detected by the Orsat analysis.)

2.49. Cities are attempting to dispose of municipal refuse and sludge from wastewater treatment plants as economically as possible. One possibility is to mix sludge with garbage from which metals, and so on, have been removed as shown in Fig. P2.49. The mixture can be turned into soil conditioner by composting.

In a typical city the garbage and refuse consists of 75% paper, cardboard, and rags; 13% garbage and garden refuse; 7% tin cans and light ferrous metals; 1.5% tramp metal (large pieces of iron and steel); 1% glass; 0.5% aluminum and nonferrous metals; and 2% material not compostible, such as rubber, heavy plastic, and cement. Glass, aluminum, tramp metals, and other materials which are not compostible are removed by manual sorting and picking. Then tin cans and light ferrous metals are removed by magnetic separation. The remaining material is milled into particles less then 1 inch in diameter and mixed with sewage sludge. The sludge provides bacteria to enhance the composting and organic nitrogen to maintain the proper carbon and nitrogen balance. After 6 days of composting, the mixture is ground and dried. The product has a texture and value comparable to peat moss.

Based on a refuse supply of 200 tons/day, answer the following questions.

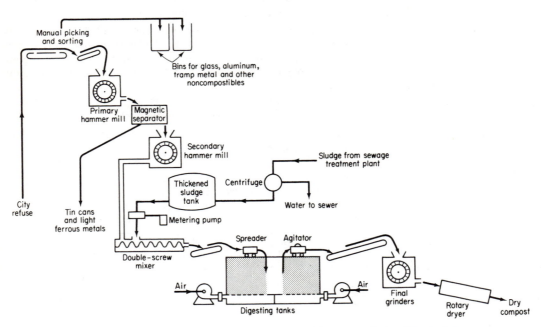

Fig. P2.49 Composting of solid wastes with sewage sludge. [Revised from *Chem. Eng.*, pp. 232–234 (November 7, 1967).]

(a) How many tons per day of (1) glass, (2) aluminum and nonferrous metals, (3) tin cans and ferrous metals, (4) tramp metal, and (5) noncompostibles are recovered per day?

(b) If the sewage sludge is 5% solids and is thickened to 30% solids, how many tons per day of the 5% sewage sludge are used if moist refuse and the 30% sludge are mixed by equal weights for composting?

(c) How much water from the sludge is discharged to the sewer?

(d) If paper, cardboard, and rags are removed for recycle prior to the first hammermill, how much sludge is needed for the case given in part (b)?

2.50.* It is desired to mix three LPG (liquefied petroleum gas) products in certain proportions in order that the final mixture will meet certain vapor-pressure specifications. These specifications will be met by a stream of composition D below. Calculate the proportions in which streams A, B, and C must be mixed to give a product with a composition of D. The values are liquid volume percent.

	Stream			
Component	A	B	C	D
C_2	5.0			1.4
C_3	90.0	10.0		31.2
iso-C_4	5.0	85.0	8.0	53.4
n-C_4		5.0	80.0	12.6
iso-C_5^+			12.0	1.4
Total	100.0	100.0	100.0	100.0

2.51. A and B are immiscible liquids, but they emulsify in one another to give uniform emulsions. See Fig. P2.51. The uniform emulsion is withdrawn from the lower layer and sent to a settler where the emulsion breaks and is separated. During the addition of 698 kg of A and 1302 kg of B to the extractor, the interface between the layers

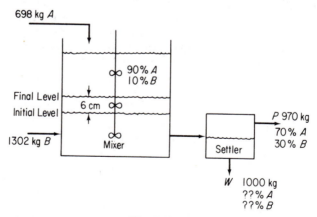

Fig. P2.51

rises to and stays at a new level of 6 cm higher than the old level. A rise of 1 cm corresponds in volume to 30 kg of A or 40 kg of B. The volumes of A and B are additive in all proportions. The top level in the extractor remains constant during the operation. What is the composition of the bottom layer from the settler?

Section 2.5

2.52. A synthesis gas is made by cracking and partially oxidizing butane (C_4H_{10}) in the presence of steam and air. The product synthesis gas has the percent composition 3.5 CO_2, 2.3 C_2H_4, 23.2 CO, 39.3 H_2, 11.6 CH_4, 1.7 C_2H_6, and 18.4 N_2. How many kilograms of air are used per kilogram of butane?

2.53. A coal analyzes 74% C and 12% ash (inert). The flue gas from the combustion of the coal analyzes 12.4% CO_2, 1.2% CO, 5.7% O_2, 80.7% N_2. Calculate:
(a) The pounds of coal fired per 100 moles of flue gas
(b) The percent excess air
(c) The pounds of air used per pound of coal
Assume that there is no nitrogen in the coal.

2.54. In an experimental apparatus air is supposed to be mixed with pure oxygen so that the methane being burned to heat the apparatus will yield combustion products that have a higher temperature than do the products for air alone. There is some question as to whether the assumed enrichment is correct. You analyze the product gases to be 22.2% CO_2, 4.4% O_2, and 73.4% N_2. Calculate the percentage of O_2 and N_2 in the oxygen-enriched air.

2.55. Wastewater flow in a sewer can be estimated by adding a known amount of tracer such as salt solution and measuring the salt concentration upstream and downstream of the point of addition. Upstream the background concentration of salt was 50 mg/liter. One hundred kilograms of a salt solution (containing 30% salt) per minute was added, and the resulting concentration of salt downstream was determined to be 0.1% salt by weight. Estimate the mass and volumetric flow rate of the wastewater in the sewer.

2.56. Your boss asks you to calculate the flow through a natural-gas pipeline. Since it is 26 in. in diameter, it is impossible to run the gas through any kind of meter or measuring device. You decide to add 100 lb of CO_2 per minute to the gas through a small $\frac{1}{2}$-in. piece of pipe, collect samples of the gas downstream, and analyze them for CO_2. Several consecutive samples after 1 hr are:

Time	Percent CO_2
1 hr, 0 min	2.0
10 min	2.2
20 min	1.9
30 min	2.1
40 min	2.0

(a) Calculate the flow of gas in pounds per minute at the point of injection.

(b) Unfortunately for you, the gas at the point of injection of CO_2 already contained 1.0% CO_2. How much was your original flow estimate in error (in percent)? [*Note:* In part (a) the natural gas is all methane, CH_4.]

2.57. How many kilograms of $CaCl_2 \cdot 6H_2O$ must be dissolved in 100 kg of water at 20°C to form a saturated solution? The solubility of $CaCl_2$ in water at 20°C is 0.744 kg of $CaCl_2$ per kilogram of water. The molecular weight of $CaCl_2$ is 111.

2.58. The nose is an extremely sensitive detector of odors; it can be matched only by the most sensitive instrumental techniques. Many smells are identified by comparison with pure compounds diluted by odor-free air until they are near the odor threshold limit. However, with the introduction of flame photometric detectors it has been possible to measure the concentration of the sulfur compounds giving rise to odors in the ppb range with the aid of preconcentration (by a factor of 10^3 to 10^4).

The main offenders are known to be H_2S, mercaptans, and disulfides, which are present in crude-oil fractions. Because of their offensive smell and low olfactory levels, these compounds give rise to the majority of complaints. Typical detectable limits for the most important sulfur compounds are:

Compound	ppm
Hydrogen sulfide, H_2S	0.001–0.014
Methyl mercaptan, CH_3SH	0.001–0.0085
Ethyl mercaptan, C_2H_5SH	0.001–0.0026
Dimethyl sulfide, $(CH_3)_2S$	0.002–0.0052

If the concentration of methyl mercaptan is measured as 0.035 ppm in a waste stream flowing at 1.7 m^3/s (specific volume = 0.021 m^3/kg mol), how many kilograms of air have to be bled into the waste stream per second to reduce the concentration of methyl mercaptan in the stream to below detectable levels?

2.59. As superintendent of a lacquer plant, your foreman brings you the following problem: he has to make up 1000 lb of an 8% nitrocellulose solution. He has available a tank of a 5.5% solution. How much dry nitrocellulose must he add to how much of the 5.5% solution in order to fill the order?

2.60. There exists some question as to whether the measuring device on a natural gas (CH_4) pipeline is operating properly. The instrument reads 14,530 kg/min. You are asked to check the flow by injecting ammonia gas into the pipeline at a constant rate and measuring the ammonia concentration at a site 10 km downstream. A check of the flow before injection of ammonia reveals no detectable trace of ammonia whatsoever. For a constant rate of injection of 72.3 kg/min for 20 min, the downstream concentration of ammonia converted to a mass fraction is 0.00382. Is the instrument for measuring flow malfunctioning?

2.61. The Clean Air Act requires automobile manufacturers to warrant their control systems as satisfying the emission standards for 50,000 mi. It requires owners to have their engine control systems serviced exactly according to manufacturers' specifications and to always use the correct gasoline. In testing an engine exhaust having a known Orsat analysis of 16.2% CO_2, 4.8% O_2, and 79% N_2 at the outlet, you find to your surprise that at the end of the muffler the Orsat analysis is 13.1 percent CO_2.

Can this discrepancy be caused by an air leak into the muffler? (Assume that the analyses are satisfactory.) If so, compute the moles of air leaking in per mole of exhaust gas leaving the engine.

2.62. In most cities, the flows in the major sanitary sewers are not known. Knowledge of existing flows, and hence available reserve sewer capacity, is essential so that planning agencies may make decisions regarding the growth that can be supported in sewered areas before expensive problems with sanitary sewers develop. Selection of equipment to be used to measure flow is based on the criteria (1) that the equipment make accurate flow measurements in sewers in wet and dry weather, including those sewers that surcharge; (2) that it be economical to purchase, install, and service; (3) that it be reasonably vandal-proof; (4) that it be reusable at other monitoring sites during the study; (5) that it maintain a minimum head loss through the measuring device to minimize unwanted effects on the hydraulic characteristics of the sewer; and (6) that it operate automatically for a minimum of 24 hr.

Where it has been either theoretically or technically impossible to use a flume and pneumatic recorder technique for flow measurement, the chemical dilution technique has been applied successfully. Figure P2.62 shows a typical installation. Rhodamine dye, used as the tracer, is dripped continuously at a constant rate into the sewer at one manhole. If 30 μg/liter of Rhodamine is detected in the flowstream at 10:30–11:00 P.M. at the downstream manhole, and the discharge rate from the Mariotte vessel is 4 lb of 50% solution per hour, what is the flow in gal/hr?

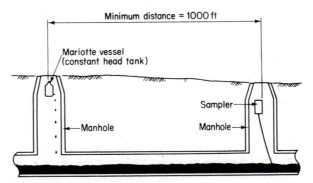

Fig. P2.62

2.63. Paper pulp is sold on the basis that it contains 12% moisture; if the moisture exceeds this value, the purchaser can deduct any charges for the excess moisture and also deduct for the freight costs of the excess moisture. A shipment of pulp became wet and was received with a moisture content of 22%. If the original price for the pulp was $40 per ton of air-dry pulp and if the freight is $1.00 per 100 lb shipped, what price should be paid per ton of pulp delivered?

2.64. To meet certain specifications, a dealer mixes bone-dry glue, selling at 25 cents per pound, with glue containing 22% moisture, selling at 14 cents per pound, so that the mixture contains 16% moisture. What should be the cost price per pound of the mixed glue?

2.65. A dairy produces casein which when wet contains 23.7% moisture. They sell this for $8.00/100 lb. They also dry this casein to produce a product containing 10% moisture. Their drying costs are $0.80/100 lb water removed. What should be the selling price of the dried casein to maintain the same margin of profit?

Section 2.6

2.66. A desert resort water system uses a secondhand reverse osmosis process to reduce the concentration of salt in the water to potable levels. Examine Fig. P2.66 for the overall process. The reverse osmosis unit produces salt only at a concentration of 100 mg/liter. Determine the value of F' in liters per second and the concentration of salt in the waste brine in milligrams per liter.

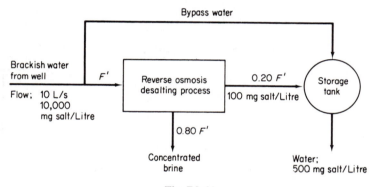

Fig. P2.66

2.67. Seawater is to be desalinized by reverse osmosis using the scheme indicated in Fig. P2.67. Use the data given in the figure to determine:

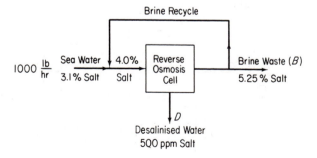

Fig. P2.67

(a) The rate of waste brine removal (B)

(b) The rate of desalinized water (called potable water) production (D)

(c) The fraction of the brine leaving the osmosis cell (which acts in essence as a separator) that is recycled

(*Note:* ppm designates parts per million.)

2.68. A solution containing 10% NaCl, 3% KCl, and 87% water is fed to the process shown in Fig. P2.68 at the rate of 18,400 kg/hr. The compositions of the streams are as follows in percent. Evaporator product, P: NaCl: 16.8, KCl: 21.6, H_2O: 61.6; Recycle, R: NaCl: 18.9. Calculate the kilograms per hour and complete compositions of every stream.

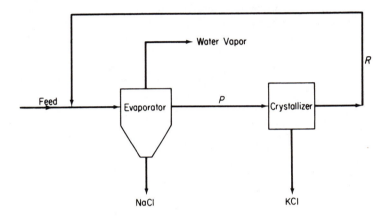

Fig. P2.68

2.69. A 50% NaCl solution is to be concentrated in a triple-effect evaporator as shown in Fig. P2.69. (Each individual evaporator is termed an *effect.*) An equal amount of water is evaporated in each effect. Determine the composition of the outlet stream from effect 2 if the internal contents of effect 2 are uniformly mixed so that the outlet stream has the same composition as the internal contents of effect 2. The steam lines in each effect are completely separate from the evaporator contents so that no mixing of the steam with the contents occurs.

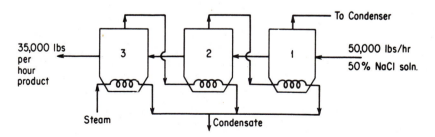

Fig. P2.69

2.70.* A natural gasoline plant at Short Junction, Oklahoma, produces gasoline by removing condensable vapors from the gas flowing out of gas wells. The gas from the well has the following composition:

Component	Mole %
CH_4	77.3
C_2H_6	14.9
C_3H_8	3.6
iso- and $n\text{-}C_4H_{10}$	1.6
C_5 and heavier	0.5
N_2	2.1
Total	100.0

This gas passes through an absorption column called a scrubber (see Fig. P2.70), where it is scrubbed with a heavy, nonvolatile oil. The gas leaving the scrubber has the following analysis:

Component	Mole %
CH_4	92.0
C_2H_6	5.5
N_2	2.5

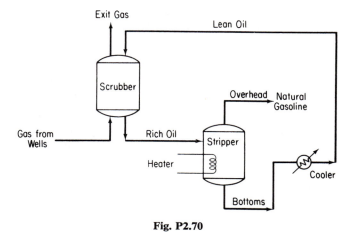

Fig. P2.70

The scrubbing oil absorbs none of the CH_4 or N_2, much of the ethane, and all of the propane and higher hydrocarbons in the gas stream. The oil stream is then sent to a stripping column, which separates the oil from the absorbed hydrocarbons. The overhead from the stripper is termed *natural gasoline*, while the bottoms from the stripper is called *lean oil*. The lean-oil stream is cooled and returned to the absorption column. Assume that there are no leaks in the system. Gas from the wells is fed to the absorption column at the rate of 52,000 lb moles/day and the flow rate of the rich oil going to the stripper is 1230 lb/min (MW = 140). Calculate:

(a) The pounds of CH_4 passing through the absorber per day

(b) The pounds of C_2H_6 absorbed from the gas stream per day

(c) The weight percentage of propane in the rich oil stream leaving the scrubber

2.71. Given that the reaction for fresh feed A and B is

$$2A + 5B \longrightarrow 3C + 6D$$

in a reactor with recycle, find the desired recycle ratio (moles recycle/moles feed) if in the fresh feed A is 20% in excess, the once-through conversion of B is 60%, and the overall conversion of B to products for the overall process is 90%. After the products exit from the reactor, a stream of pure B is separated from the products and forms the recycle.

2.72. In the operation of a synthetic ammonia plant, an excess of hydrogen is burned with air so that the burner gas contains nitrogen and hydrogen in a 1:3 mole ratio and no oxygen. Argon is also present in the burner gas since it accounts for 0.94% of air. The burner gas is fed to a converter where a 25% conversion of the N_2-H_2 mixture to ammonia is produced. The ammonia formed is separated by condensation and the unconverted gases are recycled to the converter. To prevent accumulation of argon in the system, some of the unconverted gases are vented before being recycled to the converter. The upper limit of argon in the converter is to be 4.5% of the entering gases. What percentage of the original hydrogen in converted into ammonia?

2.73. Methanol (CH_3OH) can be converted into formaldehyde (HCHO) either by oxidation to form formaldehyde and water or by direct decomposition to formaldehyde and hydrogen. Suppose that in the fresh feed to the process the ratio of methanol to oxygen is 4 moles to 1, the conversion of methanol per pass in the reactor is 50%, and all of the oxygen reacts in the reactor—none leaves with the hydrogen. After the reaction a separation process removes first all the formaldehyde and water, and then all the hydrogen is removed from the recycled methanol. Determine the flow rates of each species and the total flow rates at each point in the process.

2.74. In an attempt to provide a means of generating NO cheaply, gaseous NH_3 is burned with 20% excess O_2:

$$4NH_3 + 5O_2 \longrightarrow 4NO + 6H_2O$$

The reaction is 70% complete. The NO is separated from the unreacted NH_3, and the latter recycled as shown Fig. P2.74. Compute:

(a) The moles of NO formed per 100 moles of NH_3 fed
(b) The moles of NH_3 recycled per mole of NO formed

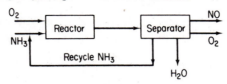

Fig. P2.74

2.75. In an ammonia converter as shown in Fig. P2.75, the fresh feed is 75.16% H_2, 24.57% N_2, and 0.27% Ar. The fresh feed is mixed with the recycle gas and enters the reactor; the gas entering the reactor is 79.52% H_2, while gas leaving the ammonia separator contain 80.01% H_2 and no ammonia. The product ammonia contains no dissolved gases.
  (a) Calculate, per 100 kg mol of fresh feed, how many kilogram moles are recycled and how many purged?
  (b) What is the percent conversion of hydrogen per pass through the reactor?

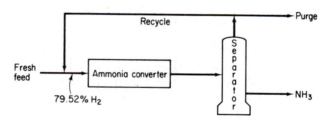

Fig. P2.75

2.76.* An isomerizer is a catalytic reactor that simply tends to rearrange isomers. The number of moles entering an isomerizer is equal to the number of moles leaving. A process, as shown in Fig. P2.76, has been designed to produce a *p*-xylene-rich product

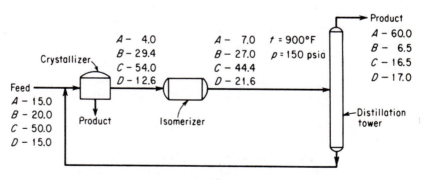

Fig. P2.76

from an aromatic feed charge. All compositions on the flow sheet are in mole percent. The components are indicated as follows:

$$A \quad \text{ethyl benzene}$$

$$B \quad o\text{-xylene}$$

$$C \quad m\text{-xylene}$$

$$D \quad p\text{-xylene}$$

Eighty percent of the ethyl benzene entering the distillation tower is removed in the top stream from the tower. The ratio of the moles fresh feed to the process as a whole to the moles of product from the crystallizer is 1.63. Find:

(a) The reflux ratio (ratio of moles of stream from the bottom of the distillation tower per mole of feed to the tower)

(b) The composition (in mole percent) of the product from the crystallizer

(c) The moles leaving the isomerizer per mole of feed

2.77. In the process sketched in Fig. P2.77, Na_2CO_3 is produced by the reaction

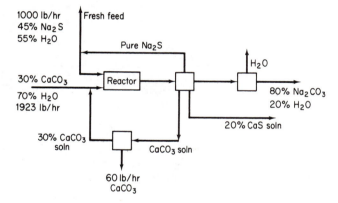

Fig. P2.77

$$Na_2S + CaCO_3 \longrightarrow Na_2CO_3 + CaS$$

The reaction is 90% complete on one pass through the reactor and the amount of $CaCO_3$ entering the reactor is 50% in excess of that needed. Calculate:

(a) The pounds of Na_2S recycled

(b) The pounds of Na_2CO_3 solution formed, per hour

2.78. Two thousand kilograms per hour of solid industrial waste containing a toxic organic compound is extracted with 20,000 kg/hr of pure solvent in a 3 unit extraction system. See Fig. P2.78. The solid waste contains 10% by weight of extractable toxic compound. On leaving units I and II, the waste retains 1.5 kg of solvent per kilogram of waste (on a toxic compound-free basis). What fraction of the toxic compound is recovered per hour? Each unit is well mixed so that the outlet and inside concentrations are identical. The solvent from I contains 0.5% by mass of toxic compound.

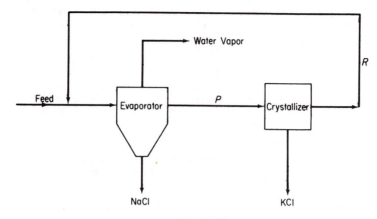

Fig. P2.78

2.79.* The process shown in Fig. P2.79 is the catalytic dehydrogenation of propane to propylene. The composition and flow rate of the recycle stream are unknown. Certain data are known for the reactor, absorber, and distillation column as follows:

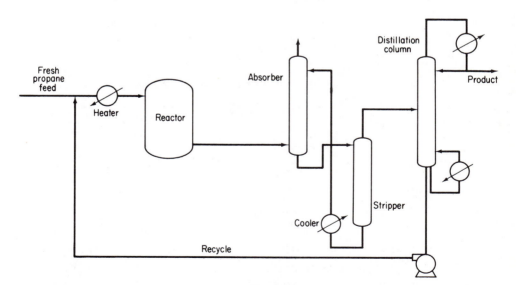

Fig. P2.79

**REACTOR EFFLUENT COMPOSITION BASED ON
100 KG MOL OF PROPANE FED TO REACTOR**

Component	Mole % in Product from Reactor
Hydrogen	25.4
Methane	3.2
Ethylene	0.3
Ethane	5.3
Propylene	21.3
Propane	44.5
Total	100.0

**ABSORBER PERFORMANCE BASED ON
100 KG MOL OF ABSORBER FEED**

Component	kg mol of Absorber Feed	Fraction of Absorber Feed Not Absorbed
Hydrogen	25.4	1.00
Methane	3.2	0.95
Ethylene	0.3	0.70
Ethane	5.3	0.60
Propylene	21.3	0.01
Propane	44.5	0.003
Total	100.0	

FRACTIONATOR PERFORMANCE

Component	Fraction of Feed Recovered in the Overhead Product
Methane	1.00
Ethane	1.00
Ethylene	1.00
Propylene	0.96
Propane	0.002

Assume that the source of all the hydrogen in the product from the reactor is the propane feed, and that any loss of mass in the reactor is due to carbon formation. Also assume that none of the hydrogen in the reactor product is absorbed in the absorber.

Calculate the flow rates in the reactor feed (after the heater), the reactor product stream, the absorber bottoms stream, the distillation column bottoms, and the recycle stream in kilograms per 100 kg of propane fed to the process. Can you calculate the composition of the recycle stream as well?

PROBLEM TO PROGRAM ON THE COMPUTER

2.1. A 15-stage extraction column similar to the one in Example 2.15 is set up to separate acetone and ethanol using two solvents, pure water and pure chloroform. The only added features to Fig. E2.15 in Example 2.15 are (1) that you must consider two solutes in each solvent instead of one solute in the two solvents and (2) the acetone–ethanol feed is introduced into stage 6. You know that the feed composition is 50 mole % ethanol and 50 mole % acetone and that the feed rate is 20 lb moles/hr. Assume that the water and chloroform are not soluble in each other. You want to have 1.50 mole % ethanol in the exit stream from the bottom of the column, and 2×10^{-5} mole % acetone in the exit stream from the top of the column. Determine the rates of flow of the two solvents. In each stage you are given a relationship between X_i^A and X_i^B ($i = 1, 2, 3$) in the two phases in terms of *mole fractions*:

$$\frac{x_i^B}{x_i^A} = \frac{\gamma_i^A}{\gamma_i^B} \tag{a}$$

Each of the γ's in phase A or phase B can be expressed in terms of some known constants and the mole fraction of the other two components. For component i,

$$\ln \gamma_i = 2x_i \sum_{j=1}^{3} x_j a_{ji} + \sum_{j=1}^{3} x_j^2 a_{ij} + \sum_{j=1}^{3} \sum_{\substack{k=1 \\ j \neq i \\ k \neq i \\ j < k}}^{3} x_j x_k a_{ijk}^*$$

$$- 2 \sum_{i=1}^{3} x_i^2 \sum_{j=1}^{3} x_j a_{ij} \tag{b}$$

$$- 2 \sum_{i=1}^{3} \sum_{j=1}^{3} \sum_{k=1}^{3} x_i x_j x_k a_{ijk}^*$$

where $a_{ijk}^* = a_{ij} + a_{ji} + a_{ik} + a_{ki} + a_{jk} + a_{kj}$. The values of the coefficients are

1: acetone		3: chloroform	
2: ethanol		4: water	
a_{11} 0.0	$a_{12} = 0.5446$	$a_{13} = 0.9417$	$a_{14} = 1.872$
$a_{21} = 0.599$	$a_{22} = 0.0$	$a_{23} = 1.61$	$a_{24} = 1.46$
$a_{31} = 0.674$	$a_{32} = 0.501$	$a_{33} = 0.0$	$a_{34} = 5.91$
$a_{41} = 1.338$	$a_{42} = 0.877$	$a_{43} = 4.76$	$a_{44} = 0.0$

The γ's are obtained for each component, 1, 2, and 3, by permuting the subscripts cyclically $i \longrightarrow j \longrightarrow k \longrightarrow i$. (*Hints:* Select some appropriate initial values for the flow rates of A and B. Divide the feed (stage 6) equally between both phases. The initial compositions can be provided by specifying both stage 1 and stage 15 and generating the remaining compositions by linear interpolation.) Be sure to include (1) one equation such as Eq. (a) for each component for each stage, (2) the correct number of component material balances for each stage (a total balance for the stage can be substituted for one component balance), and (3) $\sum x_i = 1$ for each stage for one phase (similar equations for the other phase are redundant).

INDUSTRIAL CHEMICAL DATA

HOW MADE	MAJOR END USES	MAJOR DERIVATIVES	ANNUAL PRODUCTION, PRICES (USA 1981)
	Ethylene	$CH_2=CH_2$	
Thermal (steam) or catalytic cracking of hydrocarbons ranging from natural gas-derived ethane to oil-derived gas oil (fuel oil)	Fabricated plastics 65%, antifreeze 10%, fibers 5%, solvents 5%	Polyethylenes 45%, ethylene oxide 20%, vinyl chloride 15%, styrene 10%	33×10^9 lb $.23/lb
	Caustic soda	NaOH	
Electrolysis of salt brine	Chemical manufacture 50%, pulp and paper 15%, alumina 5%, petroleum refining 5%, soap and detergents 5%		12.2×10^6 tons $270–$360 per ton 75% concentration
	Chlorine	Cl_2	
Electrolysis of salt brine; recovery from hydrochloric acid or from coproduction in making metals, caustic potash, or potassium nitrate	Chemical manufacture 50%, plastics (mostly PVC) 15%, solvents 15%, pulp and paper 15%	Ethylene dichloride 20%, other chlorinated hydrocarbons 15%	11.8×10^6 tons <$145/ton

Adapted from Chemical and Engineering News by permission of the American Chemical Society.

3

GASES, VAPORS, LIQUIDS, AND SOLIDS

In planning and decision making for our modern technology, engineers and scientists must know with reasonable accuracy the properties of the fluids and solids with which they deal. If you are engaged in the design of equipment, say the volume required for a process vessel, you need to know the specific volume or density of the gas or liquid that will be in the vessel as a function of temperature and pressure. If you are interested in predicting the possibility or extent of rainfall, you have to know something about the relation between the vapor pressure of water and the temperature. Whatever your current or future job, you need to have an awareness of the character and sources of information concerning the physical properties of fluids and solids.

Clearly, it is not possible to have reliable detailed experimental data at hand on all the useful pure compounds and mixtures that exist in the world. Consequently, in the absence of experimental information, we estimate (predict) properties based on modifications of well-established principles, such as the *ideal* gas law, or based on empirical correlations. Thus the foundation of the estimation methods ranges from quite theoretical to completely empirical and their reliability ranges from excellent to terrible.

Rather than define the states of matter, a task that is not easy to accomplish with precision in the brief space we have here, let us instead characterize the states in terms of two quantities, motion and structure of the molecules:

State	Motion of molecules	Structure of collections of molecules
Perfect gas	Extensive trajectories	None
Real gas	Extensive trajectories	Almost none
Liquid	Short distance	Related structure
Liquid crystal	Some	Some crystal structure
Amorphous solid	Little	Little
Real crystal	Almost none	Highly structured
Perfect crystal	None	Completely structured

In this chapter we first discuss ideal and real gas relationships, including some of the gas laws for pure components and mixtures of ideal gases. You will learn about methods of expressing the p–V–T properties of real gases by means of equations of state and, alternatively, by compressibility factors. Next we introduce the concepts of vaporization, condensation, and vapor pressure and illustrate how material balances are made for saturated and partially saturated gases. Finally, we examine the qualitative characteristics of gas–liquid–solid phases with the aid of diagrams. Figure 3.0 shows the interrelationships among the topics discussed in this chapter and how they relate to the making of material and energy balances.

Section 3.1   Ideal Gas Laws

In 1787, Jacques Charles, a French chemist and physicist, published his conclusions about the relationship between the volume of gases and temperature. He demonstrated that the volume of a dry gas varies directly with temperature if the pressure remains constant. Charles, Boyle, Gay-Lussac, Dalton, and Amagat, the investigators who originally developed correlating relations among gas temperature, pressure, and volume, worked at temperatures and pressures such that the average distance between the molecules was great enough to neglect the effect of the intermolecular forces and the volume of the molecules themselves. Under these conditions a gas came to be termed an *ideal* gas. More properly, an *ideal gas* is an imaginary gas which obeys exactly certain simple laws such as the laws of Boyle, Charles, Dalton, and Amagat. No real gas obeys these laws exactly over all ranges of temperature and pressure, although the "lighter" gases (hydrogen, oxygen, air, etc.) under ordinary circumstances obey the ideal gas laws with but negligible deviations. The "heavier" gases, such as sulfur dioxide and hydrocarbons, particularly at high pressures and low temperatures, deviate considerably from the ideal gas laws. Vapors, under conditions near the boiling point, deviate markedly from the

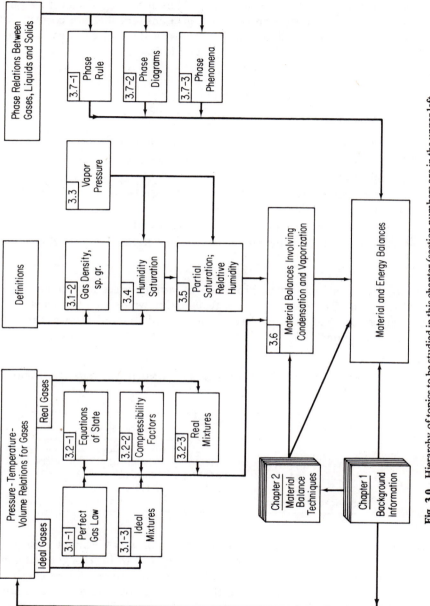

Fig. 3.0 Hierarchy of topics to be studied in this chapter (section numbers are in the upper left-hand corner of the boxes).

ideal gas laws. However, at low pressures and high temperatures, the behavior of a vapor approaches that of an ideal gas. Thus for many engineering purposes, the ideal gas laws, if properly applied, will give answers that are correct within a few percent or less. But for liquids and solids with the molecules compacted relatively close together, we do not have such general laws.

3.1-1 Calculations Using the Ideal Gas Law

Your objectives in studying this section are to be able to:

1. Write down the ideal gas law, and define all its variables and parameters and their associated dimensions.
2. Calculate the values and units of the ideal gas law constant in any set of units from the standard conditions.
3. Convert gas volumes to moles (and mass), and vice versa.
4. Use ratios of variables in the ideal gas law to calculate p, V, T, or n.

In order that the volumetric properties of various gases may be compared, several arbitrarily specified standard states (usually known as *standard conditions,* or S.C.) of temperature and pressure have been selected by custom. See Table 3.1 for the most common ones. The fact that a substance cannot exist as a gas at 0°C and 1 atm

TABLE 3.1 COMMON STANDARD CONDITIONS FOR THE IDEAL GAS

System	T	p	$\hat{V}$
SI	273.15 K	101.325 kPa	22.41 m³/kg mol
Universal scientific	0.0°C	760 mm Hg	22.41 liters/g mol
Natural gas industry	60.0°F	14.696 psia	379.4 ft³/lb mol
	(15.0°C)	(101.325 kPa)	
American engineering	32°F	1 atm	359 ft³/lb mol

is immaterial. Thus, as we see later, water vapor at 0°C cannot exist at a pressure greater than its saturation pressure of 0.61 kPa (0.18 in. Hg) without condensation occurring. However, the imaginary volume at standard conditions can be calculated and is just as useful a quantity in the calculation of volume–mole relationships as though it could exist. In the following, the symbol V will stand for total volume and the symbol $\hat{V}$ for volume per mole, or per unit mass.

EXAMPLE 3.1 Use of Standard Conditions

Calculate the volume, in cubic meters, occupied by 40 kg of CO_2 at standard conditions.

Solution

Basis: 40 kg of CO_2

$$\frac{40 \text{ kg } CO_2}{} \left| \frac{1 \text{ kg mol } CO_2}{44 \text{ kg } CO_2} \right| \frac{22.41 \text{ m}^3 CO_2}{1 \text{ kg mol } CO_2} = 20.4 \text{ m}^3 CO_2 \text{ at S.C.}$$

Notice in this problem how the information that 22.41 m³ at S.C. = 1 kg mol is applied to transform a known number of moles into an equivalent number of cubic meters.

Incidentally, whenever you use cubic measure for volume, you must establish the conditions of temperature and pressure at which the cubic measure for volume exists, since the term "m³" or "ft³," standing alone, is really not any particular *quantity* of material.

Boyle found that the volume of a gas is inversely proportional to the absolute pressure at constant temperature. Charles showed that, at constant pressure, the volume of a given mass of gas varies directly with the absolute temperature. From the work of Boyle and Charles, scientists developed the relationship now called the *perfect gas law* (or sometimes *the ideal gas law*)

$$pV = nRT \qquad (3.1)$$

In applying this equation to a process going from an initial set of conditions to a final set of conditions, you can set up ratios of similar terms which are dimensionless as follows:

$$\left(\frac{p_1}{p_2}\right)\left(\frac{V_1}{V_2}\right) = \left(\frac{n_1}{n_2}\right)\left(\frac{T_1}{T_2}\right) \qquad (3.2)$$

Here the subscripts 1 and 2 refer to the initial and final conditions. This arrangement of the perfect gas law has the convenient feature that the pressures may be expressed in any system of units you choose, such as kPa, in. Hg, mm Hg, atm, and so on, as long as the same units are used for both conditions of pressure (do not forget that the pressure must be *absolute* pressure in both cases). Similarly, the grouping together of the *absolute* temperature and the volume terms gives ratios that are dimensionless. Notice how the ideal gas constant R is eliminated in taking the ratio of the initial to the final state.

Let us see how we can apply the perfect gas law both in the form of Eq. (3.2) and Eq. (3.1) to problems.

EXAMPLE 3.2 Perfect Gas Law

An oxygen cylinder used as a standby source of oxygen contains 1.000 ft³ of O_2 at 70°F and 200 psig. What will be the volume of this O_2 in a dry-gas holder at 90°F and 4.00 in. H_2O above atmospheric? The barometer reads 29.92 in. Hg. See Fig. E3.2.

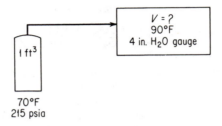

Fig. E3.2

Solution

You must first convert the temperatures and pressures into absolute units:

$$460 + 70 = 530°R$$
$$460 + 90 = 550°R$$

atmospheric pressure = 29.92 in. Hg = std atm = 14.7 psia

$$\text{initial pressure} = \frac{200 \text{ psig} + 14.7 \text{ psia}}{} \left| \frac{29.92 \text{ in. Hg}}{14.7 \text{ psia}} \right. = 437 \text{ in. Hg}$$

$$\text{final pressure} = 29.92 \text{ in. Hg} + \frac{4 \text{ in. } H_2O}{\frac{12 \text{ in. } H_2O}{\text{ft } H_2O}} \left| \frac{29.92 \text{ in. Hg}}{33.91 \text{ ft } H_2O} \right.$$

$$= 29.92 + 0.29 = 30.21 \text{ in. Hg}$$

The simplest way to proceed, now that the data are in good order, is to apply the laws of Charles and Boyle, and, in effect, to apply the perfect gas law.

From Charles' law, since the temperature *increases*, the volume *increases;* hence the ratio of the temperatures *must* be greater than 1. The pressure *decreases;* therefore, from Boyle's law, the volume will *increase;* hence the ratio of pressures will be *greater* than 1.

Basis: 1 ft³ of oxygen at 70°F and 200 psig

$$\text{final volume} = \frac{1.00 \text{ ft}^3}{} \left| \frac{550°R}{530°R} \right| \frac{437 \text{ in. Hg}}{30.21 \text{ in. Hg}}$$

$$= 15.0 \text{ ft}^3 \text{ at } 90°F \text{ and } 4 \text{ in. } H_2O \text{ gauge}$$

Formally, the same calculation can be made using Eq. (3.2),

$$V_2 = V_1 \left(\frac{p_1}{p_2} \right) \left(\frac{T_2}{T_1} \right) \qquad \text{since } n_1 = n_2$$

EXAMPLE 3.3 Perfect Gas Law

Probably the most important constituent of the atmosphere that fluctuates is the water. Rainfall, evaporation, fog, and even lightning are associated with water as a vapor or a liquid in air. To obtain some feeling for how little water vapor there is at higher altitudes, calculate the mass of 1.00 m³ of water vapor at 2.00 kPa and 23°C. Assume that water vapor is an ideal gas under these conditions.

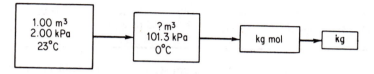

Fig. E3.3

Solution

First visualize the information available to you, and then decide how to convert in into the desired mass. If you can convert the original amount of water vapor to S.C. by use of the ideal gas law and then make use of the fact that 22.4 m³ = 1 kg mol, you can easily get the desired mass of water vapor. See Fig. E3.3.

Basis: 1.00 m³ H₂O vapor at 2.00 kPa and 23°C

$$\frac{1.00 \text{ m}^3}{} \left| \frac{2.00 \text{ kPa}}{101.3 \text{ kPa}} \right| \frac{(273 + 0) \text{ K}}{(273 + 23) \text{ K}} \left| \frac{1 \text{ kg mol}}{22.4 \text{ m}^3} \right| \frac{18 \text{ kg H}_2\text{O}}{1 \text{ kg mol}} = 1.46 \times 10^{-2} \text{ kg H}_2\text{O}$$

Notice how the entire calculation can be carried out in a single-dimensional equation.

EXAMPLE 3.4 Perfect Gas Law

You have 10 lb of CO_2 in a 20-ft³ fire extinguisher tank at 30°C. Assuming that the ideal gas law holds, what will the pressure gauge on the tank read in a test to see if the extinguisher is full? See Fig. E3.4.

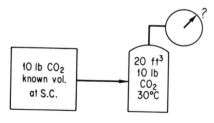

Fig. E3.4

Solution

Employing Eq. (3.2), we'can write (the subscript 1 stands for standard conditions, 2 for the conditions in the tank)

$$p_2 = p_1 \left(\frac{V_1}{V_2}\right)\left(\frac{T_2}{T_1}\right)$$

$$\underbrace{\frac{p_1}{14.7 \text{ psia}} \left| \frac{10 \text{ lb } CO_2}{} \right| \frac{1 \text{ lb mol } CO_2}{44 \text{ lb } CO_2} \left| \frac{}{1 \text{ lb mol}} \right.}_{V_1} \left| \underbrace{20 \text{ ft}^3}_{V_2} \right| \underbrace{\frac{303 \text{ K}}{273 \text{ K}}}_{\frac{T_2}{T_1}} = p_2 = 66 \text{ psia}$$

Hence the gauge on the tank will read (assuming that it reads gauge pressure and that the barometer reads 14.7 psia) $66 - 14.7 = 51.3$ psig.

We have not used the gas constant R in the solution of any of the example problems so far, but you can use Eq. (3.1) to solve for one unknown as long as all the other variables in the equation are known. However, such a calculation requires that the units of R be expressed in units corresponding to those used for the quantities p–$\hat{V}$–T. There are so many possible units you can use for each variable that a very large table of R values will be required. What do you do if a table of values of R is not handy? If you want to use R, you can always determine R from the p–$\hat{V}$–T data at standard conditions which you have already memorized and used.

EXAMPLE 3.5 Calculation of R

Find the value for the universal gas constant R for the following combinations of units:
  (a) For 1 lb mol of ideal gas when the pressure is expressed in psia, the volume is in ft^3/lb mol, and the temperature is in °R.
  (b) For 1 g mol of ideal gas when the pressure is in atm, the volume in cm^3, and the temperature in K.
  (c) For 1 kg mol of ideal gas when the pressure is in kPa, the volume is in m^3/kg mol, and the temperature is in K.

Solution

  (a) At standard conditions

$$p = 14.7 \text{ psia}$$
$$\hat{V} = 359 \text{ ft}^3/\text{lb mol}$$
$$T = 492°R$$

Then

$$R = \frac{p\hat{V}}{T} = \frac{14.7 \text{ psia}}{492°R} \left| \frac{359 \text{ ft}^3}{1 \text{ lb mol}} \right. = 10.73 \frac{(\text{psia})(\text{ft}^3)}{(°R)(\text{lb mol})}$$

  (b) Similarly, at standard conditions,

$$p = 1 \text{ atm}$$

$$\hat{V} = 22,400 \text{ cm}^3/\text{g mol}$$

$$T = 273 \text{ K}$$

$$R = \frac{p\hat{V}}{T} = \frac{1 \text{ atm}}{273 \text{ K}} \left| \frac{22,400 \text{ cm}^3}{1 \text{ g mol}} \right. = 82.06 \frac{(\text{cm}^3)(\text{atm})}{(\text{K})(\text{g mol})}$$

(c) In the SI system of units standard conditions are

$$p = 1.013 \times 10^5 \text{ Pa} \quad (\text{or N/m}^2)$$

$$\hat{V} = 22.4 \text{ m}^3/\text{kg mol}$$

$$T = 273 \text{ K}$$

$$R = \frac{p\hat{V}}{T} = \frac{1.013 \times 10^5 \text{ Pa}}{273 \text{ K}} \left| \frac{22.4 \text{ m}^3}{1 \text{ kg mol}} \right. = 8.31 \times 10^3 \frac{(\text{kPa})(\text{m}^3)}{(\text{K})(\text{kg mol})} = 8.31 \frac{\text{kJ}}{(\text{K})(\text{kg mol})}$$

To summarize, we want to emphasize that R does not have a universal value even though it is sometimes called the *universal gas constant*. The value of R depends on the units of p, $\hat{V}$, and T.

EXAMPLE 3.6 Application of the Perfect Gas Law

Calculate the volume occupied by 88 lb of CO_2 at a pressure of 32.2 ft of water and at 15°C.

Solution

See Fig. E3.6.

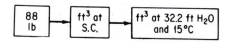

Fig. E3.6

Solution No. 1 (using Boyle's and Charles' laws):

At S.C.:

$$p = 33.91 \text{ ft H}_2\text{O}$$

$$\hat{V} = 359 \frac{\text{ft}^3}{\text{lb mol}}$$

$$T = 273 \text{ K}$$

Basis: 88 lb of CO_2

$$\frac{88 \text{ lb CO}_2}{} \left| \frac{}{\frac{44 \text{ lb CO}_2}{1 \text{ lb mol CO}_2}} \right| \frac{359 \text{ ft}^3}{1 \text{ lb mol}} \left| \frac{288}{273} \right| \frac{33.91}{32.2} = 798 \text{ ft}^3 \text{ CO}_2 \text{ at } 32.2 \text{ ft H}_2\text{O and } 15°\text{C}$$

Solution No. 2 (using the gas constant *R*):

First, the value of *R* must be obtained in the same units as the variables p, $\hat{V}$, and T. For 1 lb mol,

$$R = \frac{p\hat{V}}{T}$$

and at S.C.,

$$p = 33.91 \text{ ft } H_2O$$

$$\hat{V} = 359 \text{ ft}^3/\text{lb mol}$$

$$T = 273 \text{ K}$$

$$R = \frac{33.91 \mid 359}{\mid 273} = 44.6 \frac{(\text{ft } H_2O)(\text{ft}^3)}{(\text{lb mol})(K)}$$

Now, using Eq. (3.1), insert the given values, and perform the necessary calculations.

Basis: 88 lb of CO_2

$$V = \frac{nRT}{p} = \frac{88 \text{ lb } CO_2 \mid (44.6 \text{ ft } H_2O)(\text{ft}^3) \mid 288 \text{ K}}{44 \text{ lb } CO_2 \mid (\text{lb mol})(K) \mid 32.2 \text{ ft } H_2O}$$
$$\frac{}{\text{lb mol } CO_2 \mid}$$

$$= 798 \text{ ft}^3 \ CO_2 \text{ at } 32.2 \text{ ft } H_2O \text{ and } 15°C$$

If you will inspect both solutions closely, you will observe that in both cases the same numbers appear and that both are identical except that in the second solution using *R* two steps are used to obtain the solution.

EXAMPLE 3.7 Application of the Perfect Gas Law

One important source of emissions from gasoline-powered automobile engines that causes smog is the nitrogen oxides NO and NO_2. They are formed whether combustion is complete or not as follows. At the high temperatures that occur in an internal combustion engine during the burning process, oxygen and nitrogen combine to form nitric oxide (NO). The higher the peak temperatures and the more oxygen available, the more NO is formed. There is insufficient time for the NO to decompose back to O_2 and N_2 because the burned gases cool too rapidly during the expansion and exhaust cycles in the engine. Although both NO and nitrogen dioxide (NO_2) are significant air pollutants (together termed NO_x), the NO_2 is formed in the atmosphere as NO is oxidized.

Suppose that you collect a sample of a NO-NO_2 mixture (after having removed the other combustion gas products by various separations procedures) in a 100-cm³ standard cell at 30°C. Certainly some of the NO will have been oxidized to NO_2,

$$2NO + O_2 \longrightarrow 2NO_2$$

during the collection, storage, and processing of the combustion gases, so that measurement of NO alone will be misleading. If the standard cell contains 0.291 g of NO_2 plus NO and the pressure measured in the cell is 170 kPa, what percent of the NO + NO$_2$ is in the form of NO? See Fig. E3.7.

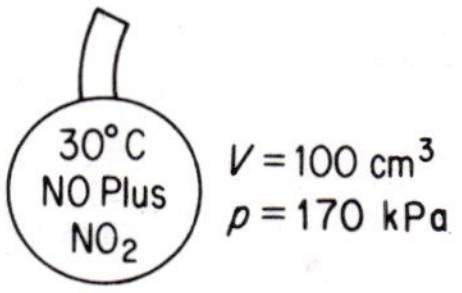

Fig. E3.7

Solution

The gas in the cell is composed partly of NO and partly of NO$_2$. We can use the ideal gas law to calculate the total gram moles present in the cell and then, by using the chemical equation and the principles of stoichiometry, compute the composition in the cell.

Basis: 100 cm³ of gas at 170 kPa and 30°C

$$R = \frac{101.3 \text{ kPa}}{273 \text{ K}} \left| \frac{22.4 \text{ L}}{1 \text{ g mol}} \right| \frac{1000 \text{ cm}^3}{1 \text{ L}} = 8.31 \times 10^3 \frac{(\text{kPa})(\text{cm}^3)}{(\text{K})(\text{g mol})}$$

$$n = \frac{pV}{RT} = \frac{170 \text{ kPa}}{8.31 \times 10^3 \dfrac{(\text{kPa})(\text{cm}^3)}{(\text{K})(\text{g mol})}} \left| \frac{100 \text{ cm}^3}{303 \text{ K}} \right. = 0.00675 \text{ g mol}$$

If the mixture is composed of NO (MW = 30) and NO$_2$ (MW = 46), because we know the total mass in the cell we can compute the fraction of, say, NO. Let x = grams of NO; then $(0.291 - x)$ = g NO$_2$.

Basis: 0.291 g of total gas

The sum of the moles is

$$\frac{x \text{ g NO}}{} \left| \frac{1 \text{ g mol NO}}{30 \text{ g NO}} + \frac{(0.291 - x) \text{ g NO}_2}{} \right| \frac{1 \text{ g mol NO}_2}{46 \text{ g NO}_2} = 0.00675$$

$$0.0333x + (0.291 - x)(0.0217) = 0.00675$$

$$x = 0.036 \text{ g}$$

The weight percent NO is

$$\frac{0.036}{0.291}(100) \simeq 12\%$$

and the mole percent NO is

$$\frac{0.036 \text{ g NO}}{0.00675 \text{ g mol total}} \left| \frac{1 \text{ g mol NO}}{30 \text{ g NO}} (100) \simeq 18\% \right.$$

Self-Assessment Test

1. Write down the ideal gas law.

2. What are the dimensions of $T, p, V, n,$ and R?

3. List the standard conditions for a gas in the SI, universal scientific, and American engineering systems of units.

4. Calculate the volume in ft³ of 10 lb mol of an ideal gas at 68°F and 30 psia.

5. A steel cylinder of volume 2 m³ contains methane gas (CH_4) at 50°C and 250 kPa absolute. How many kilograms of methane are in the cylinder?

6. What is the value of the ideal gas constant R to use if the pressure is to be expressed in atm, the temperature in kelvin, the volume in cubic feet, and the quantity of material in pound moles?

3.1-2 Gas Density and Specific Gravity

> **Your objectives in studying this section are to be able to:**
> 1. Define gas density and specific gravity.
> 2. Calculate the specific gravity of a gas even if the reference condition is not clearly specified.
> 3. Calculate the density of a gas given its specific gravity.

The density of a gas is defined as the mass per unit volume and can be expressed in kilograms per cubic meter, pounds per cubic foot, grams per liter, or other units. Inasmuch as the mass contained in a unit volume varies with the temperature and pressure, as we have previously mentioned, you should always be careful to specify these two conditions. If not otherwise specified, the densities are presumed to be at S.C. Density can be calculated by selecting a unit volume as the basis and calculating the mass of the contained gas.

EXAMPLE 3.8 Calculation of Gas Density

What is the density of N_2 at 27°C and 100 kPa expressed in:
    (a) SI units?
    (b) American engineering units?

Solution

    (a)

Basis: 1 m³ of N_2 at 27°C and 100 kPa

$$\frac{1 \text{ m}^3}{} \left| \frac{273 \text{ K}}{300 \text{ K}} \right| \frac{100 \text{ kPa}}{101.3 \text{ kPa}} \left| \frac{1 \text{ kg mol}}{22.4 \text{ m}^3} \right| \frac{28 \text{ kg}}{1 \text{ kg mol}} = 1.123 \text{ kg}$$

density = 1.123 kg/m³ of N_2 at 27°C (300 K) and 100 kPa

(b)

Basis: 1 ft³ of N_2 at 27°C and 100 kPa

$$\frac{1 \text{ ft}^3 \mid 273 \text{ K} \mid 100 \text{ kPa} \mid 1 \text{ lb mol} \mid 28 \text{ lb}}{\mid 300 \text{ K} \mid 101.3 \text{ kPa} \mid 359 \text{ ft}^3 \mid 1 \text{ lb mol}} = 0.0701 \text{ lb}$$

density = 0.0701 lb/ft³ of N_2 at 27°C (80°F) and 100 kPa (14.5 psia)

The *specific gravity* of a gas is usually defined as the ratio of the density of the gas at a desired temperature and pressure to that of air (or any specified reference gas) at a certain temperature and pressure. The use of specific gravity occasionally may be confusing because of the manner in which the values of specific gravity are reported in the literature. You must be very careful in using literature values of specific gravity. Be particularly careful to ascertain that the conditions of temperature and pressure are known both for the gas in question and for the reference gas. Among the examples below, several represent inadequate methods of expressing specific gravity.

(a) *What is the specific gravity of methane?* Actually, this question may have the same answer as the question: How many grapes are in a bunch? Unfortunately, occasionally one may see this question and the best possible answer is

$$\text{sp gr} = \frac{\text{density of methane at S.C.}}{\text{density of air at S.C.}}$$

(b) *What is the specific gravity of methane* ($H_2 = 1.00$)? Again a poor question. The notation of ($H_2 = 1.00$) means that H_2 at S.C. is used as the reference gas, but the question does not even give a hint regarding the conditions of temperature and pressure of the methane. Therefore, the best interpretation is

$$\text{sp gr} = \frac{\text{density of methane at S.C.}}{\text{density of } H_2 \text{ at S.C.}}$$

(c) *What is the specific gravity of ethane* (air = 1.00)? Same as question (b) except that in the petroleum industry the following is used:

$$\text{sp gr} = \frac{\text{density of ethane at 60°F and 760 mm Hg}}{\text{density of air at S.C. (60°F, 760 mm Hg)}}$$

(d) *What is the specific gravity of butane at 50°C and* 90 kPa? No reference gas nor state of reference gas is mentioned. However, when no reference gas is mentioned, it is taken for granted that air is the reference gas. In the case at hand the best thing to do is to assume that the reference gas and the desired gas are under the same conditions of temperature and pressure:

$$\text{sp gr} = \frac{\text{density of butane at 50°C and 90 kPa}}{\text{density of air at 50°C and 90 kPa}}$$

(e) *What is the specific gravity of* CO_2 *at 60°F and 740 mm Hg (air* = 1.00)*?*

$$\text{sp gr} = \frac{\text{density of } CO_2 \text{ at 60°F and 740 mm Hg}}{\text{density of air at S.C.}}$$

(f) *What is the specific gravity of* CO_2 *at 60°F and 740 mm Hg (ref. air at S.C.)?*

$$\text{sp gr} = \text{same as question (e)}$$

EXAMPLE 3.9 Specific Gravity of a Gas

What is the specific gravity of N_2 at 80°F and 745 mm Hg compared to
  (a) Air at S.C. (32°F and 760 mm Hg)?
  (b) Air at 80°F and 745 mm Hg?

Solution

First you must obtain the density of the N_2 and the air at their respective conditions of temperature and pressure, and then calculate the specific gravity by taking a ratio of their densities. Example 3.8 covers the calculation of the density of a gas, and therefore, to save space, no units will appear in the following calculations:
  (a)

Basis: 1 ft³ of N_2 at 80°F and 745 mm Hg

$$\frac{1}{} \left| \frac{492}{540} \right| \frac{745}{760} \left| \right| \frac{28}{359} \left| \right. = 0.0697 \text{ lb } N_2/\text{ft}^3 \text{ at 80°F, 745 mm Hg}$$

Basis: 1 ft³ of air at 32°F and 760 mm Hg

$$\frac{1}{} \left| \frac{492}{492} \right| \frac{760}{760} \left| \right| \frac{29}{359} \left| \right. = 0.0808 \text{ lb air/ft}^3 \text{ at 32°F, 760 mm Hg}$$

Therefore,

$$\text{sp. gr.} = \frac{0.0697}{0.0808} = 0.862 \frac{\text{lb } N_2/\text{ft}^3 \text{ } N_2 \text{ at 80°F, 745 mm Hg}}{\text{lb air/ft}^3 \text{ air at S.C.}}$$

Note: Specific gravity is not a dimensionless number.
  (b)

Basis: 1 ft³ of air at 80°F and 745 mm Hg

$$\frac{1}{} \left| \frac{492}{540} \right| \frac{745}{760} \left| \right| \frac{29}{359} \left| \right. = 0.0721 \text{ lb/ft}^3 \text{ at 80°F and 745 mm Hg}$$

$$(\text{sp. gr.})_{N_2} = \frac{0.0697}{0.721} = 0.967 \frac{\text{lb } N_2/\text{ft}^3 \text{ } N_2 \text{ at 80°F, 745 mm Hg}}{\text{lb air/ft}^3 \text{ air at 80°F, 745 mm Hg}}$$

$$= 0.967 \text{ lb } N_2/\text{lb air}$$

Note: You can work part (b) by dividing the unit equations instead of dividing the resulting densities:

$$
\text{sp gr} = \frac{d_{N_2}}{d_{air}} = \frac{\begin{array}{c|c|c|c|c} 1 & 492 & 745 & & 28 \\ \hline & 540 & 760 & 359 & \end{array}}{\begin{array}{c|c|c|c|c} 1 & 492 & 745 & & 29 \\ \hline & 540 & 760 & 359 & \end{array}} = \frac{28}{29} = 0.966 \text{ lb } N_2/\text{lb air}
$$

The latter calculation shows that the specific gravity is equal to the *ratio of the molecular weights* of the gases when the densities of *both* the desired gas and the reference gas are at the same temperature and pressure. This, of course, is true only for ideal gases and should be no surprise to you, since Avogadro's law in effect states that at the same temperature and pressure 1 mole of any ideal gas is contained in identical volumes.

Self-Assessment Test

1. What is the density of a gas that has a molecular weight of 0.123 kg/kg mol at 300 K and 1000 kPa?

2. What is the specific gravity of CH_4 at 70°F and 2 atm compared to air at S.C.?

3.1-3 Ideal Gas Mixtures

> *Your objectives in studying this section are to be able to:*
>
> 1. Write down and apply Dalton's law and Amagat's law.
> 2. Define and use partial pressure in gas calculations.
> 3. Show under certain assumptions that volume fraction equals the mole fraction in a gas.

In the majority of cases, as an engineer you will deal with mixtures of gases instead of individual gases. There are three ideal gas laws which can be applied successfully to gaseous mixtures:

(a) Dalton's law of partial pressures
(b) Amagat's law of partial volumes
(c) Dalton's law of the summation of partial pressures

Dalton's laws. Dalton postulated that the total pressure of a gas is equal to the sum of the pressures exerted by the individual molecules of each component gas. He went one step further to state that each individual gas of a gaseous mixture can hypothetically be considered to exert a *partial pressure*. A partial pressure is the pressure that would be obtained if this *same mass* of individual gas were alone in the *same total volume* at the *same temperature*. The sum of these partial pressures for each component in the gaseous mixture would be equal to the total pressure, or

$$p_1 + p_2 + p_3 + \cdots + p_n = p_t \tag{3.3}$$

Equation (3.3) is Dalton's law of the summation of the partial pressures.

To illustrate the significance of Eq. (3.3) and the meaning of partial pressure, suppose that you carried out the following experiment with ideal gases. Two tanks of 1.50 m^3 volume, one containing gas A at 300 mm Hg and the other gas B at 400 mm of Hg (both gases being at the same temperature of 20°C), are connected together. All the gas in B is forced into tank A isothermally. Now you have a 1.50-m^3 tank of $A + B$ at 700 mm of Hg. For this mixture (in the 1.50-m^3 tank at 20°C and a total pressure of 700 mm Hg) you could say that gas A exerts a partial pressure of 300 mm and gas B exerts a partial pressure of 400 mm. Of course you cannot put a pressure gauge on the tank and check this conclusion because the pressure gauge will read only the total pressure. These partial pressures are hypothetical pressures that the individual gases would exert and are equivalent to the pressures they actually would have if they were put into the same volume at the same temperature all by themselves.

You can surmise that, at constant volume and at constant temperature, the pressure is a function only of the number of molecules of gas present. If you divide the perfect gas law for component 1, $p_1 V_1 = n_1 R T_1$, by that for component 2, $p_2 V_2 = n_2 R T_2$, for the *same temperature and volume*, you can obtain

$$\frac{p_1}{p_2} = \frac{n_1}{n_2} \tag{3.4}$$

which shows that the ratio of the partial pressures is exactly the same numerically as the ratio of the moles of components 1 and 2. Similarly, dividing the ideal gas law for component 1 by the gas law for all the molecules, $p_t V_t = n_t R T_t$, you will get Dalton's law of partial pressures:

$$\frac{p_1}{p_t} = \frac{n_1}{n_t} = \text{mole fraction} = y_1 \tag{3.5}$$

Equation (3.5) shows that the ratio of the partial pressure of an individual component to the total pressure is exactly the same numerically as the ratio of the moles of the individual component to the total moles. With this principle under your belt, if the mole fraction of an individual gaseous component in a gaseous mixture is known and the total pressure is known, you are able to calculate the partial pressure of this component of the gas by generalizing Eq. (3.5):

$$p_i = y_i p_t \tag{3.5a}$$

where i stands for any component.

Amagat's law. Amagat's law of additive volumes is analogous to Dalton's law of additive pressures. Amagat stated that the total volume of a gaseous mixture is equal to the sum of the volumes of the individual gas components if they were to be measured at the same temperature and at the total pressure of all the molecules. The individual volumes of these individual components at the same temperature and pressure are called the *partial volumes* (or sometimes *pure component* volumes) of the individual components, and

$$V_1 + V_2 + V_3 + \cdots + V_n = V_t \tag{3.6}$$

Reasoning in the same fashion as in our explanation of partial pressures, *at the same temperature and pressure*, the partial volume is a function only of number of molecules of the individual component gas present in the gaseous mixture, or

$$\frac{V_1}{V_2} = \frac{n_1}{n_2} \tag{3.7}$$

and

$$\frac{V_1}{V_t} = \frac{n_1}{n_t} = y_1 = \text{mole fraction} \tag{3.8}$$

which shows that the ratio of the partial volumes is exactly the same, numerically, as the ratio of the moles of components 1 and 2, or the ratio of the moles of component 1 to the total moles. Equation (3.8) states the principle, presented without proof in Chap. 1, that

$$\text{volume fraction} = \text{mole fraction} = y_i \tag{3.9}$$

for an ideal gas.

EXAMPLE 3.10 Partial Pressures and Volumes

A gas-tight room has a volume of 1000 m³. This room contains air (considered to be 21% O_2 and 79% N_2) at 20°C and a total pressure of 1 atm.
    (a) What is the partial volume of O_2 in the room?
    (b) What is the partial volume of N_2 in the room?
    (c) What is the partial pressure of O_2 in the room?
    (d) What is the partial pressure of N_2 in the room?
    (e) If all of the O_2 were removed from the room by some method, what would be the subsequent total pressure in the room?

Solution

See Fig. E3.10a.

Basis: 1000 m³ of air at 20°C and 1 atm

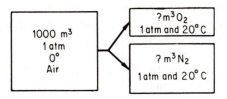

Fig. E3.10a

Partial volumes can be calculated by multiplying the total volume by the respective component mole fractions [Eq. (3.8)]:

(a) $V_{O_2} = (0.21)(1000) = \underline{\hphantom{00}}210 \text{ m}^3 \text{ O}_2$ at 20°C, 1 atm
(b) $V_{N_2} = (0.79)(1000) = \underline{\hphantom{00}}790 \text{ m}^3 \text{ N}_2$ at 20°C, 1 atm
      total volume $= \overline{1000 \text{ m}^3}$ air at 20°C, 1 atm

Note how the temperature and pressure have to be specified for the partial volumes to make them meaningful.

Partial pressures can be calculated by multiplying the total pressure by the respective component mole fractions [Eq. (3.5a)]; the basis is still the same:

(c) $p_{O_2} = (0.21)(1 \text{ atm}) = 0.21$ atm when $V = 1000 \text{ m}^3$ at 20°C
(d) $p_{N_2} = (0.79)(1 \text{ atm}) = \underline{0.79}$ atm when $V = 1000 \text{ m}^3$ at 20°C
      total pressure $= \overline{1.00}$ atm when $V = 1000 \text{ m}^3$ at 20°C

(e) If a tight room held dry air at 1 atm and all the oxygen were removed from the air by a chemical reaction, the pressure reading would fall to 0.79 atm. This is the partial pressure of the nitrogen and inert gases in the air. Alternatively, if it were possible to remove the nitrogen and inert gases and leave only the oxygen, the pressure reading would fall to 0.21 atm. In either case, you would have left in the room 1000 m³ of gas at 20°C. See Fig. E3.10b.

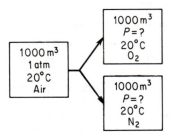

Fig. E3.10b

For use in our subsequent calculations you should clearly understand now that the original room contained:

(1)   790 m³ dry N_2 at 1 atm and 20°C
(2)   210 m³ dry O_2 at 1 atm and 20°C
(3) 1000 m³ dry air at 1 atm and 20°C
                                    (add 1 and 2)

or

(1) 1000 m³ dry O_2 at 0.21 atm and 20°C
(2) 1000 m³ dry N_2 at 0.79 atm and 20°C
(3) 1000 m³ dry air at 1.00 atm and 20°C
                                    (add 1 and 2)

EXAMPLE 3.11 Calculation of Partial Pressures from Gas Analysis

A flue gas analyzes 14.0% CO_2, 6.0% O_2, and 80.0% N_2. It is at 400°F and 765.0 mm Hg pressure. Calculate the partial pressure of each component.

Solution

Basis: 1.00 kg (or lb) mol flue gas

component	kg (or lb) mol	p (mm Hg)
CO_2	0.140	107.1
O_2	0.060	45.9
N_2	0.800	612.0
Total	1.000	765.0

On the basis of 1.00 mole of flue gas, the mole fraction y of each component, when multiplied by the total pressure, gives the partial pressure of that component.

Self-Assessment Test

1.   Write down Dalton's law and Amagat's law.

2.   A gas has the following composition at 120°F and 13.8 psia.

Component	Mole %
N_2	2
CH_4	79
C_2H_6	19

(*a*) What is the partial pressure of each component?
(*b*) What is the partial volume of each component if the total volume of the container is 2 ft³?
(*c*) What is the volume fraction of each component?

3. (**a**) If the C_2H_6 were removed from the gas in problem 2, what would be the subsequent pressure in the vessel?

(**b**) What would be the subsequent partial pressure of the N_2?

Section 3.2 Real Gas Relationships

We have said that at room temperature and pressure many gases can be assumed to act as ideal gases. However, for some gases under normal conditions, and for most gases under conditions of high pressure, values of the gas properties that you might obtain using the ideal gas law would be at wide variance with the experimental evidence. You might wonder exactly how the behavior of real gases does compare with that calculated from the ideal gas laws. In Fig. 3.1 you can see how the $p\hat{V}$ product of several gases deviates from that predicted by the ideal gas laws as the pressure increases substantially. Thus it is clear that we need some way of computing the p–V–T properties of a gas that is not ideal.

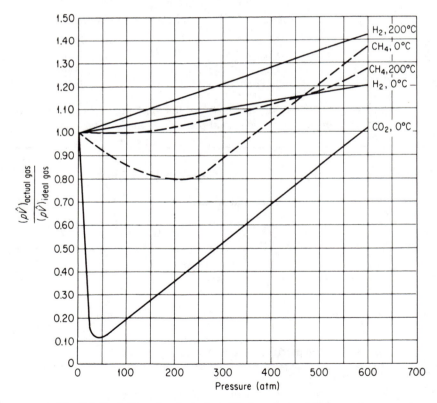

Fig. 3.1 Deviation of real gas from ideal gas laws at high pressures.

Essentially there are four methods of handling real gas calculations:

(a) Equations of state

(b) Compressibility charts

(c) Estimated properties[1]

(d) Actual experimental data

Even if experimental data are available, the other three techniques still may be quite useful for certain types of calculations. Of course, under conditions such that part of the gas liquefies, the gas laws apply only to the gas phase portion of the system—you cannot extend these real gas laws into the liquid region any more than you can apply the ideal gas laws to a liquid.

3.2-1   Equations of State

> ### *Your objectives in studying this section are to be able to:*
>
> 1. Cite two reasons for using equations of state to predict p, V, T properties of gases.
> 2. Write down van der Waals equation.
> 3. Explain what the units are for the coefficients in van der Waals equation or other equations of state.
> 4. Solve van der Waals equation for either p, V, n, or T given the values for the coefficients and the other variables.
> 5. Write down the names of five equations of state.
> 6. Solve equations of state for the value of one of the variables that can be determined explicitly given the equation, the values of the coefficients, and the values of the other variables.

Equations of state relate the p–V–T properties of a pure substance (or mixtures) by semitheoretical or empirical relations. By *property* we shall mean any measurable characteristic of a substance, such as pressure, volume, or temperature, or a characteristic that can be calculated or deduced, such as internal energy, to be discussed in Chap. 4. Experimental data for a gas, carbon dioxide, are illustrated in Fig. 3.2, which is a plot of pressure versus molal volume with absolute temperature (converted to °C) as the third parameter. For a constant temperature the perfect gas law, $p\hat{V} = RT$ (for 1 mole), is, on this plot, a rectangular hyperbola because $p\hat{V} = $ constant. The experimental data at high temperatures and low pressures is not

---

[1]Computer programs are available to estimate physical properties of compounds and mixtures based on structural group contributions and other basic parameters; refer to E. L. Meadows, *Chem. Eng. Progr.*, v. 61, p. 93 (1965), for a discussion of the principles involved.

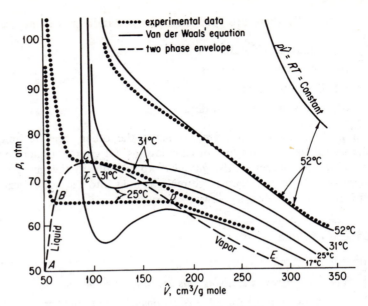

Fig. 3.2 p–V–T properties of CO_2.

shown but closely approximate the perfect gas law. However, as we approach the point C, called the critical point (discussed in Sec. 3.2-2), we see that the traces of the experimental data are quite different from the lines that would represent the relation $p\hat{V} = $ constant.

The problem, then, is to predict these experimental points by means of some equation which can, with reasonable accuracy, represent the experimental data throughout the gas phase. Over 100 equations of state have been proposed to accomplish this task, of which a few are shown in Table 3.2. For others, refer to Chao and Robinson.[2] Coefficient values for various gases can be found in Perry or Reid, Prausnitz, and Sherwood. We would not, of course, expect the function to follow the experimental data within the two-phase region outlined by the area inside the envelope A, B, C, D, and E.

Van der Waals equation. Van der Waals equation, first proposed in 1893, does not yield satisfactory predictions at low gas temperatures (see Fig. 3.2), but it does have historical interest and is one of the simplest equations of state. Thus it illustrates some of the typical theoretical development that has taken place as well as some of the computational problems involved in using equations of state. Keep in mind that some of the equations listed in Table 3.2 are completely empirical, whereas others, such as van der Waals equation, are semiempirical; that is, although they were developed from theory, the constants in the equation or portions of the equation are determined by empirical methods.

[2]K. C. Chao and R. L. Robinson, *Equations of State in Engineering and Research*, American Chemical Society, Washington, D.C., 1980.

TABLE 3.2   EQUATIONS OF STATE (FOR 1 MOLE)

Van der Waals:

$$\left(p + \frac{a}{\hat{V}^2}\right)(\hat{V} - b) = RT$$

Lorentz:

$$p = \frac{RT}{\hat{V}^2}(\hat{V} + b) - \frac{a}{\hat{V}^2}$$

Dieterici:

$$p = \frac{RT}{\hat{V} - b}e^{-a/\hat{V} RT}$$

Berthelot:

$$p = \frac{RT}{\hat{V} - b} - \frac{a}{T\hat{V}^2}$$

Redlich–Kwong:

$$\left[p + \frac{a}{T^{1/2}\hat{V}(\hat{V} + b)}\right](\hat{V} - b) = RT$$

$$a = 0.4278\frac{R^2 T_c^{2.5}}{p_c}$$

$$b = 0.0867\frac{RT_c}{p_c}$$

Kammerlingh-Onnes:

$$p\hat{V} = RT\left(1 + \frac{B}{\hat{V}} + \frac{C}{\hat{V}^2} + \cdots\right)$$

Holborn:

$$p\hat{V} = RT(1 + B'p + C'p^2 + \cdots)$$

Beattie–Bridgeman:

$$p\hat{V} = RT + \frac{\beta}{\hat{V}} + \frac{\gamma}{\hat{V}^2} + \frac{\delta}{\hat{V}^3}$$

$$\beta = RTB_0 - A_0 - \frac{Rc}{T^2}$$

$$\gamma = -RTB_0 b + aA_0 - \frac{RB_0 c}{T^2}$$

$$\delta = \frac{RB_0 bc}{T^2}$$

Benedict–Webb–Rubin:

$$p\hat{V} = RT + \frac{\beta}{\hat{V}} + \frac{\sigma}{\hat{V}^2} + \frac{\eta}{\hat{V}^4} + \frac{w}{\hat{V}^5}$$

$$\beta = RTB_0 - A_0 - \frac{C_0}{T^2}$$

$$\sigma = bRT - a + \frac{c}{T^2}\exp\left(-\frac{\gamma}{\hat{V}^2}\right)$$

$$\eta = c\gamma\exp\left(-\frac{\gamma}{\hat{V}^2}\right)$$

$$w = a\alpha$$

Peng–Robinson:

$$p = \frac{RT}{\hat{V} - b} - \frac{a\alpha}{\hat{V}(\hat{V} + b) + b(\hat{V} - b)}$$

$$a = 0.45724\left(\frac{R^2 T_c^2}{p_c}\right)$$

$$b = 0.07780\left(\frac{RT_c}{p_c}\right)$$

$$\alpha = [1 + \kappa(1 - T_r^{1/2})]^2$$

$$\kappa = 0.37464 + 1.54226\omega - 0.26992\omega^2$$

$$\omega = \text{acentric factor (see p. 238)}$$

Van der Waals tried to include in the ideal gas law the effect of the attractive forces among the molecules by adding to the pressure term, in the ideal gas law, the term

$$\frac{n^2 a}{V^2} \tag{3.10}$$

where n = number of moles
V = volume
a = a constant, different for each gas

He also tried to take into account the effect of the volume occupied by the molecules themselves by subtracting a term from the volume in the ideal gas law. These corrections led to the following equation:

$$\left(p + \frac{n^2 a}{V^2}\right)(V - nb) = nRT \tag{3.11}$$

where a and b are constants determined by fitting van der Waals equation to experimental p–V–T data, particularly the values at the critical point. Values of the van der Waals constants for a few gases are given in Table 3.3.

TABLE 3.3 CONSTANTS FOR THE VAN DER WAALS AND REDLICH–KWONG EQUATIONS CALCULATED FROM THE LISTED VALUES OF THE CRITICAL CONSTANTS

	van der Waals		Redlich–Kwong	
	$a*$ $\left[\text{atm}\left(\frac{\text{cm}^3}{\text{g mol}}\right)^2\right]$	$b\dagger$ $\left(\frac{\text{cm}^3}{\text{g mol}}\right)$	$a\ddagger$ $\left[(\text{atm})(\text{k})^{1/2}\left(\frac{\text{cm}^3}{\text{g mol}}\right)\right]$	$b\dagger$ $\left(\frac{\text{cm}^3}{\text{g mol}}\right)$
Air	1.33×10^6	36.6	15.65×10^6	25.3
Ammonia	4.19×10^6	37.3	85.00×10^6	25.7
Carbon dioxide	3.60×10^6	42.8	63.81×10^6	29.7
Ethane	5.50×10^6	65.1	97.42×10^6	45.1
Ethylene	4.48×10^6	57.2	76.92×10^6	39.9
Hydrogen	0.246×10^6	26.6	1.439×10^6	18.5
Methane	2.25×10^6	42.8	31.59×10^6	29.6
Nitrogen	1.347×10^6	38.6	15.34×10^6	26.8
Oxygen	1.36×10^6	31.9	17.12×10^6	22.1
Propane	9.24×10^6	90.7	180.5×10^6	62.7
Water vapor	5.48×10^6	30.6	140.9×10^6	21.1

*To convert to psia (ft³/lb mol)², multiply table value by 3.776×10^{-3}.
†To convert to ft³/lb mol, multiply table value by 1.60×10^{-2}.
‡To convert to psia (°R)$^{1/2}$(ft³/lb mol)², multiply table value by 2.807×10^{-3}.

Let us look at some of the computational problems that may arise when using equations of state to compute p–V–T properties by using van der Waals equation as an illustration. Van der Waals equation can easily be explicitly solved for p as follows:

$$p = \frac{nRT}{V - nb} - \frac{n^2 a}{V^2} \tag{3.12}$$

However, if you want to solve for V (or n), you can see that the equation becomes cubic in V (or n):

$$V^3 - \left(nb + \frac{nRT}{p}\right)V^2 + \frac{n^2 a}{p}V - \frac{n^3 ab}{p} = 0 \tag{3.13}$$

Consequently, to solve for V (or n) you would have to:

(a) Resort to a trial-and-error method,

(b) Plot the equation assuming various values of V (or n), and see at what value of V (or n) the curve you plot intersects the abscissa, or

(c) Use Newton's method or some other method for solving nonlinear equations (refer to Appendix L).

A computer program can be of material help in technique (c). You can obtain a close approximation to V (or n) in many cases from the ideal gas law, useful at least for the first trial, and then you can calculate a more exact value of V (or n) by finding the real root of a cubic equation.

EXAMPLE 3.12 Van der Waals' Equation

A 5.0-ft³ cylinder containing 50.0 lb of propane (C_3H_8) stands in the hot sun. A pressure gauge shows that the pressure is 665 psig. What is the temperature of the propane in the cylinder? Use van der Waals equation. See Fig. E3.12.

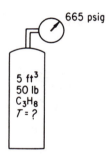

665 psig

5 ft³
50 lb
C_3H_8
$T = ?$

Fig. E3.12

Solution

Basis: 50 lb of propane

The van der Waals constants obtained from any suitable handbook or Table 3.3 are

$$a = 3.27 \times 10^4 \text{ psia} \left(\frac{\text{ft}^3}{\text{lb mol}} \right)^2$$

$$b = 1.35 \frac{\text{ft}^3}{\text{lb mol}}$$

$$\left(p + \frac{n^2 a}{V^2} \right)(V - nb) = nRT$$

All the additional information you need is as follows:

$$p = 665 \text{ psig} + 14.7 = 679.7 \text{ psia}$$

$$R \text{ in proper units is} = 10.73 \frac{(\text{psia})(\text{ft}^3)}{(\text{lb mol})(°\text{R})}$$

$$n = \frac{50 \text{ lb}}{44 \text{ lb/lb mol}} = 1.137 \text{ lb mol} \quad \text{propane}$$

$$\left[679.7 + \frac{(1.137)^2(3.27 \times 10^4)}{(5)^2} \right][5 - 1.137(1.35)] = 1.137(10.73)T$$

$$T = 673°\text{R} = 213°\text{F}$$

Can you solve the Redlich–Kwong equation for T?

Other equations of state. Table 3.2 lists a few of the many equations of state in addition to van der Waals equation. The form of these equations is of interest inasmuch as they are attempts to fit the experimental data with as few constants in the equation as possible.

The two-constant Redlich–Kwong equation appears to be quite good according to a study of Thodos and Shah.[3] Figure 3.3 compares the van der Waals and the Redlich–Kwong equations of state with experimental data.

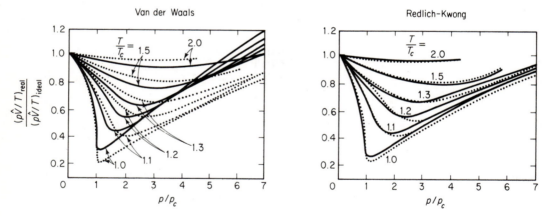

Fig. 3.3 Comparison of experimental values (dots) with predicted values (solid lines) for two equations of state.

One form of the equation known as the *virial* form is illustrated by the equations of Kammerlingh-Onnes and Holborn. These are essentially power series in $1/\hat{V}$ or in p, and the quantities B, C, D, and so on, are known as *virial coefficients*. These equations reduce to $p\hat{V} = RT$ at low pressures. The virial form of an equation of

[3]G. Thodos and K. K. Shah, *Ind. Eng. Chem.*, v. 57, p. 30 (1965).

state is the best we have today; however, it is not possible to solve theoretically by statistical thermodynamics for the constants beyond the first two. Therefore, fundamentally, these equations are semiempirical, the constants being determined by fitting the equation to experimental data. The Beattie–Bridgeman equation, which has five constants (exclusive of R), and the Benedict–Webb–Rubin equation, which has eight constants, are among the best we have at the present time. Naturally, the use of these equations involves time-consuming calculations, particularly when carried out by a hand calculator rather than on the computer, but the results will frequently be more accurate than those obtained by other methods to be described shortly.

In spite of the complications involved in their use, equations of state are important for several reasons. They permit a concise summary of a large mass of experimental data and also permit accurate interpolation between experimental data points. They provide a continuous function to facilitate calculation of physical properties involving differentiation and integration. Finally, they provide a point of departure for the treatment of thermodynamic properties of mixtures. However, for instructional purposes, and for many engineering calculations, the techniques of predicting p–V–T values discussed in the next section are considerably more convenient to use and usually have adequate accuracy.

Self-Assessment Test

1. Equations of state for gases are often used to predict p–V–T properties of gas. Cite two reasons.

2. The constants for an equation of state are usually evaluated in one of two ways.
(*a*) What are these?
(*b*) Which method would you expect to give an equation showing the least error?

3. What are some of the tests that you might apply to see how an equation of state fits P–V–T data?

4. What factors in a real gas cause the gas to behave in a nonideal manner? How are these factors taken into account in van der Waals equation?

5. What are the units of a and b in the SI system for the Redlich–Kwong equation?

6. You measure that 0.00220 lb mol of a certain gas occupies a volume of 0.95 ft^3 at 1 atm and 32°F. If the equation of state for this gas is $PV = nRT(1 + bP)$, where b is a constant, find the volume at 2 atm and 71°F.

7. Calculate the temperature of 2 g mol of a gas using van der Waals equation with $a = 1.35 \times 10^{-6}$ (m^6)(atm)(g mol^{-2}), $b = 0.0322 \times 10^{-3}$ (m^3)(g mol^{-1}) if the pressure is 100 kPa and the volume is 0.0515 m^3.

8. Calculate the pressure of 10 kg mol of ethane in a 4.86-dm^3 vessel at 300 K using two equations of state: (*a*) ideal gas and (*b*) Beattie–Bridgeman. For the latter, $A_0 = 5.880$, $a = 0.05861$, $B_0 = 0.09400$, $b = 0.01915$, and $c = 90.0 \times 10^4$. Compare with your answer the observed value of 34.0 atm.

9. The van der Waals constants for a gas are $a = 2.31 \times 10^6$ (atm)(cm³/g mol)² and $b = 44.9$ cm³/g mol. Find the volume per kilogram mole if the gas is at 90 atm and 373 K.

3.2-2 Critical State, Reduced Parameters, and Compressibility

> ***Your objectives in studying this section are to be able to:***
>
> 1. State the law of corresponding states.
> 2. Define the critical state.
> 3. Calculate the reduced temperature, reduced pressure, and reduced volume, and to use any two of these parameters to obtain the compressibility factor, z, from the Nelson and Obert charts.
> 4. Use compressibility factors and appropriate charts to predict the p–V–T behavior of a gas, or given the required data, find compressibility factors.
> 5. List Newton's corrections for H_2 and He to T_c and p_c.
> 6. Calculate the ideal critical volume and ideal reduced volume, and be able to use V_{r_i} as a parameter in the Nelson and Obert charts.

In the attempt to devise some truly universal gas law for high pressures, the idea of corresponding states was developed. Early experimenters found that at the critical point all substances are in approximately the same state of molecular dispersion. Consequently, it was felt that their thermodynamic and physical properties should be similar. The *law of corresponding states* expresses the idea that in the critical state all substances should behave alike.

Now that we have mentioned *critical state* several times, let us consider exactly what the term means. You can find many definitions, but the one most suitable for general use with pure component systems as well as with mixtures of gases is the following:

The *critical state* for the gas-liquid transition is the set of physical conditions at which the density and other properties of the liquid and vapor become identical.

Refer to Fig. 3.4. Note how as the pressure increases the specific volumes of the liquid and gas approach each other and finally become the same at the point C. This point, for a pure component (only), is the highest temperature at which liquid and vapor can exist in equilibrium. From a slightly different view a critical point is a limiting point that marks the disappearance of a state.

Consider Fig. 3.2. When gaseous carbon dioxide is compressed at 25°C its pressure rises until a certain density is reached (D). On further compression the pressure remains constant and the gas condenses, liquid and vapor being in coexist-

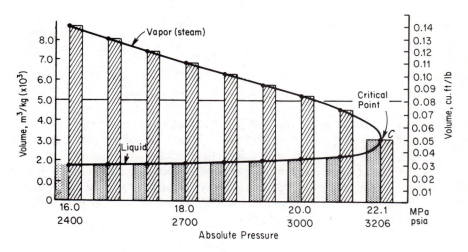

Fig. 3.4 Critical point for a pure substance (water).

ence. Only after all vapor is condensed to liquid (*B*) does the pressure rise again. If the temperature is raised a little, the vapor reaches a higher density before it starts condensing; on the other hand, the condensation is completed at lower liquid density. Thus the coexisting phases have a smaller difference in density at the higher temperature. Above 30°C this density range diminishes very rapidly with temperature and finally, just above 30°C, the two densities at which condensation is initiated and completed coincide. Vapor and liquid are no longer distinguishable; the separating meniscus disappears. Below the critical temperature, the system can exhibit two coexisting phases; above, it goes from the dilute to the dense state without a phase transition. The critical point marks the highest temperature on the coexistence curve.

You can find experimental values of the critical temperature (T_c) and the critical pressure (p_c) for various compounds in Appendix D. If you cannot find a desired critical value in this text or in a handbook, you can always consult Reid, Prausnitz, and Sherwood[4] or Hakata and Hirata,[5] which describe and evaluate methods of estimating critical constants for various compounds.

The gas–liquid transition described above is only one of several possible transitions exhibiting a critical point. Critical phenomena are observed in liquids and solids as well, as described in Sec. 3.7.

Another set of terms with which you should immediately become familiar are the *reduced* conditions. These are *corrected*, or *normalized*, conditions of temperature, pressure, and volume and are expressed mathematically as

[4]R. C. Reid, J. M. Prausnitz, and T. K. Sherwood, *The Properties of Gases and Liquids*, 3rd ed., McGraw-Hill, New York, 1977.

[5]T. Hakata and M. Hirata, *J. Chem. Eng. Japan*, v. 3, p. 5 (1970).

$$T_r = \frac{T}{T_c} \tag{3.14}$$

$$p_r = \frac{p}{p_c} \tag{3.15}$$

$$V_r = \frac{V}{V_c} \tag{3.16}$$

The idea of using the reduced variables to correlate the p–V–T properties of gases, as suggested by van der Waals, is that all substances behave alike in their reduced (i.e., their corrected) states. In particular, any substance would have the same *reduced* volume at the same *reduced* temperature and pressure. If a relationship does exist involving the reduced variables that can be used to predict p_r, T_r, and V_r, what would the equation be? Certainly, the simplest form of such an equation would be to imitate the ideal gas law, or

$$p_r \hat{V}_r = \gamma T_r \tag{3.17}$$

where γ is some constant.

Now, how does this concept work out in practice? If we plot p_r vs. T_r with a third parameter of constant V_r,

$$\frac{p_r}{T_r} = \frac{\gamma}{V_r} \tag{3.18}$$

we should get straight lines for constant V_r. That this is actually what happens can be seen from Fig. 3.5 for various hydrocarbons. Furthermore, this type of correlation works in the region where $pV = nRT$ is in great error. Also, at the critical point ($p_r = 1.0$, $T_r = 1.0$), V_r should be 1.0 as it is in the diagram.

However, the ability of Eq. (3.18) to predict gas properties breaks down for most substances as they approach the perfect gas region (the area of low pressures somewhat below $p_r = 1.0$). If this concept held in the perfect gas region for all gases, then for 1 mole

$$p\hat{V} = RT \tag{3.1a}$$

should hold as well as Eq. (3.17),

$$p_r \hat{V}_r = \gamma T_r$$

For *both* of these relations to be true would mean that

$$\frac{p\hat{V}}{RT} = \frac{p_r \hat{V}_r}{\gamma T_r} = \frac{(p/p_c)(\hat{V}/\hat{V}_c)}{\gamma(T/T_c)} \tag{3.19}$$

or

$$\frac{p_c \hat{V}_c}{RT_c} = \text{constant} \tag{3.20}$$

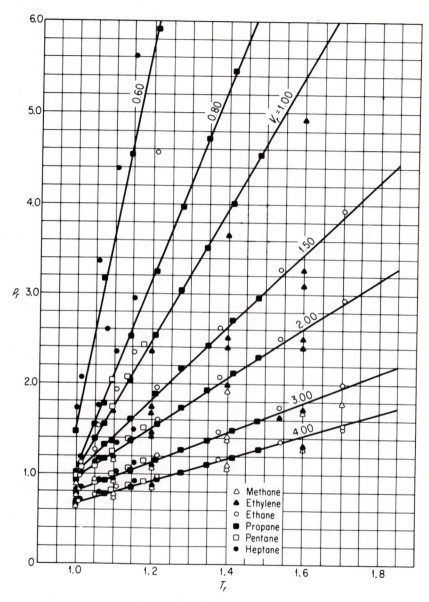

Fig. 3.5 Test of the relation $p_r/T_r = \gamma/V_r$.

for all substances. If you look at Table 3.4 you can see from the experimental data that Eq. (3.20) is not quite true, although the range of values for most substances is not too great (from 0.24 to 0.28). He (0.305) and HCN (0.197) represent the extremes of the range.

TABLE 3.4 EXPERIMENTAL VALUES OF $p_c \hat{V}_c / RT_c$ FOR VARIOUS GASES

NH_3	0.243	HCN	0.197	CH_4	0.290	C_2H_4	0.270
Ar	0.291	H_2O	0.233	C_2H_6	0.285	CH_3OH	0.222
CO_2	0.279	N_2	0.292	C_3H_8	0.277	C_2H_5OH	0.248
He	0.305	Toluene	0.270	C_6H_{14}	0.264	CCl_4	0.272

A more convenient and accurate way that has been developed to tie together the concepts of the law of corresponding states and the ideal gas law is to revert to a correction of the ideal gas law, that is, a generalized equation of state expressed in the following manner:

$$pV = znRT \qquad (3.21)$$

where the dimensionless quantity z is called the *compressibility factor* and is a function of the pressure and temperature:

$$z = \psi(p, T) \qquad (3.22)$$

One way to look at z is to consider it to be a factor that makes Eq. (3.21) an equality. If the compressibility factor is plotted for a given temperature against the pressure for different gases, we obtain something like Fig. 3.6(a). However, if the compressibility is plotted against the reduced pressure as a function of the reduced temperature,

$$z = \psi(p_r, T_r) \qquad (3.23)$$

then for most gases the compressibility values at the same reduced temperature and reduced pressure fall at about the same point, as illustrated in Fig. 3.6(b).

This permits the use of what is called a generalized compressibility factor, and Figs. 3.7 through 3.10 are the *generalized compressibility charts* or z-factor charts prepared by Nelson and Obert.[6] These charts are based on 30 gases. Figures 3.7(b) and (c) represent z for 26 gases (excluding H_2, He, NH_3, H_2O) with a maximum deviation of 1%, and H_2 and H_2O within a deviation of 1.5%. Figure 3.8 is for 26 gases and is accurate to 2.5%, while Fig. 3.9 is for nine gases and errors can be as high as 5%. For H_2 and He only, Newton's corrections to the actual critical constants are used to give pseudocritical constants

$$T'_c = T_c + 8 \text{ K} \qquad (3.24)$$

$$p'_c = p_c + 8 \text{ atm} \qquad (3.25)$$

which enable you to use Figs. 3.7 through 3.10 for these two gases as well with minimum error. Figure 3.10 is a unique chart which, by having several parameters plotted simultaneously on it, helps you avoid trial-and-error solutions or graphical solutions

[6]L. C. Nelson and E. F. Obert, *Chem. Eng.*, v. 61, no. 7, pp. 203–208 (1954).

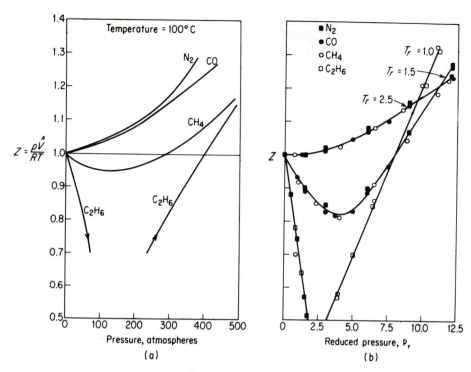

Fig. 3.6 (a) Compressibility factor as a function of temperature and pressure; (b) compressibility as a function of reduced temperature and reduced pressure.

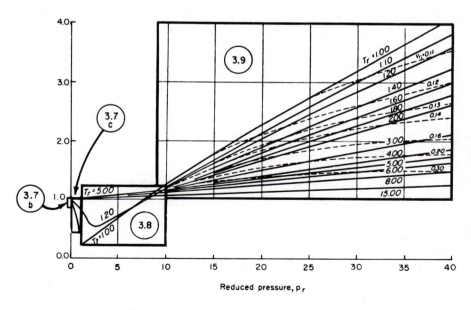

Fig. 3.7 (a) Generalized compressibility chart;

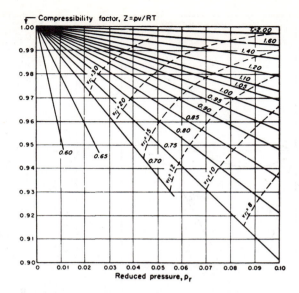

Fig. 3.7 (b) general compressibility chart, very low temperatures;

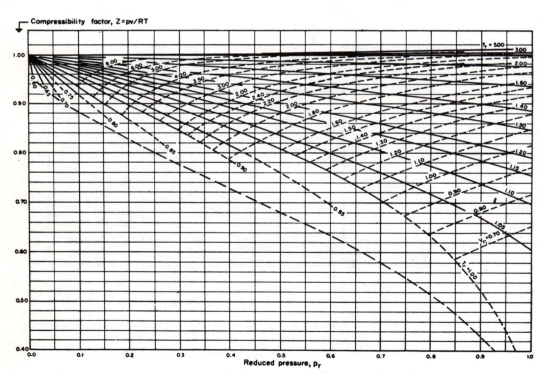

(c) general compressibility chart, low pressures.

234

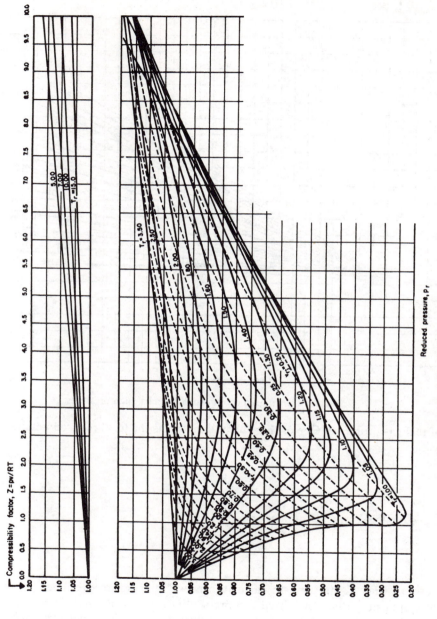

Compressibility factor, $Z = pv/RT$

Reduced pressure, p_r

Fig. 3.8 Generalized compressibility chart, medium pressures.

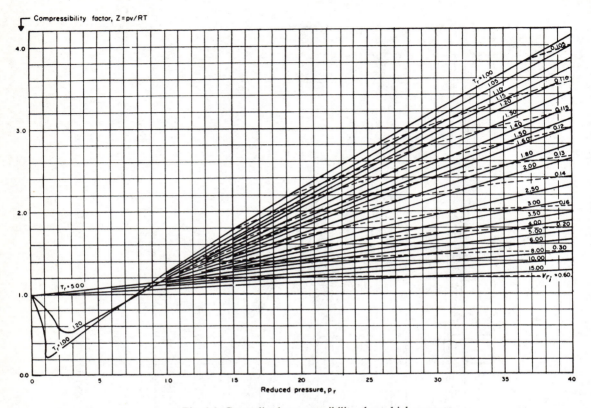

Fig. 3.9 Generalized compressibility chart, high pressures.

of real gas problems. One of these helpful factors is the ideal reduced volume defined as

$$V_{r_i} = \frac{\hat{V}}{\hat{V}_{c_i}} \qquad (3.26)$$

$\hat{V}_{c_i}$ is the ideal critical volume, or

$$\hat{V}_{c_i} = \frac{RT_c}{p_c} \qquad (3.27)$$

Both V_{r_i} and $\hat{V}_{c_i}$ are easy to calculate since T_c and p_c are presumed known. The development of the generalized compressibility chart is of considerable practical as well as pedagogical value because it enables engineering calculations to be made with considerable ease and also permits the development of thermodynamic functions for gases for which no experimental data are available. All you need to know to use these charts are the critical temperature and the critical pressure for a pure

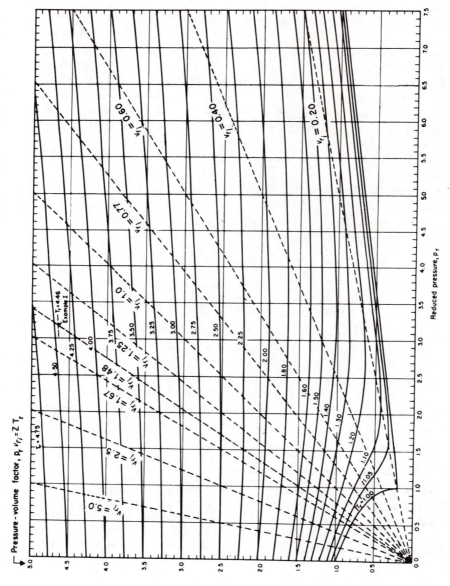

Fig. 3.10 Generalized compressibility chart, with special scales.

237

substance (or the pseudovalues for a mixture, as we shall see later). The value $z = 1$ represents ideality, and the value $z = 0.27$ is the compressibility factor at the critical point.

Quite a few authors have suggested that the generalized compressibility relation, Eq. (3.23), could provide better accuracy if another parameter for the gas were included in the argument on the right-hand side (in addition to p_r and T_r). Clearly, if a third parameter is used, adding a third dimension to p_r and T_r, you have to employ a *set* of tables or charts rather than a single table or chart. Lydersen et al.[7] developed tables of z that included as the third parameter the critical compressibility factor, $z_c = p_c \hat{V}_c / R T_c$. Pitzer[8] used a different third parameter termed the *acentric* factor, defined as being equal to $-\ln p_{rs} - 1$, where p_{rs} is the value of the reduced vapor pressure at $T_r = 0.70$.

Viswanath and Su[9] compared the z factors from the Nelson and Obert charts, the Lydersen–Greenkorn–Hougen charts, and the two-parameter (T_r and p_r) Viswanath and Su chart with the experimental z's for T_r's of 1.00 to 15.00 and p_r's of 0 to 40 for 19 gases with the following results:

	Viswanath–Su	Nelson–Obert	Lydersen–Greenkorn–Hougen
Number of points	415	355	339
Average deviation (%) in z	0.61	0.83	1.26

It is seen that all these charts yield quite reasonable values for engineering purposes. Because the increased precision that may accompany the use of a third parameter in calculating z is not necessarily for our purposes, and because the presentation of z values so as to include the third parameter is considerably more cumbersome, we shall not show the three-parameter tables here.

EXAMPLE 3.13 Use of the Compressibility Factor

In spreading liquid ammonia fertilizer, the charges for the amount of NH_3 are based on the time involved plus the pounds of NH_3 injected into the soil. After the liquid has been spread, there is still some ammonia left in the source tank (volume = 120 ft³), but in the form of a gas. Suppose that your weight tally, which is obtained by difference, shows a net weight of 125 lb of NH_3 left in the tank as a gas at 292 psig. Because the tank is sitting in the sun, the temperature in the tank is 125°F.

[7]A. L. Lydersen, R. A. Greenkorn, and O. A. Hougen, *University of Wisconsin Engineering Experimental Station Report No. 4*, Madison, Wis., 1955.

[8]K. S. Pitzer, *J. Amer. Chem. Soc.*, v. 77, p. 3427 (1955).

[9]D. S. Viswanath and G. J. Su, *AIChE J.*, v. 11, p. 202 (1965).

Your boss complains that his calculations show that the specific volume of the gas is 1.20 ft³/lb and hence that there are only 100 lb of NH_3 in the tank. Could he be correct? See Fig. E3.13.

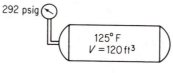

292 psig

125° F
$V = 120 \text{ ft}^3$

Fig. E3.13

Solution

Basis: 1 lb of NH_3

Apparently, your boss used the ideal gas law in calculating his figure of 1.20 ft³/lb of NH_3 gas:

$$R = 10.73 \frac{(\text{psia})(\text{ft}^3)}{(\text{lb mol})(°R)}$$

$$T = 125°F + 460 = 585°R$$

$$p = 292 + 14.7 = 307 \text{ psia}$$

$$n = \frac{1 \text{ lb}}{17 \text{ lb/lb mol}}$$

$$\hat{V} = \frac{nRT}{p} = \frac{\frac{1}{17}(10.73)(585)}{307} = 1.20 \text{ ft}^3/\text{lb}$$

However, he should have used the compressibility factor because NH_3 does not behave as an ideal gas under the observed conditions of temperature and pressure. Let us again compute the mass of gas in the tank this time using

$$pV = znRT$$

What is known and unknown in the equation?

$$p = 307 \text{ psia}$$

$$V = 120 \text{ ft}^3$$

$$z = ?$$

$$n = \tfrac{1}{17} \text{ lb mol}$$

$$T = 585°R$$

The additional information needed (taken from Appendix D) is

$$T_c = 405.4 \text{ K} \simeq 729°R$$

$$p_c = 111.3 \text{ atm} \simeq 1640 \text{ psia}$$

Then, since z is a function of T_r and p_r,

$$T_r = \frac{T}{T_c} = \frac{585°R}{729°R} = 0.803$$

$$p_r = \frac{p}{p_c} = \frac{307 \text{ psia}}{1640 \text{ psia}} = 0.187$$

From the Nelson and Obert chart, Fig. 3.7(c), you can read $z \simeq 0.845$. Now $\hat{V}$ can be calculated as

$$\hat{V} = \frac{znRT}{p} = \hat{V}_{ideal}\left(\frac{z}{z_{ideal}}\right)$$

or

$$\hat{V} = \frac{1.20 \text{ ft}^3 \text{ ideal}}{\text{lb}} \frac{0.845}{1} = 1.01 \text{ ft}^3/\text{lb NH}_3$$

$$\frac{1 \text{ lb NH}_3}{1.01 \text{ ft}^3} \frac{120 \text{ ft}^3}{} = 119 \text{ lb NH}_3$$

Certainly 119 lb is a more realistic figure than 100 lb, and it is easily possible to be in error by 6 lb if the residual weight of NH_3 in the tank is determined by difference. As a matter of interest you might look up the specific volume of NH_3 at the conditions in the tank in a handbook—you would find that $\hat{V} = 0.973 \text{ ft}^3/\text{lb}$, and hence the compressibility factor calculation yielded a volume with an error of only about 4%.

EXAMPLE 3.14 Use of the Compressibility Factor

Suppose that 3.500 kg of liquid O_2 is vaporized into a tank of 0.0284-m³ volume at −25°C. What will the pressure in the tank be; will it exceed the safety limit of the tank (100 atm)?

Solution

See Fig. E3.14.

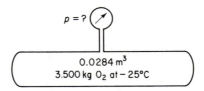

p = ?

0.0284 m³
3.500 kg O_2 at −25°C

Fig. E3.14

Basis: 3.500 kg O_2

We do know from Appendix D that

$$T_c = -118.8°C \quad \text{or} \quad -118.8 + 273.1 = 154.3 \text{ K}$$

$$p_c = 49.7 \text{ atm}$$

However, this problem cannot be worked exactly the same way as the preceding problem because we do not know the pressure of the O_2 in the tank. Thus we need to use the other parameter, V_{r_i}, that is available on the Nelson and Obert charts:

$$\hat{V} \text{ (molal volume)} = \frac{0.0284 \text{ m}^3}{3.500 \text{ kg}} \Bigg| \frac{32 \text{ kg}}{1 \text{ kg mol}} = 0.259 \text{ m}^3/\text{kg mol}$$

Note that the *molal volume must* be used in calculating V_{r_i} since $\hat{V}_{c_i}$ is a volume per mole. Next,

$$\hat{V}_{c_i} = \frac{RT_c}{p_c} = \frac{0.08206 \text{ (m}^3)(\text{atm})}{(\text{kg mol})(\text{K})} \Bigg| \frac{154.3 \text{ K}}{49.7 \text{ atm}} = 0.255 \frac{\text{m}^3}{\text{kg mol}}$$

Then

$$V_{r_i} = \frac{\hat{V}}{\hat{V}_{c_i}} = \frac{0.259}{0.255} = 1.02$$

Now we know two parameters, V_{r_i} and, with $T = -25°C + 273 = 248$ K,

$$T_r = \frac{T}{T_c} = \frac{248 \text{ K}}{154.3 \text{ K}} = 1.61$$

From the Nelson and Obert chart, Figs. 3.8 or 3.10,

$$p_r = 1.43$$

Then

$$p = p_r p_c$$
$$= 1.43(49.7) = 71.0 \text{ atm}$$

The pressure of 100 atm is not exceeded.

EXAMPLE 3.15 Use of Compressibility Factor

Repeat Example 3.12, this time using the compressibility factor, that is, the equation of state $pV = znRT$. "A 5-ft³ cylinder containing 50.0 lb of propane (C_3H_8) stands in the hot sun. A pressure gauge shows that the pressure is 665 psig. What is the temperature of the propane in the cylinder?"

Solution

See Fig. E3.15. In the equation $pV = znRT$, we know that

$$V = 5 \text{ ft}^3$$

$$n = \frac{50 \text{ lb}}{44 \text{ lb/lb mol}} = 1.135 \text{ lb mol } C_3H_8$$

$$p = 665 \text{ psig} + 14.7 \simeq 680 \text{ psia}$$

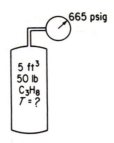

665 psig

5 ft³
50 lb
C₃H₈
T = ?

Fig. E3.15

However, z as well as T is unknown. Since z is a function of T_r and p_r, we should first calculate the latter two quantities ($T_c = 370$ K, or $666°$R, and $p_c = 42.1$ atm, or 617 psia, from Appendix D) but are stopped by the fact that T is unknown. Another parameter available for use, to avoid a trial-and-error or graphical solution, is the factor zT_r (or V_{r_i} could be used, as in the last example):

$$zT = \frac{pV}{nR} \quad \text{or} \quad zT_r = \frac{pV}{nRT_c}$$

$$zT_r = \frac{680 \text{ psia}}{\quad} \Big| \frac{5 \text{ ft}^3}{\quad} \Big| \frac{}{1.135 \text{ lb mol}} \Big| \frac{}{10.73 \frac{(\text{psia})(\text{ft}^3)}{(\text{lb mol})(°\text{R})}} \Big| 666°\text{R}$$

$$= 0.42$$

From the Nelson and Obert chart, Fig. 3.10, at

$$zT_r = 0.42$$

$$p_r = \frac{680}{617} = 1.10$$

you can find $T_r = 1.02$ and

$$T = T_r T_c = 370(1.02) = 378 \text{ K}$$
$$= 666(1.02) = 680°\text{R}$$

Self-Assessment Test

1. What is the ideal critical volume? What is the advantage of using V_{c_i}?

2. In a proposed low pollution vehicle burning H_2–O_2, the gases are to be stored in tanks at 2000 psia. The vehicle has to operate from -40 to $130°$F.
   (a) Is the ideal gas law a sufficiently good approximation for the design of these tanks?
   (b) A practical operating range requires that 3 lb_m of hydrogen be stored. How large must the hydrogen tank be if the pressure is not to exceed 2000 psia?
   (c) The H_2/O_2 ratio is 2 on a molar basis. How large must the oxygen tank be?

3. A carbon dioxide fire extinguisher has a volume of 40 liters and is to be charged to a pressure of 20 atm at a storage temperature of 20°C. Determine the mass in kilograms of CO_2 within a charged extinguisher assuming it initially contains CO_2 at 1 atm.

4. Calculate the pressure of 4.00 g mol CO_2 contained in a 6.25×10^{-3} m³ fire extinguisher at 25°C.

3.2-3   Gaseous Mixtures

Your objectives in studying this section are to be able to:

1. Use Dalton's law and an equation of state such as van der Waals' law to compute the total pressure of a gas mixture.
2. Use average constants given the desired weighting method in an equation of state.
3. Compute the mean compressibility factor.
4. Use Kay's method of pseudocritical values to calculate the pseudoreduced values and predict p, V, T, and n via the compressibility factor.

So far we have discussed only pure compounds and their p–V–T relations. Most practical problems involve gaseous mixtures and there is very little experimental data available for gaseous mixtures. Thus the question is: How can we predict p–V–T properties with reasonable accuracy for gaseous mixtures? We treated ideal gas mixtures in Sec. 3.1.3, and saw, as we shall see later (in Chap. 4 for the thermodynamic properties), that for ideal mixtures the properties of the individual components can be added together to give the desired property of the mixture. But this technique does not prove to be satisfactory for real gases. The most desirable technique would be to develop methods of calculating p–V–T properties for mixtures based solely on the properties of the pure components. Possible ways of doing this for real gases are discussed below.

Equations of state. Take, for example, van der Waals' equation, Eq. (3.12). You can compute the partial pressure of each component by van der Waals' equation and then add the individual partial pressures together according to Dalton's law to give the total pressure of the system,

For component A,

$$p_A = \frac{n_A RT}{V - n_A b_A} - \frac{n_A^2 a_A}{V^2} \tag{3.28}$$

For component B,

$$p_B = \frac{n_B RT}{V - n_B b_B} - \frac{n_B^2 a_B}{V^2} \tag{3.29}$$

and so on; *in total,*

$$p_T = RT\left(\frac{n_A}{V - n_A b_A} + \frac{n_B}{V - n_B b_B} + \cdots\right)$$

$$-\frac{1}{V^2}(n_A^2 a_A + n_B^2 a_B + \cdots) \tag{3.30}$$

This technique provides a direct solution for p_T if everything else is known; however, the solution for V or n_A from Eq. (3.30) is quite cumbersome.

Average constants. A simpler and usually equally effective way to solve gas mixture problems is to use average constants in the equation of state. The problem resolves itself into this question: How should the constants be averaged to give the least error on the whole? For van der Waals' equation, you should proceed as follows. For:

b (use linear mole fraction weights):

$$b_{\text{mixture}} = b_A y_A + b_B y_B + \cdots \tag{3.31}$$

a (use linear mole fraction weights of the square roots of a_i):

$$a_{\text{mixture}}^{1/2} = a_A^{1/2} y_A + a_B^{1/2} y_B + \cdots \tag{3.32}$$

If you are puzzled as to the reasoning behind the use of the square-root weighting for a, remember that the term in van der Waals' equation involving a is

$$\frac{n^2 a}{V^2}$$

in which the number of moles is squared. The method of average constants can be successfully applied to most of the other equations of state listed in Table 3.2. The average constants can also be calculated from the pseudoreduced constants described below. For example, for van der Waals' equation,

$$a = \frac{27}{64}\frac{R^2 T_c^2}{p_c} \quad \text{and} \quad b = \frac{RT_c}{8p_c}$$

Mean compressibility factor. Another approach toward the treatment of gaseous mixtures is to say that $pV = z_m nRT$, where z_m can be called the *mean compressibility factor.* With such a relationship the only problem is how to evaluate the mean compressibility factor satisfactorily. One obvious technique that might occur to you is to make z_m a mole average as follows:

$$z_m = z_A y_A + z_B y_B + \cdots \tag{3.33}$$

Since z is a function of both the reduced temperature and the reduced pressure, it is necessary to decide what pressure will be used to evaluate p_r.

(a) *Assume Dalton's law of partial pressures.* For each component z is evaluated at T_r and the reduced *partial pressure* for each gaseous component. The reduced partial pressure is defined as

$$p_{r_A} = \frac{p_A}{p_{c_A}} = \frac{(p_T)y_A}{p_{c_A}} \tag{3.34}$$

(b) *Assume Amagat's law of pure component volumes.* For each component z is evaluated at T_{r_t} and the reduced *total pressure* on the system.

Pseudocritical properties. Many weighting rules have been proposed to combine the critical properties of the components of a real gas mixture in order to get an effective set of critical properties (*pseudocritical*) that enable *pseudoreduced* properties of the mixture to be computed.[10] The psuedoreduced properties in turn are used in exactly the same way as the reduced properties for a pure compound, and in effect the mixture can be treated as a pure compound. In instances where you know nothing about the gas mixture, this technique is preferable to any of those discussed previously.

In Kay's method pseudocritical values for mixtures of gases are calculated on the assumption that each component in the mixture contributes to the pseudocritical value in the same proportion as the number of moles of that component. Thus the pseudocritical values are computed as follows:

$$p'_c = p_{c_A}y_A + p_{c_B}y_B + \cdots \tag{3.35}$$
$$T'_c = T_{c_A}y_A + T_{c_B}y_B + \cdots \tag{3.36}$$

where p'_c = pseudocritical pressure and T'_c = pseudocritical temperature. (It has also been found convenient in some problems to calculate similarly a weighted pseudo-ideal-critical volume V'_{c_i}.) You can see that these are linearly weighted mole average pseudocritical properties. (In Sec. 3.7 we compare the true critical point of gaseous mixtures with the pseudocritical point.) Then the respective pseudoreduced values are

$$p'_r = \frac{p}{p'_c} \tag{3.37}$$

$$T'_r = \frac{T}{T'_c} \tag{3.38}$$

If you are faced with a complicated mixture of gases whose composition is not well known, you still can estimate the pseudocritical constants from charts[11] such as shown in Fig. 3.11(a) and (b) if you know the gas specific gravity. Figure 3.11 is good only for natural gases composed mainly of methane which contain less than 5% impurities (CO_2, N_2, H_2S, etc.).

---

[10]You should still add Newton's corrections for H_2 and He.

[11]Natural Gasoline Supply Men's Association, *Engineering Data Book*, Tulsa, Okla., 1957, p. 103.

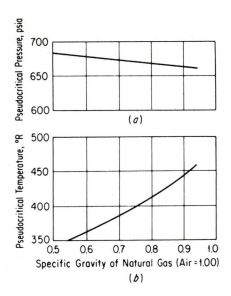

Fig. 3.11 Estimation of critical properties of natural gases.

Kay's method is known as a two-parameter rule since only p_c and T_c for each component are involved in the calculation of z. If a third parameter such as z_c, the Pitzer acentric factor, or $\hat{V}_{c_i}$ is included in the determination of the mean compressibility factor, then we would have a three-parameter rule. Reid and Leland[12] have shown that all the weighting rules can be obtained from a common base, and all reduce to Kay's rule as limiting cases with the proper assumptions. All the pseudocritical methods do not provide equal accuracy in predicting p–V–T properties, but most suffice for engineering work. Stewart et al.[13] reviewed 21 different methods of determining the pseudoreduced parameters by three-parameter rules (see Table 3.5). Although Kay's method was not the most accurate, it was easy to use and not considerably poorer than some of the more complex techniques of averaging critical properties.

In summary, we would have to state that no one method will consistently give the best results. Kay's pseudocritical method, on the average, will prove reliable, although other methods of greater accuracy (and of greater complexity) are available in the literature, as brought out in Table 3.5. Some of the longer equations of state, with average constants, are quite accurate. All these methods begin to break down near the true critical point for the mixture.

[12] R. C. Reid and T. W. Leland, *AIChEJ.*, v. 11, p. 228 (1965).

[13] W. E. Stewart, S. F. Burkhart, and David Voo, paper given at the AIChE meeting in Kansas City, Mo., May 18, 1959. Also refer to A. Satter and J. M. Campbell, *Soc. Petrol. Engr. J.*, p. 333 (December 1963); and H. E. Barner and W. C. Quinlan, *I&EC Process Design and Development*, v. 8, p. 407 (1969).

TABLE 3.5   PREDICTION OF p–V–T VALUES OF GAS DENSITY BY PSEUDOCRITICAL METHODS

Method	Percent Root-Mean-Square Deviation of Density, 35 Systems	Percent Root-Mean-Square Deviation of Density, CO_2 and H_2S Free Systems
(1) Three-parameter Kay's rule: $$T_c' = \sum T_{c_i} y_i$$ $$p_c' = \sum p_{c_i} y_i$$ $$z_c' = \sum z_{c_i} y_i$$	10.74	6.82
(2) Empirical: $$T_c' = \sum T_{c_i} y_i$$ $$\frac{T_c'}{p_c'} = \sum \frac{T_{c_i}' y_i}{p_{c_i}}$$ $$V_c' = \sum \hat{V}_{c_i} y_i$$	6.06	5.85
(3) Virial approach (Joffe's method III): $$\frac{T_c'}{p_c'} = \frac{1}{8} \sum_i \sum_j y_i y_j \left[\left(\frac{T_{c_i}}{p_{c_i}} \right)^{1/3} + \left(\frac{T_{c_j}}{p_{c_j}} \right)^{1/3} \right]$$ $$\frac{T_c'}{\sqrt{p_c'}} = \sum y_i \frac{T_{c_i}}{\sqrt{p_{c_i}}}$$ $$\hat{V}_c' = \sum \hat{V}_{c_i} y_i$$	4.98	4.24
(4) "Recommended" method: $$\frac{T_c'}{p_c'} = \frac{1}{3} \sum y_i \frac{T_{c_i}}{p_{c_i}} + \frac{2}{3} \left[\sum y_i \left(\frac{T_{c_i}}{p_{c_i}} \right)^{1/2} \right]^2$$ $$\frac{T_c'}{\sqrt{p_c'}} = \sum y_i \frac{T_{c_i}}{\sqrt{p_{c_i}}}$$ $$z_c' = \sum z_{c_i} y_i$$	4.32	3.26

EXAMPLE 3.16   p–V–T Relations for Gas Mixtures

A gaseous mixture has the following composition (in mole percent):

Methane, CH_4	20
Ethylene, C_2H_4	30
Nitrogen, N_2	50

at 90 atm pressure and 100°C. Compare the molal volume as computed by the methods of:

(a) Perfect gas law
(b) Van der Waals' equation plus Dalton's law
(c) Van der Waals' equation using averaged constants
(d) Mean compressibility factor and Dalton's law
(e) Mean compressibility factor and Amagat's law
(f) Pseudoreduced technique (Kay's method)

Solution

Basis: 1 g mol of gas mixture

Additional data needed are:

component	a $(atm)\left(\dfrac{cm^3}{g\ mol}\right)^2$	b $\left(\dfrac{cm^3}{g\ mol}\right)$	T_c (K)	p_c (atm)
CH_4	2.25×10^6	42.8	191	45.8
C_2H_4	4.48×10^6	57.2	283	50.9
N_2	1.35×10^6	38.6	126	33.5

$$R = 82.06\frac{(cm^3)(atm)}{(g\ mol)(K)}$$

(a) Perfect gas law:

$$V = \frac{nRT}{p} = \frac{1(82.06)(373)}{90} = 340\ cm^3 \text{ at 90 atm and 373 K}$$

(b) Combine van der Waals' equation and Dalton's law according to Eq. (3.30):

$$p_T = RT\left(\frac{n_{CH_4}}{V - n_{CH_4}b_{CH_4}} + \frac{n_{C_2H_4}}{V - n_{C_2H_4}b_{C_2H_4}} + \frac{n_{N_2}}{V - n_{N_2}b_{N_2}}\right)$$
$$-\frac{1}{V^2}(n_{CH_4}^2 a_{CH_4} + n_{C_2H_4}^2 a_{C_2H_4} + n_{N_2}^2 a_{N_2})$$

Substitute the numerical values (in the proper units):

$$90 = 82.06(373)\left(\frac{0.2}{V - 8.56} + \frac{0.3}{V - 17.1} + \frac{0.5}{V - 19.3}\right)$$
$$-\frac{1}{V^2}(9 \times 10^4 + 40.5 \times 10^4 + 33.8 \times 10^4) \tag{1}$$

The solution to this equation can be obtained by trial-and-error graphical means, or solution of a cubic equation. For the first trial assume ideal conditions. By plotting the right-hand side of Eq. (1) against V until the total is 90, we find the solution is at $V = 332\ cm^3$ at 90 atm and 373 K.

(c) To use average constants in van der Waals' equation, write it in the following fashion:

$$V^3 - \left(\bar{b} + \frac{RT}{p}\right)nV^2 + \left(\frac{\bar{a}}{p}\right)n^2 V - \frac{\bar{a}\bar{b}}{p}n^3 = 0$$

where $\bar{a}$ and $\bar{b}$ are the average constants.

$$(\bar{a})^{1/2} = 0.2a_{CH_4}^{1/2} + 0.3a_{C_2H_4}^{1/2} + 0.5a_{N_2}^{1/2}$$
$$\bar{a} = 2.30 \times 10^6\ atm\left(\frac{cm^3}{g\ mol}\right)^2$$
$$\bar{b} = 0.2b_{CH_4} + 0.3b_{C_2H_4} + 0.5b_{N_2} = 45.0\frac{cm^3}{g\ mol}$$
$$n = 1\ g\ mol\ \text{(the basis)}$$

$$V^3 - \left[45.0 + \frac{(82.06)(373)}{90}\right]V^2 + \left[\frac{2.30 \times 10^6}{90}\right]V - \left[\frac{(2.30 \times 10^6)(45.0)}{90}\right] = 0 \tag{2}$$

Let us work out a method of solving Eq. (2) for V different from the method used in part (b):

$$f(V) = V^3 - 385V^2 + 2.56 \times 10^4 V - 1.15 \times 10^6 = 0$$

$$-V^3 + 385V^2 = 2.56 \times 10^4 V - 1.15 \times 10^6 \tag{3}$$

Splitting this equation into two parts,

$$f_1(V) = 2.56 \times 10^4 V - 1.15 \times 10^6 \tag{4}$$

$$f_2(V) = -V^3 + 385V^2 = V^2(385 - V) \tag{5}$$

and plotting these parts (see Fig. E3.16) to see where they cross is one method of obtaining the real root of the equation. You know V is in the vicinity of 340 cm³.

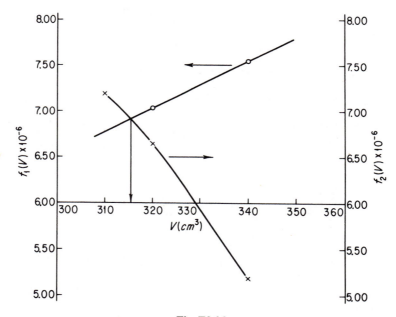

Fig. E3.16

V (cm³)	$f_1(V)$	$f_2(V)$
340	7.55×10^6	5.20×10^6
320	7.04×10^6	6.65×10^6
310	(only two points) required for straight line)	7.20×10^6

From the graph $V \simeq 316$ cm³ at 90 atm and 373 K.

(d) The table below shows how Dalton's law can be used to estimate a mean compressibility factor.

component	p_c (atm)	T_c (K)	y	$(90)y = p$	$p_r = \dfrac{p}{p_c}$	$T_r = \dfrac{373}{T_c}$
CH_4	45.8	191	0.2	18	0.394	1.95
C_2H_4	50.9	283	0.3	27	0.535	1.32
N_2	33.5	126	0.5	45	1.340	2.95

component	z	$(z)(y)$
CH_4	0.99	0.198
C_2H_4	0.93	0.279
N_2	1.00	0.500
z_{mean}	=	0.977

Then

$$V = \frac{z_{mean}RT}{p} = \frac{0.977(82.06)(373)}{90}$$

$$= 332 \text{ cm}^3 \text{ at 90 atm and 373 K}$$

(e) Combining Amagat's law and the z factor,

component	p_c (atm)	T_c (K)	y	$p_r = \dfrac{90}{p_c}$	$T_r = \dfrac{373}{T_c}$
CH_4	45.8	191	0.2	1.97	1.95
C_2H_4	50.9	283	0.3	1.78	1.32
N_2	33.5	126	0.5	2.68	2.95

component	z	$(z)(y)$
CH_4	0.97	0.194
C_2H_4	0.75	0.225
N_2	1.01	0.505
z_{mean}	=	0.924

$$V = \frac{0.924(82.06)(373)}{90} = 313 \text{ cm}^3 \text{ at 90 atm and 373 K}$$

(f) According to Kay's method, we first calculate the pseudocritical values for the mixture by Eqs. (3.35) and (3.36).

$$\qquad\qquad\qquad\qquad CH_4 \qquad\quad C_2H_4 \qquad\quad N_2$$

$$p_c' = p_{c_A}y_A + p_{c_B}y_B + p_{c_C}y_C = (45.8)(0.2) + (50.9)(0.3) + (33.5)(0.5)$$
$$= 41.2 \text{ atm}$$
$$T_c' = T_{c_A}y_A + T_{c_B}y_B + T_{c_C}y_C = (191)(0.2) + (283)(0.3) + (126)(0.5)$$
$$= 186 \text{ K}$$

Then we calculate the pseudoreduced values for the mixture by Eqs. (3.37) and (3.38):

$$p_r' = \frac{p}{p_c'} = \frac{90}{41.2} = 2.18, \qquad T_r' = \frac{T}{T_c'} = \frac{373}{186} = 2.00$$

With the aid of these two parameters we can find from Fig. 3.8 that $z = 0.965$. Thus

$$V = \frac{zRT}{p} = \frac{0.965(82.06)(373)}{90} = 328 \text{ cm}^3 \text{ at 90 atm and 373 K}$$

EXAMPLE 3.17 Use of Pseudoreduced Ideal Molal Volume

In instances where the temperature or pressure of a gas mixture is unknown, it is convenient, to avoid a trial-and-error solution using the generalized compressibility charts, to compute a pseudocritical ideal volume and a pseudoreduced ideal volume as illustrated below. Suppose we have given that the molal volume of the gas mixture in the preceding problem was 326 cm³ at 90.0 atm. What was the temperature?

Solution

Basis: 326 cm³ of gas at 90.0 atm and T K

component	T_c (K)	p_c (atm)	y	yT_c	yp_c
CH_4	190.7	45.8	0.20	38.14	9.16
C_2H_4	283.1	50.9	0.30	84.93	15.27
N_2	126.2	33.5	0.50	63.10	16.75
Total				186.2	41.2

Note we have used mole fractions as weighting factors to find an average $T_c' = 186.2$ and $p_c' = 41.2$ as in Example 3.16. Next we compute V_{c_i}':

$$\hat{V}_{c_i}' = \frac{RT_c'}{p_c'} = \frac{82.06(186.2)}{41.2} = 371.0 \text{ cm}^3/\text{g mol}$$

$$V_r' = \frac{\hat{V}}{\hat{V}_{c_i}'} = \frac{326}{371} = 0.879$$

$$p_r' = \frac{p}{p_c'} = \frac{90.0}{41.2} = 2.19$$

From Fig. 3.8 or Fig. 3.10, $T_r' = 1.98$. Then

$$T = T_c'T_r' = 186.2(1.98) = 369 \text{ K}$$

Self-Assessment Test

1. One pound mole of a mixture containing 0.400 lb mol of N_2 and 0.600 lb mol C_2H_4 at 50°C occupies a volume of 1.44 ft³. What is the pressure in the vessel? Compute your answer by:
   (a) Dalton's law plus van der Waals' equation
   (b) Averaged constants for van der Waals' equation
   (c) Mean compressibility factor (assuming Dalton's law)
   (d) Kay's method

Section 3.3 Vapor Pressure

Your objectives in studying this section are to be able to:

1. Define vapor pressure, triple point, equilibrium, dew point, bubble point, saturated, superheated, subcooled, and quality, and be able to locate the region or point in a *p–T* chart in which each term applies.

2. Calculate the vapor pressure of a substance from a vapor-pressure equation given values for the coefficients, and look up the vapor pressure in reference books.

3. Calculate the properties of a wet mixture given the temperature, pressure, and volume of the mixture.

4. Prepare a Cox chart; calculate the scaling factors for each axis so that the chart can be used to predict the vapor pressure of a liquid.

The terms *vapor* and *gas* are used very loosely. A gas that exists below its critical temperature is usually called a vapor because it can condense. If you continually compress a pure gas at constant temperature, provided that the temperature is below the critical temperature, some pressure is eventually reached at which the gas starts to condense into a liquid. Further compression does not increase the pressure but merely increases the fraction of gas that condenses. A reversal of the procedure just described will cause the liquid to be transformed into the gaseous state again. From now on, the word *vapor* will be reserved to describe a gas below its critical point in a process in which the phase change is of primary interest, while the word *gas or noncondensable gas* will be used to describe a gas above the critical point or a gas in a process in which it cannot condense.

Vaporization and condensation at constant temperature and pressure are *equilibrium* processes, and the equilibrium pressure is called the *vapor pressure*. At a given temperature there is only one pressure at which the liquid and vapor phases of a pure substance may exist in equilibrium. Either phase alone may exist, of course, over a wide range of conditions. By equilibrium we mean a state in which there is no tendency toward spontaneous change. Another way to say the same thing is to say equilibrium is a state in which all the rates of attaining and departing from the state are balanced.

You can visualize vapor pressure, vaporization, and condensation more easily with the aid of Fig. 3.12. Figure 3.12 is an expanded *p-T* diagram for pure water. For each temperature you can read the corresponding pressure at which water vapor and water liquid exist in equilibrium. You have encountered this condition of equilibrium many times—for example, in boiling. Any substance has an infinite number of boiling points, but by custom we say the "normal" boiling point is the temperature at which boiling takes place under a pressure of 1 atm (101.3 kPa, 760 mm Hg). Unless another pressure is specified, 1 atm is assumed, and the term *boiling point* is

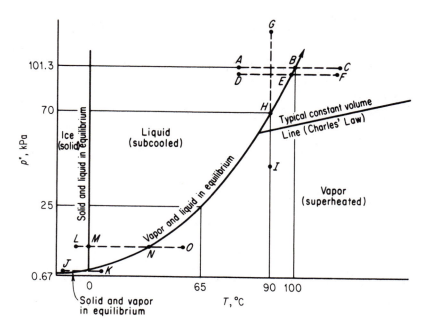

Fig. 3.12 Vapor-pressure curve for water.

taken to mean the "normal boiling point." The normal boiling point for water occurs when the vapor pressure of the water equals the pressure of the atmosphere on top of the water. A piston with a force of 101.3 kPa could just as well take the place of the atmosphere, as shown in Fig. 3.13. For example, you know that at 100°C water will boil (vaporize) and the pressure will be 101.3 kPa or 1 atm (point *B*). Suppose

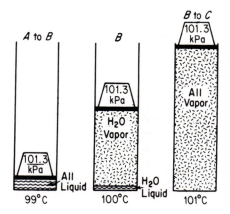

Fig. 3.13 Transformation of liquid water into water vapor at constant pressure.

that you heat water starting at 77°C (point *A*) in an open pan—what happens? We assume that the water vapor above the pan is at all times in equilibrium with the liquid water. This is a constant-pressure process since the moist air around the water in the pan acts similarly to a piston in a cylinder to keep the pressure at atmospheric pressure. As the temperature rises and the confining pressure stays constant, nothing particularly noticeable occurs until 100°C is reached, at which time the water begins to boil (i.e., evaporate). It pushes back the atmosphere and will completely change from liquid into vapor. If you heated the water in an enclosed cylinder, and if after it had all evaporated at point *B* you continued heating the water vapor formed at constant pressure, you could apply the gas laws in the temperature region *B-C* (and at higher temperatures). A reversal of this process from the temperature *C* would cause the vapor to condense at *B* to form a liquid. The temperature at point *B* would in these circumstances represent the *dew point*.

Suppose that you went to the top of Pikes Peak and repeated the experiment—what would happen then? Everything would be the same (points *D-E-F*) with the exception of the temperature at which the water would begin to boil, or condense. Since the pressure of the atmosphere at the top of Pikes Peak would presumably be lower than 101.3 kPa, the water would start to displace the air, or boil, at a lower temperature. However, water still would exert a vapor pressure of 101.3 kPa at 100°C. You can see that (a) at any given temperature water exerts its vapor pressure (at equilibrium); (b) as the temperature goes up, the vapor pressure goes up; and (c) it makes no difference whether water vaporizes into air, into a cylinder closed by a piston, or into an evacuated cylinder—at any temperature it still exerts the same vapor pressure as long as the water is in equilibrium with its vapor.

A process of vaporization or condensation at constant temperature is illustrated by the lines *G–H–I* or *I–H–G*, respectively, in Fig. 3.12. Water would vaporize or condense at constant temperature as the pressure reached point *H* on the vapor-pressure curve (also look at Fig. 3.14.)

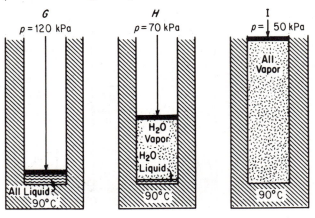

Fig. 3.14 Transformation of liquid water into water vapor at constant temperature.

Figure 3.12 also shows the *p–T* conditions at which ice (in its common form) and water vapor are in equilibrium. When the solid passes directly into the vapor phase without first becoming a liquid (line *J–K* as opposed to line *L–M–N–O*), it is said to *sublime*. Iodine crystals do this at room temperature; water sublimes only below 0°C, as when the frost disappears in the winter when the thermometer reads −6°C.

The vapor-pressure line extends well past the conditions shown in Fig. 3.12 all the way to the critical temperature and pressure (not shown). Why does it terminate at the critical temperature and pressure? Above the critical temperature, water can exist only as a gas.

A term commonly applied to the vapor-liquid portion of the vapor-pressure curve is the word *saturated*, meaning the same thing as vapor and liquid in equilibrium with each other. If a gas is just ready to start to condense its first drop of liquid, the gas is called a *saturated gas;* if a liquid is just about to vaporize, it is called a *saturated liquid*. These two conditions are also known as the *dew point* and *bubble point*, respectively.

The region to the right of the vapor-pressure curve in Fig. 3.12 is called the *superheated* region and the one to the left of the vapor-pressure curve is called the *subcooled* region. The temperatures in the superheated region, if measured as the difference $(O - N)$ between the actual temperature of the superheated vapor and the saturation temperature for the same pressure, are called *degrees of superheat*. For example, steam at 500°F and 100 psia (the saturation temperature for 100 psia is 327.8°F) has $(500 - 327.8) = 172.2$°F of superheat. Another new term you will find used frequently is the word *quality*. A *wet* vapor consists of saturated vapor and saturated liquid in equilibrium. The mass fraction of vapor is known as the quality.

EXAMPLE 3.18 Properties of Wet Vapors

The properties of a mixture of vapor and liquid in equilibrium (for a single component) can be computed from the individual properties of the saturated vapor and saturated liquid. We will use water as an example because it is easy to obtain the volumetric data we need from the steam tables (inside back cover). At 400 K and 245.6 kPa the specific volume of a wet steam mixture is 0.505 m³/kg. What is the quality of the steam?

Solution

From the steam tables the specific volumes of the saturated vapor and liquid are

$$\hat{V}_l = 0.001067 \text{ m}^3/\text{kg}, \qquad \hat{V}_g = 0.7308 \text{ m}^3/\text{kg}$$

Basis: 1 kg of wet steam mixture

Let x = mass fraction vapor.

$$\frac{0.001067 \text{ m}^3}{1 \text{ kg liquid}} \Bigg| \frac{(1-x) \text{ kg liquid}}{} + \frac{0.7308 \text{ m}^3}{1 \text{ kg vapor}} \Bigg| \frac{x \text{ kg vapor}}{} = 0.505 \text{ m}^3$$

$$0.001067 - 0.001067x + 0.7308x = 0.505$$

$$x = 0.69$$

Other properties of wet mixtures can be treated in the same manner.

3.3-1 Change of Vapor Pressure with Temperature

• A large number of experiments on many substances have shown that a plot of the vapor pressure (p^*) of a compound against temperature does not yield a straight line but a curve, as you saw in Fig. 3.12. Many types of correlations have been proposed to transform this curve to a linear form ($y = mx + b$); a plot of log (p^*) vs. ($1/T$), for moderate temperature intervals, is reasonably linear:

$$\underbrace{\log(p^*)}_{y} = \underbrace{m\left(\frac{1}{T}\right) + b}_{mx + b} \tag{3.39}$$

Equation (3.39) is derived from the Clausius–Clapeyron equation (see Chap. 4). Empirical correlations of vapor pressure are frequently given in the form of the Antoine equation (refer to Appendix G for values of the constants):

$$\log(p^*) = -\frac{A}{T + C} + B \tag{3.40}$$

where A, B, C = constants different for each substance
    T = temperature in degrees kelvin

Over very wide temperature intervals the experimental data will not prove to be exactly linear as indicated by Eq. (3.39), but have a slight tendency to curve. This curvature can be straightened out by using a special plot known as a Cox chart.[14] The log of the vapor pressure of a compound is plotted against a special nonlinear temperature scale constructed from the vapor-pressure data for water (called a *reference substance*).

As illustrated in Fig. 3.15, the temperature scale is established by recording the temperature at a given vapor pressure of water for a number of vapor pressures. The numbers on the line for water indicate the vapor pressures of water for pre-selected values (50, 100, 150, etc.) of temperature. The vapor pressures of other substances plotted on this specially prepared set of coordinates will yield straight lines over extensive temperature ranges and thus facilitate the extrapolation and

[14]E. R. Cox, *Ind. Eng. Chem.*, v. 15, p. 592 (1923).

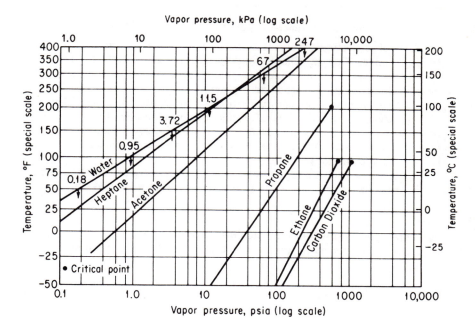

Fig. 3.15 Cox chart.

interpolation of vapor-pressure data. It has been found that lines so constructed for closely related compounds, such as hydrocarbons, all meet at a common point. Since straight lines can be obtained on a Cox chart only two sets of vapor-pressure data are needed to provide complete information about the vapor pressure of a substance over a considerable temperature range. We discuss in Chap. 4 under the topic of the Clausius–Clapeyron equation other information that can be obtained from such vapor-pressure plots.

3.3-2 Change of Vapor Pressure with Pressure

The equation for the change of vapor pressure with total pressure at constant temperature in a system is

$$\left(\frac{\partial(p^*)}{\partial p_T}\right)_T = \frac{\hat{V}_l}{\hat{V}_g} \tag{3.41}$$

where $\hat{V}$ = molal volume of saturated liquid or gas
   p_T = total pressure on the system

Under normal conditions the effect is negligible.

EXAMPLE 3.19 Extrapolation of Vapor-Pressure Data

The control of existing solvents is described in the *Federal Register*, v. 36, no. 158, August 14, 1971, under Title 42, Chapter 4, Appendix B, Section 4.0, Control of Organic Compound Emissions. Section 4.6 indicates that reductions of at least 85% can be accomplished by (a) incineration or (b) carbon adsorption.

Chlorinated solvents and many other solvents used in industrial finishing and processing, dry-cleaning plants, metal degreasing, printing operations, and so forth, can be recycled and reused by the introduction of carbon adsorption equipment. To predict the size of the adsorber, you first need to know the vapor pressure of the compound being adsorbed at the process conditions.

The vapor pressure of chlorobenzene is 400 mm Hg at 110.0°C and 5 atm at 205°C. Estimate the vapor pressure at 245°C and at the critical point (359°C).

Solution

The vapor pressures will be estimated by use of a Cox chart. See Fig. E3.19. The temperature scale is constructed by using the following data from the steam tables:

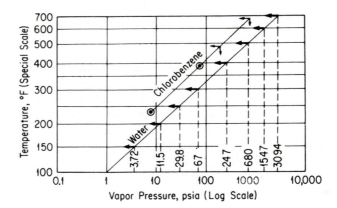

Fig. E3.19

p* H₂O (psia)	t (°F)
0.95	100
3.72	150
11.5	200
29.8	250
67.0	300
247	400
680	500
1543	600
3094	700

Vapor pressures from to 3.72 to 3094 psia are marked on the horizontal logarithmic scale as shown in Fig. E3.19. We selected the vapor pressures to correspond to even values of the temperature in °F.

Next, we draw a line representing the vapor pressure of water at any suitable angle on the graph so as to stretch the desired temperature range from the bottom to the top of the vertical axis. For each vapor pressure, the temperature is marked and a horizontal line drawn for the ordinate. This establishes the temperature scale (which looks almost logarithmic in nature).

Next we convert the two vapor pressures of chlorobenzene into psia,

$$\frac{400 \text{ mm Hg}}{} \left| \frac{14.7 \text{ psia}}{760 \text{ mm Hg}} \right. = 7.74 \text{ psia} \quad \left| \quad 110°C = 230°F \right.$$

$$\frac{5 \text{ atm Hg}}{} \left| \frac{14.7 \text{ psia}}{1 \text{ atm Hg}} \right. = 73.5 \text{ psia} \quad \left| \quad 205°C = 401°F \right.$$

and plot these two points on the graph paper. Examine the encircled dots. Next we draw a straight line between the encircled points and extrapolate to 471°F (245°C) and 678°F (359°C). At these two temperatures read off the estimated vapor pressures:

	471°F (245°C)	678°F (359°C)
Estimated:	150 psia	700 psia
Experimental:	147 psia	666 psia

Experimental values are given for comparison.

An alternative way to extrapolate vapor pressure data would be to use the Othmer[15] plot as explained in Sec. 4.4.

3.3-3 Liquid Properties

Considerable experimental data are available for liquid densities of pure compounds as a function of temperature and pressure. Gold[16] reviews 13 methods of estimating liquid densities and provides recommendations for different classes of compounds if you cannot locate experimental data. As to liquid mixtures, it is even more difficult to predict the $p–V–T$ properties of liquid mixtures than of real gas mixtures. Probably more experimental data (especially at low temperatures) are available than for gases, but less is known about the estimation of the $p–V–T$ properties of liquid mixtures. Corresponding state methods are usually used, but with less assurance than for gas mixtures.[17,18]

[15]D. F. Othmer and E. S. Yu, *Indus. Eng. Chem.*, v. 60, p. 20 (1968).

[16]P. I. Gold, *Chem. Eng.*, p. 170 (November 18, 1968).

[17]P. N. Shah and C. L. Yaws, "Densities of Liquids," *Chem. Eng.*, p. 131 (October 25, 1976).

[18]V. K. Mathur and R. N. Maddox, "Liquid Density Correlations," *Indian Chem. Eng.*, v. 18, no. 2, p. 29 (1976).

Self-Assessment Test

1. Draw a *p–T* chart for water. Label the following clearly: vapor-pressure curve, dew-point curve, saturated region, superheated region, subcooled region, and triple point. Show where evaporation, condensation, and sublimation take place by arrows.

2. Describe the state and pressure conditions of water initially at 20°F as the temperature is increased to 250°F in a fixed volume.

3. Look in Appendix J at the diagram for CO_2, Fig. J.3.
   (*a*) At what pressure is CO_2 solid in equilibrium with CO_2 liquid and vapor?
   (*b*) If the solid is placed in the atmosphere, what happens?

4. Use the Antoine equation to calculate the vapor pressure of ethanol at 50°C, and compare with the experimental value.

5. Determine the normal boiling point for benzene from the Antoine equation.

6. Prepare a Cox chart from which the vapor pressure of toluene can be predicted over the temperature range −20°C to 140°C.

Section 3.4   Saturation

> *Your objectives in studying this section are to be able to:*
>
> 1. Define saturated gas.
> 2. Calculate the partial pressure of the components of a saturated ideal gas given combinations of the temperature, pressure, volume and/or number of moles present; or calculate the partial volume; or calculate the number of moles of vapor.
> 3. Determine the condensation temperature of the vapor in a saturated gas given the pressure, volume, and/or number of moles.

How can you predict the properties of a mixture of a *pure vapor* (which can condense) *and a noncondensable gas*? A mixture containing a vapor behaves somewhat differently than does a pure component by itself. A typical example with which you are quite familiar is that of water vapor in air.

Water vapor is a gas, and like all gases its molecules are free to migrate in any direction. They will do so as long as they are not stopped by the walls of a container. Furthermore, the molecules will distribute themselves evenly throughout the entire volume of the container.

When any pure gas (or a gaseous mixture) comes in contact with a liquid, the gas will acquire vapor from the liquid. If contact is maintained for a considerable length of time, equilibrium is attained, at which time the *partial pressure of the vapor will equal the vapor pressure* of the liquid at the temperature of the system. Regardless of the duration of contact between the liquid and gas, after equilibrium is reached no more net liquid will vaporize into the gas phase. The gas is then said to be *saturated* with the particular vapor at the given temperature. We can also say that the gas mixture is at its dew point. Dew point for the mixture of pure vapor and noncondensable gas means the temperature at which the vapor just starts to condense.

Now, what do these concepts mean with respect to the quantitative measurement of gas–vapor properties. Suppose that you inject liquid water at 65°C into a cylinder of air at the same temperature, and keep the system at a constant temperature of 65°C. The pressure at the top of the cylinder will be maintained at 101.3 kPa (1 atm). What happens to the volume of the cylinder as a function of time? Figure 3.16 shows that the volume of the air plus the water vapor increases until the air is saturated with water vapor, after which stage the volume remains constant. Figure 3.17(a) indicates how the partial pressure of the water vapor increases with time until it reaches its vapor pressure of 24.9 kPa (187 mm Hg). Why does the partial pressure of the air decrease?

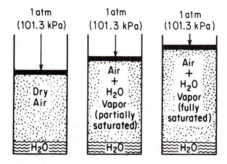

Fig. 3.16 Evaporation of water at constant pressure and temperature of 65°C.

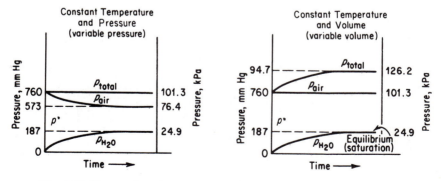

Fig. 3.17 Change of partial and total pressures on vaporization of water into air at constant temperature: (a) constant temperature and pressure (variable volume); (b) constant temperature and volume (variable pressure).

Next, suppose that you carry out a similar experiment, but maintain the volume constant and let the total pressure vary in the cylinder. Will the pressure go up or down with respect to time? What will be the asymptotic value of the partial pressure of the water vapor? The air? Look at Fig. 3.17(b) to see if your answers to these questions were correct.

Finally, is it possible to have the water evaporate into air and saturate the air, and yet maintain both a constant temperature, volume, and pressure in the cylinder? (*Hint:* What would happen if you let some of the gas–vapor mixture escape from the cylinder?)

Assuming that the ideal gas laws apply to both air and water vapor, as they do with excellent precision, we can say that the following relations hold *at saturation.* From Eq. (3.4), we know that

$$\frac{p_{air}}{p_{H_2O}} = \frac{n_{air}}{n_{H_2O}} \qquad \text{at constant temperature}$$

or, from Eq. (3.7),

$$\frac{V_{air}}{V_{H_2O}} = \frac{n_{air}}{n_{H_2O}} \qquad \text{at constant temperature}$$

Then

$$\frac{p_{air}}{p_{H_2O}} = \frac{p_{air}}{p_{total} - p_{air}} = \frac{V_{air}}{V_{total} - V_{air}} \qquad (3.42)$$

EXAMPLE 3.20 Saturation

What is the minimum number of cubic meters of dry air at 20°C and 100 kPa (750 mm Hg) that are necessary to evaporate 6 kg of alcohol if the total pressure remains constant at 100 kPa? Assume that the air is blown over the alcohol to evaporate it in such a way that the exit pressure of the air–alcohol mixture is at 100 kPa.

Solution

See Fig. E3.20. Assume that the process is isothermal. The additional data needed are

$$p^*_{alcohol} \text{ at } 20°C \text{ (68°F)} = 5.93 \text{ kPa (44.5 mm Hg)}$$

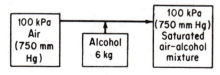

Fig. E3.20

The minimum volume of air calls for a saturated mixture; any condition less than saturated would require more air.

Basis: 6 kg of alcohol

The ratio of moles of alcohol to moles of air in the final gaseous mixture is the same as the ratio of the partial pressures of these two substances. Since we know the moles of alcohol, we can find the number of moles of air:

$$\frac{p_{\text{alcohol}}}{p_{\text{air}}} = \frac{n_{\text{alcohol}}}{n_{\text{air}}}$$

From Dalton's law,

$$p_{\text{air}} = p_{\text{total}} - p_{\text{alcohol}}$$
$$p_{\text{alcohol}} = 5.93 \text{ kPa (44.5 mm Hg)}$$
$$p_{\text{air}} = (100 - 5.93) \text{ kPa} = 94.1 \text{ kPa or } (750 - 44.5) \text{ mm Hg} = 705.5 \text{ mm Hg}$$

$$\frac{6 \text{ kg alcohol}}{} \left| \frac{1 \text{ kg mol alcohol}}{46 \text{ kg alcohol}} \right| \frac{94.1 \text{ kg mol air}}{5.93 \text{ kg mol alcohol}}$$

$$\left| \frac{22.4 \text{ m}^3}{1 \text{ kg mol}} \right| \frac{293 \text{ K}}{273 \text{ K}} \left| \frac{101.3 \text{ kPa}}{100 \text{ kPa}} \right| = 50.4 \text{ m}^3 \text{ air at } 20°C \text{ and } 100 \text{ kPa (750 mm Hg)}$$

EXAMPLE 3.21   Saturation

A telescopic gas holder contains 10,000 ft³ of saturated gas at 80°F and a pressure of 6.0 in. H_2O above atmospheric. The barometer reads 28.46 in. Hg. Calculate the weight of water vapor in the gas.

Solution

The total pressure of the saturated gas must be calculated first, say, in inches of mercury. See Fig. E3.21.

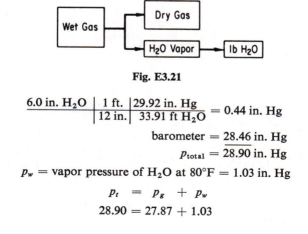

Fig. E3.21

$$\frac{6.0 \text{ in. } H_2O}{} \left| \frac{1 \text{ ft.}}{12 \text{ in.}} \right| \frac{29.92 \text{ in. Hg}}{33.91 \text{ ft } H_2O} = 0.44 \text{ in. Hg}$$

$$\text{barometer} = 28.46 \text{ in. Hg}$$
$$p_{\text{total}} = \overline{28.90 \text{ in. Hg}}$$

$$p_w = \text{vapor pressure of } H_2O \text{ at } 80°F = 1.03 \text{ in. Hg}$$

$$p_t = p_g + p_w$$
$$28.90 = 27.87 + 1.03$$

There exists

$$10,000 \text{ ft}^3 \text{ wet gas} \quad \text{at} \quad 28.90 \text{ in. Hg and } 80°\text{F}$$

$$10,000 \text{ ft}^3 \text{ dry gas} \quad \text{at} \quad 27.87 \text{ in. Hg and } 80°\text{F}$$

$$10,000 \text{ ft}^3 \text{ water vapor} \quad \text{at} \quad 1.03 \text{ in. Hg and } 80°\text{F}$$

$$°\text{R} = 80 + 460 = 540°\text{R}$$

From the volume of water vapor present at known conditions, you can calculate the volume at S.C. and hence the pound moles and finally the pounds of water.

Basis: 10,000 ft³ of water vapor at 1.03 in. Hg and 80°F

$$\frac{10{,}000 \text{ ft}^3 \text{ H}_2\text{O vapor}}{} \left| \frac{492°\text{R}}{540°\text{R}} \right| \frac{1.03 \text{ in. Hg}}{29.92 \text{ in. Hg}} \left| \frac{1 \text{ lb mol}}{359 \text{ ft}^3} \right| \frac{18 \text{ lb H}_2\text{O}}{1 \text{ lb mol H}_2\text{O}}$$

$$= 15.7 \text{ lb H}_2\text{O}$$

EXAMPLE 3.22 Smokestack Emission and Pollution

A local pollution-solutions group has reported the Simtron Co. boiler plant as being an air polluter and has provided as proof photographs of heavy smokestack emissions on 20 different days in January and February. As the chief engineer for the Simtron Co., you know that your plant is not a source of pollution because you burn natural gas (essentially methane) and your boiler plant is operating correctly. Your boss believes the pollution-solutions group has made an error in identifying the stack—it must belong to the company next door that burns coal. Is he correct? Is the pollution-solutions group correct? See Fig. E3.22.

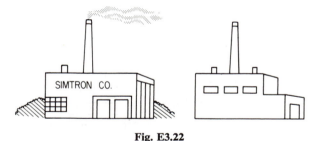

Fig. E3.22

Solution

Methane (CH_4) contains 2 kg mol of H_2 per kilogram mole of C, while coal (see Section 1.7) contains 71 kg of C per 5.6 kg of H_2 in 100 kg of coal. The coal analysis is equivalent to

$$\frac{71 \text{ kg C}}{} \left| \frac{1 \text{ kg mol C}}{12 \text{ kg C}} \right. = 5.9 \text{ kg mol C} \qquad \frac{5.6 \text{ kg H}_2}{} \left| \frac{\text{kg mol H}_2}{2.02 \text{ kg H}_2} \right. = 2.8 \text{ kg mol H}_2$$

or a ratio of 2.8/5.9 = 0.47 kg mol of H_2/kg mol of C. Suppose that each fuel burns with 40% excess air and that combustion is complete. We can compute the mole fraction of water vapor in each stack gas.

Basis: 1 kg mol of C

Natural Gas

$$CH_4 + 2O_2 \longrightarrow CO_2 + 2H_2O$$

fuel composition	kg mol	composition of combustion gases (kg mol)			
		CO_2	excess O_2	N_2	H_2O
C	1.0	1.0			
H_2	2.0				2.0
Air			0.80	10.5	
Total		1.0	0.80	10.5	2.0

Excess O_2: $2(0.40) = 0.80$
N_2: $(2)(1.40)(79/21) = 10.5$
The total kilogram moles of gas produced are 14.3 and the mole fraction H_2O is

$$\frac{2.0}{14.3} \simeq 0.14$$

Cool

$$C + O_2 \longrightarrow CO_2 \qquad H_2 + \tfrac{1}{2}O_2 \longrightarrow H_2O$$

fuel composition	kg mol	composition of gas produced (kg mol)			
		CO_2	excess O_2	N_2	H_2O
C	1	1			
H_2	0.47				0.47
Air			0.49	6.5	
Total		1	0.49	6.5	0.47

Excess O_2: $\underset{C}{1(0.40) = 0.40} \quad \underset{H_2}{(0.47)(1/2)(0.40) = 0.094}$
N_2: $1.40(79/21)(1 + 0.47(1/2)) = 6.5$
The total kilogram moles of gas produced are 8.46 and the mole fraction H_2O is

$$\frac{0.47}{8.46} \simeq 0.056$$

If the barometric pressure is 100 kPa, the stack gas becomes saturated at

	natural gas	coal
Pressure:	$100(0.14) = 14$ kPa (2.03 psia)	$100(0.056) = 5.6$ kPa (0.81 psia)
Equivalent temperature:	52.5°C (126°F)	35°C (95°F)

By mixing the stack gas with air, the mole fraction water vapor is reduced, and hence the condensation temperature is reduced. However, for equivalent dilution, the coal-burning plant will always have a lower condensation temperature. Thus, on cold winter days, the condensation of water vapor will occur more often and to a greater extent from power plants firing natural gas than from those using other fuels.

The public, unfortunately, sometimes concludes that all the emissions they perceive are pollution. Natural gas could appear to the public to be a greater pollutant than either oil or coal when, in fact, the emissions are just water vapor. The sulfur content of coal and oil can be released as sulfur dioxide to the atmosphere, and the polluting capacities of coal and oil are much greater than natural gas when all three are being burned properly. The sulfur contents as delivered to the consumers are as follows: natural gas, $4 \times 10^{-4}\%$ (as added mercaptans); number 6 fuel oil, up to 2.6%; and coal, from 0.5 to 5%.

What additional steps would you take to resolve the questions that were originally posed?

3.4-1   Gas–Liquid Equilibria for Multicomponent Systems

When a multicomponent gas is in equilibrium with a multicomponent liquid phase, more than one vapor can condense. Two useful linear equations exist to relate the mole fraction of one component in the gas phase to the mole fraction of the same component in the liquid phase.

Raoult's law. Used primarily for a component whose mole fraction approaches unity or for solutions of components quite similar in chemical nature, such as straight-chain hydrocarbons. Let the subscript i denote component, p_i be the partial pressure of component i in the gas phases, y_i be the gas-phase mole fraction, and x_i be the liquid-phase mole fraction. Then:

$$p_i = p_i^* x_i \tag{3.43}$$

Note that in the limit where $x_i \equiv 1$, $p_i \equiv p_i^*$. Often an equilibrium constant K_i is defined from Eq. (3.43) as follows by assuming that Dalton's law applied to the gas phase ($p_i = p_t y_i$):

$$K_i = \frac{y_i}{x_i} = \frac{p_i^*}{p_t} \tag{3.43a}$$

Most solutions used in industrial practice cannot be represented by Eq. (3.43) for wide ranges of x_i, hence Eq. (3.43) is treated as an ideal relationship and is modified by correction factors termed *activity coefficients* both for the gas and liquid phases, much as we use z to correct the ideal gas law. Activity factors and values of K_i can be found in numerous textbooks and handbooks.

Henry's law. Used primarily for a component whose mole fraction approaches zero, such as a dilute gas dissolved in a liquid:

$$p_i = H_i x_i \tag{3.44}$$

where H_i is the *Henry's law constant*. Note that in the limit where $x_i \equiv 0$, $p_i \equiv 0$. What would be the equilibrium constant K_i based on Henry's law if Dalton's law applied to the gas phase? Values of H_i can be found in Perry and many other handbooks.

3.4-2   Liquid–Liquid and Liquid–Solid Equilibria for Multicomponent Systems

Liquid–liquid equilibria are treated essentially in the same way as nonideal gas–liquid equilibria with activity-coefficient corrections in both phases. For liquid–solid equilibria, engineers rely almost entirely on recorded experimental data, as prediction of the liquid-phase composition is quite difficult.

EXAMPLE 3.23   Application of Raoult's Law

Suppose that a liquid mixture of n-hexane and n-octane is to be vaporized. What is the composition of the initial vapor formed if the total pressure is 1 atm?

Solution

The answer, of course, depends on the temperature of vaporization. We will assume that Raoult's law applies (as it does for such like compounds). You have to look up the coefficients of the Antoine equation to obtain the vapor pressures of the two components:

$$\log_{10} p^* = A - \frac{B}{C + t}$$

where p^* is in mm Hg and t is in °C:

	A	B	C
n-hexane (C_6):	6.87776	1171.530	224.366
n-octane (C_8):	6.92374	1355.126	209.517

Basis: 1 kg mol of liquid

We know that

$$p_t y_{C_6} = p_{C_6} = p_{C_6}^* x_{C_6} \qquad p_t y_{C_8} = p_{C_8} = p_{C_8}^* x_{C_8}$$

and

$$p_{C_6} + p_{C_8} = p_t \qquad x_{C_6} + x_{C_8} = 1 \qquad y_{C_6} + y_{C_8} = 1$$

We can combine these equations to get

$$y_{C_6} = \left(\frac{p_{C_6}^*}{p_t} \right) x_{C_6} \tag{a}$$

and

$$\frac{p_t - p_{C_8}^*}{p_{C_6}^* - p_{C_8}^*} = x_{C_6} \tag{b}$$

One way to proceed is to assume a temperature, then calculate x_{C_6} from Eq. (b) and y_{C_6} from Eq. (a). For example, at 121°C,

$$x_{C_6} = \frac{760 - 666}{3059 - 666} = 0.0393 \qquad y_{C_6} = \frac{3059}{760}(0.0393) = 0.158$$

Self-Assessment Test

1. What does the term "saturated gas" mean?

2. If a container with a volumetric ratio of air to liquid water of 5 is heated to 60°C and equilibrium is reached, will there still be liquid water present? at 125°C?

3. A mixture of air and benzene contains 10 mole % benzene at 43°C and 105 kPa pressure. At what temperature does the first liquid form? What is the liquid?

4. The dew point of water in atmospheric air is 82°F. What is the mole fraction of water vapor in the air if the barometric pressure is 750 mm Hg?

5. Ten pounds of $KClO_3$ is completely decomposed and the oxygen evolved collected over water at 80°F. The barometer reads 29.7 in. Hg. What weight of saturated oxygen is obtained?

6. If a gas is saturated with water vapor, describe the state of the water vapor and the air if it is:
   (*a*) Heated at constant pressure
   (*b*) Cooled at constant pressure
   (*c*) Expanded at constant temperature
   (*d*) Compressed at constant temperature

Section 3.5   Partial Saturation and Humidity

Your objectives in studying this section are to be able to:

1. List four ways of expressing partial saturation, and given the partial saturation in one form, calculate the corresponding values in the other three forms as well as the dew point.
2. Define relative saturation (humidity), molal saturation (humidity), absolute saturation (humidity), and humidity by formulas involving partial pressures of the gas components.

In Sec. 3.4 we dealt with mixtures of gas and vapor in which the gas was saturated with the vapor. Often, the contact time required between the gas and liquid for equilibrium (or saturation) to be attained is too long, and the gas is not compeltely saturated with the vapor. Then the vapor is not in equilibrium with a liquid phase, and the partial pressure of the vapor is less than the vapor pressure of the liquid at the given temperature. This condition is called *partial saturation*. What we have is simply a mixture of two or more gases that obey the real gas laws. What distinguishes this case from the previous examples for gas mixtures is that under suitable conditions it is possible to condense part of one of the gaseous components. In Fig. 3.18 you can see how the partial pressure of the water vapor in a gaseous mixture at constant volume obeys the ideal gas laws as the temperature drops until saturation is reached,

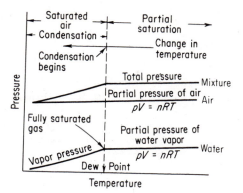

Fig. 3.18 Transformation of a partially saturated water vapor–air mixture into a saturated mixture as the temperature is lowered (volume = constant).

at which time the water vapor starts to condense. Until these conditions are achieved, you can confidently apply the gas laws to the mixture.

Several ways exist to express the concentration of a vapor in a gas mixture. You sometimes encounter mass or mole fraction (or percent), but more frequently one of the following:

(a) Relative saturation (relative humidity); Sec. 3.5.1

(b) Molal saturation (molal humidity); Sec. 3.5.2

(c) "Absolute" saturation ("absolute" humidity) or percent saturation (percent humidity); Secs. 3.5.3 and 5.3

(d) Humidity; Secs. 3.5.2 and 5.3

When the vapor is water vapor and the gas is air, the special term *humidity* applies. For other gases or vapors, the term *saturation* is used.

3.5-1 Relative Saturation

Relative saturation is defined as

$$\mathcal{RS} = \frac{p_{\text{vapor}}}{p_{\text{satd}}} = \text{relative saturation} \qquad (3.45)$$

where p_{vapor} = partial pressure of the vapor in the gas mixture

p_{satd} = partial pressure of the vapor in the gas mixture *if* the gas were saturated at the given temperature of the mixture (i.e., the vapor pressure of the vapor component)

Then, for brevity, if the subscript 1 denotes vapor,

$$\mathscr{RS} = \frac{p_1}{p_1^*} = \frac{p_1/p_t}{p_1^*/p_t} = \frac{V_1/V_t}{V_{\text{satd}}/V_t} = \frac{n_1}{n_{\text{satd}}} = \frac{\text{mass}_1}{\text{mass}_{\text{satd}}} \qquad (3.46)$$

You can see that relative saturation, in effect, represents the fractional approach to total saturation, as shown in Fig. 3.19. If you listen to the radio or TV and hear the

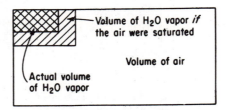

Fig. 3.19 Partially saturated gas with the water and air separated conceptually.

announcer say that the temperature is 20°C and the *relative humidity* is 60%, he or she implies that

$$\frac{p_{\text{H}_2\text{O}}}{p_{\text{H}_2\text{O}}^*}(100) = \% \; \mathscr{RH} \qquad (3.47)$$

with both the $p_{\text{H}_2\text{O}}$ and the $p_{\text{H}_2\text{O}}^*$ being measured at 20°C. Zero percent relative saturation means no vapor in the gas. What does 100% relative saturation mean? It means that the partial pressure of the vapor is the same as the vapor pressure of the substance that is the vapor.

EXAMPLE 3.24 Relative Humidity

The weatherman reported on the radio this morning that the temperature this afternoon would reach 94°F, the relative humidity would be 43%, the barometer 29.67 in. Hg, partly cloudy to clear, with the wind from SSE at 8 mi/hr. How many pounds of water vapor would be in 1 mi³ of afternoon air? What would be the dew point of this air?

Solution

The vapor pressure of water at 94°F is 1.61 in. Hg. We can calculate the partial pressure of the water vapor in the air from the given percent relative humidity; from this point forward, the problem is the same as the examples in the preceding section.

$$p_w = (1.61 \text{ in. Hg})(0.43) = 0.69 \text{ in. Hg}$$

$$p_{\text{air}} = p_t - p_w = 29.67 - 0.69 = 28.98 \text{ in. Hg}$$

Basis: 1 mi³ water vapor at 94°F and 0.69 in. Hg

$$\frac{1 \text{ mi}^3}{} \left| \left(\frac{5280 \text{ ft}}{1 \text{ mi}}\right)^3 \right| \frac{492°\text{R}}{554°\text{R}} \left| \frac{0.69 \text{ in. Hg}}{29.92 \text{ in. Hg}} \right| \frac{1 \text{ lb mol}}{359 \text{ ft}^3} \left| \frac{18 \text{ lb H}_2\text{O}}{1 \text{ lb mol}} \right| = 1.51 \times 10^8 \text{ lb H}_2\text{O}$$

Now the dew point is the temperature at which the water vapor in the air will first condense on cooling at *constant total pressure and composition*. As the gas is cooled you can see from Eq. (3.47) that the percent relative humidity increases since the partial pressure of the water vapor is constant while the vapor pressure of water decreases with temperature. When the percent relative humidity reaches 100%,

$$100 \frac{p_1}{p_1^*} = 100\% \quad \text{or} \quad p_1 = p_1^*$$

the water vapor will start to condense. This means that at the dew point the vapor pressure of water will be 0.69 in. Hg. From the steam tables you can see that this corresponds to a temperature of about 69°F.

3.5-2 Molal Saturation

Another way to express vapor concentration in a gas is to use the ratio of the moles of vapor to the moles of vapor-free gas:

$$\frac{n_{\text{vapor}}}{n_{\text{vapor-free gas}}} = \text{molal saturation} \tag{3.48}$$

If subscripts 1 and 2 represent the vapor and the dry gas, respectively, then for a binary system,

$$p_1 + p_2 = p_t \tag{3.49}$$

$$n_1 + n_2 = n_t \tag{3.50}$$

$$\frac{n_1}{n_2} = \frac{p_1}{p_2} = \frac{V_1}{V_2} = \frac{n_1}{n_t - n_1} = \frac{p_1}{p_t - p_1} = \frac{V_1}{V_t - V_1} \tag{3.51}$$

By multiplying by the appropriate molecular weights, you can find the mass of vapor per mass of dry gas:

$$\frac{(n_{\text{vapor}})(\text{mol. wt}_{\text{vapor}})}{(n_{\text{dry gas}})(\text{mol. wt}_{\text{dry gas}})} = \frac{\text{mass}_{\text{vapor}}}{\text{mass}_{\text{dry gas}}} \tag{3.52}$$

The special term *humidity* ($\mathcal{H}$) refers to the mass of water vapor per mass of bone-dry air and is used in connection with the humidity charts in Sec. 5.3.

3.5-3 "Absolute" Saturation (Humidity); Percentage Saturation (Humidity)

"Absolute" saturation is defined as the ratio of the moles of vapor per mole of *vapor-free* gas to the moles of vapor *that would be present* per mole of *vapor-free* gas if the mixture were completely saturated at the existing temperature and total pressure:

$$\mathfrak{a} \mathcal{S} = \text{``absolute saturation''} = \frac{\left(\dfrac{\text{moles vapor}}{\text{moles vapor-free gas}}\right)_{\text{actual}}}{\left(\dfrac{\text{moles vapor}}{\text{moles vapor-free gas}}\right)_{\text{saturated}}} \qquad (3.53)$$

Using the subscripts 1 for vapor and 2 for vapor-free gas,

$$\text{percent absolute saturation} = \frac{\left(\dfrac{n_1}{n_2}\right)_{\text{actual}}}{\left(\dfrac{n_1}{n_2}\right)_{\text{saturated}}}(100) = \frac{\left(\dfrac{p_1}{p_2}\right)_{\text{actual}}}{\left(\dfrac{p_1}{p_2}\right)_{\text{saturated}}}(100) \qquad (3.54)$$

Since p_1 saturated $= p_1^*$ and $p_t = p_1 + p_2$,

$$\text{percent absolute saturation} = 100\,\frac{\dfrac{p_1}{p_t - p_1}}{\dfrac{p_1^*}{p_t - p_1^*}} = \frac{p_1}{p_1^*}\left(\frac{p_t - p_1^*}{p_t - p_1}\right)100 \qquad (3.55)$$

Now you will recall that $p_1/p_1^* = $ relative saturation. Therefore,

$$\text{percent absolute saturation} = (\text{relative saturation})\left(\frac{p_t - p_1^*}{p_t - p_1}\right)100 \qquad (3.56)$$

Percent absolute saturation is always less than relative saturation except at saturated conditions (or at zero percent saturation) when percent absolute saturation = percent relative saturation.

EXAMPLE 3.25 Partial Saturation

Helium contains 12% by volume of ethyl acetate. Calculate (a) the percent relative saturation and (b) the percent absolute saturation of the mixture at a temperature of 30°C and a pressure of 98 kPa.

Solution

The additional data needed are

$$p_{\text{EtAc}}^* \text{ at } 30°C = 15.9 \text{ kPa (119 mm Hg)} \qquad \text{(from any suitable handbook)}$$

Using Dalton's laws, we obtain

$$p_{\text{EtAc}} = p_t y_{\text{EtAc}} = p_t\left(\frac{n_{\text{EtAc}}}{n_t}\right) = p_t\left(\frac{V_{\text{EtAc}}}{V_t}\right)$$

$$= (98)(0.12) = 11.76 \text{ kPa}$$

$$p_{\text{He}} = p_t - p_{\text{EtAc}}$$

$$= 98 - 11.76 = 86.2 \text{ kPa}$$

At 30°C:

(a) Percent relative saturation =

$$100 \frac{p_{EtAc}}{p^*_{EtAc}} = 100 \frac{11.76}{15.9} = 74.0\%$$

(b) Percent absolute saturation =

$$100 \frac{\dfrac{p_{EtAc}}{p_t - p_{EtAc}}}{\dfrac{p^*_{EtAc}}{p_t - p^*_{EtAc}}} = \frac{\dfrac{11.76}{98 - 11.76}}{\dfrac{15.9}{98 - 15.9}} 100 = \frac{\dfrac{11.76}{86.2}}{\dfrac{15.9}{82.1}} 100$$

$$= 70.0\%$$

EXAMPLE 3.26 Partial Saturation

A mixture of ethyl acetate vapor and air has a relative saturation of 50% at 30°C and a total pressure of 740 mm Hg. Calculate (a) the analysis of the vapor and (b) the molal saturation.

Solution

From Example 3.25, the vapor pressure of ethyl acetate at 30°C is 119 mm Hg.

$$\% \text{ relative saturation} = 50 = \frac{p_{EtAc}}{p^*_{EtAc}} 100$$

From this relation, the p_{EtAc} is

$$p_{EtAc} = 0.50(119) = 59.5 \text{ mm Hg}$$

(a)
$$\frac{n_{EtAc}}{n_t} = \frac{p_{EtAc}}{p_t} = \frac{59.5}{740} = 0.0805$$

Hence the vapor analyzes 8.05% EtAc and 91.95% air.

(b) Molal saturation is

$$\frac{n_{EtAc}}{n_{air}} = \frac{p_{EtAc}}{p_{air}} = \frac{p_{EtAc}}{p_t - p_{EtAc}} = \frac{59.5}{740 - 59.5}$$

$$= 0.0876 \frac{\text{mol EtAc}}{\text{mol air}}$$

EXAMPLE 3.27 Partial Saturation

The percent absolute humidity of air at 86°F and a total pressure of 750 mm Hg is 20%. Calculate the percent relative humidity and the partial pressure of the water vapor in the air. What is the dew point of the air?

Solution

Data from the steam tables are

$$p^*_{H_2O} \text{ at } 86°F = 31.8 \text{ mm Hg}$$

To get the relative humidity, $p_{H_2O}/p^*_{H_2O}$, we need to find the partial pressure of the water vapor in the air. This may be obtained from

$$\mathcal{RH} = 20 = \frac{\dfrac{p_{H_2O}}{p_t - p_{H_2O}}}{\dfrac{p^*_{H_2O}}{p_t - p^*_{H_2O}}} \, 100 = \frac{\dfrac{p_{H_2O}}{750 - p_{H_2O}}}{\dfrac{31.8}{750 - 31.8}} \, 100$$

This equation can be solved for p_{H_2O}:

$$0.00886 = \frac{p_{H_2O}}{750 - p_{H_2O}}$$

(a) $$p_{H_2O} = 6.7 \text{ mm Hg}$$

(b) $$\% \, \mathcal{RH} = 100\frac{6.7}{31.8} = 21.1\%$$

(c)   The dew point is the temperature at which the water vapor in the air would commence to condense. This would be at the vapor pressure of 6.7 mm, or about 41°F.

Self-Assessment Test

1. A mixture of air and benzene is found to have a 50% relative saturation at 27°C and an absolute pressure of 110 kPa. What is the mole fraction of benzene in the air?

2. A TV announcer says that the dew point is 92°F. If you compress the air to 110°F and 2 psig, what is the percent absolute humidity?

3. Nine hundred forty-seven cubic feet of wet air at 70°F and 29.2 in. Hg are dehydrated. If the water removed contains 0.94 lb of H_2O, what was the relative humidity of the wet air?

Section 3.6   Material Balances Involving Condensation and Vaporization

> **Your objective in studying this section is to be able to:**
>
> 1. Solve material balance problems involving a vapor (in a noncondensable gas) which may condense and may or may not vaporize.

The solution of material balance problems involving partial saturation, condensation, and vaporization will now be illustrated. Remember the drying problems in Chap. 2? They included water and some bone-dry material, as shown at the top of Fig. 3.20. To complete the diagram, we add the air that is used to remove the water from the material being dried.

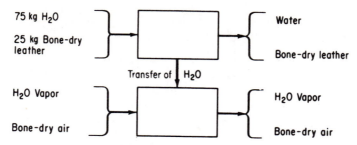

Fig. 3.20 Complete drying schematic.

You can analyze material balance problems involving water vapor in air in exactly the same fashion as you analyzed the material balance problems for the drying of leather (or paper, etc.), depending on the information provided and sought. (Humidity and saturation problems that include the use of energy balances and humidity charts are presented in Chap. 5.)

In connection with the examples that follow, we again stress that if you know the dew point of a gas mixture, you automatically know the partial pressure of the water vapor in the gas mixture. When a partially saturated gas is cooled at constant pressure, as in the cooling of air containing some water vapor at atmospheric pressure, the volume of the mixture may change slightly, but the partial pressures of the air and water vapor remain constant until the dew point is reached. At this point water begins to condense; the air remains saturated as the temperature is lowered. All that happens is that more water goes from the vapor into the liquid phase. At the time the cooling is stopped, the air is still saturated, and at its dew point.

The material balance problems that follow are solved using the 10-step strategy presented in Table 2.1.

EXAMPLE 3.28 Material Balance with Condensation

If the atmosphere in the afternoon during a humid period is at 90°F and 80 ℛℋℭ (barometer reads 738 mm Hg) while at night it is at 68°F (barometer reads 745 mm Hg), what percent of the water in the afternoon air is deposited as dew?

Solution

Assume a closed system.

Steps 1, 2, 3, and 4: The data are shown in Fig. E3.28. The check marks indicate that the compositions of the streams can be calculated from the information given.

If any dew is deposited at night, the air must be fully saturated at 68°F. You can easily check the dew point of the day air and find that it is 83°F.

$$\text{vapor pressure } H_2O \text{ at } 90°F = 36.1 \text{ mm Hg}$$
$$\text{vapor pressure } H_2O \text{ at } 68°F = 17.5 \text{ mm Hg}$$

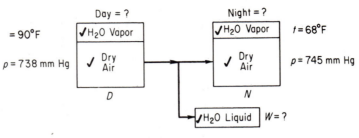

Fig. E3.28

The partial pressure of the water vapor in the day air is

$$(36.1 \text{ mm Hg})(0.80) = 28.9 \text{ mm Hg} \leftrightsquigarrow 83°F$$

What is the partial pressure of the water vapor in the night air?

$$(17.5 \text{ mm Hg})(1.00) = 17.5 \text{ mm Hg}$$

Next, we calculate all the partial pressures in the D and N streams, the results of which in effect give us the compositions of the two streams.

$$p_t = p_{air} + p_w \qquad \text{so that}$$

$$\text{day:} \quad p_{air} = 738 - 28.9 = 709.1 \text{ mm Hg}$$

$$\text{night:} \quad p_{air} = 745 - 17.5 = 727.5 \text{ mm Hg}$$

Step 5: If we take a basis, we have only two stream variables unknown, so that two component material balances can be made; hence the problem has a unique solution.

Step 6: As a basis you could select 1 ft³ of wet air, 1 lb mol wet (or dry) air, or many other suitable bases. The simplest basis to take is

Basis: 738 lb mol of moist day air

because then the moles = the partial pressures.

Step 4 Repeated:

component	initial mixture (day) lb mol = partial pressures		final mixture (night) lb mol = partial pressures
Air	709.1	← tie component →	727.5
H₂O	28.9		17.5
Total	738.0		745.0

Steps 7, 8, and 9: Because a tie component, the air, exists, the solution of the problem is now quite simple.

On the basis of 738 lb mol of moist day air, we have the following water in the night air:

Air balance: $\dfrac{17.5 \text{ lb mol } H_2O \text{ in night air}}{727.5 \text{ lb mol air}} \left| \dfrac{709.1 \text{ lb mol air}}{} \right. = 17.1 \text{ lb mol } H_2O$

Water balance: $28.9 - 17.1 = 11.8 \text{ lb mol } H_2O$ deposited as dew

$100\dfrac{11.8}{28.9} = 41\%$ of water in day air deposited as dew

Step 10: This problem could also be solved on the basis of

Basis: 1 lb mol of bone-dry air (BDA)

$$\underset{\textbf{initial}}{} \qquad - \qquad \underset{\textbf{final}}{} \qquad = \textbf{change}$$

$$\dfrac{28.8 \text{ lb mol } H_2O}{709.2 \text{ lb mol BDA}} - \dfrac{17.5 \text{ lb mol } H_2O}{727.5 \text{ lb mol BDA}} = \text{change}$$

$$0.0406 \qquad - \qquad 0.0241 \qquad = 0.0165 \dfrac{\text{lb mol } H_2O}{\text{lb mol BDA}}$$

$$\dfrac{0.0165}{0.0406} \, 100 = 41\% \text{ of water vapor deposited as dew}$$

Note especially that the operation

$$\underset{\text{initial}}{} \qquad \underset{\text{final}}{}$$

$$\dfrac{28.8}{738} - \dfrac{17.5}{745}$$

is meaningless since two different bases are involved—the wet initial air and the wet final air.

EXAMPLE 3.29 Dehydration

By absorption in silica gel you are able to remove all (0.33 kg) of the H_2O from moist air at 15°C and 98.6 kPa. The same air measures 1000 m³ at 20°C and 108.0 kPa when dry. What was the relative humidity of the moist air?

Solution

Steps 1, 2, 3, and 4: Figure E3.29 contains the known data. The check marks indicate that the moist air and other compositions can be calculated from the data given.

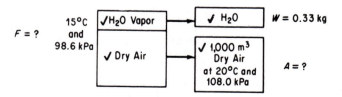

Fig. E3.29

Step 5: Values of two of the variables are unknown and two component material balances can be written; hence the problem has a unique solution.

Step 6: Either the *W* or the *A* stream can serve as the basis.

Basis: 1000 m³ bone-dry air (BDA) at 20°C and 108.0 kPa

Steps, 7, 8, and 9: Determine the kg mol of BDA.

$$\frac{1000 \text{ m}^3 \text{ BDA}}{} \left| \frac{273 \text{ K}}{293 \text{ K}} \right| \frac{108.0 \text{ kPa}}{101.3 \text{ kPa}} \left| \frac{1 \text{ kg mol}}{22.4 \text{ m}^3} \right. = 44.35 \text{ kg mol BDA}$$

Thus the wet air (*F*) has the composition (and partial pressures)

component	kg mol	partial pressure (kPa)
BDA	44.35	(44.35/44.68)(98.6) = 97.9
H₂O	0.33	(0.33/44.68)(98.6) = 0.7
Total	44.68	98.6

Saturated air at 15°C has a vapor pressure of 1.70 kPa (12.78 mm Hg), so that the percent relative humidity of the moist air was

$$\frac{p_{H_2O}}{p_{H_2O}^*} = \frac{0.7}{1.70}(100) = 41\%$$

EXAMPLE 3.30 Humidification

One thousand cubic meters of moist air at 101 kPa and 22°C and with a dew point of 11°C enters a process. The air leaves the process at 98 kPa with a dew point of 58°C. How many kilograms of water vapor are added to each kilogram of wet air entering the process?

Solution

Steps 1, 2, 3, and 4: The known data appear in Fig. E3.30. Additional data needed are

dew-point temp. (°C)	$p_{H_2O}^*$ (mm Hg)	$p_{H_2O}^*$ (kPa)
11	9.84	1.31†
58	136.1	18.1†

†These give the partial pressures of the water vapor in the initial and final gas mixtures.

Step 3 Continued: In effect, all the compositions are known as indicated by the check marks in the figure. The partial pressures of the water vapor are the pressures at the dew point in each case.

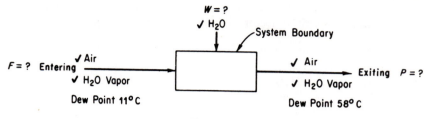

Fig. E3.30

component	initial moles	initial = initial partial pressures (kPa)	final partial pressures (kPa)
Water vapor	1.31	tie	18.1
BDA*	99.7	← component →	79.9
Total	101.0		98.0

*Bone-dry air.

Step 5: F is known; hence only W and P are unknown. Two component balances can be written; hence the problem has a unique solution.

Steps 6, 7, and 8: We have a tie component of bone-dry air (BDA) to tie F to P, and could then get W by subtraction; hence a good basis is the one implied in step 3:

Basis: 101 kg mol of moist air entering

Step 9: In the exit gas we can say that there are

18.1 kg mol water vapor/79.9 kg mol BDA

18.1 kg mol water vapor/98.0 kg mol moist air

79.9 kg mol BDA/98.0 kg mol moist air

Since we have a tie element of bone-dry air and want to know how many moles of water vapor are present in the final air on our basis of 101 kg mol entering wet air, we write

BDA balance: $\dfrac{99.7 \text{ kg mol BDA in} \mid 18.1 \text{ kg mol vapor in exit}}{79.9 \text{ kg mol BDA in exit}}$

$= 22.6 \text{ kg mol vapor in exit}$

H₂O balance: 22.6 kg mol exit vapor − 1.31 kg mol entering vapor

$= 21.3 \text{ kg mol of water vapor added}$

Therefore, our answer should be

$$\frac{21.3 \text{ kg mol vapor added}}{101 \text{ kg mol wet gas in}} = \frac{0.211 \text{ kg mol vapor added}}{1 \text{ kg mol wet gas in}}$$

Next we convert to weights from moles. The molecular weight of water is 18; the average molecular weight of the original wet air has to be calculated:

component	kg mol	mol. wt.	kg
BDA	99.7	29	2891
H_2O	1.31	18	23.6
Total	101.0		2915

$$\frac{2915 \text{ kg}}{101 \text{ kg mol}} = 28.9$$

$$\frac{0.213 \text{ kg mol } H_2O}{1 \text{ kg mol wet gas}} \left| \frac{18 \text{ kg}}{1 \text{ kg mol } H_2O} \right| \frac{1 \text{ kg mol wet gas}}{28.9 \text{ kg}} = 0.133 \frac{\text{kg } H_2O}{\text{kg wet gas}}$$

Comments:

(a) If the original problem had asked for the moles of exiting wet gas, we could have used the tie component in the following manner:

$$\frac{99.7 \text{ mol BDA in}}{} \left| \frac{98.0 \text{ mol wet gas out}}{79.9 \text{ mol BDA out}} \right. = 122.3 \text{ mol wet gas out}$$

(b) If a basis of 1 mol of bone-dry air had been used, the following calculation would apply:

Basis: 1 mol BDA

in: $\dfrac{\text{mol vapor}}{\text{mol BDA}} = \dfrac{1.31}{99.7} = 0.0131$

out: $\dfrac{\text{mol vapor}}{\text{mol BDA}} = \dfrac{18.1}{79.9} = 0.227$

Since there are more moles of vapor in the final air (per mole of bone-dry air), we know that vapor must have been added:

$$\text{moles vapor added} = 0.227 - 0.0131 = 0.214 \text{ mole}$$

The remainder of the calculations would be the same.

We have gone over a number of examples of condensation and vaporization, and you have seen how a given amount of air at atmospheric pressure can hold only a certain maximum amount of water vapor. This amount depends on the temperature of the air, and any decrease in the temperature will lower the water-bearing capacity of the air.

An increase in pressure also will accomplish the same effect. If a pound of saturated air at 75°F is isothermally compressed (with a reduction in volume, of course), liquid water will be deposited out of the air just like water being squeezed out of a wet sponge (Fig. 3.21). This process has been previously described in the $p-T$ diagram for water (Fig. 3.12), by line $I-H-G$.

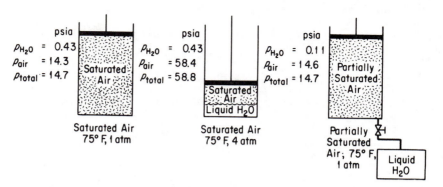

Fig. 3.21 Effect of an increase of pressure on saturated air and a return to the initial pressure.

For example, if a pound of saturated air at 75°F and 1 atm is compressed isothermally to 4 atm (58.8 psia), almost three-fourths of the original content of water vapor now will be in the form of liquid, and the air has a dew point of 75°F at 4 atm. Remove the liquid water, expand the air isothermally back to 1 atm, and you will find that the dew point has been lowered to about 36°F. Mathematically (1 = state at 1 atm, 4 = state at 4 atm) with $z = 1.00$ for both components:

For saturated air at 75°F and 4 atm:

$$\left(\frac{n_{H_2O}}{n_{air}}\right)_4 = \left(\frac{p^*_{H_2O}}{p_{air}}\right)_4 = \frac{0.43}{58.4}$$

For the same air saturated at 75°F and 1 atm:

$$\left(\frac{n_{H_2O}}{n_{air}}\right)_1 = \left(\frac{p^*_{H_2O}}{p_{air}}\right)_1 = \frac{0.43}{14.3}$$

Since the air is the tie element in the process,

$$\left(\frac{n_4}{n_1}\right)_{H_2O} = \frac{\dfrac{0.45}{58.4}}{\dfrac{0.43}{14.3}} = \frac{14.3}{58.4} = 0.245$$

24.5% of the original water will remain as vapor.

After the air–water vapor mixture is returned to a total pressure of 1 atm, the following two familiar equations now apply:

$$p_{H_2O} + p_{air} = 14.7$$

$$\frac{p_{H_2O}}{p_{air}} = \frac{n_{H_2O}}{n_{air}} = \frac{0.43}{58.4} = 0.00737$$

From these two relations you can find that

$$p_{H_2O} = 0.108 \text{ psia}$$
$$p_{air} = 14.6$$
$$\overline{p_{total} = 14.7 \text{ psia}}$$

The pressure of the water vapor represents a dew point of about 36°F, and a relative humidity of

$$(100)\,\frac{p_{H_2O}}{p_{H_2O}^*} = \frac{0.108}{0.43}\,(100) = 25\%$$

Self-Assessment Test

1. Gas from a synthetic gas plant analyzes (on a dry basis) 4.8% CO_2, 0.2% O_2, 25.4% CO, 0.4% C_2H_4, 12.2% H_2, 3.6% CH_4, and 53.4% N_2. The coal used analyzes 70.0% C, 6.5% H, 16.0% O, and 7.5% ash. The entering air for combustion has a partial pressure of water equal to 2.67 kPa. The barometer reads 101 kPa. Records show that 465 kg of steam is supplied to the combustion vessel per metric ton of coal fired. Calculate the dew point of the exit gas.

2. A liquid solution of pharmaceutical material to be dried is sprayed into a stream of hot gas. The water evaporated from the solution leaves with the exit gases. The solid is recovered by means of cyclone separators. Operating data are:

   Inlet air: 100,000 ft³/hr, 600°F, 780 mm Hg, humidity of 0.00505 lb H_2O/lb dry air

   Inlet solution: 300 lb/hr, 15% solids, 70°F

   Outlet air: 200°F, 760 mm Hg, dew point 100°F

   Outlet solid: 130°F

   Calculate the composition of the outlet solid—it is not entirely dry.

3. A gas leaves a solvent recovery system saturated with benzene at 50°C and 750 mm Hg. The gas analyzes, on a benzene-free basis, 15% CO, 4% O_2, and the remainder nitrogen. This mixture is compressed to 3 atm and is subsequently cooled to 20°C. Calculate the percent benzene condensed in the process. What is the relative saturation of the final gas?

Section 3.7   Phase Phenomena

> *Your objectives in studying this section are to be able to:*
>
> 1. Write down the phase rule, define each parameter in the phase rule, and be able to apply the phase rule to determine the degrees of freedom, or the number of components, or the phases that exist in a system.

2. Sketch (roughly) the p–V–T surface and the two-dimensional projections thereof for water.

3. Predict the p–V–T behavior of a pure component or a mixture given the p–V–T diagram, and identify a specified state as residing in the solid, liquid, or vapor single-phase region or the solid plus liquid, solid plus vapor, or liquid plus vapor two-phase region.

4. Completely describe, from a phase diagram, all changes that occur on cooling or heating a given composition of a two-component system; give all phases present at each temperature and approximate amounts of each phase; also, give the degrees of freedom at any point in the cooling or heating process.

5. Locate a reference containing the properties of a pure substance in your library.

We now consider briefly some of the qualitative characteristics of vapors, liquids, and solids.

3.7-1   The Phase Rule

You will find the phase rule a useful guide in establishing how many properties, such as pressure and temperature, have to be specified to definitely fix all the remaining properties and number of phases that can coexist for any physical system. The rule can be applied only to systems in *equilibrium*. It says that

$$F = C - \mathcal{P} + 2 \tag{3.57}$$

where F = number of degrees of freedom (i.e., the number of independent properties that have to be specified to determine all the intensive properties of each phase of the system of interest)

C = number of components in the system; for circumstances involving chemical reactions, C is *not* identical to the number of chemical compounds in the system but is equal to the number of chemical compounds less the number of independent-reaction and other equilibrium relationships among these compounds (refer to Appendix L for details)

$\mathcal{P}$ = number of phases that can exist in the system; a phase is a homogeneous quantity of material such as a gas, a pure liquid, a solution, or a homogeneous solid

Variables of the kind with which the phase rule is concerned are called *phase-rule variables*, and they are *intensive* properties of the system. By this we mean properties that do not depend on the quantity of material present. If you think about the properties we have employed so far in this text, you have the feeling that pressure and temperature are independent of the amount of material present. So is concentration, but what about volume? The total volume of a system is called an *extensive* property because it does depend on how much material you have; the specific volume, on the other hand, the cubic meter per kilogram, for example, is an intensive property because it is independent of the amount of material present. In Chap. 4 we take up additional intensive properties, such as internal energy and enthalpy. You should remember that the specific (per unit mass) values of these quantities are intensive properties; the total quantities are extensive properties.

An example will clarify the use of these terms in the phase rule. You will remember for a pure gas that we had to specify three of the four variables in the ideal gas equation $pV = nRT$ in order to be able to determine the remaining one unknown. You might conclude that $F = 3$. If we apply the phase rule, for a single phase $\mathcal{P} = 1$, and for a pure gas $C = 1$, so that

$$F = C - \mathcal{P} + 2 = 1 - 1 + 2 = 2 \qquad \text{variables to be specified}$$

How can we reconcile this apparent paradox with our previous statement? Since the phase rule is concerned with intensive properties only, the following are phase-rule variables in the ideal gas law:

$$\left.\begin{array}{l} P \\ \hat{V} \text{ (specific molar volume)} \\ T \end{array}\right\} \quad \text{3 intensive properties}$$

Thus the ideal gas law should be written

$$p\hat{V} = RT$$

and once two of the intensive variables are fixed (so that F actually equals 2), the third is also automatically fixed.

An *invariant* system is one in which no variation of conditions is possible without one phase disappearing. An example with which you may be familiar is the ice–water–water vapor system, which exists at only one temperature (0.01°C):

$$F = C - \mathcal{P} + 2 = 1 - 3 + 2 = 0$$

With all three phases present, none of the physical conditions can be varied without the loss of one phase. As a corollary, if the three phases are present, the temperature, the specific volume, and so on, must always be fixed at the same values. This phenomenon is useful in calibrating thermometers and other instruments.

A complete discussion of the significance of the term C and the other terms in the phase rule is beyond the scope of this text; for further information read one of the general references on the phase rule listed at the end of the chapter or consult an advanced book on thermodynamics or physical chemistry. We are now going to consider how phase phenomena can be illustrated by means of diagrams.

3.7-2 Phase Phenomena of Pure Components

If we are to clearly understand phase phenomena, the properties of a substance should be shown in three dimensions, particularly if there can be more than one phase in the region for which the p–$\hat{V}$–T properties are to be presented. We previously overcame the handicap of using two dimensions to present three-dimensional data by showing, on the two-dimensional graph, lines of constant value for the other properties of the system (see Fig. 3.2, which has p and $\hat{V}$ as axes, and lines of constant temperature as the third parameter). In this section we present a more elaborate picture of these p–$\hat{V}$–T properties. Figure 3.22 illustrates the relationships between the three-dimensional presentations of p–$\hat{V}$–T data for water and the conventional two-dimensional diagrams you usually encounter. Be certain that you view Fig. 3.22 as a *surface* and not as solid figures; only the p–$\hat{V}$–T points on the surface exist, and no point exists above or below the surface.

You can see that the two-dimensional diagrams are really projections of the three-dimensional surfaces onto suitable axes. On the p–$\hat{V}$ diagram, you will observe lines of constant temperature called *isothermal lines*, and on the p–T diagram you will see lines of constant volume, called *isometric lines* or *isochores*. On the T–$\hat{V}$ diagram, we might have put lines of constant pressure known as *isobaric lines*, but these are not shown since they would obscure the more important features which are shown. The two-phase vapor–liquid region is represented by the heavy envelope in the p–$\hat{V}$ and the T–$\hat{V}$ diagrams. On the p–T diagram this two-phase region appears only as a single curve because you are looking at the three-dimensional diagram from the side. The p–T diagram shows what we have so far called the *vapor-pressure curve*, but examination of the associated diagrams show that the "curve" is really a surface on which two phases coexist, each having the properties found at the appropriate bounding line for the surface. A point on the surface shows merely the overall composition of one phase in equilibrium with another. For example, at point A we have liquid and at point B we have vapor, and in between them no single phase exists. If we added the liquid at A to the vapor at B, we would have a two-phase mixture with the gross properties shown by point M. You will notice that in the p–$\hat{V}$ diagram the lines of constant temperature and pressure become horizontal in the two-phase region, because if you apply the phase rule

$$F = C - \mathcal{P} + 2 = 1 - 2 + 2 = 1$$

only one variable can change—the specific volume. In the two-phase region, where condensation or evaporation takes place, the pressure and temperature remain constant. Do you recall our discussion of boiling water?

Other features illustrated on the diagrams are the critical point, the critical isotherm, the triple point in the p–T diagram, which actually becomes a triple-point line in the p–$\hat{V}$ and T–$\hat{V}$ diagrams; and the existence of two solid phases and a liquid phase in equilibrium, such as solid ice I, solid ice III, and liquid. The temperature at the freezing point for water (i.e., the point where ice, water, air, and water vapor

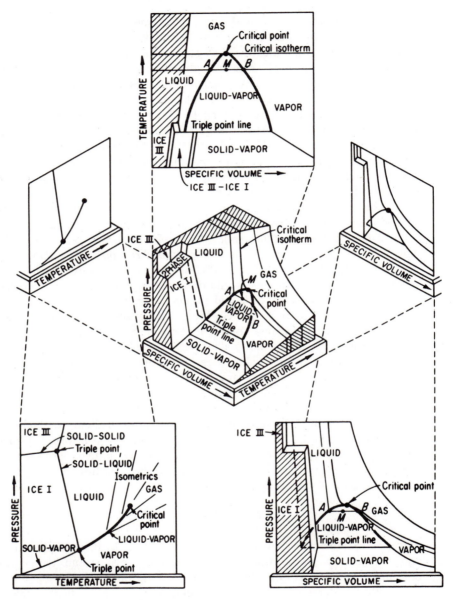

Fig. 3.22 p–$\hat{V}$–T surface and projections for H_2O.

are in equilibrium) is not quite the same temperature as that at the triple point, which is the point where ice, water, and water vapor are in equilibrium. The presence of air at 1 atm lowers the freezing temperature by about 0.01°C, so that the triple point is at 0.00602 atm (0.61 kPa) and 0.01°C. Another triple point for the water system exists at 2200 atm and 20°C between ice III, ice I, and liquid water.

3.7-3   Phase Phenomena of Mixtures

Presentation of phase phenomena for mixtures that are completely miscible in the liquid state involves some rather complex reductions of three-, four-, and higher-dimensional diagrams into two dimensions. We shall restrict ourselves to two-component systems because, although the ideas discussed here are applicable to any number of components, the graphical presentation of more complex systems in an elementary text such as this is probably more confusing than helpful. We shall use hydrocarbons for most of our examples of two-component systems since a vast amount of experimental work has been reported for these compounds.

It would be convenient if the critical temperature of a mixture were the mole weight average of the critical temperatures of its pure components, and the critical pressure of a mixture were simply a mole weight average of the critical pressures of the pure components (the concept used in Kay's rule), but these maxims simply are not true, as shown in Fig. 3.23. The pseudocritical temperature falls on the dashed line between the critical temperatures of CO_2 and SO_2, whereas the actual critical point for the mixture lies somewhere else. The dashed line in Fig. 3.25 illustrates (for another system) the three-dimensional aspects of the locus of the actual critical points.

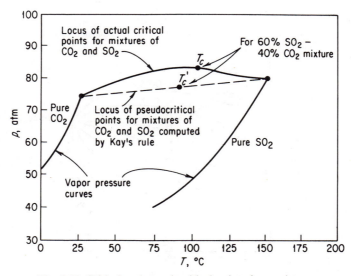

Fig. 3.23 Critical and pseudocritical points for a mixture.

Figure 3.24 is a p–T diagram for a 25% methane/75% butane mixture. You can compare this figure with the one for a pure substance, water, as in Fig. 3.12; the p–T curve for pure butane or pure methane would look just like the diagram for water (excluding the solid phases). For a mixture of a fixed composition, we have only three variables to consider, p, $\hat{V}$, and T. Lines of constant specific volume are shown outside the envelope of the two-phase region. Inside the envelope are shown lines of constant fractions of liquid, starting at the high pressures, where 100% liquid exists (zero percent vapor), and ranging down to 0% liquid and 100% vapor at the low pressures. The 100% saturated liquid line is the bubble-point line and the 100% saturated vapor line is the dew-point line.

The critical point is shown as point C. You will note that this is not the point of maximum temperature at which vapor and liquid can exist in equilibrium; the latter is the maximum cricondentherm, point D. Neither is it the point of maximum pressure at which vapor and liquid can exist in equilibrium because that point is the maximum cricondenbar (point E). The maximum temperature at which liquid and vapor can exist in equilibrium for this mixture is about $30°$ higher than the critical temperature, and the maximum pressure is 60 or 70 psia greater than the critical pressure. You can clearly see, therefore, that the best definition of the critical point was the one which stated that the density and other properties of the liquid and vapor become identical at the critical point. This definition holds for single components as well as mixtures.

Figure 3.24 represented a two-component system with a fixed overall composition. You might wonder what a diagram would look like if we were to try to show systems of several compositions on one page. This has been done in Fig. 3.25. Here we have a composite p–T diagram, which is somewhat awkward to visualize, but

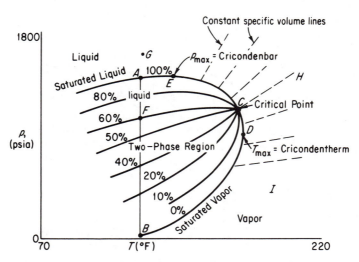

Fig. 3.24 p–T diagram for binary system: 25% CH_4–75% C_4H_{10}.

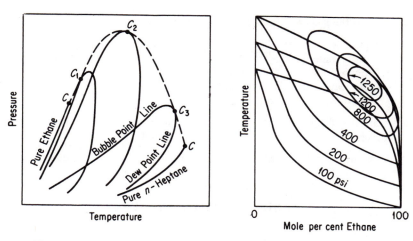

Fig. 3.25 p–T and T–x diagrams for ethane–heptane mixtures of different compositions.

represents the bubble-point and dew-point curves for various mixtures of ethane and heptane. These curves in essence are intersections of surfaces in the composition coordinate sliced out of a three-dimensional system and are stacked one in front of the other, although in two dimensions it appears that they intersect one another. The vapor-pressure curves for the two pure components are at the extreme sides of the diagrams as single curves (as you might expect). Each of the loops represents the two-phase area for a system of a specific composition. An infinite number of these surfaces are possible, of course. The dashed line indicates the envelope of the critical points for each possible composition. Although this line appears to be two-dimensional in Fig. 3.25, it actually is a three-dimensional line of which only the projection is shown in the figure.

Another way to illustrate the phase phenomena for the two-component systems we have been discussing is to use pressure–composition diagrams at constant temperature or, alternatively, to use temperature–composition diagrams at constant pressure. A temperature–composition diagram with pressure as the third parameter is illustrated in Fig. 3.25 for the ethane–heptane system.

More complicated phase diagrams for multicomponent systems (ternary and higher) can be found in the references at the end of the chapter.

Self-Assessment Test

1. Is the critical point a single phase? If not, what phases are present? Repeat for the triple point (for water).

2. What does a p–V–T value lying above the surface in Fig. 3.23 mean? Below the surface?

3. Can a liquid be changed to a vapor without passing through a distinct discontinuity in properties? Can a solid be transformed similarly to vapor?

4. Sketch the p–V, p–T, and T–V diagrams for water.

5. A vessel contains air: $N_2(g)$, $O_2(g)$, and $Ar(g)$.
   (*a*) How many phases, components, and degrees of freedom are there according to the phase rule?
   (*b*) Repeat for a vessel one-third filled with liquid ethanol and two-thirds filled with N_2 plus ethanol vapor.

6. Can the following system exist at equilibrium: $H_2O(s)$, $H_2O(l)$, $H_2O(g)$, decane(s), decane(l), decane(g)? (*Hint:* Decane is insoluble in water.)

7. How many independent properties are required to fix the equilibrium state of a pure compound?

SUPPLEMENTARY REFERENCES

General

1. American Petroleum Institute, Division of Refining, *Technical Data Book—Petroleum Refining*, API, New York, 1970.
2. Angus, S., B. Armstrong, and K. M. de Reuck, eds., *International Thermodynamic Tables of the Fluid State*, IUPAC Chemical Data Series, 16, Pergamon Press, Oxford, 1976, and periodically thereafter.
3. Boublik, T., V. Fried, and E. Hala, *The Vapour Pressures of Pure Substances*, Elsevier, Amsterdam, 1973.
4. *CALPHAD* (a journal), published quarterly since 1979, Pergamon Press, Oxford.
5. Coward, I., S. E. Gale, and D. R. Webb, "Process Engineering Calculations with Equations of State," *Trans. Inst. Chem. Eng.*, v. 56, pp. 19–27 (1978).
6. Edmister, Wayne C., *Applied Hydrocarbon Thermodynamics*, Gulf Publications Co., Houston, Tex., 1961, Chaps. 1–4.
7. Kehianian, H. V., ed., *Selected Data on Mixtures, Series A*, Thermodynamics Research Center, Texas A & M University, College Station, Tex., 1976 and following.
8. Ohe, S., *Computer Aided Data Book of Vapor Pressure*, Data Book Publishing Co., Tokyo, 1977.
9. Reid, R. C., J. M. Prausnitz, and T. K. Sherwood, *The Properties of Gases and Liquids*, 3rd ed., McGraw-Hill, New York, 1977.
10. Sterbacek, Z., B. Biskup, and P. Tausk, *Calculation of Properties Using Corresponding State Methods*, Elsevier, Amsterdam, 1979.
11. Vargaftik, N. B., *Tables on the Thermophysical Properties of Liquids and Gases*, 2nd ed., Halsted Press (Wiley), New York, 1975.
12. Wexler, A., *Humidity and Moisture*, 4 vols., Van Nostrand Reinhold, New York, 1965.
13. Williams, E. T., and R. C. Johnson, *Stoichiometry for Chemical Engineers*, McGraw-Hill, New York, 1958.

References on Phase Phenomena

1. Domb, C., and M. S. Green, eds., *Phase Transitions and Critical Phenomena*. Vol. 2, Academic Press, New York, 1972.
2. Ricci, J. E., *The Phase Rule and Heterogeneous Equilibrium*, Van Nostrand Reinhold, New York, 1951.

3. Stanley, H. E., *Introduction to Phase Transitions and Critical Phenomena*, Oxford University Press, London, 1972.

PROBLEMS[19]

Section 3.1

3.1. One liter of a gas is under a pressure of 780 mm Hg. What will be its volume at standard pressure, the temperature remaining constant?

3.2. A gas occupying a volume of 1 m³ under standard pressure is expanded to 1.200 m³, the temperature remaining constant. What is the new pressure?

3.3. A gas of 0.200 m³ is at 15.7°C. Find its volume at 0°C, the pressure remaining constant.

3.4. Under standard conditions, a gas measures 10.0 liters in volume. What is its volume at 92°F and 29.4 in. Hg?

3.5. A gas measured 150 cm³ at 17.5°C, and because of a change in temperature, the pressure remaining constant, the volume decreased to 125 cm³. What was the new temperature?

3.6. The pressure on a confined gas, at 60°F, was 792 mm Hg. If the pressure later registered 820 mm Hg, what was the temperature then, the volume remaining unchanged?

3.7. Since 1917, Goodyear has built over 300 airships, most of them military. One of the newest is the *America*, which has a bag 192 ft long, is 50 ft in diameter, and holds about 202,000 ft³ of helium. Her twin 210-hp engines produce a crusing speed of 30 to 35 mi/hr and a top speed of 50 mi/hr. The 23-ft gondola will hold the pilot and and six passengers. Blimps originally were made of rubber-impregnated cotton, but the bag of the *America* is made of a two-ply fabric of Dacron polyester fiber coated with neoprene. The bag's outer surface is covered with an aluminized coat of Hypalon synthetic rubber. Assuming that the bag size cited is at 1 atm and 25°C, estimate the temperature increase or decrease in the bag at a height of 1000 m (where the pressure is 740 mm Hg) if the bag volume does not change. If the temperature remains 25°C, explain how you might estimate the volume change in the bag.

3.8. Argon gas occupies a volume of 200 cm³ under a pressure of 780 mm Hg. The temperature remaining constant, what pressure must be applied if the volume becomes 400 cm³?

3.9. Compressed air from a tank at 125 psig is used to inflate to 30 psig a tire that has a volume of 0.50 ft³, the temperature remaining constant. What volume of air is taken from the tank? What volume is this at standard pressure?

3.10. A recent newspaper report states: "Home meters for fuel gas measure the volume of gas usage based on a standard temperature, usually 60 degrees. But gas contracts when it's cold and expands when warm. East Ohio Gas Co. figures that in chilly Cleveland, the homeowner with an outdoor meter gets more gas than the meter

[19]An asterisk designates problems appropriate for computer solution. Refer also to the computer problems at the end of the chapter.

says he does, so that's built into the company's gas rates. The guy who loses is the one with an indoor meter: If his home stays at 60 degrees or over, he'll pay for more gas than he gets. (Several companies make temperature-compensating meters, but they cost more and aren't widely used. Not surprisingly, they are sold mainly to utilities in the North.)"

Suppose that the outside temperature drops from 60°F to 10°F. What is the percentage increase in the mass of the gas passed by a noncompensated outdoor meter that operates at constant pressure? Assume that the gas is CH_4.

3.11. The highest temperature of a gas holder in summer is 42°C, and the lowest in winter is −38°C. Calculate how many more kilograms of methane may be contained by the gas holder of volume equal to 2000 m³ at the lowest winter temperature compared to the highest summer temperature. Assume that the pressure in the gas holder is maintained at 780 mm Hg in both cases.

3.12. The water tank on a boat has a volume of 3.00 m³, and is tested for leaks by attaching it to a helium tank (after initial evacuation of the water tank). At 35°C and an absolute pressure of 205 kPa, how many kilograms of helium are needed to fill the water tank?

3.13. A cylinder of nitrogen contains 1.00 m³ of gas at 20°C and atmospheric pressure. If the valve on the cylinder is opened and the cylinder is heated to 100°C, calculate the fraction of the nitrogen that leaves the cylinder.

3.14. One of the experiments in the fuel-testing laboratory has been giving some trouble because a particular barometer gives erroneous readings owing to the presence of a small amount of air above the mercury column. At a pressure of 755 mm Hg the barometer reads 748 mm Hg, and at 740 the reading is 736. What will the barometer read when the actual pressure is 760 mm Hg?

3.15. Explain whether or not the statement is correct, and, if not, modify the statement to make it correct:
(a) Pressure is how much a fluid weighs.
(b) If more of a gas is pumped into a closed drum, the volume of gas in the drum increases.
(c) If more of a gas is pumped out of a closed drum, the weight of gas in the drum decreases.

3.16. Two tanks are initially sealed off from one another by means of valve *A*. Tank I initially contains 1.00 ft³ of air at 100 psia and 150°F. Tank II initially contains a nitrogen–oxygen mixture containing 95 mole % nitrogen at 200 psia and 200°F. Valve *A* is then opened allowing the contents of the two tanks to mix. After complete mixing had been effected, the gas was found to contain 85 mole % nitrogen. Calculate the volume of tank II. See Fig. P3.16.

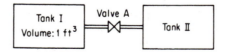

Fig. P3.16

3.17. From the known standard conditions, calculate the value of the gas law constant R in the following sets of units:

(a) cal/(g mol)(K)

(b) Btu/(lb mol)(°R)

(c) (psia)(ft³)/(lb mol)(°R)

(d) J/(g mol)(K)

(e) (cm³)(atm)/(g mol)(K)

(f) (ft³)(atm)/(lb mol)(°R)

3.18. A gas-fired furnace uses 3000 ft³/hr of gas at a pressure of 30.0 in. Hg and at a temperature of 100°F. The gas is purchased on a metered basis referred to 30.0 in. Hg and 60°F. Your accounting department wants to know if there is any difference between these two measurements. What percentage increase or decrease, if any, in volume is there? Base your percentage on the conditions of the gas as used in the furnace.

3.19. How many kilograms of CS_2 must be burned with the stoichiometric quantity of O_2 ($CS_2 + 3O_2 \rightarrow CO_2 + 2SO_2$) to yield 200 m³ of gas at 300°C and 100 kPa?

3.20. How much mercuric oxide must be heated to liberate a liter of oxygen measured at 25°C and 765 mm Hg pressure?

3.21. Three kilogram moles of C_2H_6 is burned with 12 kg mol of O_2. Eighty percent of the ethane burns to CO_2 and H_2O, whereas the remainder does not burn. What is the composition of the resulting gases, and what is the final volume of the gases (in m³) if they are at 200°C and 100 kPa?

3.22. An industrial fuel gas consists of 40% CO and 60% CH_4; 121 ft³/min of the gas, measured at 500 psia and 90°F, is completely burned with 25% excess air that enters at 70°F and 770 mm Hg. The resultant gases leave the furnace at 570°F and 770 mm Hg. The resultant gases leave the furnace at 570°F and 770 mm Hg. Compute the composition and number of cubic feet per minute of the hot gases leaving the furnace.

3.23. A medium-Btu gas analyzes 6.4% CO_2, 0.1% O_2, 39% CO, 51.8% H_2, 0.6% CH_4, and 2.1% N_2. It enters the combustion chamber at 90°F and a pressure of 35.0 in. Hg. It is burned with 40% excess air (dry) at 70°F and the atmospheric pressure of 29.4 in. Hg, but 10% of the CO remains unburned. How many cubic feet of air are supplied per cubic foot of entering gas?

3.24. Turbojet aircraft operating under most conditions produce lower concentrations of pollutants than do motor vehicles. At idle, CO and unburned hydrocarbons are higher than in motor vehicles at idle, but in the operating mode the emission index (grams of pollutant per kilogram of fuel) is:

	CO	Hydrocarbons	NO_x
Turbojet	8.7	0.16	2.7
Automotive piston	300	55	27

Given that the Orsat analysis from a turbojet shows 12.2% CO_2, 0.4% CO, and 6.2% O_2, compute the net hydrogen/carbon ratio in the fuel (the fuel contains negligible sulfur and nitrogen). Also compute the cubic meters of air used at 27°C and 101.4 kPa/kg of fuel burned.

3.25. In the first stage of the manufacture of sulfuric acid by the contact process, iron pyrites (FeS_2) is burned in dry air, the iron being oxidized to Fe_2O_3. Sufficient air is supplied so that it is 40% in excess of that required if all the sulfur were oxidized to sulfur trioxide. Of the pyrites charged, 15% is lost by falling through the grate with the cinders and is not burned.

Calculate the cubic meters of air at 30°C and 150 kPa to be used per 100 kg of pyrites charged.

3.26. Pine wood has the following composition: 50.31% C, 6.20% H_2, 43.08% O_2, and 0.41% ash.
(a) Calculate the cubic feet of air at 76°F and 29.4 in. Hg necessary for complete combustion per pound of wood.
(b) If 30% excess air were used, calculate the cubic feet of dry gas at 600°F and 29.4 in. Hg produced per pound of wood.
(c) Calculate the Orsat analysis of the flue gas for parts (a) and (b).

3.27. In a test on an oil-fired boiler, it is not possible to measure the amount of oil burned, but the air used is determined by inserting a venturi meter in the air line. It is found that 5000 ft³/min of air at 80°F and 10 psig is used. The dry gas analyzes 10.7% CO_2, 0.55% CO, 4.75% O_2, and 84.0% N_2. If the oil is assumed to be all hydrocarbon, calculate the gallons per hour of oil burned. The specific gravity of the oil is 0.94.

3.28. Methane is flowing in a pipeline at the rate of 20.7 m³/min at 30°C and 250 kPa. To check the flow rate, He is to be introduced at 25°C and 300 kPa. How many cubic centimeters per minute of He must be introduced if the minimum detectable downstream concentration of the He in the methane is 10 ppm?

3.29. A hydrogen-free coke analyzes 4.2% moisture, 10.3 ash, and 85.5% carbon. It is burned giving a stack gas that analyzes 13.6% CO_2, 1.5% CO, 6.5% O_2, and 78.4% N_2, on a dry basis. Calculate the following:
(a) The percentage of excess air used
(b) The cubic feet of air at 80°F and 740 mm entering per pound of carbon burned
(c) Same as part (b) per pound of coke burned
(d) The cubic feet of dry flue gas at 690°F per pound of coke
(e) The cubic feet of wet flue gas at S.C. per pound of coke

3.30. A furnace produces a flue gas that contains 16.8% CO_2. The flue gas is drawn through a waste heat boiler and on exiting it is found to contain 15.2% CO_2. Calculate the cubic meters of air that have leaked into the system per cubic meter of flue gas.

3.31. For the manufacture of dry ice a furnace produces 1.31×10^5 ft³/hr (at 750°F and 1 atm) of a flue gas that contains 16.8% CO_2. It passes through a waste heat boiler to the absorbers, at which point it contains 15.2% CO_2. Calculate the cubic feet of air that have leaked into the system per hour if the air is at 70°F and 1 atm.

3.32. A rigid closed vessel having a volume of 1 ft³ contains NH_3 gas at 330°C and 30 psia. Into the closed vessel is pumped 0.35 ft³ of HCl gas measured at 200°F and 20 psia. NH_4Cl is formed according to the reaction

$$NH_3 + HCl \longrightarrow NH_4Cl$$

Assume that the reaction goes to completion and that the vapor pressure of NH_4Cl at 330°C is 610.6 mm Hg.
   (a) How much NH_4Cl will be formed?
   (b) Assuming that some NH_4Cl exists as a solid, what will be the final pressure in the closed vessel if the final temperature is 330°C?

3.33. Gas at 15°C and 105 kPa is flowing through an irregular duct. To determine the rate of flow of the gas, CO_2 is passed into the gas stream. The gas analyzes 1.2% CO_2 by volume before, and 3.4% CO_2 after addition. The CO_2 tank is placed on a scale and found to lose 2.27 kg in 10 min. What is the rate of flow of the entering gas in cubic meters per minute?

3.34. Gas at 60°F and 42.1 in. Hg is flowing through an irregular duct. To determine the rate of flow of the gas, He is passed into the gas stream. The gas analyzes 1.0% He by volume before, and 1.4% He after addition. The He tank is placed on a scale and is observed to lose 10 lb in 30 min. What is the rate of flow of the gas in the duct in cubic feet per minute?

3.35. Automobiles built in 1975 and thereafter must be capable of markedly lower emissions than any built heretofore in the United States. To measure the exhaust emissions of a vehicle, it is driven through a prescribed speed–time pattern on a dynamometer. A fraction of the exhaust is collected in a bag, and at the end of the test, pollutant concentrations in the bag are measured with special electronic gas-analyzing equipment to give the emissions in grams per mile of travel. A tank presumably containing propane is placed on a scale and fed to a new type of engine that burns propane instead of gasoline. The gas collected in the bag analyzes (on a dry basis) 10.1% CO_2, 1.1% CO, 4.7% O_2, and 84.1% N_2. During the test the cylinder loses 0.50 kg of gas while the bag picks up 1.73 m³ of gas at 15°C and 108 kPa. Because of some confusion in the stockroom, the tank was not labeled, and there is some question as to whether the hydrocarbon burned was really propane or perhaps another hydrocarbon. Determine whether or not an error has been made.

3.36. A gas containing only C and H on complete combustion yields 0.302 kg mol of CO_2 and 0.309 kg mol of H_2O. What is the C/H ratio in the gas? The specific gravity of the gas at 20°C and 97.3 kPa compared to dry air at 80°C and 30.78 in. Hg is 1.631. Calculate the molecular formula of the gas.

3.37. A gaseous mixture consisting of 50 mole % hydrogen and 50 mole % acetaldehyde (C_2H_4O) is initially contained in a rigid vessel at a total pressure of 760 mm Hg abs. The formation of ethanol (C_2H_6O) occurs according to

$$C_2H_4O + H_2 \longrightarrow C_2H_6O$$

After a time it was noted that the total pressure in the rigid vessel had dropped to

700 mm Hg abs. Calculate the degree of completion of the reaction using the following assumptions: (1) all reactants and products are in the gaseous state; and (2) the vessel and its contents were at the same temperature when the two pressures were measured.

3.38. The working of an old oil well yields 2.13×10^4 ft³ (at 20°C and 1 atm) of gas per hour of the following composition:

$$
\begin{array}{ll}
CO_2 & 4.0\% \\
CH_4 & 72.0\% \\
C_2H_6 & 13.0\% \\
N_2 & 11.0\%
\end{array}
$$

What is the specific gravity of the gas compared to pure methane?

3.39. What is the density of propane gas (C_3H_8) in kg per cubic meter at 200 kPa and 40°C? What is the specific gravity of propane?

3.40. What is the specific gravity of gas (C_3H_8) at 100°F and 800 mm Hg relative to air at 60°F and 760 mm Hg?

3.41 A glass weighing cell is used to determine the density of a gas. For a certain determination, the data are as follows:

weight of the cell full of air in air = 18.602 g

weight of the evacuated cell in air = 18.294 g

weight of the cell filled with sample gas in air = 18.345 g

Calculate the density of the sample gas in grams per cubic centimeter.

3.42. What is the mass of 1 m³ of H_2 at 5°C and 110 kPa? What is the specific gravity of this H_2 compared to air at 5°C and 110 kPa?

3.43. Gas from the Padna Field, Louisiana, is reported to have the following components and volume percent composition. What is:
(a) The mole percent?
(b) The weight percent of each component in the gas?
(c) The apparent molecular weight of the gas?
(d) Its specific gravity?

Component	Percent	Component	Percent
Methane	87.09	Pentanes	0.46
Ethane	4.42	Hexanes	0.29
Propane	1.60	Heptanes	0.06
Isobutane	0.40	Nitrogen	4.76
Normal butane	0.5	Carbon dioxide	0.40
		Total	100.00

3.44.   A natural gas from a gas well has the following composition:

Component	Percent	Mol. Wt.
CH_4	60	16
C_2H_6	16	30
C_3H_8	10	44
C_4H_{10}	14	58

(a) What is the composition in weight percent?
(b) What is the composition in mole percent?
(c) How many cubic feet will be occupied by 100 lb of the gas at 70°F and 74 cm Hg?
(d) What is the density of the gas in pounds per cubic foot at 70°F and 740 mm Hg?
(e) What is the specific gravity of the gas?

3.45.   A gas used to extinguish fires is composed of 80% CO_2 and 20% N_2. It is stored in a 2-m³ tank at 200 kPa and 25°C. What is the partial pressure of the CO_2 in the tank in kilopascals? What is the partial volume of the N_2 in the tank in cubic meters?

3.46.   The contents of a gas cylinder are found to contain 20% CO_2, 60% O_2, and 20% N_2 at a pressure of 740 mm Hg and at 20°C. What are the partial pressures of each of the components? If the temperature is raised to 40°C, will the partial pressures change? If so, what will they be?

3.47.   A tank composed of air augmented by 10 ppm of SO_2 is used to study the reactions involved in the acid-rain problem. What is the partial pressure of the SO_2 in the tank? Of the air?

3.48.   A liter of oxygen at 760 mm is forced into a vessel containing a liter of nitrogen at 760 mm. What will be the resulting pressure? What assumptions are necessary for your answer?

3.49.   You have a flat tire that contains only air at 130 kPa. To inflate the tire you use a portable CO_2 extinguisher and an adapter nozzle. The pressure increases to 180 kPa in the tire. Assume that the gases can be treated as ideal gases, and calculate the partial pressures of the O_2, the N_2, and the CO_2, respectively, in the tire.

3.50.   A mixture of 15 lb N_2 and 20 lb H_2 is at a pressure of 50 psig and a temperature of 60°F. Determine the following:
(a) The partial pressure of each component
(b) The partial (or pure component) volumes
(c) The specific volume of the mixture

Section 3.2

3.51.   You drive up to a gas distributor, and obtain a tankful of gaseous propane. The temperature of the gas (after filling) is 5°C (why?), the tank volume is 0.150 m³, and the weight of the propane delivered (final tank weight less initial tank weight) is 1.20 kg. What is the pressure of the gas in the tank?

3.52.   What pressure would be developed if 100 ft³ of ammonia at 20 atm and 400°F were compressed into a volume of 5.0 ft³, the final temperature being 350°F?

3.53. A 40-kg block of ice is put into a 0.3-m³ container and heated to 900 K. What is the pressure in the container?

3.54. One pound of carbon dioxide has a volume of 0.15 ft³ at a pressure of 100 atm. What is the temperature of the gas?

3.55. A block of dry ice weighing 50 lb is dropped into an empty steel bomb, the volume of which is 5.0 ft³. The bomb is closed and heated until the pressure gauge reads 1600 psig. What was the temperature of the gas in the bomb?

3.56. Calculate the specific volume of CO_2 at 600 atm and 40°C.

3.57. One-half of a cubic meter of CO_2 gas is held at a constant pressure of 3690kPa and is heated from 36°C to 77°C. Compare the volumes calculated for the gas after heating by considering it (**a**) as a nonideal gas and (**b**) as an ideal gas.

3.58. A tank contains compressed N_2 at a temperature of 25°C and at a pressure of 3×10^3 kPa gauge. The tank has a capacity of 3 m³. How many kilograms of N_2 are in the tank?

3.59. Ethylene at 500 atm pressure and a temperature of 100°C is contained in a cylinder of internal volume of 1.0 ft³. How many pounds of C_2H_4 are in the cylinder?

3.60. A steel cylinder contains ethylene (C_2H_4) at 200 psig. The cylinder and gas weigh 222 lb. The supplier refills the cylinder with ethylene until the pressure reaches 1000 psig, at which time the cylinder and gas weigh 250 lb. The temperature is constant at 25°C. Calculate the charge to be made for the ethylene if the ethylene is sold at $0.41 per pound, and what the weight of the cylinder is for use in billing the freight charges. Also find the volume of the empty cylinder in cubic feet.

3.61.*[20] A large gas storage tank contains 10,000 lb of ethylene at a pressure of 1050 psig. One-half or 5000 lb of this gas is needed for polyethylene production by a batch process. The gas leaves the tank through a large pipe. The only indication of how much ethylene has been removed is the pressure in the tank. At what pressure should removal be stopped in order to deliver exactly the necessary 5000 lb? Assume that the temperature is constant at 24°C throughout the process.

3.62. Polyethylene can be made from ethylene in a very high pressure process. The compression of ethylene is carried out by piston-type compressors, as sketched in Fig. P3.62. The ethylene enters at 1000 psia and is compressed to 20,000 psia. (Assume

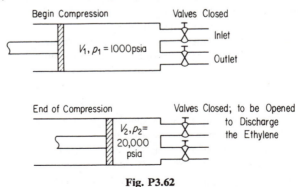

Fig. P3.62

[20]If equations of state are used.

that the temperature is constant at 25°C.) Calculate the following:

(a) The compression ratio (i.e., V_1/V_2)

(b) The density of the high-pressure ethylene

3.63.[*21]A steel cylinder contains ethylene (C_2H_4) at 10^4 kPa gauge. The weight of the cylinder and gas is 70 kg. Ethylene is removed from the cylinder until the gauge pressure measured falls to one-third of the original reading. The cylinder and gas now weigh 52 kg. The temperature is constant at 25°C. Calculate:

(a) The fraction of the original gas (i.e., at 10^4 kPa) that remains in the cylinder at the lower pressure

(b) The volume of the cylinder in cubic meters

3.64. "After 58 years of lying 2 miles under the Atlantic Ocean, the luxury liner *Titanic* may once more ride the ocean surface. A 21-man international team led by Douglas Woolley of Baldock, England, hopes to refloat the 66,000-ton wreck using plastic pontoons filled with hydrogen produced by electrolysis of seawater at the site.

"The team will use remote controlled equipment to circumvent the main problem—a pressure of more than 5000 psig at a depth of 2 miles. In Mr. Woolley's plan, the television-controlled deep-sea device will be used to break in some of the ship's more than 2000 portholes to attach lines connected to large bags or pontoons. Each of the 200 nylon-reinforced polyvinyl chloride pontoons will be linked to the ship by 10 2-inch-diameter nylon ropes. Each rope will support about 200 tons, for a total lifting capacity of about 400,000 tons.

"A ship-borne 200-volt 20-megawatt generator will provide power to electrolyze the seawater inside the cylindrical pontoons. Current will flow from an external anode through an aperture on the pontoon's underside to the internal cathodes. Mr. Woolley estimates it will take a week to generate the 85,000 cubic yards of hydrogen required." [*Chem. Eng. News*, p. 22 (October 5, 1970).]

(a) How many cubic feet of hydrogen will be produced at 5000 psig and 35°F in 1 week by the generator if all the 20 MW of electricity are used to produce gas?

(b) Comment on the proposed lifting method. (*Hint:* Write the equation for the generation of the hydrogen. Also, what happens as the pontoons rise?)

3.65. Gauges used on cylinders containing compressed gases frequently are calibrated to show both the pressure in the cylinder and the volumetric contents of the cylinder. You have a Matheson Co. size 1A cylinder containing methane gas. You have a gauge for a 1A hydrogen cylinder, but not a gauge for methane. When you put the hydrogen gauge on the methane cylinder, it registers 2000 psig at 77°F and a capacity of 200 ft³ (as hydrogen at 77°F and 1 atm pressure). Prepare a calibration so that the cubic feet as hydrogen can be converted to cubic feet as methane.

3.66. When a scuba diver goes to the dive shop to have his or her scuba tanks filled, the tank is connected to a compressor and filled to about 2100 psia while immersed in a tank of water. (Why immerse the tank in water? So that the compression of air into the tank will be approximately isothermal.)

Suppose that the tank is filled without inserting it into a water bath, and air at 27°C is compressed very rapidly from 1 atm absolute to the same final pressure. The final temperature would be about 700°C. Compute the fractional increase or decrease in the final quantity of air in the tank relative to the isothermal case:

[21]If equations of state are used.

(a) Assuming that the air behaves as an ideal gas

(b) Assuming that the air behaves as a real gas

Treat air as a single component with $p_c = 37.2$ atm and $T_c = 132.5$ K.

3.67. A sample of natural gas taken at 3500 kPa absolute and 120°C is separated by chromatography at standard conditions. It was found by calculation that the grams of each component in the gas were:

Component	g
Methane (CH_4)	100
Ethane (C_2H_6)	240
Propane (C_3H_8)	150
Nitrogen (N_2)	50
Total	540

What was the density of the original gas sample?

3.68. In a miscible displacement project, natural gas consisting of 90 mole % methane and 10% propane is combined with pure propane to make up a mixture consisting of 50 mole % propane. This gas is injected into the oil reservoir in sufficient quantities to displace 32,000 ft³/day of oil at reservoir conditions of 1500 psia and 220°F. What would be the volume per day of the 50% gas mixture at the surface conditions of 175.0 psia and 100°F? What volumes of (a) natural gas and (b) propane must be metered separately per day (at 175.0 psia and 100°F) in order to produce the 50% propane mixture? Solve this problem by using for the gases:

(1) The ideal gas law

(2) The compressibility factor (with Kay's method for mixtures)

(3) The compressibility factor (with the mean compressibility for mixtures computed according to Dalton's law)

3.69.* A gaseous mixture has the following composition (in mole percent):

$$C_2H_4 \quad 57$$
$$Ar \quad 40$$
$$He \quad 3$$

at 120 atm pressure and 25°C. Compare the experimental volume of 0.14 liter/g mol with that computed by the following:

(a) Van der Waals' equation plus Dalton's law

(b) Van der Waals' equation using averaged constants

(c) Mean compressibility factor and Dalton's law

(d) Mean compressibility factor and Amagat's law

(e) Compressibility factor using pseudoreduced conditions (Kay's method)

(f) Perfect gas law

3.70. A gas analyzes 60% methane and 40% ethylene by volume. It is desired to store 28 lb of this gas mixture in a cylinder having a capacity of 1.82 ft³ at a maximum temperature of 112°F. Calculate the pressure inside the cylinder:

(a) Assuming that the mixture obeys the ideal gas laws

(b) Using the compressibility factor determined from the pseudocritical point of the mixture

How many pounds of this mixture can be stored in the cylinder at 112°F if the maximum allowable pressure is 1250 psig?

3.71. You are in charge of a pilot plant using an inert atmosphere composed of 60% ethylene (C_2H_4) and 40% argon (Ar). How big a cylinder (or how many) must be purchased if you are to use 300 ft³ of gas measured at the pilot plant conditions of 100 atm and 300°F?

Cylinder Type	Cost	Pressure (psig)	lb Gas
1A	$40.25	2000	62
2	32.60	1500	47
3	25.50	1500	35

State any additional assumptions. You can buy only one type of cylinder.

Section 3.3

3.72. Prepare a Cox chart for:
(a) Acetic acid vapor
(b) Heptane
(c) Ammonia
(d) Methanol

from 0°C to the critical point (for each substance). Compare the estimated vapor pressure at the critical point with the critical pressure.

3.73.* Estimate the vapor pressure of ethyl bromide at 125°C from vapor-pressure data taken from a handbook or a journal.

3.74.* From the following data, estimate the vapor pressure of sulfur dioxide at 100°C; the actual vapor pressure is about 29 atm.

t (°C)	−10	6.3	32.1	55.5
p^* (atm)	1	2	5	10

3.75. Plot the vapor pressure of benzene (in mm Hg) over the range of temperature −40 to 80°C in each of the following ways. Obtain the necessary data from the *Chemical Engineers' Handbook*. Indicate all plotted points by small circles.
(a) p vs. t in °C
(b) $\log p$ vs. t in °C (use semilog paper)
(c) $\log p$ vs. $1/T$, where T is in Kelvin (use semilog paper)
(d) Use water as the reference substance; plot the vapor pressure on a Cox chart.
Plots (b) and (c) can be on the same piece of paper if labeled properly.

3.76. Use the Antoine equation, Eq. (3.40), to estimate the vapor pressure of sulfur dioxide at −10°C and compare with the experimental value (taken from a handbook).

Section 3.4

3.77. It is desired to represent the vapor pressure of an organic substance by an equation of the form

$$p^* = p_0^* e^{m/T}$$

where p^* is the vapor pressure, T the absolute temperature, and m and p_0^* are unknown constants. If 100 m³ of dry air (measured at standard conditions) is required to vaporize 0.5 kg mol at a temperature of 290 K and a pressure of 1 atm and only 50 m³ are required for the 0.5 kg mol at 340 K and 1 atm, calculate the constants p_0^* and m.

3.78. Fifty cubic feet of air saturated with water 90.0°F and 29.80 in. Hg is dehydrated. Compute the volume of the dry air and the pounds of moisture removed.

3.79. A mixture of air and benzene contains 10 mole % benzene at 38°C and 790 mm Hg pressure absolute. The vapor pressure of benzene is given as

$$\log_{10} p^* = 6.906 - \frac{1211}{220.8 + t}$$

where p^* is the vapor pressure in mm Hg and t is in °C. What is the dew point of the mixture?

3.80. Suppose that a vessel of dry nitrogen at 70.0°F and 29.90 in. Hg is saturated thoroughly with water. What will be the pressure in the vessel after saturation if the temperature is still 70.0°F?

3.81. How many grams of sodium peroxide containing 80% Na_2O_2 must be taken to produce 1500 cm³ of oxygen measured saturated with water vapor at 15°C and 100 kPa pressure?

$$2\,Na_2O_2 + 2\,H_2O \longrightarrow 4\,NaOH + O_2$$

3.82. Carbon disulfide (CS_2) at 20°C has a vapor pressure of 352 mm Hg. Dry air is bubbled through the CS_2 at 20°C until 2.00 kg of CS_2 is evaporated. What was the volume of the dry air required to evaporate this CS_2 (assuming that the air is saturated) if the air was initially at 20°C and 10 atm and the final pressure on the air–CS_2 vapor mixture is 750 mm Hg?

3.83. The dew point of water in atmospheric air is 82°F. What is the mole fraction of water vapor in the air if the barometric pressure is 750 mm Hg?

3.84. Dry air at 20°C and 100 kPa absolute is bubbled through benezene (C_6H_6) at 30°C, the saturated air leaving at 30°C and 100 kPa absolute. How many kilograms of benzene are evaporated by 30.0 m³ entering air? Show units.

3.85. A mixture of acetylene (C_2H_2) with an excess of oxygen measured 350 ft³ at 25°C and 745 mm Hg pressure. After explosion, the volume of the dry gaseous product was 300 ft³ at 60°C and the partial pressure of the dry product was 745 mm Hg. Calculate the partial volumes of acetylene and of oxygen in the original mixture. Assume that the final gas is saturated and that only enough water is formed to saturate the gas.

3.86. The vapor pressure of hexane (C_6H_{14}) at $-20°C$ is 14.1 mm Hg. Dry air at this temperature is saturated with the vapor under a total pressure of 760 mm Hg.
(a) What is the percent excess air for combustion?
(b) What is the flue-gas analysis if complete combustion occurs?

3.87. Liquid hexane is insoluble in water; hence each substance exerts its vapor pressure independently of the other. Compute the partial pressure of (a) the hexane and (b) the water if the gas mixture, containing 150,000 kg of hexane (C_6H_{14}) and 4200 kg of water, is initially at 100°C and 200 kPa absolute. To what temperature must the mixture be lowered before the water starts to condense out, if the pressure remains constant? Has hexane started to condense at this temperature? Support your answers with the necessary data and calculations.

3.88. A solution containing 12 wt % of dissolved nonvolatile solid is fed to a flash distillation unit. The molecular weight of the solid is 123.0. The effective vapor pressure of the solution is equal to the mole fraction of water,

$$p = p^*x$$

where p = effective vapor pressure of the solution
$\quad\quad x$ = mole fraction of water
$\quad p^*$ = vapor pressure of pure water

The pressure in the flash distillation unit is 1.121 psia and the temperature is 110°F. Calculate the pounds of pure water obtained in the vapor stream per 100 lb of feed solution and the weight percent of the dissolved nonvolatile solid leaving in the liquid stream.

Section 3.5

3.89. What is the mass of 3.00 m³ of air at 27°C and a total pressure of 115 kPa absolute if the relative humidity of the air is 65%?

3.90. The Environmental Protection Agency has promulgated a national ambient air quality standard for hydrocarbons: 160 $\mu g/m^3$ is the maximum 3-hr concentration not to be exceeded more than once a year. It was arrived at by considering the role of hydrocarbons in the formation of photochemical smog. Suppose that in an exhaust gas benzene vapor is mixed with air at 25°C such that the partial pressure of the benzene vapor is 2.20 mm Hg. The total pressure is 800 mm Hg. Calculate:
(a) The moles of benzene vapor per mole of gas (total)
(b) The moles of benzene per mole of benzene free gas
(c) The weight of benzene per unit weight of benzene-free gas
(d) The relative saturation
(e) The percent saturation
(f) The micrograms of benzene per cubic meter
(g) The grams of benzene per cubic foot
Does the exhaust gas concentration exceed the national quality standard?

3.91. Sufficient oxygen at 21°C and 100 kPa absolute with 80% absolute saturation of water vapor is supplied to react with sewage to reduce the concentration of biological material to below the quality control limits. Calculate the following for the supplied gas:

(a) The molal saturation

(b) kg water/kg bone-dry oxygen

(c) Relative saturation

(d) Dew point

3.92. Under what circumstances can the relative humidity and percent absolute humidity be equal?

3.93. Ethyl ether (mol. wt. = 74) is mixed with H_2 at 10°C in such proportions that the partial pressure of the vapor is 150 mm Hg. The total pressure is 715 mm Hg. Calculate:

(a) Relative saturation

(b) Moles ether per mole of vapor-free gas

(c) Weight of ether per unit weight of vapor-free gas

(d) Percentage saturation

(e) Percent ether by volume

3.94. One hundred kilograms of water vapor at 200°C and 120 kPa is cooled until one-half of the water is condensed. What is the temperature of the remaining water vapor after the condensation takes place?

3.95. Thermal pollution is the introduction of waste heat into the environment in such a way as to adversely affect environmental quality. Most thermal pollution results from the discharge of cooling water into the surroundings. It has been suggested that power plants use cooling towers and recycle water rather than dump water into the streams and rivers. In a proposed cooling tower, air enters and passes through baffles over which warm water from the heat exchanger falls. The air enters at a temperature of 80°F and leaves at a temperature of 70°F. The partial pressure of the water vapor in the air entering is 5 mm Hg and the partial pressure of the water vapor in the air leaving the tower is 18 mm Hg. The total pressure is 740 mm Hg. Calculate:

(a) The relative humidity of the air-water vapor mixture entering and of the mixture leaving the tower

(b) The percentage composition by volume of the moist air entering and of that leaving

(c) The percentage composition by weight of the moist air entering and of that leaving

(d) The percent absolute humidity of the moist air entering and leaving

(e) The pounds of water vapor per 1000 ft³ of mixture both entering and leaving

(f) The pounds of water vapor per 1000 ft³ of vapor-free air both entering and leaving

(g) The weight of water evaporated if 800,000 ft³ of air (at 740 mm and 80°F) enters the cooling tower per day

3.96. Air at 70°F, 760 mm Hg absolute, and 90% relative humidity is compressed to 150 psig during which the temperature increases to 120°F. In a subsequent step the compressed air is expanded until the pressure becomes 25 psig, and in so doing the temperature drops to 80°F. The air is supplied at a rate of 10,000 ft³/hr (at the initial conditions of 70°F, 760 mm, and 90% RH).

(a) Does any water condense during the compression step?

(b) If water does condense, how much?

(c) Assume that if condensate is formed, it is separated from the air. Compute the

following quantities for the resulting air at the final conditions (i.e., 25 psig, 80°F).
(1) Percent relative humidity
(2) Dew point
(3) Molal humidity

3.97. On Thursday the temperature was 90°F, and the dew point was 70°F. At 2 P.M. the barometer read 29.83 in. Hg, but owing to an approaching storm it dropped by 5 P.M. to 29.08 in. Hg, with no other changes. What change occurred in (a) the relative humidity and (b) the percent absolute humidity, between 2 and 5 P.M.?

3.98. If a gas at 60°C and 101.5 kPa absolute has a molal humidity of 0.03, calculate:
(a) The percentage humidity
(b) The relative humidity
(c) The dew point of the gas (in °C)

3.99. A chemist determines the humidity of the air by absorbing the water vapor in P_2O_5 and weighing the water found in a definite volume of air. Her setup looks like that in Fig. P3.99. A total of 1.153 ft³ of air is drawn through the wet-test meter. The thermometer on the wet-test meter reads 86°F and the manometer shows that the pressure within is 0.4 in. H_2O less than the outside pressure. What is the molal humidity of the air? What is its percent absolute humidity? Assume that all the water in the air is removed in the P_2O_5 tube.

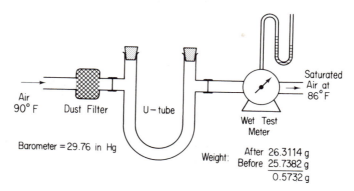

Fig. P3.99

Section 3.6

3.100. Methane gas contains CS_2 vapor in an amount such that the relative saturation at 35°C is 85%. To what temperature must the gas mixture be cooled to condense out 60% of the CS_2 by volume? The total pressure is constant at 750 mm Hg.

3.101. A process must be supplied with air and water vapor in small amounts. It is calculated that 0.04 kg of water vapor should accompany each kilogram of dry air into the process. The air enters at 1.00 atm. The engineers decide to accomplish this by saturating the air with water. The atmospheric air is at 25°C, 1 atm, and 55% humidity. The air is heated and then bubbled through water in an insulated tank.
(a) To what temperature was the air heated?

(b) How many cubic meters of entering air were required to evaporate 0.0190 m³ of water?

3.102. One gallon of benzene (C_6H_6) vaporizes in a room 20 × 20 × 9 ft at a constant barometric pressure of 750 mm and 70°F. The lower explosive limit for benzene in air is 1.4%. Has this been exceeded?

3.103. One thousand pounds of a slurry containing 10% by weight of $CaCO_3$ is to be filtered on a rotary vacuum filter. The filter cake from the filter contains 60% water. This cake is then placed into a drier and dried to a moisture content of 9.09 lb of H_2O/100 lb of $CaCO_3$. If the humidity of the air entering the drier is 0.005 lb of water per pound of dry air and the humidity of the air leaving the drier is 0.015 lb of water per pound of dry air, Calculate:
(a) Pounds of water removed by the filter
(b) Pounds of dry air needed in the drier

3.104. Around airports jet aircraft can become major contributors to pollution, and as aircraft activity increases and public concern brings other sources under control, the relative contribution of aircraft to pollution could go up. Recently, federal-, state-, and local-government pressure has speeded the introduction of new combustors in aircraft. In a test for a supersonic aircraft fuel with the average composition $C_{1.20}H_{4.40}$, the fuel is completely burned with the exact stochiometric amount of air required. The air is supplied at 25°C and 101 kPa, with an absolute humidity of 80%. The combustion products leave at 480°C and 106 kPa pressure and are passed through a heat exchanger from which they emerge at 57°C and 100 kPa pressure.
(a) For the entering air, compute:
(1) The dew point
(2) The molal humidity
(3) The relative humidity
(b) How much water is condensed in the heat exchanger per kilogram of gas burned, and hence must be removed as liquid water?

3.105. Air at 25°C and 100 kPa has a dew point of 16°C.
(a) Compute the percent relative humidity for the wet air.
(b) If the air is compressed isothermally (i.e., the temperature remains at 25°C), at what pressure (in kPa) will water vapor first start to condense?
(c) If you want to remove 50% of the initial moisture in the air (at a constant pressure of 100 kPa), to what temperature should you cool the air?

3.106. Air at 80°F and 1 atm has a dew point of 60°F.
(a) What is the percent relative humidity?
(b) To what pressure must this air be compressed to cause condensation to start (the temperature remains at 80°F)?
(c) To what temperature must this air be cooled to remove 25% of the moisture (the pressure stays constant at 1 atm)?
(d) What would be the percent relative humidity of this air after heating to 150°F (at constant pressure)?

3.107. In the summer when the temperature reaches 32.0°C, you measure the dew point of the air as 21.0°C. The barometric pressure is 100.9 kPa. At night the air cools to a temperature of 10.0°C and the barometric pressure remains unchanged. Calculate:
(a) The volume (in m³) of night air per cubic meter of day air

(b) The grams of water deposited out at night per cubic meter of day air

3.108. Oxalic acid (H_2C_2O) is burned with 248% excess air, 65% of the carbon burning to CO. Calculate:

(a) The Orsat gas analysis

(b) The volume of air at 90°F and 785 mm Hg used per pound of oxalic acid burned

(c) The volume of stack gases at 725°F and 785 mm Hg per pound of oxalic acid burned

(d) The dew point of the stack gas

3.109. A synthesis gas analyzing 4.0% CO_2, 40.0% CO, 26.0% H_2, and 30.0% N_2 is completely burned using 30% excess air. Calculate:

(a) The flue-gas analysis on a dry basis

(b) The cubic meter of flue gas (wet) at 270°C and 99.5 kPa per cubic meter of synthesis gas at 15.0°C and 101 kPa

3.110. Air is humidified in the spray chamber shown in Fig. P3.110. Calculate how much water must be added per hour to the tower to process 10,000 ft³/hr of air metered at the entrance conditions.

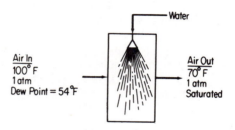

Fig. P3.110

3.111. A flue gas from a furnace leaves at 315°C and has an Orsat analysis of 16.7% CO_2, 4.1% O_2, and 79.2% N_2. It is cooled in a spray cooler and passes under slight suction through a duct to an absorption system at 32.0°C to remove CO_2 for the manufacture of dry ice. The gas at the entrance to the absorber analyzes 14.6% CO_2, 6.2% O_2, and 79.2% N_2, due to air leaking into the system. Calculate the cubic meters of air leaked in per cubic meter of gas to the absorber, both measured at the same temperature and pressure.

3.112.* A certain gas contains moisture, and you have to remove this by compression and cooling so that the gas will finally contain not more than 1% moisture (by volume). You decide to cool the final gas down to 21°C.

(a) Determine the minimum final pressure needed.

(b) If the cost of the compression equipment is

$$\text{cost in } \$ = (\text{pressure in psia})^{1.40}$$

and the cost of the cooling equipment is

$$\text{cost in } \$ = (350 - \text{temp. K})^{1.9}$$

is 21°C the best temperature to use?

3.113. A low-energy gas from in situ combustion of shale oil has the following composition: 10.0% CO_2, 20.0% CO, 20.0% H_2, 2.0% CH_4, and 48.0% N_2. On combustion the Orsat analysis of the flue gas gives 14.3% CO_2, 1.0% CO, 4.3% O_2, and 80.4% N_2. Calculate:

(a) The percent excess air used in the combustion of the low-energy gas

(b) The dew point of the flue gas if burned with air entering at 25°C and 30% relative humidity

The barometer reads 100 kPa.

3.114. Soybean flakes from an extraction process are reduced from 0.96 lb of C_2HCl_3 per pound of dry flakes to 0.05 lb of C_2HCl_3 per pound of dry flakes in a desolventizer by a stream of N_2 which vaporizes the C_2HCl_3. The entering N_2 contains C_2HCl_3 such that its dew point is 30°C. The N_2 leaves at 90°C with a relative saturation of 60%. The pressure in the desolventizer is 760 mm, and 1000 lb/hr of dry flakes pass through the drier.

(a) Compute the volume of N_2 plus C_2HCl_3 leaving the desolventizer at 90°C and 760 mm Hg in cubic feet per minute.

(b) The N_2 leaving the desolventizer is compressed and cooled to 40°C, thus condensing out the C_2HCl_3 picked up in the desolventizer. What must the pressure in the condenser be if the gas is to have a dew point of 30°C at the pressure of the desolventizer?

3.115. A synthesis gas is made by partial oxidation of butane (C_4H_{10}) in the presence of steam and air. The product synthesis gas has the percent composition 3.5% CO_2, 2.3% C_2H_4, 23.2% CO, 39.3% H_2, 11.6% CH_4, 1.7% C_2H_6, and 18.4% N_2. How many cubic meters of air (at 40% relative humidity, 30°C, 104 kPa) are used per cubic meter of butane (at 25°C, 120 kPa, dry)?

3.116. A wet sewage sludge contains 50% by weight of water. A centrifuging step removes water at a rate of 100 lb/hr. The sludge is dried further by air. Use the data in Fig. P3.116 to determine how much moist air (in cubic feet per hour) is required for the process shown.

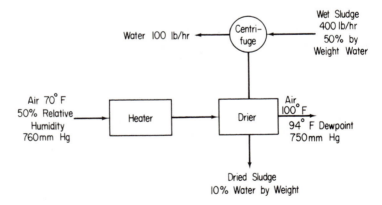

Fig. P3.116

3.117. A wet yellowcake sludge (uranium) contains 50% by weight of water. This sludge is first centrifuged and 0.1 kg of water is removed per kilogram of wet sludge feed. The yellowcake is dried further using air so that the final product contains 10% by weight of water. The air for drying is heated, passed into an oven drier, and vented back into the atmosphere. On a day when the atmospheric pressure is 760 mm Hg, the temperature is 70°F, and the relative humidity is 50%, calculate the cubic meters of air required to dry 1 kg of wet yellowcake feed. The air vented from the oven drier is at 100°F and 780 mm Hg. It has a dew point of 94°F. The data are:

Temperature (°F)	Vapor Pressure of Water (mm Hg)
70	18.76
80	26.21
90	36.09
94	40.37
100	49.07

3.118. Moist air is partially dehydrated and cooled before it is passed through a refrigerator room maintained at 0°F, to prevent excessive ice formation on the cooling coils. The cool air is passed through the room at the rate of 20,000 ft³/day measured at the entrance temperature (40°F) and 800 mmHg pressure. At the end of 30 days the refrigerator room must be warmed to remove the ice from the coils. How many pounds of water are removed? See Fig. P3.118.

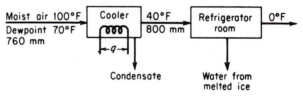

Fig. P3.118

3.119. Refer to the process flow diagram (Fig. P3.119) for a process that produces maleic anhydride by the partial oxidation of benzene. The moles of O_2 fed to the reactor per mole of pure benzene fed to the reactor is 18.0. All the maleic acid produced in the reactor is removed with water in the bottom stream from the water scrubber. All the C_6H_6, O_2, CO_2, and N_2 leaving the reactor leave in the stream from the top of the water scrubber, saturated with H_2O. Originally, the benzene contains trace amounts of a nonvolatile contaminant that would inhibit the reaction. This contaminant is removed by steam distillation in the steam still. The steam still contains liquid phases of both benzene and water (benzene is completely insoluble in water). The benzene phase is 80% by weight, and the water phase is 20% by weight of the total of the two liquid phases in the still. Other process conditions are given in the flow sheet. Use the following vapor-pressure data:

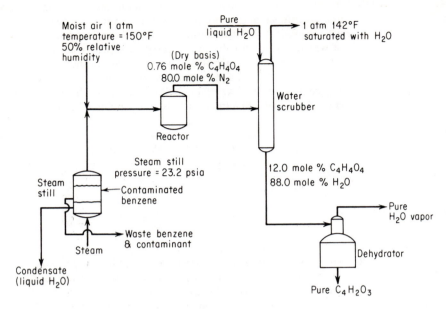

Fig. P3.119

Temperature (°F)	Benzene (psia)	Water (psia)
110	4.045	1.275
120	5.028	1.692
130	6.195	2.223
140	7.570	2.889
150	9.178	3.718
160	11.047	4.741
170	13.205	5.992
180	15.681	7.510
190	18.508	9.339
200	21.715	11.526

The reactions are

$$C_6H_6 + 4\tfrac{1}{2}O_2 \longrightarrow \begin{matrix} CH-C \\ \| \\ CH-C \end{matrix}\begin{matrix} O \\ OH \\ O \\ OH \end{matrix} + 2CO_2 + H_2O \qquad (1)$$

$$C_6H_6 + 7\tfrac{1}{2}O_2 \longrightarrow 6CO_2 + 3H_2O \qquad (2)$$

Calculate:

(a) The moles of benzene undergoing reaction (2) per mole of benzene feed to the reactor

(b) The pounds of H_2O removed in the top stream from the dehydrator per pound mole of benzene feed to the reactor

(c) The composition (mole percent, wet basis) of the gases leaving the top of the water scrubber

(d) The pounds of pure liquid H_2O added to the top of the water scrubber per pound mole of benzene feed to the reactor

Section 3.7

3.120. When production is taken from a gas reservoir, is it possible to decrease the pressure at the well and recover liquids as well as gas? Explain using the phase diagram in Fig. P3.120 showing the p–T properties of a two-component gas condensate.

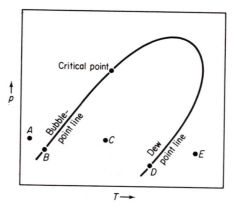

Fig. P3.120

3.121. At the critical point, how many degrees of freedom exist according to the phase rule? How many degrees of freedom exist at each of the points in Fig. P3.120 marked A, B, C, D and E? Show your computations.

3.122. Water is pumped into a pipe at the rate of 250 kg/hr at 950 kPa and 175°C. Because of friction in the pipe, pressure losses due to pipe fittings and valves, the pressure at the exit of the pipe drops to 760 kPa. Is the entering water liquid, vapor, or a mixture of both? Determine the fraction of each. Is the exit water liquid, vapor, or both? Determine the fraction of each. From the steam tables or similar source, determine the density of the overall mixture in the pipe at both locations. How many degrees of freedom exist according to the phase rule at each location?

3.123. Examine the volume pressure chart for propane in Appendix J. If a cylinder contains propane with a volume of 2.1 ft³/lb mol at a pressure of 700 psia, is the propane all liquid, all vapor, or a mixture of the two phases?

3.124. Is carbon dioxide at 31°C and 70 atm a two-phase or a single-phase system? Repeat for 25°C.

PROBLEMS TO PROGRAM ON THE COMPUTER

3.1. You need to fill a reactor with 8.00 lb of NH_3 gas to a pressure of 150 psia at a temperature of 250°F. Apply van der Waals' equation to find the volume of the reactor. Prepare a computer program to solve Eq. (3.13) for V. How do you select the initial estimate of V to start the calculations?

3.2. An experiment was designed to test the validity of Charles' law using the expansion of a fixed amount of air in a balloon in a flask of water. The data taken were as follows at 751 mm Hg constant pressure:

Temp. (°C)	Incremental Vol. of Balloon (cm³)	Temp. (°C)	Incremental Vol. of Balloon (cm³)
22	0.0	44	4.2
23	0.3	47	4.9
24	0.45	50	5.5
25	0.8	53	6.5
26	0.8	55	7.0
27	1.2	59	9.0
29	2.0	61	10.0
34	3.0	63	11.9
37	3.2	65	13.5
39	3.8	67	14.7
43	3.8		

Ascertain how well the ideal gas law (Charles' law) is obeyed. A least-squares computer code should be used to evaluate the experimental data. What was the initial volume of gas (at 22°C)? Assume that the pressure on the air in the balloon is constant.

3.3. Use the Benedict–Webb–Rubin (BWR) equation of state (see Table 3.2)

$$p = RT\rho + \left(B_0 RT - A_0 - \frac{C_0}{T^2}\right)\rho^2 + (bRT - a)\rho^3$$
$$+ a\alpha\rho^6 + \frac{c\rho^3}{T^2}(1 + \gamma\rho^2)\exp(-\gamma\rho^2)$$

where ρ is the density, to predict the density of a mixture of two nonideal gases. Let the pressure be in atmospheres and calculate the density in both gram moles per liter and pound moles per cubic foot for selected pairs of p and T. Use the following rule illustrated for A_0 to obtain the coefficients A_0, B_0, C_0, a, b, c, α, and γ for the mixture:

$$A_0 = \left(\sum_{i=1}^{n} x_i A_{0i}^{1/2}\right)^2$$

where n is the number of components in the mixture and x_i is the mole fraction. Values of the BWR constants for 38 pure components have been tabulated by H. W. Cooper and J. C. Goldfrank, *Hydrocarbon Processing*, v. 46, no. 12, p. 141 (1967). For reduced temperatures above 0.6 and reduced specific volumes less than 0.5, the BWR equation should be accurate to within 1%.

4

ENERGY
BALANCES

It was not long ago that resources and reserves of petroleum and natural gas seemed to be adequate for many decades at stable prices. But no longer. The world energy supply–demand picture for these two most convenient sources of energy has changed so that alternative sources that at one time appeared to be uneconomical now have become not only technically but economically feasible. Figure 4.1(a) and (b) show the major uses and sources of energy in the United States over a 30-year time span. Figure 4.2 illustrates the energy balance for a typical chemical industry showing the inputs and outputs of energy.

In the short run only small changes can be made in the supply and usage of energy. But in the long run, only four alternative sources appear to be capable of replacing oil and gas as the primary energy supply for the future:

(a) *Coal:* a life expectancy of 100 to 400 years.
(b) *Fission:* a life expectancy of 25 to 100 years, but with breeder reactors a life expectancy of perhaps 500 years.
(c) *Solar:* infinite expectancy, but very capital intensive, and thus can only be utilized when higher costs for competing sources arrive.
(d) *Fusion:* potentially extensive life but not yet technically feasible.

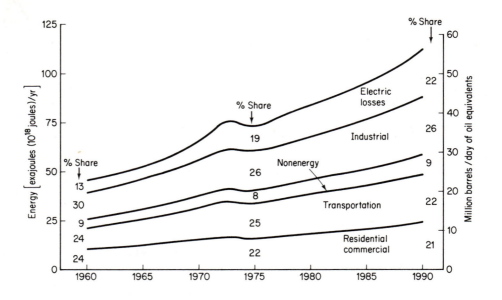

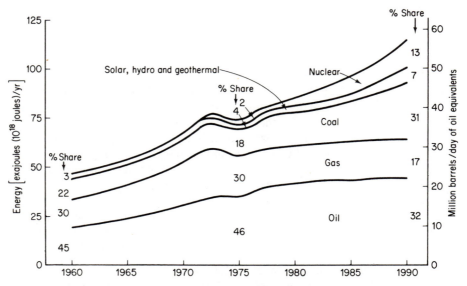

Fig. 4.1 (a) United States energy demand by consuming sector; (b) United States energy supply by source.

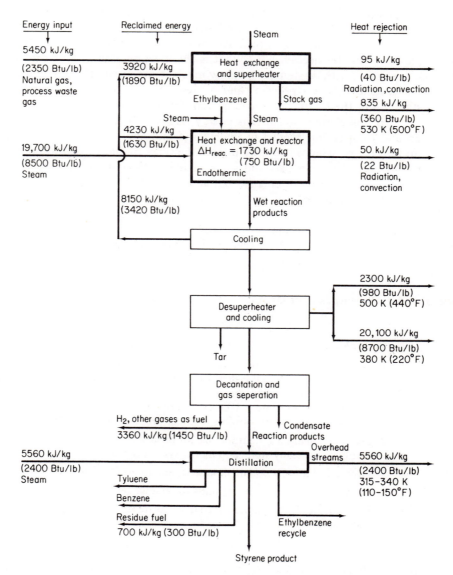

Fig. 4.2 Energy balance in styrene production in the United States: 40% conversion of ethylbenzene to products, 90% selectivity to styrene. (From J. T. Reding and B. P. Shepherd, *Energy Consumption: The Chemical Industry*, Report EPA-650/2-75-032a, Environmental Protection Agency, Washington, D.C., April 1975.)

To provide publicly acceptable, effective, and yet economical conversion of our resources into energy and to properly utilize the energy so generated, you must understand the basic principles underlying the generation, uses, and transformation of energy in its different forms. The answers to questions such as "Is thermal pollution inherently necessary"; "What is the most economic source of fuel"; "What can be done with waste heat"; "How much steam at what temperature and pressure are needed to heat a process" and related questions can only arise from an understanding of the treatment of energy transfer by natural processes or machines.

In this chapter we discuss energy balances together with the accessory background information needed to understand and apply them correctly. Before taking up the energy balance itself, we discuss the types of energy that are of major interest to engineers and scientists and some of the methods that are used to measure and evaluate these forms of energy. Our main attention will be devoted to heat, work, enthalpy, and internal energy. Next, the energy balance will be described and applied to practical and hypothetical problems. Finally, we shall discuss the energy balance with reaction and how energy generation fits into industrial process calculations. Figure 4.0 shows the relationships among the topics discussed in this chapter and in previous chapters.

Section 4.1   Concepts and Units

> *Your objectives in studying this section are to be able to:*
>
> 1. Define or explain the following terms: energy, system, thermochemical calorie, closed system, nonflow system, open system, flow system, surroundings, property, extensive property, intensive property, state, heat, work, kinetic energy, potential energy, internal energy, enthalpy, initial state, final state, point (state) function, state variable, cyclical process, and path function.
> 2. Select and define a system suitable for problem solution, either closed or open, steady or unsteady state, and fix the system boundary.
> 3. Distinguish among potential, kinetic, and internal energy.
> 4. Convert energy in one set of units to another set.
> 5. List and apply the equations used to calculate kinetic energy, potential energy, and work.

As you already know, energy exists in many different forms. In Sec. 4.5 we assemble the various types of energy in the form of an energy balance. Before doing so, we need to clearly distinguish between the common types of energy, describe the units used to express energy, and learn how to calculate the values of various forms of energy.

What units are associated with energy? If you have forgotten, refer back to Table 1.1. No confusion exists with respect to the joule, but it is necessary to be

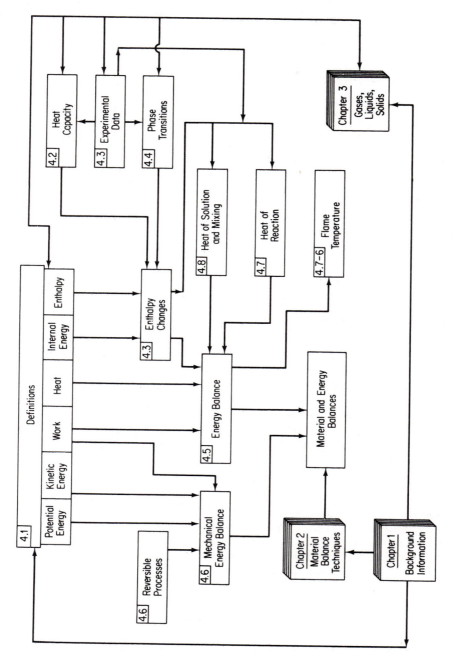

Fig. 4.0 Hierarchy of topics to be studied in this chapter (section numbers are in the upper left-hand corner of the boxes).

careful in specifying what type of calorie or British thermal unit (Btu) is under consideration (there are four or five common kinds). For example, the type of calorie that is found in tables of thermochemical properties of substances is the *thermochemical calorie* (equal to 4.184 J), whereas a second type of calorie is the *International Steam Table (I.T.) calorie* (equal to 4.1867 J). In this text 1 cal will be equal to 4.184 J and 1 Btu will be equal to 1.055×10^3 J, but be careful when selecting values from references to check on the specifications of the energy unit.

In this section we characterize and define several common forms of energy. Unfortunately, many of the terms described below are used loosely in our ordinary conversation and writing, and thus have different connotations than those presented below. You may have the impression that you understand the terms from long acquaintance—be sure that you really do.

Certain terms that have been described in earlier chapters occur repeatedly in this chapter; these terms are summarized below with some elaboration in view of their importance.

System. Any arbitrarily specified mass of material or segment of apparatus to which we would like to devote our attention. A system must be defined by surrounding it with a system boundary. A system enclosed by a boundary that prohibits the transfer of mass across the boundary is termed a *closed* system, or *nonflow* system, in distinction to an *open* system, or *flow* system, in which the exchange of mass is permitted. All the mass or apparatus external to the defined system is termed the *surroundings*. Reexamine some of the example problems in Chap. 2 for illustrations of the location of system boundaries. You should always draw similar boundaries in the solution of your problems, since this will fix clearly the system and surroundings.

Property. A characteristic of material that can be measured, such as pressure, volume, or temperature—or calculated, if not directly measured, such as certain types of energy. The properties of a system are dependent on its condition at any given time and not on what has happened to the system in the past.

An *extensive property* (variable, parameter) is one whose value is the sum of the values of each of the subsystems comprising the whole system. For example, a gaseous system can be divided into two subsystems which have volumes or masses different from the original system. Consequently, mass or volume is an extensive property.

An *intensive property* (variable, parameter) is one whose value is not additive and does not vary with the quantity of material in the subsystem. For example, temperature, pressure, density (mass per volume), and so on, do not change if the system is sliced in half or if the halves are put together.

Two properties are *independent* of each other if at least one variation of state for the system can be found in which one property varies while the other remains fixed. The number of independent intensive properties necessary and sufficient to fix the state of the system can be ascertained from the phase rule of Sec. 3.7.1.

State. Material with a given set of properties at a given time. The state of a system does not depend on the shape or configuration of the system but only on its intensive properties.

Next we shall take up some new concepts, although they may not be entirely new, because you perhaps have encountered them before.

Heat. In a discussion of *heat* we enter an area in which our everyday use of the term may cause confusion, since we are going to use heat in a very restricted sense when we apply the laws governing energy changes. Heat (Q) is commonly defined as that part of the total energy flow across a system boundary that is caused by a temperature difference between the system and the surroundings. Heat may be exchanged by conduction, convection, or radiation. A more effective general qualitative definition is given by Callen[1]:

> A macroscopic observation [of a system] is a kind of "hazy" observation which discerns gross features but not fine detail. . . . Of the enormous number of atomic coordinates [which can exist], a very few with unique symmetric properties survive the statistical averaging associated with a transition to the macroscopic description [and are macroscopically observable]. Certain of these surviving coordinates are mechanical in nature [such as volume]. Others are electrical in nature [such as dipole moments]. . . . It is equally possible to transfer energy to the hidden atomic modes of motion [of the atom] as well as to those modes which happen to be macroscopically observable. An energy transfer to the hidden atomic modes is called heat.

To evaluate heat transfer *quantitatively*, unless given a priori, you must apply the energy balance that is discussed in Sec. 4.5, evaluate all the terms except Q and then solve for Q. Heat transfer can be *estimated* for engineering purposes by many empirical relations, which can be found in books treating heat transfer or transport processes.

Work. Work (W) is commonly defined as energy transferred between the system and its surroundings by means of a vector force acting through a vector displacement on the system boundaries

$$W = \int F \, dl$$

where F is the direction of dl. However, this definition is not always useful inasmuch as:

(a) The displacement may not be easy to define.

(b) The product of $F \, dl$ does not always result in an equal amount of work.

(c) Work can be exchanged without a force acting on the system boundaries (such as through magnetic or electric effects).

Since heat and work are by definition mutually exclusive exchanges of energy between the system and the surroundings, we shall qualitatively classify work as energy that can be transferred to or from a mechanical state, or mode, of the system,

[1] H. B. Callen, *Thermodynamics*, Wiley, New York, 1960, p. 7.

while heat is the transfer of energy to atomic or molecular states, or modes, which are not macroscopically observable. To measure or evaluate work *quantitatively* by a mechanical device is difficult, so that, unless one of the energy balances in Sec. 4.5 or 4.6 can be applied, in many instances the value of the work done must be given a priori, or measured by experiment.

Kinetic energy. Kinetic energy (K) is the energy a system possesses because of its velocity relative to the surroundings. Kinetic energy may be calculated from the relation

$$K = \tfrac{1}{2}mv^2 \tag{4.1a}$$

or

$$\hat{K} = \tfrac{1}{2}v^2 \tag{4.1b}$$

where the superscript caret (^) refers to the energy per unit mass (or sometimes per mole) and not the total kinetic energy as in Eq. (4.1a).

EXAMPLE 4.1 Calculation of Kinetic Energy

Water is pumped from a storage tank into a tube of 3.00 cm inner diameter at the rate of 0.001 m³/s. See Fig. E4.1. What is the specific kinetic energy of the water?

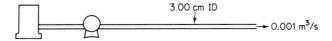

3.00 cm ID

0.001 m³/s

Fig. E4.1

Solution

$$\text{Assume that } \rho = \frac{1000 \text{ kg}}{\text{m}^3} \qquad r = \tfrac{1}{2}(3.00) = 1.50 \text{ cm}$$

$$v = \frac{0.001 \text{ m}^3}{\text{s}} \left| \frac{1}{\pi(1.50)^2 \text{ cm}^2} \right| \left(\frac{100 \text{ cm}}{1 \text{ m}}\right)^2 = 1.415 \text{ m/s}$$

$$\hat{K} = \frac{1}{2} \left(\frac{1.415 \text{ m}}{\text{s}}\right)^2 \left| \frac{1 \text{ N}}{\frac{(\text{kg})(\text{m})}{\text{s}^2}} \right| \frac{1 \text{ J}}{1 \text{ (N)(m)}} = 1.00 \text{ J/kg}$$

Potential energy. Potential energy (P) is energy the system possesses because of the body force exerted on its mass by a gravitational field with respect to a reference surface. Potential energy can be calculated from

$$P = mgh \tag{4.2a}$$

or

$$\hat{P} = gh \tag{4.2b}$$

where the symbol (^) again means potential energy per unit mass (or sometimes per mole).

EXAMPLE 4.2 Calculation of Potential Energy

Water is pumped from one reservoir to another 300 ft away, as shown in Fig. E4.2. The water level in the second reservoir is 40 ft above the water level of the first reservoir. What is the increase in specific potential energy of the water in Btu/lb$_m$?

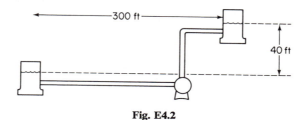

Fig. E4.2

Solution

Let the water level in the first reservoir be the reference plane. Then $h = 40$ ft.

$$\hat{P} = \frac{32.2 \text{ ft}}{s^2} \left| \frac{40 \text{ ft}}{} \right| \frac{}{\frac{32.2 \text{ (lb}_m)(\text{ft})}{(\text{lb}_f)(s^2)}} \left| \frac{1 \text{ Btu}}{778.2 \text{ (ft)(lb}_f)} \right| = 0.0514 \text{ Btu/lb}_m$$

Internal energy. Internal energy (U) is a macroscopic measure of the molecular, atomic, and subatomic energies, all of which follow definite microscopic conservation rules for dynamic systems. Because no instruments exist with which to measure internal energy directly on a macroscopic scale, internal energy must be calculated from certain other variables that can be measured macroscopically, such as pressure, volume, temperature, and composition.

To calculate the internal energy per unit mass ($\hat{U}$) from the variables that can be measured, we make use of a special property of internal energy, namely, that it is an exact differential (because it is a *point* or *state* property, a matter to be described shortly) and, for a pure component, can be expressed in terms of the temperature and specific volume alone. If we say that

$$\hat{U} = \hat{U}(T, \hat{V})$$

by taking the total derivative, we find that

$$d\hat{U} = \left(\frac{\partial \hat{U}}{\partial T}\right)_{\hat{V}} dT + \left(\frac{\partial \hat{U}}{\partial \hat{V}}\right)_T d\hat{V} \tag{4.3}$$

By definition $(\partial \hat{U}/\partial T)_{\hat{V}}$ is the heat capacity at constant volume, given the special symbol C_v. For most practical purposes in this text the term $(\partial \hat{U}/\partial \hat{V})_T$ is so small that the second term on the right-hand side of Eq. (4.3) can be neglected. Consequently, changes in the internal energy can be computed by integrating Eq. (4.3) as follows:

$$\hat{U}_2 - \hat{U}_1 = \int_{T_1}^{T_2} C_v \, dT \tag{4.4}$$

Note that you can only calculate *differences* in internal energy, or calculate the internal energy relative to a reference state, *but not absolute values* of internal energy. Instead of using Eq. (4.4), internal energy changes are usually calculated from enthalpy values, the next topic of discussion.

Enthalpy. In applying the energy balance you will encounter a variable which is given the symbol H and the name *enthalpy* (pronounced en'-thal-py). This variable is defined as the combination of two variables which will appear very often in the energy balance:

$$H = U + pV \tag{4.5}$$

where p is the pressure and V is the volume. (The term "enthalpy" has replaced the now obsolete terms "heat content" or "total heat," to eliminate any connection whatsoever with heat as defined earlier.)

To calculate the enthalpy per unit mass, we use the property that the enthalpy is an exact differential. For a pure substance, the enthalpy can be expressed in terms of the temperature and pressure (a more convenient variable than volume). If we let

$$\hat{H} = \hat{H}(T, p)$$

by taking the total derivative of $\hat{H}$, we can form an expression corresponding to Eq. (4.3):

$$d\hat{H} = \left(\frac{d\hat{H}}{\partial T}\right)_p dT + \left(\frac{\partial \hat{H}}{\partial p}\right)_T dp \tag{4.6}$$

By definition $(\partial \hat{H}/\partial T)_p$ is the heat capacity at constant pressure, given the special symbol C_p. For most practical purposes $(\partial \hat{H}/\partial p)_T$ is so small at modest pressures that the second term on the right-hand side of Eq. (4.6) can be neglected. Changes in enthalpy can then be calculated by integration of Eq. (4.6) as follows:

$$\hat{H}_2 - \hat{H}_1 = \int_{T_1}^{T_2} C_p \, dT \tag{4.7}$$

However, in high-pressure processes the second term on the right-hand side of Eq. (4.6) cannot necessarily be neglected, but must be evaluated from experimental data. One property of ideal gases that should be noted is that their enthalpies and internal energies are functions of temperature only and are not influenced by changes in pressure or specific volume.

As with internal energy, *enthalpy has no absolute value;* only changes in enthalpy can be calculated. Often you will use a reference set of conditions (perhaps implicitly) in computing enthalpy changes. For example, the reference conditions used in the steam tables are liquid water at 0°C (32°F) and its vapor pressure. This does not mean that the enthalpy is actually zero under these conditions but merely that the enthalpy has arbitrarily been assigned a value of zero at these conditions. In computing enthalpy changes, the reference conditions cancel out as can be seen from the following:

initial state of system   *final state of system*

$$\text{enthalpy} = \hat{H}_1 - \hat{H}_{\text{ref}} \qquad\qquad \text{enthalpy} = \hat{H}_2 - \hat{H}_{\text{ref}}$$

$$\text{net enthalphy change} = (\hat{H}_2 - \hat{H}_{\text{ref}}) - (\hat{H}_1 - \hat{H}_{\text{ref}}) = \hat{H}_2 - \hat{H}_1$$

Point, or state, functions. The variables enthalpy and internal energy are called *point functions*, or *state variables*, meaning that their values depend *only* on the state of the material (temperature, pressure, phase, and composition) and *not* on how the material reached that state. Figure 4.3 illustrates the concept of a state variable.

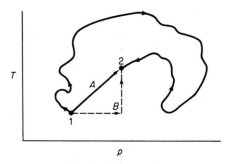

Fig. 4.3 Point function.

In proceeding from state 1 to state 2, the actual process conditions of temperature and pressure are shown by the wiggly line. However, you may calculate $\Delta\hat{H}$ by route A or B, or any other route, and still obtain the same net enthalpy change as for the route shown by the wiggly line. The change of enthalpy depends only on the initial and final states of the system. A process that proceeds first at constant pressure and then at constant temperature from 1 to 2 will yield exactly the same $\Delta\hat{H}$ as one that takes place first at constant temperature and then at constant pressure as long as the end point is the same. The concept of the point function is the same as that of an airplane passenger who plans to go straight to Chicago from New York but is detoured because of bad weather by way of Cincinnati. His trip costs him the same whatever way he flies, and he eventually arrives at his destination. The fuel consumption of the plane may vary considerably, and in analogous fashion heat (Q) or work (W), the two "path" functions with which we deal, may vary depending on the specific path chosen, while $\Delta\hat{H}$ is the same regardless of path. If the plane were turned back by mechanical problems and landed at New York, the passenger might be irate, but at least could get a refund. Thus $\Delta\hat{H} = 0$ if a cyclical process is involved which goes from state 1 to 2 and back to state 1 again, or

$$\oint d\hat{H} = 0$$

All the intensive properties we shall work with, such as $\hat{P}, T, \hat{U}, p, \hat{H}$, and so on, are state variables and depend only on the state of the substance of interest, so we can say, for example, that

$$\oint dT = 0$$

$$\oint d\hat{U} = 0$$

Always keep in mind that the values for a difference in a point function can be calculated by taking the value in the final state and subtracting the value in the initial state, regardless of the actual path.

Self-Assessment Test

1. Contrast the following property classifications: extensive–intensive, measurable–unmeasurable, state–path.

2. Define heat and work.

3. Consider the hot-water heater in your house. Classify each case below as an open system, closed system, neither, or both.
   (a) The tank is being filled with cold water.
   (b) Hot water is being drawn from the tank.
   (c) The tank leaks.
   (d) The heater is turned on to heat the water.
   (e) The tank is full and the heater is turned off.

4. The units of potential energy or kinetic energy are (select all the correct expressions):
   (a) $(ft)(lb_f)$
   (b) $(ft)(lb_m)$
   (c) $(ft)(lb_f)/(lb_m)$
   (d) $(ft)(lb_m)/(lb_f)$
   (e) $(ft)(lb_f)/(hr)$
   (f) $(ft)(lb_m)/(hr)$

5. Review the selection of a system and surroundings by reading two or three examples in Secs. 2.3 to 2.5, covering up the solution, and designating the system. Compare with the system shown in the example.

6. Will the kinetic energy per unit mass of an incompressible fluid flowing in a pipe increase, decrease, or remain the same if the pipe diameter is increased at some place in the line?

7. A 100-kg ball initially on the top of a 5-m ladder is dropped and hits the ground. With reference to the ground:
   (*a*) What is the initial kinetic and potential energy of the ball?
   (*b*) What is the final kinetic and potential energy of the ball?
   (*c*) What is the change in kinetic and potential energy for the process?
   (*d*) If all the initial potential energy were somehow converted to heat, how many calories would this amount to? how many Btu?

8. In expanding a balloon, two types of work are done by the air (the system). One is stretching the balloon $dW = \sigma \, dA$, where σ is the surface tension of the balloon. The other is the work of pushing back the atmosphere. If the balloon is spherical and expands from a diameter of 1 m to 1.5 m, what is the work done in pushing back the atmosphere?

Section 4.2   Heat Capacity

> *Your objectives in studying this section are to be able to:*
>
> 1. Define heat capacity.
> 2. Convert an expression for the heat capacity from one set of units to another.
> 3. Look up from a reference source an equation that expresses the heat capacity as a function of temperature, and compute the heat capacity at a given temperature.
> 4. Estimate the value of the heat capacity for solids and liquids.

Before formulating the general energy balance, we shall discuss in some detail the calculation of enthalpy changes and provide some typical examples of such calculations. Our discussion will be initiated with consideration of the heat capacity C_p.

The two heat capacities have been defined as

(a)
$$C_p = \left(\frac{\partial \hat{H}}{\partial T}\right)_p$$

(b)
$$C_v = \left(\frac{\partial \hat{U}}{\partial T}\right)_v$$

To give these two quantities some physical meaning, you can think of them as representing the amount of energy required to increase the temperature of a substance by 1 degree, energy that might be provided by heat transfer in certain specialized processes. To determine from experiments values of C_p (or C_v), the enthalpy (or internal energy) change must first be calculated from an energy balance, and then the heat capacity evaluated from Eq. (4.7) [or (4.4)].

Suppose that you want to calculate the heat capacity for steam at 10.0 kPa (45.8°C). You can determine the enthalpy change for essentially constant pressure from the steam tables as

$$H_{47.7^{\circ}C} - H_{43.8^{\circ}C} = (2588.1 - 2581.1) \text{ kJ/kg} = 7.00 \text{ kJ/kg}$$

and if you assume C_p is essentially constant over the small temperature range indicated by the subscripts, then

$$C_p \cong \frac{\Delta H}{\Delta t} = \frac{7.00 \text{ kJ}}{\text{kg}} \bigg| \frac{}{3.9 \Delta^{\circ}C} = 1.79 \frac{\text{kJ}}{(\text{kg})(\Delta^{\circ}C)}$$

What are the units of the heat capacity? From the definitions of heat capacity you can see that the units are (energy)/(temperature difference) (mass or moles). The common units found in engineering practice are (we suppress the Δ symbol)

$$\frac{\text{J}}{(\text{kg mol})(\text{K})} \qquad \frac{\text{cal}}{(\text{g mol})(^{\circ}C)} \qquad \frac{\text{Btu}}{(\text{lb mol})(^{\circ}F)}$$

Because of the definition of the calorie or Btu, the heat capacity can be expressed in certain different systems of units and still have the same numerical value; for example, heat capacity may be expressed in the units of

$$\frac{\text{cal}}{(\text{g mol})(^{\circ}C)} = \frac{\text{kcal}}{(\text{kg mol})(^{\circ}C)} = \frac{\text{Btu}}{(\text{lb mol})(^{\circ}F)}$$

and still have the same numerical value.

Alternatively, heat capacity may be in terms of

$$\frac{\text{cal}}{(\text{g})(^{\circ}C)} = \frac{\text{Btu}}{(\text{lb})(^{\circ}F)} \qquad \text{or} \qquad \frac{\text{J}}{(\text{kg})(\text{K})}$$

Note that

$$\frac{1 \text{ Btu}}{(\text{lb})(^{\circ}F)} = \frac{4.184 \text{ J}}{(\text{g})(\text{K})}$$

and that the heat capacity of water in the SI system is 4184 J/(kg)(K). These relations are worth memorizing.

Figure 4.4 illustrates the behavior of the heat capacity of a pure substance over a wide range of absolute temperatures. Observe that at zero degrees absolute the heat capacity is zero. As the temperature rises, the heat capacity also increases until a certain temperature is reached at which a phase transition takes place. The phase transitions are shown also on a related p–T diagram in Fig. 4.5 for water. The phase transition may take place between two solid states, or between a solid and a liquid state, or between a solid and a gaseous state, or between a liquid and a gaseous state. Figure 4.4 shows first a transition between solid state I and solid state II, then the transition between solid state II and the liquid state, and finally the transition between the liquid and the gaseous state. Note that the heat capacity is a continuous

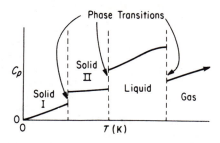

Fig. 4.4 Heat capacity as a function of temperature for a pure substance.

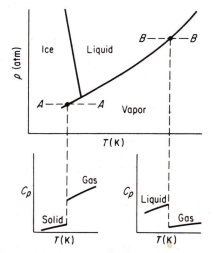

Fig. 4.5 Heat capacity and phase transitions.

function *only* in the region between the phase transitions; consequently, it is not possible to have a heat capacity equation for a substance that will go from absolute zero up to any desired temperature. What an engineer does is to determine experimentally the heat capacity between the temperatures at which the phase transitions occur, fit the data with an equation, and then determine a new heat capacity equation for the next range of temperatures between the succeeding phase transitions.

Experimental evidence indicates that the heat capacity of a substance is not constant with temperature, although at times we may assume that it is constant in order to get approximate results. For the ideal monoatomic gas, of course, the heat capacity at constant pressure is constant even though the temperature varies (see Table 4.1). For typical real gases, see Fig. 4.6; the heat capacities shown are for pure components. For ideal mixtures, the heat capacities of the individual components may be computed separately and each component handled as if it were alone (see Sec. 4.8 for additional details regarding mixtures).

TABLE 4.1 HEAT CAPACITIES OF IDEAL GASES

	Approximate Heat Capacity*, C_p	
Type of Molecule	High Temperature (Translational, Rotational, and Vibrational Degrees of Freedom)	Room Temperature (Translational and Rotational Degrees of Freedom Only)
Monoatomic	$\frac{5}{2}R$	$\frac{5}{2}R$
Polyatomic, linear	$(3n - \frac{3}{2})R$	$\frac{7}{2}R$
Polyatomic, nonlinear	$(3n - 2)R$	$4R$

*n, number of atoms per molecule; R, gas constant defined in Sec. 3.1.

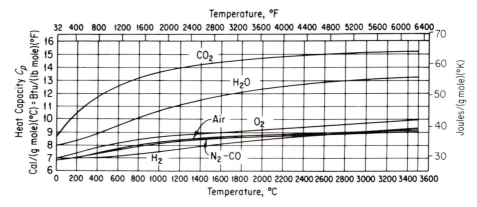

Fig. 4.6 Heat capacity curves for the combustion gases.

Most of the equations for the heat capacities of solids, liquids, and gases are empirical. We usually express the heat capacity at constant pressure C_p as a function of temperature in a power series, with constants a, b, c, and so on; for example,

$$C_p = a + bT$$

or

$$C_p = a + bT + cT^2$$

where the temperature may be expressed in degrees Celsius, degrees Fahrenheit, degrees Rankine, or degrees kelvin. If C_p is expressed in the form of

$$C_p = a + bT + cT^{-1/2}$$
$$C_p = a + bT - cT^{-2}$$

or a form such that you divide by T, then it is necessary to use degrees kelvin or degrees Rankine in the heat capacity equations, because if degrees Celsius or Fahrenheit were to be used, you might divide at some point in the temperature range by zero. Since these heat capacity equations are valid only over moderate temperature ranges, it is possible to have equations of different types represent the experimental heat capacity data with almost equal accuracy. The task of fitting heat capacity equations to heat capacity data is greatly simplified by the use of digital computers, which can determine the constants of best fit by means of a standard prepared program and at the same time determine how precise the predicted heat capacities are. Heat capacity information can be found in Appendix E. The change of C_p with pressure at high pressures is beyond the scope of our work here. Details can be found in several of the references listed at the end of the chapter.

Specific heat is a term often considered synonymous with heat capacity, but this connotation has arisen from loose usage. In principle, specific heat is the ratio of the heat capacity of a substance to the heat capacity of a reference substance, such as

$$\frac{C_{p_A}}{C_{p_{H_2O}}} = \frac{\text{Btu}/(\text{lb}_A)(°F)}{\text{Btu}/(\text{lb}_{H_2O})(°F)}$$

where the reference substance temperature must be specified. Because water has a heat capacity of 1.00 at about 17°C, numerical values of specific heats and heat capacities in the American engineering and thermochemical systems are about the same, although their units are not. An even more confusing issue is that by "specific heat" many engineers simply mean a heat capacity based on a gram or pound.

EXAMPLE 4.3 Heat Capacity Equation

The heat capacity equation for CO_2 gas is

$$C_p = 6.393 + 10.100T \times 10^{-3} - 3.405T^2 \times 10^{-6}$$

with C_p expressed in cal/(g mol)(K) and T in K. Convert this equation into a form so that the heat capacity will be expressed over the entire temperature range in

(a) Cal/(g mol)(°C) with T in °C and ΔT in Δ°C
(b) Btu/(lb mol)(°F) with T in °F and ΔT in Δ°F
(c) Cal/(g mol)(K) with T in °F and ΔT in ΔK
(d) J/(kg mol)(K) with T in K and ΔT in ΔK

Solution

Changing a heat capacity equation from one set of units to another is merely a problem in the conversion of units. Each term in the heat capacity equation must have the same units as the left-hand side of the equation. To avoid confusion in the conversion, you must remember to distinguish between the temperature symbols that represent temperature and the temperature symbols that represent temperature difference even though the same symbol often is used for each concept. In the conversions below we shall distinguish between the temperature and the temperature difference symbols for clarity.

(a) The heat capacity equation with T in °C and ΔT in Δ°C is

$$C_p\frac{\text{cal}}{(\text{g mol})(\Delta°\text{C})} = 6.393\frac{\text{cal}}{(\text{g mol})(\Delta\text{K})}\left|\frac{1\,\Delta\text{K}}{1\,\Delta°\text{C}}\right.$$

$$+ 10.1000 \times 10^{-3}\frac{\text{cal}}{(\text{g mol})(\Delta\text{K})(\text{K})}\left|\frac{1\,\Delta\text{K}}{1\,\Delta°\text{C}}\right|\frac{(T°_\text{C}+273)\,\text{K}}{}$$

$$- 3.405 \times 10^{-6}\frac{\text{cal}}{(\text{g mol})(\Delta\text{K})(\text{K})^2}\left|\frac{1\,\Delta\text{K}}{1\,\Delta°\text{C}}\right|\frac{(T°_\text{C}+273)^2\,\text{K}^2}{}$$

$$= 6.393 + 10.100 \times 10^{-3}T°_\text{C} + 2.757 - 3.405 \times 10^{-6}T°_\text{C}^2$$

$$- 1.860 \times 10^{-3}T°_\text{C} - 0.254$$

$$= 8.896 + 8.240 \times 10^{-3}T°_\text{C} - 3.405 \times 10^{-6}T°_\text{C}^2$$

(b)

$$C_p\frac{\text{Btu}}{(\text{lb mol})(\Delta°\text{F})} = 6.393\frac{\text{cal}}{(\text{g mol})(\Delta\text{K})}\left|\frac{454\,\text{g mol}}{1\,\text{lb mol}}\right|\frac{1\,\text{Btu}}{252\,\text{cal}}\left|\frac{1\,\Delta\text{K}}{1.8\,\Delta°\text{F}}\right.$$

$$+ 10.100 \times 10^{-3}\frac{\text{cal}}{(\text{g mol})(\Delta\text{K})(\text{K})}\left|\frac{454\,\text{g mol}}{1\,\text{lb mol}}\right|\frac{1\,\text{Btu}}{252\,\text{cal}}\left|\frac{1\,\Delta\text{K}}{1.8\,\Delta°\text{F}}\right|\frac{\left(273 + \frac{T°_\text{F}-32}{1.8}\right)\,\text{K}}{}$$

$$- 3.405 \times 10^{-6}\frac{\text{cal}}{(\text{g mol})(\Delta\text{K})(\text{K})^2}\left|\frac{454\,\text{g mol}}{1\,\text{lb mol}}\right|\frac{1\,\text{Btu}}{252\,\text{cal}}\left|\frac{1\,\Delta\text{K}}{1.8\,\Delta°\text{F}}\right|\frac{\left(273 + \frac{T°_\text{F}-32}{1.8}\right)^2\,\text{K}^2}{}$$

$$= 6.393 + 10.100 \times 10^{-3}\,[273 + (T°_\text{F}-32)/1.8]$$

$$- 3.405 \times 10^{-6}[273 + (T°_\text{F}-32)/1.8]^2$$

$$= 6.393 + 2.575 + 5.61 \times 10^{-3}T°_\text{F} - 0.222$$

$$- 0.964 \times 10^{-3}T°_\text{F} - 1.05 \times 10^{-6}T°_\text{F}^2$$

$$= 8.746 + 4.646 \times 10^{-3}T°_\text{F} - 10.5 \times 10^{-6}T°_\text{F}^2$$

Note: In Sec. 1.4 we developed and made use of the expression

$$T_\text{K} = 273 + \frac{T°_\text{F} - 32}{1.8}$$

If the equation for part (a) were available, the solution to part (b) would be simplified. For example, we use

$$C_p\frac{\text{cal}}{(\text{g mol})(\Delta°\text{C})} = 8.896 + 8.240 \times 10^{-3}T°_\text{C} - 3.405 \times 10^{-6}T°_\text{C}$$

and convert it into

$$C_p\frac{\text{Btu}}{(\text{lb mol})(\Delta°\text{F})} = \frac{8.896\,\text{cal}}{(\text{g mol})(\Delta°\text{C})}\left|\frac{454\,\text{g mol}}{\text{lb mol}}\right|\frac{1\,\text{Btu}}{252\,\text{cal}}\left|\frac{1\,\Delta°\text{C}}{1.8\,\Delta°\text{F}}\right.$$

$$+ 8.240 \times 10^{-3}\frac{\text{cal}}{(\text{g mol})(\Delta°\text{C})(°\text{C})}\left|\frac{454\,\text{g mol}}{1\,\text{lb mol}}\right|\frac{1\,\text{Btu}}{252\,\text{cal}}\left|\frac{1\,\Delta°\text{C}}{1.8\,\Delta°\text{F}}\right|\frac{[(T°_\text{F}-32)/1.8]°\text{C}}{}$$

$$- 3.405 \times 10^{-6}\frac{\text{cal}}{(\text{g mol})(\Delta°\text{C})(°\text{C}^2)}\left|\frac{454\,\text{g mol}}{1\,\text{lb mol}}\right|\frac{1\,\text{Btu}}{252\,\text{cal}}\left|\frac{1\,\Delta°\text{C}}{1.8\,\Delta°\text{F}}\right|\frac{[(T°_\text{F}-32)/1.8]^2°\text{C}^2}{}$$

$$= 8.746 + 4.646 \times 10^{-3}T°_\text{F} - 1.05 \times 10^{-6}T°_\text{F}^2$$

(c) Since

$$\frac{Btu}{(lb\ mol)(\Delta°F)} = \frac{cal}{(g\ mol)(\Delta K)}$$

the equation developed in part (b) gives the heat capacity in the desired units.

(d)

$$C_p \frac{J}{(kg\ mol)(\Delta K)} = 6.393 \frac{cal}{(g\ mol)(\Delta K)} \left| \frac{4.184\ J}{1\ cal} \right| \frac{1000\ g}{1\ kg}$$

$$+ 10.1000 \times 10^{-3} \frac{cal}{(g\ mol)(\Delta K)(K)} \left| \frac{4.184\ J}{1\ cal} \right| \frac{1000 g}{1\ kg} \left| T_K \right.$$

$$- 3.405 \times 10^{-6} \frac{cal}{(g\ mol)(K)^2(\Delta K)} \left| \frac{4.184\ J}{1\ cal} \right| \frac{1000\ g}{1\ kg} \left| T_K^2 \right.$$

$$= 2.90 \times 10^4 + 42.27 T_K - 1.425 \times 10^{-2} T_K^2$$

EXAMPLE 4.4 Heat Capacity of the Ideal Gas

Show that $C_p = C_v + \hat{R}$ for the ideal monoatomic gas.

Solution

The heat capacity at constant volume is defined as

$$C_v = \left(\frac{\partial \hat{U}}{\partial T} \right)_v \tag{a}$$

For any gas,

$$C_p = \left(\frac{\partial \hat{H}}{\partial T} \right)_p = \left[\frac{\partial \hat{U} + \partial(p\hat{V})}{\partial T} \right]_p = \left[\frac{\partial \hat{U} + p\partial \hat{V}}{\partial T} \right]_p$$

$$= \left(\frac{\partial \hat{U}}{\partial T} \right)_p + p \left(\frac{\partial \hat{V}}{\partial T} \right)_p \tag{b}$$

For the *ideal gas*, since $\hat{U}$ is a function of temperature only,

$$\left(\frac{\partial \hat{U}}{\partial T} \right)_p = \left(\frac{\partial \hat{U}}{\partial T} \right)_v = C_v \tag{c}$$

and from $p\hat{V} = \hat{R}T$ we can calculate

$$\left(\frac{\partial \hat{V}}{\partial T} \right)_p = \frac{\hat{R}}{p} \tag{d}$$

so that

$$C_p = C_v + \hat{R}$$

4.2-1 Estimation of Heat Capacities

We now mention a few ways by which to *estimate* heat capacities of solids, liquids, and gases. For the most accurate results, you should employ actual experimental heat capacity data or equations derived from such data in your calculations. However, if experimental data are not available, there are a number of equations that you may use which give estimates of values for the heat capacities.

 Solids. Only very rough approximations of solid heat capacities can be made. *Kopp's rule* (1864) should only be used as a last resort when experimental data cannot be located or new experiments carried out. Kopp's rule states that at room temperature the sum of the heat capacities of the individual elements is approximately equal to the heat capacity of a solid compound. For elements below potassium, numbers have been assigned from experimental data for the heat capacity for each element as shown in Table 4.2. For liquids Kopp's rule can be applied with a modified series of values for the various elements, as shown also in Table 4.2.

TABLE 4.2 VALUES FOR MODIFIED KOPP'S RULE: ATOMIC HEAT CAPACITY AT 20°C [CAL/(G ATOM)(°C)]

Element	Solids	Liquids
C	1.8	2.8
H	2.3	4.3
B	2.7	4.7
Si	3.8	5.8
O	4.0	6.0
F	5.0	7.0
P or S	5.4	7.4
All others	6.2	8.0

EXAMPLE 4.5 Use of Kopp's Rule

Determine the heat capacity at room temperature of $Na_2SO_4 \cdot 10H_2O$.

$$\text{Basis: 1 g mol } Na_2SO_4 \cdot 10H_2O$$

$$
\begin{array}{llr}
\text{Na:} & 2 \times 6.2 = & 12.4 \\
\text{S:} & 1 \times 5.4 = & 5.4 \\
\text{O:} & 14 \times 4.0 = & 56.0 \\
\text{H:} & 20 \times 2.3 = & \underline{46.0} \\
& & 119.8 \text{ cal/(g mol)(°C)}
\end{array}
$$

The experimental value is about 141 cal/(g mol)(°C), so you can see that Kopp's rule does not yield very accurate estimates.

Liquids

Aqueous solutions. For the special but very important case of aqueous solutions, a rough rule in the absence of experimental data is to use the heat capacity of the water only. For example, a 21.6% solution of NaCl is assumed to have a heat capacity of 0.784 cal/(g)(°C); the experimental value at 25°C is 0.806 cal/(g)(°C).

Hydrocarbons. An equation for the heat capacity of liquid hydrocarbons and petroleum products recommended by Fallon and Watson[2] is

$$C_p = [(0.355 + 0.128 \times 10^{-2} \,°\text{API}) + (0.503 + 0.117 \times 10^{-2} \,°\text{API})$$
$$\times 10^{-3} t](0.05K + 0.41)$$

where $°\text{API} = \dfrac{141.5}{\text{sp gr } (60°\text{F}/60°\text{F})} - 131.5$ and is a measure of specific gravity

$t = °\text{F}$

$K =$ Universal Oil Products characterization factor, which has been related to six easily applied laboratory tests; this factor is not a fundamental characteristic but is easy to determine experimentally (values of K range from 10.1 to 13.0 and are discussed in Appendix K)

$C_p = \text{Btu}/(\text{lb})(°\text{F})$

Organic liquids. A simple and reasonably accurate relation between C_p in cal/(g)(°C) at 25°C and molecular weight is

$$C_p = kM^a$$

where M is the molecular weight and k and a are constants. Pachaiyappan et al.[3] give a number of values of the set (k, a). For example

compounds	k	a
Alcohols	0.85	−0.1
Acids	0.91	−0.152
Ketones	0.587	−0.0135
Esters	0.60	−0.0573
Hydrocarbons, aliphatic	0.873	−0.113

Reid and San Jose[4] review a number of estimation methods for liquids and indicate their precision.

Gases and vapors

Petroleum vapors. The heat capacity of petroleum vapors can be estimated from[5]

[2]J. F. Fallon and K. M. Watson, *Natl. Petrol. News, Tech. Sec.* (June 7, 1944).

[3]V. Pachaiyappan, S. H. Ibrahim, and N. R. Kuloor, *Chem. Eng.*, p. 241 (October 9, 1967).

[4]R. C. Reid and J. L. San Jose, *Chem. Eng.*, pp. 67–71 (December 20, 1976).

[5]W. H. Bahlke and W. B. Kay, *Ind. Eng. Chem.*, v. 21, p. 942 (1929).

$$C_p = \frac{(4.0 - s)(T + 670)}{6450}$$

where C_p is in Btu/(lb)($°F$), T is in $°F$, and s is the specific gravity at $60°F/60°F$, with air as the reference gas.

Kothari–Doraiswamy equation. Kothari and Doraiswamy[6] recommend plotting

$$C_p = a + b \log_{10} T_r$$

Given two values of C_p at known temperatures, values of C_p can be predicted at other temperatures with good precision. The most accurate methods of estimating heat capacities of vapors are those of Dobratz[7] based on spectroscopic data and generalized correlations based on reduced properties[8].

Additional methods of estimating solid and liquid heat capacities may be found in Reid et al., who compare various techniques we do not have the space to discuss and make recommendations as to their use.

Self-Assessment Test

1. A problem indicates that the enthalpy of a compound can be predicted by an empirical equation H (J/g) $= -30.2 + 4.25T + 0.001T^2$, where T is in kelvin. What is the heat capacity at constant pressure for the compound?

2. What is the heat capacity at constant pressure at room temperature of O_2 if the O_2 is assumed to be an ideal gas?

3. A heat capacity equation in cal/(g mol)(K) for ammonia gas is

   $$C_p = 8.4017 + 0.70601 \times 10^{-2}T + 0.10567 \times 10^{-5}T^2 - 1.5981 \times 10^{-9}T^3$$

   where T is in $°C$. What are the units of each of the coefficients in the equation?

4. Calculate the heat capacity (C_p) of N_2 gas at 1 atm and 500 K.

5. Estimate the heat capacity (C_p) of acetic acid by **(a)** Kopp's rule and **(b)** Pachaiyappan's constants. Compare with the experimental value at room temperature.

6. Convert the following equation for the heat capacity of carbon monoxide gas, where C_p is in Btu/(lb mol)($°F$) and T is in $°F$:

   $$C_p = 6.865 + 0.08024 \times 10^{-2}T - 0.007367 \times 10^{-5}T^2$$

   to yield C_p in J/(kg mol)(K) with T in kelvin.

[6]M. S. Kothari and L. K. Doraiswamy, *Hydrocarbon Processing and Petroleum Refiner*, v. 43, no. 3, p. 133 (1964).

[7]C. J. Dobratz, *Ind. Eng. Chem.*, v. 33, p. 759 (1941).

[8]R. C. Reid, J. M. Prausnitz, and T. K. Sherwood, *The Properties of Gases and Liquids*, 3rd ed., McGraw-Hill, New York, 1977.

Section 4.3 Calculation of Enthalpy Changes (without Change of Phase)

> **Your objectives in studying this section are to be able to:**
>
> 1. Define mean heat capacity and compute a value given the reference state and final state.
> 2. Calculate an enthalpy change given (a) the heat capacity equation or (b) mean heat capacities, plus the initial and final states of the material.
> 3. Look up the enthalpy values in a reference table (especially the steam tables), data base, or property chart, and compute the enthalpy change given the initial and final states of the material.

Now that we have examined sources for heat capacity values, we turn to the details of the calculation of enthalpy changes. If we omit for the moment consideration of phase changes and examine solely the problem of how to calculate enthalpy changes from heat capacity data, we can see that in general if we use Eq. (4.7), $\Delta\hat{H}$ is the area under the curve in Fig. 4.7:

$$\int_{H_1}^{H_2} d\hat{H} = \Delta\hat{H} = \int_{T_1}^{T_2} C_p \, dT$$

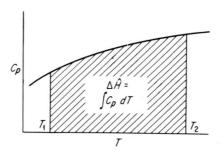

Fig. 4.7 Calculation of enthalpy change.

(You will find details of the technique of graphical integration in Example 6.6.) If our heat capacity is expressed in the form $C_p = a + bT + cT^2$, then

$$\Delta\hat{H} = \int_{T_1}^{T_2} (a + bT + cT^2) \, dT = a(T_2 - T_1) + \frac{b}{2}(T_2^2 - T_1^2)$$
$$+ \frac{c}{3}(T_2^3 - T_1^3) \tag{4.8}$$

If a different functional form of the heat capacity is available, the integration result will no doubt be different.

We can define a mean heat capacity, which, if known, provides a quick and convenient way of calculating enthalpy changes. The *mean heat capacity* C_{p_m} is defined as the enthalpy change divided by the temperature difference for this change, or

$$C_{p_m} = \frac{\hat{H}_2 - \hat{H}_1}{T_2 - T_1} \tag{4.9}$$

Then it is possible to compute an enthalpy change, if C_{p_m} is available, as

$$\Delta \hat{H} = C_{p_m} \Delta T = C_{p_m}(T_2 - T_1) \tag{4.10}$$

where C_{p_m} is identified with the interval between T_1 and T_2.

If the heat capacity can be expressed as a power series, $C_p = a + bT + cT^2$, then C_{p_m} is

$$
C_{p_m} = \frac{\int_{T_1}^{T_2} C_p \, dT}{\int_{T_1}^{T_2} dT} = \frac{\int_{T_1}^{T_2} (a + bT + cT^2) \, dT}{T_2 - T_1}
$$

$$
= \frac{a(T_2 - T_1) + (b/2)(T_2^2 - T_1^2) + (c/3)(T_2^3 - T_1^3)}{T_2 - T_1} \tag{4.11}
$$

By choosing 0°F or 0°C as the reference temperature (T_1) for C_{p_m}, you can simplify the expression for C_{p_m}. Tables 4.3a and 4.3b lists mean heat capacities for combustion gases based on the same reference point (0°C) for two different temperature scales.

TABLE 4.3A   MEAN HEAT CAPACITIES OF COMBUSTION GASES* [J/(G MOL)(K)]†

°C	N_2	O_2	Air	H_2	CO	CO_2	H_2O
0	29.116	29.241	29.062	28.610	29.120	35.961	33.476
18	29.120	29.279	29.074	28.693	29.124	36.425	33.509
25	29.120	29.296	29.074	28.718	29.129	36.467	33.522
100	29.141	29.526	29.141	28.978	29.175	38.166	33.727
200	29.225	29.932	29.292	29.099	29.497	40.124	34.099
300	29.384	30.438	29.513	29.149	29.526	41.852	34.543
400	29.601	30.877	29.781	29.216	29.790	43.346	35.049
500	29.865	31.333	30.082	29.279	30.108	44.685	35.593
600	30.154	31.760	30.400	29.350	30.430	45.877	36.166
700	30.446	32.149	30.710	29.438	30.756	46.948	36.756
800	30.752	32.501	31.020	29.547	31.078	47.910	37.354
900	31.045	32.823	31.317	29.677	31.384	48.869	37.948
1000	31.304	33.120	31.585	29.823	31.672	49.580	38.534
1100	31.593	33.388	31.865	29.978	31.944	50.291	39.112
1200	31.840	33.635	32.108	30.145	32.166	50.919	39.672
1300	32.070	33.865	32.338	30.237	32.434	51.546	40.220
1400	32.292	34.082	32.555	30.375	32.656	52.048	40.752
1500	32.505	34.275	32.760	30.526	32.865	52.551	41.254

SOURCE: Page 30 of Reference 16 in Table 4.5.

*Reference temperature, 0°C; pressure, 1 atm.

†To convert to cal/(g mol)(°C), multiply by 0.2390.

TABLE 4.3B   MEAN HEAT CAPACITIES OF COMBUSTION GASES*
[BTU/LB MOL)(°F)]

°F	N_2	O_2	Air	H_2	CO	CO_2	H_2O
32	6.959	6.989	6.946	6.838	6.960	8.595	8.001
60	6.960	6.996	6.948	6.855	6.961	8.682	8.008
77	6.960	7.002	6.949	6.864	6.962	8.715	8.012
100	6.961	7.010	6.952	6.876	6.964	8.793	8.019
200	6.964	7.052	6.963	6.921	6.972	9.091	8.055
300	6.970	7.102	6.978	6.936	6.987	9.362	8.101
400	6.984	7.159	7.001	6.942	7.007	9.612	8.154
500	7.002	7.220	7.028	6.961	7.033	9.844	8.210
600	7.026	7.283	7.060	6.964	7.065	10.060	8.274
700	7.055	7.347	7.096	6.978	7.101	10.262	8.341
800	7.087	7.409	7.134	6.981	7.140	10.450	8.411
900	7.122	7.470	7.174	6.984	7.182	10.626	8.484
1000	7.158	7.529	7.214	6.989	7.224	10.792	8.558
1100	7.197	7.584	7.256	6.993	7.268	10.948	8.634
1200	7.236	7.637	7.298	7.004	7.312	11.094	8.712
1300	7.277	7.688	7.341	7.013	7.355	11.232	8.790
1400	7.317	7.736	7.382	7.032	7.398	11.362	8.870
1500	7.356	7.781	7.422	7.054	7.439	11.484	8.950
1600	7.395	7.824	7.461	7.061	7.480	11.60	9.029
1700	7.433	7.865	7.500	7.073	7.519	11.71	9.107
1800	7.471	7.904	7.537	7.081	7.558	11.81	9.185
1900	7.507	7.941	7.573	7.093	7.595	11.91	9.263
2000	7.542	7.976	7.608	7.114	7.631	12.01	9.339

SOURCE: Page 31 of Reference 16 in Table 4.5.
*Reference temperature, 32°F; pressure, 1 atm.

EXAMPLE 4.6   Calculation of $\Delta\hat{H}$ Using Mean Heat Capacity

Calculate the enthalpy change for 1 kg mol of nitrogen (N_2) which is heated at constant pressure (100 kPa) from 18°C to 1100°C.

Solution

The enthalpy data will be taken from Table 4.3a; reference conditions, 0°C:

$$C_{pm} = 31.593 \text{ kJ/(kg mol)(°C) at } 1100°C$$
$$C_{pm} = 29.120 \text{ kJ/(kg mol)(°C) at } 18°C$$

Basis: 1 kg mol of N_2

$$\Delta\hat{H}_{1100-18} = \Delta\hat{H}_{1100} - \Delta\hat{H}_{18}$$

$$= \frac{31.593 \text{ kJ}}{\text{(kg mol)(°C)}} \Bigg| (1100 - 0)°C - \frac{29.120 \text{ kJ}}{\text{(kg mol)(°C)}} \Bigg| (18 - 0)°C$$

$$= 34.228 \text{ kJ/kg mol}$$

If the reference conditions are not zero (0°C, 0°F etc.), such as for Table 4.3b where the reference temperature is 32°F, the $\Delta\hat{H}_{1100-18}$ would be

$$C_{pm_{1100}}(T_{1100} - T_{ref}) - C_{pm_{18}}(T_{18} - T_{ref})$$

and *not* $C_{pm_{1100}}T_{1100} - C_{pm_{18}}T_{18}$.

EXAMPLE 4.7 Calculation of $\Delta\hat{H}$ Using Heat Capacity Equations

The conversion of solid wastes to innocuous gases can be accomplished in incinerators in an environmentally acceptable fashion. However, the hot exhaust gases must be cooled or diluted with air. An economic feasibility study indicates that solid municipal waste can be burned to a gas of the following composition (on a dry basis):

CO_2	9.2%
CO	1.5%
O_2	7.3%
N_2	82.0%
	100.0%

What is the enthalpy difference for this gas between the bottom and the top of the stack if the temperature at the bottom of the stack is 550°F and the temperature at the top is 200°F? Ignore the water vapor in the gas. Because these are ideal gases, you can neglect any energy effects resulting from the mixing of the gaseous components.

Solution

Heat capacity equations from Table E.1 in Appendix E [T in °F; C_p = Btu/(lb mol)(°F)] are

$$N_2: \quad C_p = 6.895 + 0.7624 \times 10^{-3}T - 0.7009 \times 10^{-7}T^2$$

$$O_2: \quad C_p = 7.104 + 0.7851 \times 10^{-3}T - 0.5528 \times 10^{-7}T^2$$

$$CO_2: \quad C_p = 8.448 + 5.757 \times 10^{-3}T - 21.59 \times 10^{-7}T^2 + 3.049 \times 10^{-10}T^3$$

$$CO: \quad C_p = 6.865 + 0.8024 \times 10^{-3}T - 0.7367 \times 10^{-7}T^2$$

Basis: 1.00 lb mol of gas

By multiplying these equations by the respective mole fraction of each component, and then adding them together, you can save time in the integration.

$$N_2: \quad 0.82(6.895 + 0.7624 \times 10^{-3}T - 0.7009 \times 10^{-7}T^2)$$

$$O_2: \quad 0.073(7.104 + 0.7851 \times 10^{-3}T - 0.5528 \times 10^{-7}T^2)$$

$$CO_2: \quad 0.092(8.448 + 5.757 \times 10^{-3}T - 21.59 \times 10^{-7}T^2 + 3.059 \times 10^{-10}T^3)$$

$$CO: \quad 0.015(6.865 + 0.8024 \times 10^{-3}T - 0.7367 \times 10^{-7}T^2)$$

$$C_{p_{net}} = 7.052 + 1.2243 \times 10^{-3}T - 2.6124 \times 10^{-7}T^2 + 0.2814 \times 10^{-10}T^2$$

$$\Delta\hat{H} = \int_{550}^{200} C_p \, dT = \int_{550}^{200} (7.052 + 1.2243 \times 10^{-3}T - 2.6124 \times 10^{-7}T^2$$
$$+ 0.2814 \times 10^{-10}T^3) \, dT$$

$$= 7.053[(200) - (550)] + \frac{1.224 \times 10^{-3}}{2}[(200)^2 - (550)^2]$$

$$- \frac{2.6124 \times 10^{-7}}{3}[(200)^3 - (550)^3]$$

$$+ \frac{0.2814 \times 10^{-10}}{4}[(200)^4 - (550)^4]$$

$$= -2469 - 160.6 + 13.8 - 0.633$$

$$= -2616 \text{ Btu/lb mol gas}$$

Note in Example 4.7 that the heat capacity of each component in the mixture has been assumed to be independent of the other components. Will this asumption always be true?

Next, we discuss briefly the use of enthalpy tables and charts to calculate enthalpy changes. The simplest and quickest method of computing enthalpy changes is to use tabulated enthalpy data available in the literature, in reference books, or in a computer data base. Thus rather than using integrated heat capacity equations or mean heat capacity data, you should first see if you can find enthalpy data given directly as $\Delta \hat{H}$ as a function of temperature. Tables 4.4a and 4.4b list typical enthalpy

TABLE 4.4A   ENTHALPIES OF COMBUSTION GASES* (BTU/LB MOL)

°R	N₂	O₂	Air	H₂	CO	CO₂	H₂O
492	0.0	0.0	0.0	0.0	0.0	0.0	0.0
500	55.67	55.93	55.57	57.74	55.68	68.95	64.02
520	194.9	195.9	194.6	191.9	194.9	243.1	224.2
537	313.2	315.1	312.7	308.9	313.3	392.2	360.5
600	751.9	758.8	751.2	744.4	752.4	963	867.5
700	1450	1471	1450	1433	1451	1914	1679
800	2150	2194	2153	2122	2154	2915	2501
900	2852	2931	2861	2825	2863	3961	3336
1000	3565	3680	3579	3511	3580	5046	4184
1100	4285	4443	4306	4210	4304	6167	5047
1200	5005	5219	5035	4917	5038	7320	5925
1300	5741	6007	5780	5630	5783	8502	6819
1400	6495	6804	6540	6369	6536	9710	7730
1500	7231	7612	7289	7069	7299	10942	8657
1600	8004	8427	8068	7789	8072	12200	9602
1700	8774	9251	8847	8499	8853	13470	10562
1800	9539	10081	9623	9219	9643	14760	11540
1900	10335	10918	10425	9942	10440	16070	12530
2000	11127	11760	11224	10689	11243	17390	13550
2100	11927	12610	12030	11615	12050	18730	14570
2200	12730	13460	12840	12160	12870	20070	15610
2300	13540	14320	13660	12890	13690	21430	16660
2400	14350	15180	14480	13650	14520	22800	17730
2500	15170	16040	15300	14400	15350	24180	18810

SOURCE: Page 30 of Reference 16 in Table 4.5.
*Pressure, 1 atm.

TABLE 4.4B   ENTHALPIES OF COMBUSTION GASES (J/G MOL)*

K	N_2	O_2	Air	H_2	CO	CO_2	H_2O
273	0	0	0	0	0	0	0
291	524	527	523	516	525	655	603
298	728	732	726	718	728	912	837
300	786	790	784	763	786	986	905
400	3695	3752	3696	3655	3699	4903	4284
500	6644	6811	6660	6589	6652	9204	7752
600	9627	9970	9673	9518	9665	13807	11326
700	12652	13225	12736	12459	12748	18656	15016
800	15756	16564	15878	15413	15899	23710	18823
900	18961	19970	19116	18384	19125	28936	22760
1000	22171	23434	22367	21388	22413	34308	26823
1100	25472	26940	25698	24426	25760	39802	31011
1200	28819	30492	29078	27509	29154	45404	35312
1300	32216	34078	32501	30626	32593	51090	39722
1400	35639	37693	35953	33789	36070	56860	44237
1500	39145	41337	39463	36994	39576	62676	48848
1750	47940	50555	48325	45275	48459	77445	60751
2000	56902	59914	57320	53680	57488	92466	73136
2250	65981	69454	66441	62341	66567	107738	85855
2500	75060	79119	75646	71211	75772	123176	98867
2750	84265	88910	84935	80290	85018	138699	112089
3000	93512	98826	94265	89453	94265	154347	125520
3500	112131	119034	113135	108030	112968	185895	152799
4000	130875	141410	132172	127528	131796	217777	180414

*To convert to cal/g mol multiply by 0.2390.

data for the combustion gases. Some sources of enthalpy data are listed in Table 4.5. The most common source of enthaply data for water is the steam tables which are reproduced on a sheet that can be found inside the back cover. The reference by Kobe and associates (Table 4.5, Reference 16) lists heat capacity equations, enthalpy values, and many other thermodynamic functions for over 100 compounds of commercial importance. Some of this information can be found in Appendix D. Computer tapes can be purchased providing information on the physical properties of large numbers of compounds as described in Reference 17 in Table 4.5, or from the Institution of Chemical Engineers, London. Immediate access to computer-based information systems via the telephone can be obtained from computer service bureaus.

If you remember that enthalpy values are all relative to some reference state, you can make enthalpy difference calculations merely by subtracting the initial enthalpy from the final enthalpy for any two sets of conditions. Of course, if the enthalpy data are not available, you will have to rely on calculations involving heat capacities or mean heat capacities.

Another convenient way to store and retrieve enthalpy data is from a chart

TABLE 4.5   SOURCES OF ENTHALPY DATA

1. American Petroleum Institute, Division of Refining, *Technical Data Book—Petroleum Refining*, 2nd ed., API, Washington, D.C., 1970.
2. American Petroleum Institute, Research Project 44, *Selected Values of Physical and Thermodynamic Properties of Hydrocarbons and Related Compounds*, API, Washington, D.C., 1953. See also Reference 23.
3. American Society of Mechanical Engineers, *Thermodynamic Data for Waste Incineration*, Book No. H00141, American Society of Mechanical Engineers, New York, 1979.
4. Barner, H. E., and R. V. Scheuerman, *Handbook of Thermochemical Data for Compounds and Aqueous Species*, Wiley–Interscience, New York, 1978.
5. *Bulletin of Chemical Thermodynamics*, published by Department of Chemistry, Oklahoma State University, Stillwater, Okla., 74074. (Index, bibliography, reports.)
6. Chermin, H. A. G., *Hydrocarbon Processing and Petroleum Refining*, Parts 26–32, 1961. (Continuation of Kobe compendium for gases; see Reference 16.)
7. Egloff, G., *Physical Constants of Hydrocarbons*, 5 vols., Van Nostrand Reinhold, New York, 1939–1953.
8. Engineering Sciences Data Unit Ltd., London. (Extensive series of property data.)
9. Fratzscher, W., et al., "The Acquisition, Collection, and Tabulation of Substance Data on Fluid Systems for Calculations in Chemical Engineering," *Intl. Chem. Eng.*, v. 20, no. 1, pp. 19–28 (1980). (A list of references for data.)
10. Hamblin, F. D., *Abridged Thermodynamic and Thermochemical Tables (S.I. Units)*, Pergamon Press, Elmsford, N.Y. (paperback edition available), 1972.
11. Haywood, R. W., *Thermodynamic Tables in SI (Metric) Units*, 2nd ed., Cambridge University Press, Cambridge, 1972.
12. Irvine, T. F., and P. E. Liley, eds., *Microcomputer Program for the Thermodynamic Properties of Air/Steam*, Rumford Publishing Co., 1980. (For personal computers.)
13. *Journal of Physical and Chemical Reference Data*, American Chemical Society, 1972 and following.
14. Keenan, J. H., et al., *Steam Tables*, Wiley, New York, 1978. (SI units.) Also J. H. Keenan, and F. G. Keyes, *Thermodynamic Properties of Steam*, Wiley, New York, 1936.
15. Kelley, K. K., various *U.S. Bureau of Mines Bulletins* on the Thermodynamic Properties of Substances, particularly minerals; 1935–1962.
16. Kobe, K. A., et al. "Thermochemistry of Petrochemicals," Reprint No. 44 from the *Petroleum Refiner*. A collection of a series of 24 articles covering 105 different substances that appeared in the *Petroleum Refiner*.
17. Meadows, E. L., "Estimating Physical Properties," *Chem. Eng. Progr.*, v. 61, p. 93 (1965). (A.I.Ch.E. Machine Computation Committee report on the preparation of thermodynamic tables on tapes.)
18. McBride, B. J., and S. Gordon, *Fortran IV Program for Calculation of Thermodynamic Data*, NASA Report TND-4097, 1967.
19. Naumov, G. B., *Handbook of Thermodynamic Data* (translation into English of *Handbook of Thermodynamic Data*, Atomizdat, Moscow, 1971). (Available as NTIS PB-226722/7.)
20. Perry, J. H., *Chemical Engineers' Handbook*, 5th ed., McGraw-Hill, New York, 1973.
21. Rossini, F. K., et al., "Tables of Selected Values of Chemical Thermodynamic Properties," *National Bureau of Standards (U.S.)*, *Circ.* 500, 1952. (Revisions are being issued periodically under the Technical Note 270 series by other authors.)
22. Stull, D. R., and H. Prophet, *JANAF Thermochemical Tables*, 2nd ed., U.S. Government Printing Office, Washington, D.C., No. 0303–0872, C13, 48: 37 (AD-732 043), 1971. (Data for over 1000 compounds.)

TABLE 4.5 (Continued)

23. Thermodynamics Research Center, Texas A & M University. Continuation of API Project 44 plus extensive additional data. Write Data Distribution Office, TRC, College Station, Tex. 77843.

24. Yaws, C. L., "Physical and Thermodynamic Properties" (Graphs), *Chem. Eng.*, Initiated with Part I in the June 10, 1974, issue. Subsequent issues: July 3, Aug. 19, Sept. 30, Oct. 28, Nov. 25, Dec. 23, 1974; Jan. 20, Feb. 17, Mar. 31, May 12, July 21, Sept. 1, Sept. 29, 1975. See "Physical Properties: A Guide . . .," published 1977, McGraw-Hill, New York.

25. Zeise, H., *Thermodynamik*, Band III/I Taballen, S. Hirzel Verlag, Leipzig, 1954. (Tables of heat capacities, enthalpies, entropies, free energies, and equilibrium constants.)

---

or diagram. Figure 4.8 is an example; others are given in Appendix J. Reynolds[9] has an excellent collection of charts of 40 substances in his volume. Other sources of such charts are listed in Table 4.6. Many forms of charts are available with coordinate axes such as

$$p \text{ vs. } \hat{H}$$

$$p \text{ vs. } \hat{V}$$

$$p \text{ vs. } T$$

$$\hat{H} \text{ vs. } \hat{S}[10]$$

Since a chart has only two dimensions, the other parameters of interest have to be plotted as lines of constant value across the face of the chart. Recall, for example, that in the p–$\hat{V}$ diagram for CO_2, Fig. 3.2, lines of constant temperature were shown as parameters. Similarly, on a chart with pressure and enthalpy as the axes, lines of constant volume and/or temperature might be drawn.

How many properties have to be specified for a single component gas to definitely fix the state of the gas? Because we know from our previous discussion in Sec. 3.7 that specifying two properties for a pure gas will ensure that all the other properties will have definite values, any two properties can be chosen at will. Since a particular state for a gas can be defined by any two independent properties, a two-dimensional thermodynamic chart can be seen to be a handy way to present many combinations of physical properties.

Although diagrams and charts are helpful in that they portray the relations among the phases and give approximate values for the physical properties of a substance, to get more precise data (unless the chart is quite large), usually you will find it best to see if tables are available which list the physical properties you want, or equations for computer calculations.

Enthalpies and other thermodynamic properties can be *estimated* by generalized

---

[9]W. C. Reynolds, *Thermodynamic Properties in SI*, Department of Mechanical Engineering, Stanford University, Stanford, Calif. 94305 (1979).

[10]S is *entropy*—this diagram is called a *Mollier* diagram.

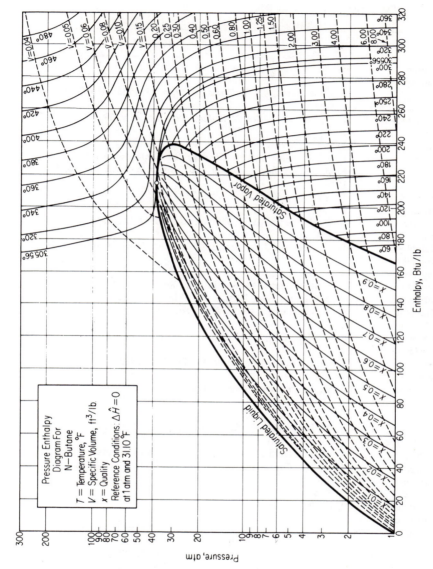

Fig. 4.8 Pressure enthalpy diagram for *n*-butane.

Pressure Enthalpy
Diagram For
N—Butane

T = Temperature, °F
V = Specific Volume, ft³/lb
x = Quality
Reference Conditions: $\Delta \hat{H} = 0$
at 1 atm and 31.10 °F

Enthalpy, Btu/lb

Pressure, atm

Saturated Vapor

Saturated Liquid

TABLE 4.6 THERMODYNAMIC CHARTS SHOWING ENTHALPY DATA FOR PURE COMPOUNDS*

Compound	Reference†
Acetone	2
Acetylene	1
Air	V. C. Williams, *AIChE Trans.*, v. 39, p. 93 (1943); *AIChE J.*, v. 1, p. 302 (1955).
Benzene	1
1, 3-Butadiene	C. H. Meyers, *J. Res. Natl. Bur. Std., Ser. A*, v. 39, p. 507 (1947).
i-Butane	1, 3
n-Butane	1, 3, 4
n-Butanol	L. W. Shemilt, in *Proceedings of the Conference on Thermodynamic Transport Properties of Fluids*, London, 1957, Institute of Mechanical Engineers, London, 1958.
t-Butanol	F. Maslan, *AIChE J.*, v. 7, p. 172 (1961).
n-Butene	1
Chlorine	R. E. Hulme and A. B. Tilman, *Chem. Eng.* (January 1949).
Ethane	1, 3, 4
Ethanol	R. C. Reid and J. M. Smith, *Chem. Eng. Progr.*, v. 47, p. 415 (1951).
Ethyl ether	2
Ethylene	1, 3
Ethylene oxide	J. E. Mock and J. M. Smith, *Ind. Eng. Chem.*, v. 42, p. 2125 (1950).
Fatty acids	J. D. Chase, *Chem. Eng.*, p. 107 (Mar. 24, 1980).
n-Heptane	E. B. Stuart et al., *Chem. Eng. Progr.*, v. 46, p. 311 (1950).
n-Hexane	1
Hydrogen sulfide	J. R. West, *Chem. Eng. Progr.*, v. 44, p. 287 (1948).
Isopropyl ether	2
Mercury	General Electric Co. Report GET-1879A, 1949.
Methane	1, 3, 4
Methanol	J. M. Smith, *Chem. Eng. Progr.*, v. 44, p. 52 (1948).
Methyl ethyl ketone	2
Monomethyl hydrazine	F. Bizjak and D. F. Stai, *AIAA J.*, v. 2, p. 954 (1964).
Neon	Cryogenic Data Center, National Bureau of Standards, Boulder, Colo.
Nitrogen	G. S. Lin, *Chem. Eng. Progr.*, v. 59, no. 11, p. 69 (1963).
n-Pentane	1, 3, 4
Propane	1, 3, 4
n-Propanol	Shemilt (*see n*-Butanol).
Propylene	1, 3

*For mixtures, see V. F. Lesavage et al., *Ind. Eng. Chem.*, v. 59, no. 11, p. 35 (1967).

†1. L. N. Cajar et al., *Thermodynamic Properties and Reduced Correlations for Gases*, Gulf Publishing Co., Houston, Tex., 1967. (Series of articles which appeared in the magazine *Hydrocarbon Processing* from 1962 to 1965.)
2. P. T. Eubank and J. M. Smith, *J. Chem. Eng. Data*, v. 7, p. 75 (1962).
3. W. C. Edmister, *Applied Hydrocarbon Thermodynamics*, Gulf Publishing Co., Houston, Tex., 1961.
4. K. E. Starling et al., *Hydrocarbon Processing*, 1971 and following. *Note:* Charts available separately from Gulf Publishing Co., Houston, Tex.

TABLE 4.6 **(Continued)**

Refrigerant 245	R. L. Shank, *J. Chem. Eng. Data*, v. 12, p. 474 (1967).
Sulfur dioxide	J. R. West and G. P. Giusti, *J. Phys. Colloid Chem.*, v. 54, p. 601 (1950).
Combustion gases	H. C. Hottel, G. C. Williams, and C. N. Satterfield, *Thermodynamic Charts for Combustion Processes* (I) Text, (II) Charts, Wiley, New York, 1949.
Hydrocarbons	3

methods based on the theory of corresponding states or additive bond contributions.[11-13]

EXAMPLE 4.8 Calculation of Enthalpy Change Using Tabulated Enthalpy Values

Recalculate Example 4.6 using data from enthalpy tables.

Solution

For the same change as in Example 4.6, the data from Table 4.4b are (ref. temp. $= 273$ K)

$$\text{at } 1100°C \backsim 1373 \text{ K}: \quad \Delta\hat{H} = 34,715 \text{ kJ/kg mol (by extrapolation)}$$
$$\text{at } 18°C \backsim 291 \text{ K}: \qquad \Delta\hat{H} = 524 \text{ kJ/kg mol}$$
$$\text{Basis: 1 kg mol of } N_2$$
$$\Delta\hat{H} = 34,715 - 524 = 34,191 \text{ kJ/kg mol}$$

EXAMPLE 4.9 Use of the Steam Tables

What is the enthalpy change in British termal units when 1 gal of water is heated from 60° to 1150°F and 240 psig?

Solution

$$\text{Basis: 1 lb of } H_2O \text{ at } 60°F$$

From steam tables (ref. temp. $= 32°F$):

$$\hat{H} = 28.07 \text{ Btu/lb} \qquad \text{at } 60°F$$
$$\hat{H} = 1604.5 \text{ Btu/lb} \qquad \text{at } 1150°F \text{ and } 240 \text{ psig (254.7 psia)}$$
$$\Delta\hat{H} = (1604.5 - 28.07) = 1576.4 \text{ Btu/lb}$$
$$\Delta H = 1576(8.345) = 13,150 \text{ Btu/gal}$$

[11]E. J. Janz, *Estimation of Thermodynamic Properties of Organic Compounds*, Academic Press, New York, 1967.

[12]E. K. Landis, M. T. Cannon, and L. N. Canjar, *Hydrocarbon Processing*, v. 44, p. 154 (1965).

[13]R. R. Tarakad and R. P. Palmer, "A Comparison of Enthalpy Prediction Methods," *AIChE J.*, v. 22, p. 409 (1976).

Note: The enthalpy values which have been used for liquid water as taken from the steam tables are for the saturated liquid under its own vapor pressure. Since the enthalpy of liquid water changes negligibly with pressure, no loss of accuracy is encountered for engineering purposes if the initial pressure on the water is not stated.

Self-Assessment Test

1. Calculate the enthalpy change in 24 g of N_2 if heated from 300 K to 1500 K at constant pressure.

2. Air conditioners use Freon-12 as the working fluid. A charging unit when full holds 10 kg of saturated liquid at 20°C. If the unit contains 3 kg at 20°C, what is the quality? Use the following extract from the Freon-12 table in solving the problem.

Satn. Pres., p_s (kPa)	Satn. Temp., t_s (°C)	Volume (cm³/g)		Enthalpy (J/g)	
		v_l	v_g	h_l	h_g
500	15.6	0.7438	34.82	50.66	194.0
600	22	0.7566	29.13	56.80	196.6

3. You are told that 4.3 kg of water at 200 kPa occupies 4.3, 43, 430, 4300, and 43,000 liters. State for each case whether the water is in the solid, liquid, liquid–vapor, or vapor regions.

4. Use the table for the mean heat capacities of H_2 to compute the enthalpy change of 1 lb mol from 1300°F to 500°F. Check your answer by using the table of enthalpies itself.

5. Use 100°F as the reference temperature and compute a mean heat capacity for acetylene gas at 200°F.

6. Two hundred pounds of a 35°API distillate is heated from 130°F to 275°F. Estimate the enthalpy change (in Btu).

7. Water at 400 kPa and 500 K is cooled to 200 kPa and 400 K. What is the enthalpy change? Use the steam tables.

Section 4.4 Enthalpy Changes for Phase Transitions

Your objectives in studying this section are to be able to:

1. Estimate the heat of fusion or heat of vaporization from empirical formulas, or look up the value in a reference table.
2. Estimate the heat of vaporization from the Clausius–Clapeyron equation, the Othmer plot, or from reference tables.
3. Calculate an enthalpy change of a substance including the phase transition.

In making enthalpy calculations, we noted in Fig. 4.4 that the heat capacity data are discontinuous at the points of a phase transition. The name usually given to the enthalpy changes for these phase transitions is *latent heat* changes, latent meaning "hidden" in the sense that the substance (e.g., water) can absorb a large amount of heat without any noticeable increase in temperature. Unfortunately, the word *heat* is still associated with these enthalpy changes for historical reasons, although they have nothing directly to do with heat as defined in Sec. 4.1. Ice at 0°C can absorb energy amounting to 335 J/g without undergoing a temperature rise or a pressure change. The enthalpy change from a solid to a liquid is called the *heat of fusion*, and the enthaply change for the phase transition between solid and liquid water is thus 335 J/g. The *heat of vaporization* is the enthalpy change for the phase transition between liquid and vapor, and we also have the *heat of sublimation*, which is the enthalpy change for the transition directly from solid to vapor. Dry ice (solid CO_2) at room temperature and pressure sublimes. The $\Delta \hat{H}$ for the phase change from gas to liquid is called the *heat of condensation*. You can find experimental values of latent heats in the references in Table 4.5, and a brief tabulation is listed in Appendix D. The symbols used for latent heat changes vary, but you usually find one or more of the following employed: $\Delta \hat{H}$, L, λ, Λ. Keep in mind that the enthalpy changes for vaporization given in the steam tables are for water under its vapor pressure at the indicated temperature.

In the absence of experimental values for the latent heats of transition, the following approximate methods will provide a rough *estimate* of the *molar latent heats*. Reid, Prausnitz and Sherwood, *The Properties of Gases and Liquids*, give many more methods.[14]

4.4-1   Heat of Fusion

The heat of fusion for many elements and compounds can be expressed as

$$\frac{\Delta \hat{H}_f}{T_f} = \text{constant} = \begin{cases} 2\text{--}3 & \text{for elements} \\ 5\text{--}7 & \text{for inorganic compounds} \\ 9\text{--}11 & \text{for organic compounds} \end{cases} \qquad (4.12)$$

$$\text{where } \Delta \hat{H}_f = \text{molar heat of fusion, cal/g mol}$$
$$T_f = \text{melting point, K}$$

4.4-2   Heat of Vaporization

Typical computer data bases such as PPDS (Physical Properties Data Service from the Institution of Chemical Engineers) use the three techniques for estimating the enthalpies of vaporization described below.

[14]See also S. H. Fishtine, *Hydrocarbon Processing*, v. 45, no. 4, p. 173 (1965), and P. I. Gold, *Chem. Eng.*, p. 109 (February 24, 1969).

Clausius–Clapeyron equation. The Clapeyron equation itself is an exact thermo-dynamic relationship between the slope of the vapor-pressure curve and the molar heat of vaporization and the other variables listed below:

$$\frac{dp^*}{dT} = \frac{\Delta \hat{H}_v}{T(\hat{V}_g - \hat{V}_l)} \tag{4.13}$$

where p^* = vapor pressure
T = absolute temperature
$\Delta \hat{H}_v$ = molar heat of vaporization at T
$\hat{V}_l$ = molar volume of gas or liquid as indicated by the subscript g or l

Any consistent set of units may be used.

If experimental vapor-pressure data are aviailable for a span of temperatures, or a correlation is available, dp^*/dT can be evaluated in the vicinity of T. Furthermore, $(\hat{V}_g - \hat{V}_l)$ can be estimated solely from $\hat{V}_g$ if we neglect V_l; hence for a nonideal gas

$$\frac{dp^*}{dT} = -\frac{\Delta \hat{H}_v}{z(RT^2/p^*)} \tag{4.14}$$

Refer to Edmonds[15] for the specific relation to use data points at equal temperature intervals to evaluate ΔH_v.

Another variation of using Eq. (4.13) is as follows. Assume that:

(a) $\hat{V}_l$ is negligible in comparison with $\hat{V}_g$

(b) The ideal gas law is applicable for the vapor:

$$\hat{V}_g = RT/p^*$$

$$\frac{dp^*}{p^*} = \frac{\Delta \hat{H}_v \, dT}{RT^2}$$

Rearrange the form to

$$\frac{d \ln p^*}{d(1/T)} = 2.303 \frac{d \log_{10} p^*}{d(1/T)} = -\frac{\Delta \hat{H}_v}{R} \tag{4.15}$$

You can plot the $\log_{10} p^*$ vs. $1/T$ and obtain the slope $-(\Delta \hat{H}_v/2.303R)$.

If we further assume that $\Delta \hat{H}_v$ is constant over the temperature range of interest, integration of Eq. (4.15) yields an indefinite integral

$$\log_{10} p^* = -\frac{\Delta \hat{H}_v}{2.303RT} + B \tag{4.16}$$

---

[15]B. Edmonds, "Enthalpies of Vaporization," *Chem. Eng.* p. 109 (February 1979); p. 179, (March 1979); p. 357 (May 1979).

or a definite one

$$\log_{10} \frac{p_1^*}{p_2^*} = \frac{\Delta \hat{H}_v}{2.303R} \left(\frac{1}{T_2} - \frac{1}{T_1} \right)$$ (4.17)

Either of these equations can be used graphically or analytically to obtain $\Delta \hat{H}_v$ for a short temperature interval.

EXAMPLE 4.10 Heat of Vaporization from the Clausius–Clapeyron Equation

Estimate the heat of vaporization of isobutyric acid at 200°C.

Solution

The vapor-pressure data for isobutyric acid (from Perry's *Handbook*) are

pressure (mm Hg)	temp. (°C)	pressure (atm)	temp. (°C)
100	98.0	1	154.5
200	115.8	2	179.8
400	134.5	5	217.0
760	154.5	10	250.0

Basis- 1 g mol of isobutyric acid

Since $\Delta \hat{H}_v$ remains essentially constant for short temperature intervals, the heat of vaporization can be estimated from Eq. (4.17) and the vapor-pressure data at 179.8°C and 217.0°C.

$$179.8°C \approx 452.8 \text{ K} \qquad 217°C \approx 490 \text{ K}$$

$$\log_{10} \frac{2}{5} = \frac{\Delta \hat{H}_v}{(2.303)(8.314)} \left(\frac{1}{490} - \frac{1}{452.8} \right)$$

$$\Delta \hat{H}_v = 45{,}442 \text{ J/g mol} = 19{,}555 \text{ Btu/lb mol}$$

Reduced form of the Clapeyron equation. This is an equation in terms of the reduced pressure and temperature which gives good results:

$$d \ln p^* = -\frac{\Delta \hat{H}_v}{zRT_c} d\left(\frac{1}{T_r} \right)$$

or

$$\frac{\Delta \hat{H}_v}{zRT_c} = -\frac{d \ln p^*}{d(1/T_r)}$$

If we compute the right-hand side, say, from the Antoine equation (see Appendix G), we get

$$\frac{\Delta \hat{H}_v}{zRT_c} = \frac{B}{T_c} \left[\frac{T_r}{T_r + (C/T_c)} \right]^2$$ (4.18)

Prediction using enthalpy of vaporization at the normal boiling point. As an example, Watson[16] found empirically that

$$\frac{\Delta \hat{H}_{v_2}}{\Delta \hat{H}_{v_1}} = \left(\frac{1 - T_{r_2}}{1 - T_{r_1}}\right)^{0.38}$$

where $\Delta \hat{H}_{v_2}$ = heat of vaporization of a pure liquid at T_2
        $\Delta \hat{H}_{v_1}$ = heat of vaporization of the same liquid at T_1

4.4-3 Reference Substance Plots

A number of graphical techniques have been proposed to estimate the molal heat of vaporization of a liquid at any temperature by comparing the $\Delta \hat{H}_v$ for the unknown liquid with that of a known liquid such as water. Two of these methods are described below.

Duhring plot. The temperature of the wanted compound A is plotted against the temperature of the known (reference) liquid at *equal vapor pressure*. For example, if the temperatures of A (isobutyric acid) and the reference substance (water) are determined at 760, 400, and 200 mm Hg pressure, then a plot of the temperatures of A vs. the temperatures of the reference substance will be approximately a straight line over a wide temperature range.

Othmer plot. The Othmer plot[17] is based on the same concepts as the Duhring plot except that the *logarithms* of vapor pressures are plotted against each other at *equal temperature*. As illustrated in Fig. 4.9, a plot of the $\log_{10}(p_A^*)$ against $\log_{10}$

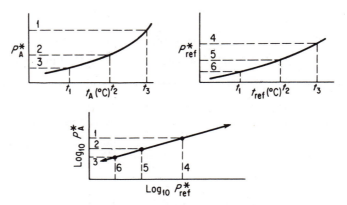

Fig. 4.9 Othmer plot and vapor-pressure curves.

[16]K. M. Watson, *Ind. Eng. Chem.*, v. 23, p. 360 (1931); v. 35, p. 398 (1943).

[17]D. F. Othmer, *Ind. Eng. Chem.*, v. 32, p. 841 (1940). For a complete review of the technique and a comparative statistical analysis among various predictive methods, refer to D. F. Othmer and H. N. Huang, *Ind. Eng. Chem.*, v. 57, p. 40 (1965); D. F. Othmer and E. S. Yu, *Ind. Eng. Chem.*, v. 60, no. 1, p. 22 (1968); and D. F. Othmer and H. T. Chem, *Ind. Eng. Chem.*, v. 60, no. 4, p. 39 (1968).

(p^*_{ref}) chosen at the same temperature yields a straight line over a very wide temperature range.

To indicate how to apply the Othmer plot to estimate the heat of vaporization of compound A, we apply the Clapeyron equation to each substance and take the ratio:

$$\frac{d(\ln p^*_A)}{d(\ln p^*_{ref})} = \frac{\dfrac{-\Delta \hat{H}_{v_A}}{R \, d(1/T_A)}}{\dfrac{-\Delta \hat{H}_{v_{ref}}}{R \, d(1/T_{ref})}} \qquad (4.19)$$

By choosing values of vapor pressure at equal temperature ($T_A = T_{ref}$) in Eq. (4.19), we obtain

$$\frac{d \ln p^*_A}{d \ln p^*_{ref}} = \frac{\Delta \hat{H}_{v_A}}{\Delta \hat{H}_{v_{ref}}} = m = \text{slope of Othmer plot as in Fig. 4.9} \qquad (4.20)$$

The Othmer plot works well because the errors inherent in the assumptions made in deriving Eq. (4.20) cancel out to a considerable extent. At very high pressures the Othmer plot is not too effective. Incidentally, this type of plot has been applied to estimate a wide variety of thermodynamic and transport relations, such as equilibrium constants, diffusion coefficients, solubility relations, ionization and dissociation constants, and so on, with considerable success.

Othmer recommended another type of relationship that can be used to estimate the heats of vaporization:

$$\ln p^*_{r_A} = \frac{\Delta \hat{H}_{v_A}}{\Delta H_{v_{ref}}} \left(\frac{T_{c_{ref}}}{T_{c_A}} \right) \ln p^*_{r_{ref}} \qquad (4.21)$$

where p^*_r is the reduced vapor pressure. Equation (4.21) is effective and gives a straight line through the point $p^*_{c_A} = 1$ and $T_{c_A} = 1$. The Gordon method plots the log of the vapor pressure of A versus that of the reference substance at equal *reduced temperature*.

EXAMPLE 4.11 Use of Othmer Plot

Repeat the calculation for the heat of vaporization of isobutyric acid at 200°C using an Othmer plot.

Solution

Data are as follows:

temp. (°C)	p^*_{iso} (atm)	$\log p^*_{iso}$	$p^*_{H_2O}$ (atm)	$\log p^*_{H_2O}$
154.5	1	0	5.28	0.723
179.8	2	0.3010	9.87	0.994
217.0	5	0.698	21.6	1.334
250.0	10	1.00	39.1	1.592

From Othmer plot, Fig. E4.11:

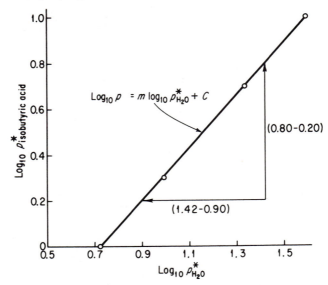

Fig. E4.11

$$\text{slope} = \frac{\Delta \hat{H}_{v_{\text{iso}}}}{\Delta \hat{H}_{v_{\text{H}_2\text{O}}}} = \frac{0.80 - 0.20}{1.42 - 0.90} = 1.15$$

At 200°C (392°F), from the steam tables,

$$\Delta \hat{H}_{v_{\text{H}_2\text{O}}} = 34{,}895 \text{ kJ/kg mol} \quad (1938.6 \text{ kJ/kg})$$

Then

$$\Delta \hat{H}_{v_{\text{iso}}} = (34{,}895)(1.15) = 40{,}130 \text{ kJ/kg mol} = 17{,}270 \text{ Btu/lb mol}$$

The answer in this case is lower than that in Example 4.10. Without the experimental value, it is difficult to say what the proper answer is. However, at the normal boiling point (154°C), $\Delta \hat{H}_{v_{\text{iso}}}$ is known to be 17,700 Btu/lb mol, and $\Delta \hat{H}_{v_{\text{iso}}}$ should be lower at 200°C, so that 17,270 Btu/lb mol seems to be the more satisfactory value.

We are now equipped to include in the calculation of an enthalpy change the occurrence of a phase change. Because enthalpy is a state variable, any arbitrary path from the initial to the final state will suffice for the actual computation. Simply choose the simplest possible path.

EXAMPLE 4.12 Calculation of Enthalpy Change Including a Phase Change

What is the enthalpy change of 1 kg of water from ice at 0°C to vapor at 120°C and 100 kPa?

Solution

We choose a path for the enthalpy change as shown in Fig. E4.12 by the solid arrows; an alternative path is shown by the dashed arrows. We need the $\Delta\hat{H}_{\text{fusion}}$ and $\Delta\hat{H}_{\text{vaporization}}$ data from Appendix D:

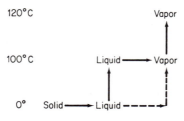

Fig. E4.12

$$\Delta\hat{H}_{\text{fusion}} = 334 \text{ kJ/kg} \qquad \Delta\hat{H}_{\text{vaporization}} = 2255 \text{ kJ/kg} \quad \text{at } 100°\text{C}$$

and will use average heat capacities (for convenience in the calculations)

$$C_{p_{\text{liquid}}} = 4.18 \text{ J/(g)(°C)} \qquad C_{p_{\text{vapor}}} = 1.88 \text{ J/(g)(°C)}$$

$$\Delta\hat{H}_{\text{overall}} = \underset{\text{at }0°\text{C}}{\Delta\hat{H}_{\text{fusion}}} + \underset{0°\text{C}\rightarrow100°\text{C}}{\Delta\hat{H}_{\text{liquid}}} + \underset{\text{at }100°\text{C}}{\Delta\hat{H}_{\text{vaporization}}} + \underset{100\rightarrow120°\text{C}}{\Delta\hat{H}_{\text{vapor}}}$$

$$= 334 + 4.18(100) + 2255 + 1.88(20) = 3045 \text{ J/g}$$

How can you check that the answer is correct? (*Hint:* What enthalpy change do you calculate from data in the steam tables?)

Self-Assessment Test

1. Calculate the enthalpy change in Btu/lb of benzene as it goes from a vapor at 300°F and 1 atm to a solid at 0°F and 1 atm.

2. What is enthalpy change that occurs when 5 lb of water is heated from ice at 32°F to vapor at 250°F and 1 atm? Use the steam tables.

3. Estimate the heat of vaporization of water at 450 K from Watson's relation given that $\Delta H = 2256.1$ kJ/kg at 373.14 K, and compare with experimental data.

4. Prepare an Othmer chart to predict the heat of vaporization of methyl alcohol at 90°C. Use water as the reference substance. Repeat using the Clausius–Clapeyron equation.

Section 4.5   The General Energy Balance

Scientists did not begin to write energy balances for physical systems prior to the latter half of the nineteenth century. Before 1850 they were not sure what energy was or even if it was important. But in the 1850s the concepts of energy and the energy balance became clearly formulated. We do not have the space here to outline the historical development of the energy balance and of special cases of it, but it truly makes a most interesting story and can be found elsewhere.[18-23] Today we consider the energy balance to be so fundamental a physical principle that we invent new kinds of energy to make sure that the equation indeed does balance. Equation (4.22) as written below is a generalization of the results of numerous experiments on relatively simple special cases. We universally believe the equation is valid because we cannot find exceptions to it in practice, taking into account the precision of the measurements.

It is necessary to keep in mind two important points as you read what follows. First, we examine only systems that are homogeneous, not charged, and without surface effects, in order to make the energy balance as simple as possible. Second, the energy balance is developed and applied from the macroscopic viewpoint (overall about the system) rather than from a microscopic viewpoint (i.e., an elemental volume within the system).

The concept of the macroscopic energy balance is similar to the concept of the macroscopic material balance, namely,

$$
\begin{Bmatrix} \text{Accumulation of} \\ \text{energy within the} \\ \text{system} \end{Bmatrix} = \begin{Bmatrix} \text{transfer of energy} \\ \text{into system through} \\ \text{system boundary} \end{Bmatrix} - \begin{Bmatrix} \text{transfer of energy out} \\ \text{of system through} \\ \text{system boundary} \end{Bmatrix}
$$
$$
+ \begin{Bmatrix} \text{energy genera-} \\ \text{tion within} \\ \text{system} \end{Bmatrix} - \begin{Bmatrix} \text{energy con-} \\ \text{sumption} \\ \text{within system} \end{Bmatrix}
\tag{4.22}
$$

Equation (4.22) can be applied to a single piece of equipment or to a complex plant such as that shown in Fig. 4.10.

While the formulation of the energy balance in words as outlined in Eq. (4.22) is easily understood and rigorous, you will discover in later courses that to express each term of (4.22) in mathematical notation may require certain simplifications to be introduced, a discussion of which is beyond our scope here, but they have a quite minor influence on our final balance.

In what follows we first treat systems in which no material flows in and out. Do you remember the name for such a system? Then we examine systems in which material does flow in and out. What is the name for such a system? Finally, in Sec.

[18]E. J. Hoffman, *The Concept of Energy: An Inquiry into Origins and Applications*, Ann Arbor Science Publishers, Ann Arbor, Mich., 1977.

[19]V. V. Raman, "Where Credit Is Due—The Energy Conservation Principle," *The Physics Teacher*, p. 80 (February 1965).

[20]D. Roller, *The Early Development of the Concepts of Temperature and Heat*, Harvard University Press, Cambridge, Mass., 1950.

[21]T. M. Brown, *Amer. J. Physics*, v. 33, no. 10, p. 1 (1965).

[22]J. Zernike, *Chem. Weekblad*, v. 61, pp. 270–284, 277–279 (1965).

[23]L. K. Nash, *J. Chem. Educ.*, v. 42, p. 64 (1965) (a resource paper).

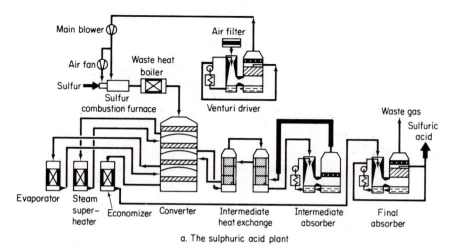

a. The sulphuric acid plant

b. The energy flows based on
1000 tons/day of acid production

Fig. 4.10 Energy balance in a double catalysis sulfuric acid plant. The thickness of the arrows is proportional to the magnitude of the energy transfers: (a) the sulfuric acid plant; (b) the energy flows based on 1000 tons/day of acid production. [From U. Sander, *Chem. Eng. Progr.*, p. 61 (March 1977).]

4.7 we discuss energy balances in which chemical reactions take place. Thus the energy generation and consumption terms will not play a role in Section 4.5 unless sources such as external electric and magnetic fields, radioactive decay, or the slowing down of neutrons enter into the process.

4.5-1   Energy Balances for Closed Systems (without Chemical Reaction)

> ***Your objectives in studying this section are to be able to:***
>
> 1. Write down the general energy balance in words (Eq. 4.22).
> 2. Write down the energy balance for a closed system in symbols (Eq. 4.23), and apply it to solve energy balance problems.
> 3. Cite the signs for work and heat entering and leaving the system.
> 4. Calculate the total energy associated with the mass of the system or any of its components (internal energy, kinetic energy, potential energy).

Figure 4.11 shows the various types of energy to be accounted for in Eq. (4.22), and Table 4.7 lists the specific individual terms to be employed in Eq. (4.22). As to the notation, the subscripts t_1 and t_2 refer to the initial and final time periods over which the accumulation is to be evaluated, with $t_2 > t_1$. The superscript caret ($^\wedge$) means that the symbol stands for energy per unit mass; without the caret, the symbol means energy of the total mass present. Other notation is evident from Table 4.7 and can be found in the notation list at the end of the book.

As is commonly done, we have split the total energy (E) associated with the mass in the system into three categories: internal energy (U), kinetic energy (K), and potential energy (P). Energy transported across the system boundary can be

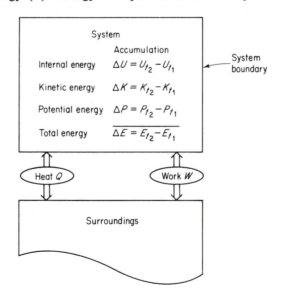

Fig. 4.11 Terms in the energy balance for a closed system.

**TABLE 4.7 LIST OF SYMBOLS TO BE USED
IN THE GENERAL ENERGY BALANCE**

Accumulation term		
Type of energy	*A time t_1*	*At time t_2*
Internal	U_{t_1}	U_{t_2}
Kinetic	K_{t_1} $\left.\right\}$ E_{t_1}	K_{t_2} $\left.\right\}$ E_{t_2}
Potential	P_t	P_{t_2}
Mass	m_{t_1}	m_{t_2}
Energy accompanying mass transport		
Type of energy	*Transport in*	*Transport out*
Internal	U_1	U_2
Kinetic	K_1	K_2
Potential	P_1	P_2
Mass	m_1	m_2
Net heat input to system	Q	
Net work done by system on surroundings		
Mechanical work or work by moving parts	W	
Work to introduce material into system, less work recovered on removing material from system	$(p_2\hat{V}_2)m_2 - (p_1\hat{V}_1)m_1$	

transferred by two modes: heat (Q) and work (W). (We discuss energy transferred with mass flow in Sec. 4.5.2.) Note that Q and W here are defined as the *net* transfer of heat and work, respectively, between the system and the surroundings, and equal the integral of the net rate of flow of heat or work over the time interval t_1 to t_2:

$$Q = \int_{t_1}^{t_2} \dot{Q} \, dt \qquad W = \int_{t_1}^{t_2} \dot{W} \, dt$$

Equation (4.22) when translated into the notation listed in Table 4.7 becomes

$$\Delta E = E_{t_2} - E_{t_1} = Q - W \qquad (4.23)$$

where Δ = difference operator signifying *final* minus *initial* in time

Q = heat absorbed *by* the system *from* the surroundings (by definition Q is *positive for heat entering the system*)

W = mechanical work done *by* the system *on* the surroundings (by definition W is *positive for work going from the system to the surroundings*)

Keep in mind that a system may do work, or have work done on it, without some obvious mechanical device such as a pump, shaft, and so on, being present. Often the nature of the work is implied rather than explicitly stated. For example, a cylinder filled with gas enclosed by a movable piston implies that the surrounding atmosphere can do work on the piston or the reverse; a batch fuel cell does no mechanical work, unless it produces bubbles, but does deliver a current at a potential difference; and so forth.

Now for a word of warning: Be certain you use *consistent units* for all terms; in the American engineering system the use of foot-pound, for example, and Btu in different places in Eq. (4.23) is a common error for the beginner.

Let us look now at some examples of applications of the energy balance for closed systems. Remember to follow the checklist presented in Chap. 2 in analyzing the problem.

EXAMPLE 4.13 Application of the Energy Balance

Changes in the heat input that spacecraft encounter constitute a threat both to any occupants as well as to the payload of instruments. If a satellite passes into the earth's shadow, the heat flux that it receives may be as little as 10% of that in direct sunlight. Consider the case in which a satellite leaves the earth's shadow and heats up. For simplicity we select just the air in the spacecraft as the system. It holds 4.00 kg of air at 20°C (the air has an internal energy of 8.00×10^5 J/kg with reference to fixed datum conditions). Energy from the sun's radiation enters the air as heat until the internal energy is 10.04×10^5 J/kg.

(a) How much heat has been transferred to the air?
(b) If some machinery in the spacecraft does 0.110×10^5 J of work *on* the air over the same interval, how would your answer to part (a) change?

Solution

(a)

Steps 1, 2, 3, and 4: The air is chosen as the system, and the process is clearly a closed or batch system. Everything outside the air is the surroundings. See Fig. E4.13. We know U_{t_2}, U_{t_1}, and $W = 0$ for part (a) and a W is also given for part (b). Potential and kinetic energy quantities are zero.

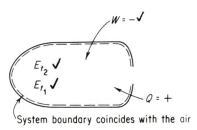

Fig. E4.13

Step 5: We have one unknown, Q, and can make one energy balance; hence the problem has a unique solution.

Step 6:   Basis: 1 kg of air

Steps 7, 8, and 9:

$$\Delta \hat{U} = \hat{U}_{t_2} - \hat{U}_{t_1} = \hat{Q} - \hat{W}$$
$$\hat{U}_{t_2} - \hat{U}_{t_1} = (10.04 \times 10^5) - (8.0 \times 10^5) = 2.04 \times 10^5 \text{ J/kg}$$

so that $\hat{Q} = 2.04 \times 10^5$ J/kg. Note that the sign of Q is positive, indicating that heat has been added to the system.

$$\text{Basis: 4.00 kg of air}$$

$$Q = \frac{2.04 \times 10^5 \text{ J}}{\text{kg}} \Bigg| \frac{4.00 \text{ kg}}{} = 8.16 \times 10^5 \text{ J}$$

(b) If work is done on the air (by compression or otherwise), the work term is not zero but

$$W = -0.110 \times 10^5 \text{ J}$$

Note that the sign on the work is negative. Why? Work is done *on* the gas. In this second case,

$$Q = \Delta U + W = (8.16 \times 10^5) - (0.11 \times 10^5) = 8.05 \times 10^5 \text{ J}$$

Step 10: Less heat is required than before to reach the same value for the internal energy because some work has been done on the gas.

EXAMPLE 4.14 Energy Balance

Ten pounds of water at 35°F, 4.00 lb of ice at 32°F, and 6.00 lb of steam at 250°F and 20 psia are mixed together. What is the final temperature of the mixture? How much steam condenses?

Solution

Steps 1, 2, 3, and 4: We can assume that the overall batch process takes place with $Q = 0$ and $W = 0$ if we define the system as in Fig. E4.14. Let t_2 be the final temperature. The system

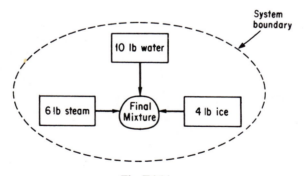

Fig. E4.14

consists of 20 lb of H_2O in various phases. ΔK and ΔP equal 0. The check marks in the figure indicate that the composition of the system is known—all water. The energy balance reduces to (superscripts are s = steam, w = water, i = ice, f = final mixture)

$$\Delta U = 20\hat{U}^f_{t_2} - (6\hat{U}^s_{t_1} + 10\hat{U}^w_{t_1} + 4\hat{U}^i_{t_1}) = 0 \qquad\qquad\text{(a)}$$

Step 5: We have one unknown value, $\hat{U}^f_{t_2}$, a function of the temperature only, and one equation. Hence the problem has a unique solution.

Step 6:   Basis: 20 lb of water

Steps 7 and 8: Because the steam tables will be the source of the values for the various forms of water, we need to express Eq. (a) in terms of specific enthalpies. Then a direct solution can be used. If $\hat{U}$ is replaced with $\hat{H} - p\hat{V}$ for convenience in the calculations, we get

$$20\hat{H}_{t_2}^f - (6\hat{H}_{t_1}^s + 10\hat{H}_{t_1}^w + 4\hat{H}_{t_1}^i)$$
$$= 20(p\hat{V})_{t_2}^f - 6(p\hat{V})_{t_1}^s - 10(p\hat{V})_{t_1}^w - 4(p\hat{V})_{t_1}^i \tag{b}$$

The last two terms on the right-hand side of Eq. (b) cannot be more than 1 Btu at the very most and can safely be neglected. Be sure to check this assumption!

Step 9: Because of the phase changes that take place as well as the nonlinearity of the heat capacities as a function of temperature, it is not possible to replace the enthalpies in Eq. (b) with functions of temperature and get a linear algebraic equation that is easy to solve. The simplest way to solve the problem is to assume the final conditions of temperature and pressure and check the assumption by Eq. (c):

$$20\hat{H}_{t_2}^f - 20(p\hat{V})_{t_2}^f = (6\hat{H}_{t_1}^s + 10\hat{H}_{t_1}^w + 4\hat{H}_{t_1}^i) - 6(p\hat{V})_{t_1}^s \tag{c}$$

If the equation does not balance, a new assumption can be made, leading to a solution to the problem by a series of iterative calculations:

$$\text{Basis: }\begin{cases} 4 \text{ lb of ice at } 32°F \\ 10 \text{ lb of } H_2O \text{ at } 35°F \\ 6 \text{ lb of steam at } 250°F \text{ and } 20 \text{ psia} \end{cases}$$

Since the enthalpy change corresponding to the heat of condensation of the steam is quite large, we might expect that the final temperature of the mixture would be near or at the saturation temperature of the 20 psia steam. Let us assume that $T_{\text{final}} = 228°F$ (the saturated temperature of 20 psia of steam). We shall also assume that all the steam condenses as a first guess.

Using the steam tables (ref.: zero enthalpy at 32°F, saturated liquid), we can calculate the following quantities to test our assumption:

Btu

$$6\,\Delta\hat{H}_{t_1}^s = \frac{6 \text{ lb}}{}\bigg|\frac{1168.0 \text{ Btu}}{\text{lb}} \qquad\qquad = 7008.0$$

$$10\,\Delta\hat{H}_{t_1}^w = \frac{10 \text{ lb}}{}\bigg|\frac{3.02 \text{ Btu}}{\text{lb}} \qquad\qquad = 30.2$$

The heat of fusion of ice is 143.6 Btu/lb:

$$4\,\Delta\hat{H}_{t_1}^i = -\frac{4 \text{ lb}}{}\bigg|\frac{143.6 \text{ Btu}}{\text{lb}} \qquad\qquad = -574.4$$

$$-6(p\hat{V})_{t_1}^s = -\frac{6 \text{ lb}_m}{}\bigg|\frac{20 \text{ lb}_f}{\text{in.}^2}\bigg|\left(\frac{12 \text{ in.}}{1 \text{ ft}}\right)^2\bigg|\frac{20.81 \text{ ft}^3}{\text{lb}_m}\bigg|\frac{1 \text{ Btu}}{778 \text{ (ft)(lb}_f)} = -462.2$$

$$\text{total right-hand side of Eq. (c)} \qquad\qquad = \overline{6001.6}$$

The left-hand side of Eq. (c) is

$$20 \, \Delta \hat{H}_{t_2}^f = \frac{20 \text{ lb}}{} \, \Bigg| \, \frac{196.2 \text{ Btu}}{\text{lb}} \qquad\qquad\qquad\qquad\qquad = 3924.0$$

liquid

$$-20(p\hat{V})_{t_2}^f = -\frac{20 \text{ lb}_m}{} \, \Bigg| \, \frac{19.7 \text{ lb}_f}{\text{in.}^2} \, \Bigg| \, \left(\frac{12 \text{ in.}}{1 \text{ ft}}\right)^2 \, \Bigg| \, \frac{0.0168 \text{ ft}^3}{} \, \Bigg| \, \frac{1 \text{ Btu}}{778 \text{ (ft)(lb}_f)} = -1.2$$

$$\text{total left-hand side of Eq. (c):} \qquad\qquad\qquad\qquad = \overline{3922.8}$$

Evidently, our assumption that all the steam condenses is wrong, since the value of the left-hand side of Eq. (c) is too small. With this initial calculation we can now see that the final temperature of the mixture will be 228°F and that a saturated steam–liquid water mixture exists in the vessel. The next question is: How much steam condenses?

Let $S =$ the amount of steam at 228°F:

$$S[\hat{H}_{t_2} - p\hat{V}]_{\text{vapor}}^f + (20 - S)[\hat{H}_{t_2} - p\hat{V}]_{\text{liquid}}^f = 6001.6$$

$$(\Delta \hat{H}_{t_1})_{\text{vapor}}^f = 1156.1 \text{ Btu/lb}$$

$$(\Delta \hat{H}_{t_2})_{\text{liquid}}^f = 196.2 \text{ Btu/lb}$$

$$(p\hat{V})_{\text{vapor}}^f = (19.7)(12)^2(20.08)/(778) = 73.2 \text{ Btu/lb}$$

$$(p\hat{V})_{\text{liquid}}^f \simeq 0.0$$

$$S(1156.1 - 73.2) + (20 - S)(196.2) = 6001.6$$

$$S = 2.34 \text{ lb}$$

In effect, $6.00 - 2.34 = 3.66$ lb of steam can be said to condense.

If there had been only 1 lb of steam instead of 6 lb to start with, then on the first calculation the right-hand side of Eq. (c) would have been

$$1168.0 + 30.2 - 574.4 - (1/6)(453.2) = 548.3 \text{ Btu}$$

Hence lower temperatures would have been chosen until the left-hand side of Eq. (c) was small enough. Only liquid water would be present.

EXAMPLE 4.15 Application of the Energy Balance

The peaks of Mount Everest, Mount Fujiyama, Mount Washington, and other mountains are usually covered with clouds even when the surrounding area has nice weather. It is clear that water condenses because the temperature is lower, but why doesn't water condense elsewhere at the same altitude? Is the cloud cap stuck by some force? See Fig. E4.15.

Solution

Steps 1, 2, 3, 4, and 7: Only two compounds are involved, water vapor and air. As the system we will pick a reasonably sized "box" of air—see Fig. E4.15. We can assume that the contents of the box are a gas until condensation takes place.

The formation of cloud caps can be predicted with the aid of the general energy balance. As a unit mass of warm moist air moves up the side of a mountain with only modest breezes,

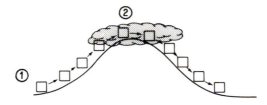

Fig. E4.15

it cannot exchange energy by way of heat with the surrounding air ($Q = 0$), nor does it exchange a significant amount of water vapor, so that in effect the system is a closed one. Both kinetic and potential energy effects can be neglected, so that Eq. (4.23) reduces to

$$\Delta \hat{U} = -\hat{W}$$

If we calculate $\hat{W}$, we can then calculate $\Delta \hat{U}$, and knowing $\hat{U}_{t_1}$ can calculate $\hat{U}_{t_2}$, and hence T_2 at the top of the mountain.

Step 5: Only one equation is needed as long as we work solely with a gas phase.

Step 6:   Basis: 1 kg mol of moist air

Steps 8 and 9: We can calculate

$$\Delta \hat{U} = \int_{T_1}^{T_2} \hat{C}_v \, dT \simeq \hat{C}_v(T_2 - T_1)$$

and

$$\hat{W} = \int_{V_1}^{V_2} p \, d\hat{V}$$

Assume that the air acts as an ideal gas as it expands, so that $p\hat{V} = \hat{R}T$. Then, from information as to the pressure change with altitude plus the fact that $C_v = 1046$ J/(kg mol) (K), by trial and error the change in temperature can be determined as the air rises. The cooling is roughly $0.5°$ to $0.7°C/100$ m of elevation. During the descent of the air on the other side of the mountain, the cloud will evaporate. You can also calculate how high a peak has to be for a cloud to form for a given relative humidity and temperature of air at the bottom of the peak or, alternatively, given the peak height, what the highest relative humidity can be for a cloud not to form. As an example of the latter, for Mt. Washington (altitude 1947 m), the maximum percent relative humidity of $21°C$ ($70°F$) air at sea level is 40%.

A more complicated but related problem that we do not have the space to go into is the formation of contrails by supersonic aircraft.

Self-Assessment Test

1. Liquid oxygen is stored in a 14,000-liter storage tank. When charged, the tank contains 13,000 liters of liquid in equilibrium with its vapor at 1 atm pressure. What is the **(a)** temperature, **(b)** mass, and **(c)** quality of the oxygen in the vessel? The pressure relief valve of the storage tank is set at 2.5 atm. If heat leaks into the oxygen tank at the rate of 5.0×10^6 J/hr, **(d)** when will the pressure relief valve operate, and **(e)** what will be the temperature in the storage tank at that time?

    Data: at 1 atm, saturated, $V_l = 0.0281$ liter/g mol, $V_g = 7.15$ liter/g mol, $\hat{H} = -133.5$ J/g; at 2.5 atm, saturated, $\hat{H} = -116.6$ J/g.

2. Suppose that you fill an insulated Thermos to 95% of the volume with ice and water at equilibrium and securely seal the opening.

    (a) Will the pressure in the Thermos go up, down, or remain the same after 2 hr?

    (b) After 2 weeks?

    (c) For the case in which after filling and sealing, the Thermos is shaken vigorously, what will happen to the pressure?

3. An 0.25-liter container initially filled with 0.225 kg of water at a pressure of 20 atm is cooled until the pressure inside the container is 100 kPa.

    (a) What are the initial and final temperatures of the water?

    (b) How much heat was transferred from the water to to reach the final state?

4. Compute ΔE in Btu for 2 moles of an ideal monoatomic gas heated from 60°F to 160°F.

5. In a shock tube experiment, the gas (air) is held at room temperature at 15 atm in a volume of 0.350 ft³ by a metal seal. When the seal is broken, the air rushes down the evacuated tube, which has a volume of 20 ft³. The tube is insulated. In the experiment:

    (a) What is the work done by the air?

    (b) What is the heat transferred to the air?

    (c) What is the internal energy change of the air?

    (d) What is the final temperature of the air after 3 min?

    (e) What is the final pressure of the air?

6. A reservoir of air of volume 0.5 m³ at 25°C and 300 kPa is used to raise a mass of 1000 kg. If the process occurs fast enough so heat transfer is negligible, how high will the mass raise up? Neglect the volume of the line to the lift cylinder.

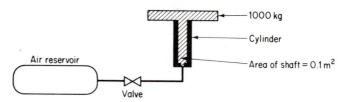

4.5-2 Energy Balances for Open Systems
(without Chemical Reaction)

> *Your objectives in studying this section are to be able to:*
>
> 1. Write down the energy balance for an open system [Eq. (4.24)] in words and symbols, explain each term, and apply the equation to solve energy balance problems using the 10-step strategy plus two new steps.
>
> 2. Explain the "pV work" (flow work) concept.

3. Define isothermal, adiabatic, isobaric, isometric, and sensible heat.

4. Make the necessary assumptions and approximations to simplify and solve the energy balance for open systems.

We now apply Eq. (4.22) to systems in which mass as well as energy can pass through the system boundaries. Figure 4.12 illustrates the general case. All the notation was listed in Table 4.7. Keep in mind that the symbol m represents a *steady* flow of mass. We give an alternative formulation of the energy balance in terms of differentials in Chap. 6, where emphasis is placed on the instantaneous rate of change of energy of a system rather than just the initial and final states of the system; the formulation in this chapter can be considered to be the result of integrating the balance in Chap. 6. The energy balance for the open system is

$$\underset{\text{accumulation}}{m_{t_2}(\hat{U} + \hat{K} + \hat{P})_{t_2} - m_{t_1}(\hat{U} + \hat{K} + \hat{P})_{t_1}} = \underset{\substack{\text{transfer in by} \\ \text{mass flow}}}{(\hat{U}_1 + \hat{K}_1 + \hat{P}_1)m_1}$$

$$\underset{\substack{\text{transfer out by} \\ \text{mass flow}}}{- (\hat{U}_2 + \hat{K}_2 + \hat{P}_2)m_2} + \underset{\substack{\text{net transfer} \\ \text{by heat}}}{Q} \underset{\text{net transfer by work}}{- W + p_1\hat{V}_1 m_1 - p_2\hat{V}_2 m_2}$$

(4.24)

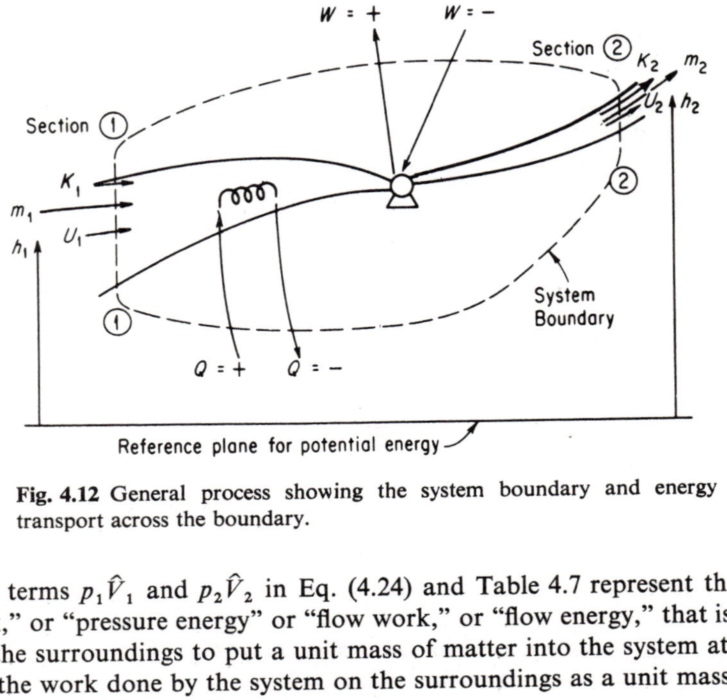

Fig. 4.12 General process showing the system boundary and energy transport across the boundary.

The terms $p_1\hat{V}_1$ and $p_2\hat{V}_2$ in Eq. (4.24) and Table 4.7 represent the so-called "pV work," or "pressure energy" or "flow work," or "flow energy," that is, the work done by the surroundings to put a unit mass of matter into the system at ① in Fig. 4.12 and the work done by the system on the surroundings as a unit mass leaves the system at ②. Work is defined as

$$W = \int_0^l F\,dl$$

Because the pressures at the entrance and exit to the system are constant for differential displacements of mass, the work done by the surroundings on the system adds energy to the system at ①:

$$W_1 = \int_0^l F_1\,dl = \int_0^{\hat{V}_1} p_1\,d\hat{V} = p_1(\hat{V}_1 - 0) = p_1\hat{V}_1$$

where $\hat{V}$ is the volume per unit mass. Similarly, the work done by the fluid on the surroundings as the fluid leaves the system is $W_2 = p_2\hat{V}_2$, a term that has to be subtracted (why?) from the right-hand side of Eq. (4.24).

In practice we introduce the expression $\Delta\hat{U} + \Delta p\hat{V} = \Delta\hat{H}$ in Eq. (4.24) to replace U with H in order to reduce the energy balance to a form easier to memorize:

$$\Delta E = E_{t_2} - E_{t_1} = -\Delta[(\hat{H} + \hat{K} + \hat{P})m] + Q - W \qquad (4.24a)$$

In Eq. (4.24a) the delta symbol (Δ) has two different meanings:

(a) In ΔE, Δ means final minus initial in time.

(b) In ΔH, and so on, Δ means out minus in.

Such usage is perhaps initially confusing, but is very common; hence you might as well become accustomed to it. A more rigorous derivation of Eq. (4.24) from a microscopic balance may be found in Slattery.[24]

If there is more than one input and output stream to the system, Eq. (4.24a) would become

$$E_{t_2} - E_{t_1} = \sum_{\text{in}} m_i(\hat{H}_i + \hat{K}_i + \hat{P}_i) - \sum_{\text{out}} m_o(\hat{H}_o + \hat{K}_o + \hat{P}_o) + Q - W \qquad (4.25)$$

Here the subscript (o) designates an output stream and the subscript (i) designates an input stream.

In most problems you do not have to use all the terms of the general energy balance equation because certain terms may be zero or may be so small that they can be neglected in comparison with the other terms. Several special cases can be deduced from the general energy balance of considerable industrial importance by introducing certain simplifying assumptions:

(a) No mass transfer (*closed* or *batch* system) ($m_1 = m_2 = 0$):

$$\Delta E = Q - W \qquad (4.23)$$

[24]J. C. Slattery, *Momentum, Energy, and Mass Transfer in Continua*, McGraw-Hill, New York, 1972.

[Equation (4.23) is known as the *first law of thermodynamics* for a closed system.]

(b) No accumulation ($\Delta E = 0$), no mass transfer ($m_1 = m_2 = 0$):

$$Q = W \tag{4.26}$$

(c) No accumulation ($\Delta E = 0$), but with mass flow:

$$Q - W = \Delta[(\hat{H} + \hat{K} + \hat{P})m] \tag{4.27}$$

(d) No accumulation, $Q = 0$, $W = 0$, $\hat{K} = 0$, $\hat{P} = 0$:

$$\Delta\hat{H} = 0 \tag{4.28}$$

[Equation (4.28) is called the "enthalpy balance."]

Take, for example, the flow system shown in Fig. 4.13. Overall, between sections 1 and 5, we would find that $\Delta P = 0$. In fact, the only portion of the system where ΔP would be of concern would be between section 4 and some other section. Between 3 and 4, ΔP may be consequential, but between 2 and 4 it may be negligible in comparison with the work introduced by the pump. Between section 3 and any further downstream point, both Q and W are zero. After reading the problem statements in the examples below but before continuing to read the solution, you should try to apply Eq. (4.24a) yourself to test your ability to simplify the energy balance for particular cases.

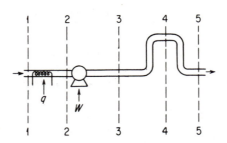

Fig. 4.13 Flow system.

Some special process names associated with energy balance problems are worth remembering:

(a) *Isothermal ($dT = 0$)*:   constant-temperature process

(b) *Isobaric ($dp = 0$)*:   constant-pressure process

(c) *Isometric* or *isochoric ($dV = 0$)*:   constant-volume process

(d) *Adiabatic* ($Q = 0$): no heat interchange (i.e., an insulated system). If we inquire as to the circumstances under which a process can be called adiabatic, one of the following is most likely:

(1) The system is insulated.

(2) Q is very small in relation to the other terms in the energy equation and may be neglected.

(3) The process takes place so fast that there is no time for heat to be transferred.

Occasionally in a problem you will encounter a special term called *sensible heat*. Sensible heat is the enthalpy difference (normally for a gas) between some reference temperature and the temperature of the material under consideration, excluding any enthalpy differences for phase changes that we have previously termed latent heats.

One further remark that we should make is that the energy balance we have presented has included only the most commonly used energy terms. If a change in surface energy, rotational energy, or some other form of energy is important, these more obscure energy terms can be incorporated in an appropriate energy expression by separating the energy of special interest from the term in which it presently is incorporated in Eq. (4.24a). As an example, kinetic energy might be split into linear kinetic energy (translation) and angular kinetic energy (rotation).

To assist you in solving problems involving energy balances, two additional steps should be added to your mental checklist for analyzing problems (see Table 2.1):

(a) *Step 3a*: Always write down the general energy balance, Eq. (4.24) or (4.24a), below your sketch. By this step you will be certain not to neglect any of the terms in your analysis.

(b) *Step 3b:* Examine each term in the general energy balance and eliminate all terms that are zero or can be neglected. Write down why you do so.

We now examine some applications of the general energy balance.

EXAMPLE 4.16 Application of the Energy Balance

Air is being compressed from 100 kPa and 255 K (where it has an enthalpy of 489 kJ/kg) to 1000 kPa and 278 K (where it has an enthalpy of 509 kJ/kg). The exit velocity of the air from the compressor is 60 m/s. What is the power required (in kW) for the compressor if the load is 100 kg/hr of air?

Solution

Steps 1, 2, and 3: Figure E4.16 shows the known quantities. The process is clearly a flow process or open system. Let us assume that the entering velocity of the air is zero.

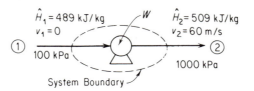

Fig. E4.16

Steps 3a, 3b, and 7:

$$\Delta E = -\Delta[(\hat{H} + \hat{K} + \hat{P})m] + Q - W$$

Equation (4.24a) can be used with the assumptions that there is no accumulation of air in the compressor ($\Delta E = 0$ and $m_1 = m_2$), $Q = 0$ (no heat exchange), and no potential energy terms are of importance. We can calculate $\hat{K}_1$ and $\hat{K}_2$ from the given data for the velocities. $\hat{H}_1$ and $\hat{H}_2$ are known.

Steps 4 and 5: Only W is unknown; hence a unique solution can be obtained.

Step 6:   Basis: 1 kg of air

Steps 8 and 9: $\hat{W}$ can be obtained directly.

$$\hat{W} = -\Delta(\hat{H} + \hat{K})$$
$$\Delta\hat{H} = 509 - 489 = 20 \text{ kJ/kg}$$

$$\Delta\hat{K} = \frac{\Delta v^2}{2} = \frac{1}{2} \left| \frac{(60\text{m})^2}{\text{s}^2} \right| \frac{1 \text{ kJ}}{\frac{1000 \text{ (kg)(m}^2)}{(\text{s})^2}} = 1.80 \text{ kJ/kg}$$

$$\hat{W} = -(20 + 1.8) = -21.8 \text{ kJ/kg}$$

(*Note:* The minus sign indicates work is done on the air.)
    To convert to power (work/time),

Basis: 100 kg of air/hr

$$kW = \frac{21.8 \text{ kJ}}{\text{kg}} \left| \frac{100 \text{ kg}}{\text{hr}} \right| \frac{1 \text{ kW}}{\frac{1 \text{ kJ}}{\text{s}}} \left| \frac{1 \text{ hr}}{3600 \text{ s}} \right. = 0.61 \text{ kW}$$

EXAMPLE 4.17 Application of the Energy Balance

Water is being pumped from the bottom of a well 15 ft deep at the rate of 200 gal/hr into a vented storage tank to maintain a level of water in a tank 165 ft above the ground. To prevent freezing in the winter a small heater puts 30,000 Btu/hr into the water during its transfer from the well to the storage tank. Heat is lost from the whole system at the constant rate of 25,000 Btu/hr. What is the temperature of the water as it enters the storage tank, assuming that the well water is at 35°F? A 2-hp pump is being used to pump the water. About 55% of the rated horsepower goes into the work of pumping and the rest is dissipated as heat to the atmosphere.

Solution

Steps 1, 2, and 3: Let the system consist of the well inlet, the piping, and the outlet at the storage tank. It is a flow process since material continually enters and leaves the system. See Fig. E4.17. The net Q is known, and W is known.

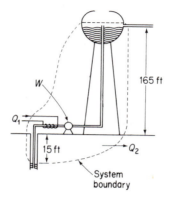

Fig. E4.17

Steps 3a, 3b, and 7: The general energy balance is

$$\Delta E = -\Delta[(\hat{H} + \hat{K} + \hat{P})m] + Q - W$$

There is no accumulation of mass in the system, and no accumulation of energy if the water stays at a constant temperature in the tank. $\Delta \hat{P}$ can be calculated. The water initially has a velocity of zero at the well inlet and the water leaving the tank also has a velocity of essentially zero. The kinetic energy terms in any case may be considered negligible in comparison with the other energy terms in the equation. Thus

$$\Delta \hat{K} \simeq 0 \tag{a}$$
$$Q - W = \Delta \hat{H} + \Delta \hat{P}$$

Steps 4, 5, and 7: We now have established that all the terms in the energy balance can be evaluated except for one, $\Delta \hat{H}$; hence the energy balance can be solved for $\Delta \hat{H}$, and the exit temperature of the water calculated from $\Delta \hat{H}$.

Step 6:   Basis: 1 hr of operation

Steps 8 and 9: The total amount of water pumped is

$$\frac{200 \text{ gal}}{\text{hr}} \left| \frac{8.33 \text{ lb}}{1 \text{ gal}} \right. = 1666 \text{ lb/hr}$$

The potential energy change is

$$\Delta P = m\,\Delta\hat{P} = mg\,\Delta h = \frac{1666\ \text{lb}_m}{\text{hr}}\left|\frac{32.2\ \text{ft}}{\text{s}^2}\right|\frac{180\ \text{ft}}{\left|\frac{32.2\ (\text{ft})(\text{lb}_m)}{(\text{s}^2)(\text{lb}_f)}\right|} = 300{,}000\ (\text{ft})(\text{lb}_f)/\text{hr}$$

or

$$\frac{300{,}000\ (\text{ft})(\text{lb}_f)/\text{hr}}{778\ (\text{ft})(\text{lb}_f)/\text{Btu}} = 386\ \text{Btu/hr}$$

The heat lost by the system is 25,000 Btu/hr while the heater puts 30,000 Btu/hr into the system; hence the net heat exchange is

$$Q = 30{,}000 - 25{,}000 = 5000\ \text{Btu/hr}$$

The rate of work being done on the water by the pump is

$$W = -\frac{2\ \text{hp}}{}\left|\frac{0.55}{}\right|\frac{33{,}000\ (\text{ft})(\text{lb}_f)}{(\text{min})(\text{hp})}\left|\frac{60\ \text{min}}{\text{hr}}\right|\frac{\text{Btu}}{778\ (\text{ft})(\text{lb}_f)}$$
$$= -2800\ \text{Btu/hr}$$

ΔH can be calculated from Eq. (a):

$$5000 - (-2800) = \Delta H + 386$$
$$\Delta H = 7414\ \text{Btu/hr}$$

We know from Eq. (4.7) that, for an incompressible fluid such as water,

$$\Delta\hat{H} = \int_{T_1}^{T_2} C_p\,dT$$

Because the temperature range considered is small, the heat capacity of liquid water may be assumed to be constant and equal to 1.0 Btu/(lb)(°F) for the problem. Thus

$$\Delta\hat{H} = C_p\Delta T = (1.0)(\Delta T)$$
$$\text{Basis: 1 lb of } H_2O$$

Now

$$\Delta\hat{H} = \frac{7414\ \text{Btu}}{\text{hr}}\left|\frac{1\ \text{hr}}{1666\ \text{lb}}\right| = 4.45\ \text{Btu/lb}$$
$$= (1.0)(\Delta T)$$
$$\Delta T \simeq 4.5°F \text{ temperature rise. Hence, } T = 39.5°F$$

EXAMPLE 4.18 Energy Balance

Steam that is used to heat a batch reaction vessel enters the steam chest, which is segregated from the reactants, at 250°C saturated and is completely condensed. The reaction absorbs 2300 kJ/kg of material in the reactor. Heat loss from the steam chest to the surroundings is

1.5 kJ/s. The reactants are placed in the vessel at 20°C and at the end of the reaction the material is at 100°C. If the charge consists of 150 kg of material and both products and reactants have an average heat capacity of $C_p = 3.26$ J/(g)(K), how many kilograms of steam are needed per kilogram of charge? The charge remains in the reaction vessel for 1 hr.

Solution

Steps 1, 2, and 3: Figure E4.18a defines the system and lists the known quantities.

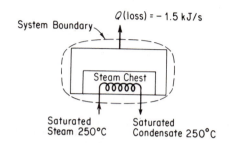

Fig. E4.18a

Steps 3a, 3b, and 7: The energy balance is

$$\Delta E = -\Delta[(\hat{H} + \hat{K} + \hat{P})m] + Q - W$$

Analysis of the process reveals that no potential energy, kinetic energy, or work terms are significant in the energy balance. The steam flows in and out, while the reacting material stays within the system. Thus there is an accumulation of energy for the material being processed but no accumulation of mass or energy associated with the steam.

With the system comprised of the reaction vessel plus the steam chest, the energy balance reduces to

$$(U_{t_2} - U_{t_1})_{\text{material}} = -\Delta H_{\text{steam}} + Q \qquad \text{(a)}$$

However, since enthalpy is more convenient to work with than internal energy, we let $\hat{U} = \hat{H} - p\hat{V}$,

$$[(\hat{H}_{t_2} - \hat{H}_{t_1}) - (p_{t_2}\hat{V}_{t_2} - p_{t_1}\hat{V}_{t_1})]_{\text{material}}m = -\Delta H_{\text{steam}} + Q$$

Since no information is available concerning the $p\hat{V}$ difference and it would no doubt be negligible if we could calculate it, it will be neglected, and then

$$(\hat{H}_{t_2} - \hat{H}_{t_1})_{\text{material}}m = -\Delta H_{\text{steam}} + Q \qquad \text{(b)}$$

Steps 4 and 5: We can calculate ΔH of the material; hence ΔH_{steam} can be calculated, and knowing the specific enthalpy of condensation, we can compute the kilograms of steam needed.

Step 6:   Basis: 1 hr of operation

Steps 8 and 9:

(a) The heat loss is given as $Q = -1.50$ kJ/s or

$$\frac{-1.50 \text{ kJ}}{\text{s}} \left| \frac{3600 \text{ s}}{1 \text{ hr}} \right| \frac{1 \text{ hr}}{} = -5400 \text{ kJ}$$

(b) The specific enthalpy change for the steam can be determined from the steam tables. The $\Delta \hat{H}_{vap}$ of saturated steam at 250°C is 1701 kJ/kg, so that

$$\Delta \hat{H}_{steam} = -1701 \text{ kJ/kg}$$

(c) The enthalpy of the initial reactants, with reference to zero enthalpy at 20°C, is

$$\Delta H_{t_1} = m(\Delta \hat{H}) = m \int_{20}^{20} C_p \, dT = 0$$

Note that the selection of zero enthalpy at 20°C makes the calculation of the enthalpy of the reactants at the start quite easy. The enthalpy of the final products relative to zero enthalpy at 20°C is

$$\Delta H_{t_2} = m \int_{20}^{100} C_p \, dT = mC_p(100 - 20)$$

$$= \frac{150 \text{ kg}}{} \left| \frac{3.26 \text{ J}}{(g)(°C)} \right| \frac{(100 - 20)°C}{} \left| \frac{1000 \text{ g}}{\text{kg}} \right| = 39{,}120 \text{ kJ}$$

$$\Delta H_{material} = 39{,}120 \text{ kJ}$$

(d) In addition to the changes in enthalpy occurring in the material entering and leaving the system, the reaction absorbs 2300 kJ/kg. This quantity of energy can be conveniently thought of as an energy loss term, or alternatively as an adjustment to the enthalpy of the reactants.

$$\Delta H = \frac{-2300 \text{ kJ}}{\text{kg}} \left| \frac{150 \text{ kg}}{} \right. = -345{,}000 \text{ kJ}$$

Introduction of all these values into Eq. (b) gives

$$39{,}120 \text{ kJ} = \left(1701 \frac{\text{kJ}}{\text{kg steam}}\right)(S \text{ kg steam}) - 5400 \text{ kJ} - 345{,}000 \text{ kJ} \qquad \text{(c)}$$

from which the kilograms of steam per hour, S, can be calculated as

$$S = \frac{389{,}500 \text{ kJ}}{} \left| \frac{1 \text{ kg steam}}{1701 \text{ kJ}} \right. = 229 \text{ kg steam}$$

or

$$\frac{389{,}500 \text{ kJ}}{} \left| \frac{1 \text{ kg steam}}{1701 \text{ kJ}} \right| \frac{1}{150 \text{ kg charge}} = 1.53 \frac{\text{kg steam}}{\text{kg charge}}$$

If the system had been chosen to be everything but the steam coils and lines, we would have a situation as shown in Fig. E4.18b. Under these circumstances we could talk about the heat transferred into the reactor from the steam chest. From a balance on the steam chest (no accumulation),

$$Q_{\text{system II}} = \Delta H_{\text{steam}}$$

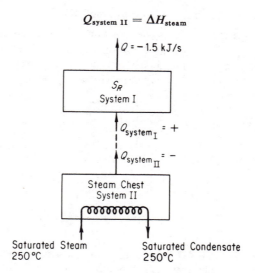

Fig. E4.18b

Both have negative values, but remember that

$$Q_{\text{system I}} = -Q_{\text{system II}}$$

so that the value of $Q_{\text{system I}}$ is plus.

The energy balance on system I reduces to

$$(\hat{H}_{t_2} - \hat{H}_{t_1})m = Q \tag{d}$$

where $Q = Q_{\text{system chest}} + (-5400 \text{ kJ})$. The steam used can be calculated from Eq. (d); the numbers will be identical to the calculation made for the original system.

Self-Assessment Test

1. In a refinery a condenser is to cool 1000 lb/hr of benzene that enters at 1 atm, 200°F, and leaves at 171°F. Assume negligible heat loss to the surroundings. How many pounds of cooling water are required per hour if the cooling water enters at 80°F and leaves at 100°F?

2. In a steady-state process, 10 g mol/s of O_2 at 100°C and 10 g mol/s of nitrogen at 150°C are mixed in a vessel which has a heat loss to the surroundings equal to $209(t - 25)$ J/s, where t is the temperature of the gas mixture in °C. Calculate the gas temperature of the exit stream in °C. Use the following heat capacity equations:

$$O_2: \quad C_p = 6.15 + 3.1 \times 10^{-3}T$$
$$N_2: \quad C_p = 6.5 + 1.25 \times 10^{-3}T$$

where T is in K and C_p is in cal/(g mol)(K).

3. An exhaust fan in a constant-area well-insulated duct delivers air at an exit velocity of 1.5 m/s at a pressure differential of 6 cm H_2O. Thermometers show the inlet and exit temperatures of the air to be 21.1°C and 22.8°C, respectively. The duct area is 0.60 m². Determine the actual power requirement for the fan.

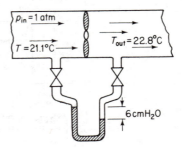

4. A water system is fed from a very large tank, large enough so that the water level in the tank is essentially constant. A pump delivers 3000 gal/min in a 12-in.-ID pipe to users 40 ft below the tank level. The work delivered to water is 1.52 hp. If the exit velocity of the water is 1.5 ft/s and the water temperature in the reservoir is the same as in the exit water, estimate the heat loss per second from the pipeline by the water in transit.

Section 4.6   Reversible Processes and the Mechanical Energy Balance

Your objectives in studying this section are to be able to:

1. Define a reversible process.
2. Identify a process as reversible or irreversible given a description of the process.
3. Define efficiency and apply the concept to calculate the work for an irreversible process.
4. Write down the steady-state mechanical energy balance for an open system and apply it to a problem in which all terms but one can be evaluated.

The energy balance of Sec. 4.3 is concerned with various classes of energy without inquiring into how "useful" each form of energy is to human beings. Our experience with machines and thermal processes indicates that some types of energy cannot be transformed completely into other types, and that energy in one state cannot be transformed to another state without the addition of extra work or heat. For example, internal energy cannot be completely converted into mechanical work. To account for these limitations on energy utilization, the second law of thermodynamics was eventually developed as a general principle.

As applied to fluid dynamic problems, one of the consequences of the second' law of thermodynamics is that two categories of energy of different "quality" can be envisioned:

(a) The so-called *mechanical* forms of energy, such as kinetic energy, potential energy, and work, which are *completely* convertible by an *ideal* (frictionless, reversible) engine from one form to another within the class.

(b) Other forms of energy, such as internal energy and heat, which are not so freely convertible.

Of course, in any real process with friction, viscous effects, mixing of components, and other dissipative phenomena taking place which prevent the complete conversion of one form of mechanical energy to another, allowances will have to be made in making a balance on mechanical energy for these "losses" in quality.

To describe the loss or transfer of energy of the freely convertible kind to energy of lesser quality, a concept termed *irreversiblity* has been coined. A process that operates without the degradation of any of the convertible types of energy is termed a *reversible* process; one that does not is an irreversible process. A reversible process proceeds under conditions of differentially balanced forces. Only an infinitesimal change is required to reverse the process, a concept that leads to the name "reversible."

Take, for example, the piston shown in Fig. 4.14. During the expansion process the piston moves the distance x and the volume of gas confined in the piston increases from V_1 to V_2. Two forces act on the reversible (frictionless) piston; one is the force exerted by the gas, equal to the pressure times the area of the piston, and the other is the force on the piston shaft and head from outside. If the force exerted by the gas equals the force F, nothing happens. If F is greater than the force of the gas, the gas will be compressed, whereas if F is less than the force of the gas, the gas will expand.

In the latter case, the work done by the expanding gas and the piston will be $W = \int F \, dx$. The work of the gas would be $W = \int p \, dV$ if the process were reversible,

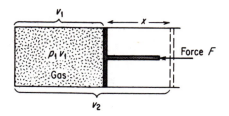

Fig. 4.14 Gas expansion.

that is, if the force F divided by the piston area were differentially less at all times than the pressure of the gas, and if the piston were frictionless; but in a real process some of the work done by the gas is dissipated by viscous effects, and the piston will not be frictionless so that the work as measured by $\int F\,dx$ will be less than $\int p\,dV$. For these two integrals to be equal, none of the available energy of the system can be degraded to heat or internal energy. To achieve such a situation we have to ensure that the movement of the piston is frictionless and that the motion of the piston proceeds under only a differential imbalance of forces so no shock or turbulence is present. Naturally, such a process would take a long time to complete.

No real process which involves friction, shock, finite temperature differences, unrestricted expansion, viscous dissipation, or mixing can be reversible, so we see that the dropping of a weight from a tower, the operation of a heat exchanger, and the combustion of gases in a furnace are all irreversible processes. Since practically all real processes are irreversible you may wonder why we bother paying attention to the theoretical reversible process. The reason is that it represents the best that can be done, the ideal case, and gives us a measure of our maximum accomplishment. In a real process we cannot do as well and so are less effective. A goal is provided by which to measure our effectiveness. And then, many processes are not too irreversible, so that there is not a big discrepancy between the practical and the ideal process.

Given the concept of an ideal (reversible process) and knowing the work in an actual process, two ways in which we can define efficiency are

$$\text{efficiency} = \eta_1 = \frac{\text{actual work output for the process}}{\text{work output for a reversible process}} \tag{4.29a}$$

and

$$\text{efficiency} = \eta_2 = \frac{\text{work input for a reversible process}}{\text{actual work input for the process}} \tag{4.29b}$$

These definitions provide a good way to compare process performance for energy conservation (but not the only way).

Now let us look at a general balance for mechanical energy. A balance on mechanical energy can be written on a microscopic basis for an elemental volume by taking the scalar product of the local velocity and the equation of motion.[25] After integration over the entire volume of the system the *steady-state mechanical energy balance* (for a system with mass interchange with the surroundings) becomes, on a per unit mass basis,

$$\Delta(\hat{K} + \hat{P}) + \int_{p_1}^{p_2} \hat{V}\,dp + \hat{W} + \hat{E}_v = 0 \tag{4.30}$$

where E_v represents the loss of mechanical energy, that is, the irreversible conversion *by the flowing fluid* of mechanical energy to internal energy, a term which must in

[25]Slattery, *Momentum, Energy, and Mass Transfer.*

each individual process be evaluated by experiment (or, as occurs in practice, by use of already existing experimental results for a similar process). Equation (4.30) is sometimes called the Bernoulli equation, especially for the reversible process for which $\hat{E}_v = 0$. The mechanical energy balance is best applied in fluid-flow calculations when the kinetic and potential energy terms and the work are of importance, and the friction losses can be evaluated from handbooks with the aid of *friction factors* or *orifice coefficients*.

EXAMPLE 4.19 Application of the Mechanical Energy Balance

Calculate the work per minute required to pump 1 lb of water per minute from 100 psia and 80°F to 1000 psia and 100°F. The exit stream is 10 ft above the entrance stream.

Solution

Steps 1, 2, 3, and 4:

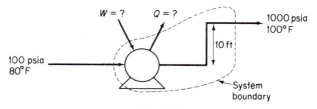

Fig. E4.19

Steps 3a, 3b, and 7: The general mechanical energy balance is

$$\Delta(\hat{K} + \hat{P}) + \int_{p_1}^{p_2} \hat{V}\, dp + \hat{W} + \hat{E}_v = 0 \tag{a}$$

We will assume that $\Delta\hat{K}$ is insignificant, and also preliminarily assume that the process is reversible so that $E_v = 0$, and the pump is 100% efficient. (Subsequently, we will consider what to do if the process is not reversible.) Equation (a) reduces to

$$\hat{W} = -\int_{p_1}^{p_2} \hat{V}\, dp - \Delta\hat{P} \tag{b}$$

Step 5 and 8: From the steam tables, the specific volume of liquid water is 0.01608 ft³/lb$_m$ at 80°F and 0.01613 ft³/lb$_m$ at 100°F. For all practical purposes the water is incompressible, and the specific volume can be taken to be 0.0161 ft³/lb$_m$. We have only one unknown in Eq. (b): $\hat{W}$.

Step 6:   Basis: 1 min of operation

Step 9:

$$\Delta\dot{P} = \frac{1\ \text{lb}_m}{\text{min}} \left|\frac{10\ \text{ft}}{}\right| \frac{32.2\ \text{ft}}{\text{sec}^2} \left|\frac{1\ \text{Btu}}{\dfrac{32.2\ \text{(ft)(lb}_m)}{\text{(lb}_f)(\text{sec}^2)}}\right| \frac{1\ \text{Btu}}{778\ \text{(ft)(lb}_f)} = 0.013\ \frac{\text{Btu}}{\text{min}}$$

$$\frac{1 \text{ lb}_m}{\text{min}} \int_{100}^{1000} 0.0161 dp = \frac{1 \text{ lb}_m}{\text{min}} \left| \frac{0.0161 \text{ ft}^3}{\text{lb}_m} \right| \frac{(1000 - 100) \text{ lb}_f}{\text{in.}^2} \left| \left(\frac{12 \text{ in.}}{1 \text{ ft}}\right)^2 \right| \frac{1 \text{ Btu}}{778 \text{ (ft)(lb}_f)}$$

$$= 2.68 \frac{\text{Btu}}{\text{min}}$$

$$\dot{W} = -2.68 - 0.013 = -2.69 \frac{\text{Btu}}{\text{min}}$$

About the same value can be calculated using Eq. (4.24) if $\hat{Q} = \hat{K} = 0$, because the enthalpy change[26] for a reversible process for 1 lb of water going from 100 psia and 100°F to 1000 psia is 2.70 Btu. Make the computation yourself. However, usually the enthalpy data for liquids other than water are missing, or not of sufficient accuracy to be valid, which forces an engineer to turn to the mechanical energy balance.

We might now well inquire for the purpose of purchasing a pump-motor, say, as to what the work would be for a real process, instead of the fictitious reversible process assumed above. First, you would need to know the efficiency of the combined pump and motor so that the actual input from the surroundings (the electric connection) to the system would be known. Second, the friction losses in the pipe, valves, and fittings must be estimated so that the term $\hat{E}_v$ could be reintroduced into Eq. (4.30). Suppose, for the purposes of illustration, that $\hat{E}_v$ was estimated to be, from an appropriate handbook, 320 (ft)(lb$_f$)/lb$_m$ and the motor-pump efficiency was 60% (based on 100% efficiency for a reversible pump-motor). Then,

$$\dot{E}_v = \frac{320 \text{ (ft)(lb}_f)}{1 \text{ lb}_m} \left| \frac{1 \text{ Btu}}{778 \text{ (ft)(lb}_f)} \right| \frac{1 \text{ lb}_m}{\text{min}} = 0.41 \text{ Btu/min}$$

$$\dot{W} = -(2.68 + 0.013 + 0.41) = -3.10 \text{ Btu/min}$$

Remember that the minus sign indicates that work is done on the sytsem. The pump motor must have the capacity

$$\frac{3.10 \text{ Btu}}{1 \text{ min}} \left| \frac{1}{0.60} \right| \frac{1 \text{ min}}{60 \text{ sec}} \left| \frac{1.415 \text{ hp}}{1 \text{ Btu/sec}} \right| = 0.122 \text{ hp}$$

EXAMPLE 4.20 Calculation of Work for a Batch Process

One kilogram mole of N_2 is in a horizontal cylinder at 1000 kPa and 20°C. A 6-cm^2 piston of 2-kg mass seals the cylinder and is fixed by a pin. The pin is released and the N_2 volume is doubled, at which time the piston is stopped again. What is the work done by the gas in this process?

Solution

Steps 1, 2, 3, and 4: Draw a picture. See Fig. E4.20. From the ideal gas law we can compute

$$\hat{V} = \frac{RT}{p} = \frac{8.31 \text{ (kPa)(m}^3)}{\text{(K)(kg mol)}} \left| \frac{293 \text{ K}}{1000 \text{ kPa}} \right| = 2.44 \text{ m}^3/\text{kg mol}$$

[26]From J. H. Keenan and F. G. Keyes, *Thermodynamic Properties of Steam*, Wiley, New York, 1936, Table 3.

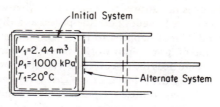

Fig. E4.20

Steps 5, 6, and 7:

(a) Is the process a flow or nonflow process? Since no material leaves or enters the cylinder, let us analyze the process as a nonflow system.

(b) This process is definitely irreversible because of friction between the piston (and the gas) and the walls of the cylinder, turbulence in the gas, the large pressure drops, and so on.

(c) In selecting a system to work with, we can choose either the gas itself or the gas plus the piston. Let us initially select the N_2 gas as the system, with the surroundings being everything else, including the piston and the cylinder.

(d) Since the gas expansion is a nonflow process, Eq. (4.30) does not apply. Even if you had an expression for the unsteady-state mechanical energy balance (which we have not discussed since it involves the concept of entropy), it would not be possible to calculate the work done by the gas because the process is irreversible and the E_v term would not be known. As might be expected, a direct application of $W = \int F \, dl$ fails for this problem because the pressure of the gas does not necessarily equal the force per unit area applied by the piston to the gas; that is, the work cannot be obtained from $W = \int p \, dV$.

(e) Let us change our system to include the gas, the piston, and the cylinder. With this choice, the system expansion is still irreversible, but the pressure in the surroundings can be assumed to be atmospheric pressure and is constant. The work done in pushing back the atmosphere probably is done almost reversibly from the viewpoint of the surroundings and can be closely estimated by calculating the work done on the surroundings:

$$W = \int_{l_1}^{l_2} F \, dl = \int_{V_1}^{V_2} p \, dV = p \, \Delta V$$

ΔV of the gas is $2V_1 - V_1 = V_1 = 2.44$ m³, so that the volume change of the surroundings is -2.44 m³. Then the work done on the surroundings when the surroundings are the system is

Basis: 2.44 m³ at 101.3 kPa and 20°C

$$W = \frac{101.3 \text{ kPa}}{} \left| \frac{-2.44 \text{ m}^3}{} \right| \frac{1 \text{ (N)(m}^{-2})}{1 \text{ Pa}} \left| \frac{1 \text{ J}}{1 \text{ (N)(m)}} \right| = -247 \text{ kJ}$$

The minus sign indicates that work is done on the surroundings. From this we know that the work done by the alternative system is

$$W_{\text{system}} = -W_{\text{surroundings}} = -(-247) = 247 \text{ kJ}$$

(f) This still does not answer the question of what work was done by the gas, because our system now includes both the piston and the cylinder as well as the gas. You may wonder what happens to the energy transferred from the gas to the piston itself. Some of it goes into work against the atmosphere, which we have just calculated; where does the rest of it go? This energy can go into raising the temperature of the piston or the cylinder or the gas. With time, part of it can be transferred to the surroundings to raise the temperature of the surroundings. From our macroscopic viewpoint we cannot say specifically what happens to this "lost" energy.

(g) You should also note that if the surroundings were a vacuum instead of air, then *no work* would be done by the system, consisting of the piston plus the cylinder plus the gas, although the gas itself probably still would do some work.

EXAMPLE 4.21 Evaporation

How much work is done by 1 liter (1 dm³) of liquid water when it evaporates from an open vessel at 25°C?

Solution

Steps 1, 2, 3, and 4: Does the water do work in evaporating? Certainly! It does work against the atmosphere. Furthermore, the process, diagrammed in Fig. E4.21, is a reversible one because the evaporation takes place at constant temperature and pressure, and presumably the conditions in the atmosphere immediately above the open portion of the vessel are in equilibrium with the water surface. Assume that the atmospheric pressure is 100 kPa.

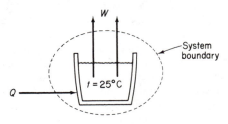

Fig. E4.21

Steps 3a, 3b, and 7: The general energy balance

$$\Delta E = -\Delta[(\hat{H} + \hat{K} + \hat{P})m] + Q - W$$

will not be useful in solving this problem because Q is unknown. However, imagine that an expansible bag is placed over the open face so that the system now becomes a closed system. Equation (4.30) will not apply (why?), but because of the special conditions established for this problem, the work is

$$W = \int_{V_1}^{V_2} p \, dV = p \, \Delta V$$

and W can be computed as shown in Example 4.20.

Self-Assessment Test

1. Which process will yield more work: (1) expansion of a piston against constant pressure or (2) reversible expansion of a piston?

2. Differentiate between thermal and mechanical energy.

3. Define a reversible process.

4. Find Q, W, ΔE, and ΔH for the reversible compression of 3 moles of an ideal gas from a volume of 100 dm³ to 2.4 dm³ at a constant T of 300 K.

5. Water is pumped from a very large storage reservoir as shown in the figure at the rate of 2000 gal/min. Determine the minimum power (i.e., for a reversible process) required by the pump in horsepower.

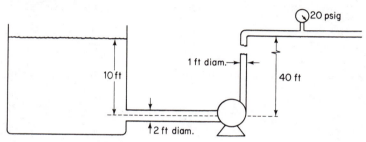

Section 4.7   Energy Balances with Chemical Reaction

So far we have not discussed how to treat energy balances in which chemical reactions occur. We do so in this section. For any given process, when a reaction takes place, the energy exchange with the surroundings will be different from the same process without a reaction. As you know, the observed heat transfer that occurs in a closed system (with zero work) as a result of a reaction represents the energy associated with the rearrangement of the bonds holding together the atoms of the reacting molecules. For an exothermic reaction, the energy required to hold the products of the reaction together is less than that required to hold the reactants together and the surplus energy is released. The reverse is true of an endothermic reaction. The energy change that appears directly as a result of a reaction is termed the *heat of reaction*, ΔH_{rxn}, a term that is a legacy of the days of the caloric theory. Note that "heat of reaction" is actually an *enthalpy change* and not heat transfer.

4.7-1   Standard Heat of Formation

> *Your objectives in studying this section are to be able to:*
>
> 1. Compute heats of formation from experimental data for the enthalpy change (including phase changes) of a process with a reaction taking place.

2. Look up heats of formation in reference tables for a given compound.
3. List the standard conventions and reference states used for reactions associated with the standard heat of formation.
4. Calculate the standard heat of reaction from tabulated standard heats of formation for a given reaction.

To take account of possible energy changes caused by a reaction, in the energy balance we incorporate in the enthalpy of each individual constituent an additional quantity termed the standard heat (really enthalpy) of formation, $\Delta \hat{H}_f^\circ$, a quantity that is discussed in detail below. (The superscript $^\circ$ denotes "standard state" and the subscript f denotes "formation.") Thus for the case of a single species A without any pressure effect on the enthalpy and omitting phase changes,

$$\Delta \hat{H}_A = \Delta \hat{H}_{fA}^\circ + \int_{T_{\text{ref}}}^{T} C_{pA} \, dT \tag{4.31}$$

For several species in which there is no energy change on mixing, we would have

$$\Delta H_{\text{mixture}} = \sum_{i=1}^{s} n_i \, \Delta \hat{H}_{fi}^\circ + \sum_{i=1}^{s} \int_{T_{\text{ref}}}^{T} n_i C_{p_i} \, dT \tag{4.32}$$

where i designates each species, n_i is the number of moles of species i, and s is the total number of species.

If a mixture enters and leaves a system without a reaction taking place, we would find that the same species entered and left so that the enthalpy change in either the flow or the accumulation terms in Eq. (4.24) would not be any different with the modification described above than what we have used before, because the terms that account for the heat of reaction in the energy balance would cancel. For example, for the case of two species in a flow system, the output enthalpy would be

$$\Delta H_{\text{output}} = \underbrace{n_1 \, \Delta \hat{H}_{f1}^\circ + n_2 \, \Delta \hat{H}_{f2}^\circ}_{\text{"heat of formation"}} + \underbrace{\int_{T_{\text{ref}}}^{T_{\text{out}}} (n_1 C_{p1} + n_2 C_{p2}) \, dT}_{\text{"sensible heat"}}$$

and the input enthalpy would be

$$\Delta H_{\text{input}} = \underbrace{n_1 \, \Delta \hat{H}_{f1}^\circ + n_2 \, \Delta \hat{H}_{f2}^\circ}_{\text{"heat of formation"}} + \underbrace{\int_{T_{\text{ref}}}^{T_{\text{in}}} (n_1 C_{p1} + n_2 C_{p2}) \, dT}_{\text{"sensible heat"}}$$

Without reaction, note that $\Delta H_{\text{output}} - \Delta H_{\text{input}}$ would only involve the sensible heat terms that we have described before; the heat-of-formation terms would be exactly the same in each equation and would cancel out.

However, if a reaction takes place, the species that enter and leave will differ, and the terms involving the heat of formation will not cancel. For example, suppose that species 1 and 2 enter the system, react, and species 3 and 4 leave. Then

$$\Delta H_{\text{out}} - \Delta H_{\text{in}} = (n_3 \, \Delta \hat{H}^\circ_{f3} + n_4 \, \Delta \hat{H}^\circ_{f4}) - (n_1 \, \Delta \hat{H}^\circ_{f1} + n_2 \, \Delta \hat{H}^\circ_{f2})$$

$$+ \int_{T_{\text{ref}}}^{T_{\text{out}}} (n_3 C_{p3} + n_4 C_{p4}) \, dT - \int_{T_{\text{ref}}}^{T_{\text{in}}} (n_1 C_{p1} + n_2 C_{p2}) \, dT \qquad (4.33)$$

You will be retrieving information on heats of formation from reference tables. The values in the tables have been determined from innumerable experiments. To determine the values of the standard heats (enthalpies) of formation, the experimenter usually selects either a simple flow process without kinetic energy, potential energy, or work effects (a flow calorimeter), or a simple batch process (a bomb calorimeter), in which to conduct the reaction. Consider an experiment in a flow process in which the summation of sensible heat terms on the right-hand side of Eq. (4.33) is zero and no work is done. The steady-state (no accumulation term) version of Eq. (4.24) reduces to

$$Q = \Delta[\hat{H}m] = \Delta H_{\text{rxn}}$$

with

$$\Delta H^\circ_{\text{rxn}} \equiv (\sum_{\text{products}} n_i \, \Delta \hat{H}^\circ_{fi} - \sum_{\text{reactants}} n_i \, \Delta \hat{H}^\circ_{fi}) \qquad (4.34)$$

where $\Delta H^\circ_{\text{rxn}}$ is the symbol used for the heat of reaction at the standard state. We observe that the enthalpy change caused by the reaction in the prescribed type of experiment appears as heat transferred to or from the system, and the value of Q that is measured is the heat transferred from the reaction and is equivalent to the heat of reaction at the standard state.

A flow calorimeter in which $Q = 0$, $W = 0$, and the temperature change is measured between the input and output streams can also be used to compute the heat of reaction, because the left-hand side of Eq. (4.33) is zero. Why? Refer to Sec. 4.7.3 for an analysis of the batch bomb calorimeter of constant volume.

The calculations that we shall make here will all be for low pressures, and, although the effect of pressure upon heats of reaction is relatively negligible under most conditions, if exceedingly high pressures are encountered, you should make the necessary corrections as explained in most texts on thermochemistry.

There are certain conventions and symbols which you should always keep in mind concerning thermochemical calculations if you are to avoid difficulty later on. These conventions can be summarized as follows:

(a) The reactants are shown on the left-hand side of the chemical equation, and the products are shown on the right: for example,

$$CH_4(g) + H_2O(l) \longrightarrow CO(g) + 3H_2(g)$$

(b) The conditions of phase, temperature, and pressure must be specified unless the last two are the standard conditions, as presumed in the example above, when only the phase is required. This is particularly important for compounds such as H_2O, which can exist as more than one phase under common conditions. If the reaction takes place at other than standard conditions, you might write

$$CH_4(g, 1.5 \text{ atm}) + H_2O(l) \xrightarrow{50°C} CO(g, 3 \text{ atm}) + H_2(g, 3 \text{ atm})$$

(c) Unless otherwise specified, the heats of reaction, the enthalpy changes, and all the constituents are at the standard state of 25°C (77°F) and 101.3 kPa (1 atm) total pressure. The heat of reaction under these conditions is called the *standard heat of reaction*, and is distinguished by the superscript ° symbol.

(d) Unless the amounts of material reacting are stated, it is assumed that the quantities reacting are the stoichiometric amounts shown in the chemical equation.

Data to compute the standard heat of reaction are reported and tabulated in two different but essentially equivalent forms:

(a) Standard heats of formation
(b) Standard heats of combustion

We first describe the standard heat of formation ($\Delta \hat{H}_f°$) and then the standard heat of combustion ($\Delta \hat{H}_c°$). The units of both quantities are usually tabulated as energy per mole, such as J/g mol, kJ/g mol, kcal/g mol, or Btu/lb mol. The "per mole" refers to the specified reference substance in the related stoichiometric equation.

The *standard heat of formation* is the special heat of reaction for the formation of a compound from the elements. The initial reactants and final products must be stable and at 25°C and 1 atm. The reaction does not necessarily represent a real reaction that would proceed at constant temperature but can be a fictitious process for the formation of a compound from the elements. By defining the heat of formation as zero in the standard state for each *element*, it is possible to design a system to express the heats of formation for all *compounds* at 25°C and 1 atm. If you use the conventions discussed above, then the thermochemical calculations will all be consistent, and you should not experience any confusion as to signs. Consequently, with the use of well-defined standard heats of formation, it is not necessary to record the experimental heats of reaction for every reaction that can take place. That would be an impossible task.

To sum up, a formation reaction is understood by convention to be a reaction that forms 1 mole of compound from the elements that make it up. Standard heats of reaction at 25°C and 1 atm for any reaction can be calculated from the tabulated (or experimental) standard heats of formation values by using Eq. (4.34), because the standard heat of formation is a state (point) function.

EXAMPLE 4.22 Heats of Formation from Reference Data

What is the standard heat of formation of HCl(g)?

Solution

Look in Appendix F.

tabulated data	$\frac{1}{2}H_2(g) + \frac{1}{2}Cl_2(g) \longrightarrow HCl(g)$		
$\Delta \hat{H}_f^\circ \left(\dfrac{J}{g\ mol}\right)$	0	0	$-92{,}312$

In the reaction at 25°C and 1 atm, both $H_2(g)$ and $Cl_2(g)$ would be assigned $\Delta \hat{H}_f^\circ$ values of 0, and the value shown in Appendix F for HCl(g) of $-92{,}312$ J/g mol is the standard heat of formation for this compound (as well as the standard heat of reaction for the reaction as written above). The value tabulated in Appendix F might actually be determined by carrying out the reaction shown for HCl(g) and measuring the energy liberated in a calorimeter, or by some other more convenient method.

EXAMPLE 4.23 Determination of Heats of Formation from Experimental Data

Suppose that you want to find the standard heat of formation of CO from experimental data. Can you prepare pure CO from C and O_2 and measure the heat transfer? This would be far too difficult. It would be easier experimentally to find the heat of reaction for the two reactions shown below and add them as follows:

Basis: 1 g mol of CO

$\Delta \hat{H}_{rxn}^\circ (experimental)$

A:	$C(\beta) + O_2(g) \longrightarrow CO_2(g)$	-393.51 kJ/g mol
B:	$CO(g) + \frac{1}{2}O_2(g) \longrightarrow CO_2(g)$	-282.99 kJ/g mol
$A - B$:	$C(\beta) + \frac{1}{2}O_2(g) \longrightarrow CO(g)$	

$$\Delta \hat{H}_{rxn\ A-B}^\circ = (-393.51) - (-282.99) = \Delta \hat{H}_f^\circ = -110.52 \text{ kJ/g mol}$$

The energy change for the overall reaction scheme is the desired heat of formation per mole of CO(g). See Fig. E4.23.

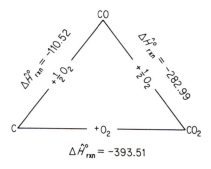

Fig. E4.23

EXAMPLE 4.24 Calculation of the Heat of Reaction from Standard Heats of Formation

Calculate $\Delta \hat{H}^{\circ}_{\text{rxn}}$ for the following reaction of 4 g mol of NH_3:

$$4NH_3(g) + 5O_2(g) \longrightarrow 4NO(g) + 6H_2O(g)$$

Solution

Basis: 4 g mol of NH_3

tabulated data	$NH_3(g)$	$O_2(g)$	$NO(g)$	$H_2O(g)$
$\Delta \hat{H}^{\circ}_f$ per mole at 25°C and 1 atm (kcal/g mol)	−11.04	0	+21.60	−57.80

We shall use Eq. (4.34) to calculate $\Delta H^{\circ}_{\text{rxn}}$ for 4 g mol of NH_3:

$$\Delta H^{\circ}_{\text{rxn}} = [4(21.60) + 6(-57.80)] - [5(0) + 4(-11.04)]$$

$$= -216.24 \text{ kcal/4 g mol } NH_3, \quad \text{or} \quad \Delta H^{\circ}_{\text{rxn}} = -54.0 \text{ kcal/g mol } NH_3$$

EXAMPLE 4.25 Heat of Formation Including a Phase Change

If the standard heat of formation for $H_2O(l)$ is −285.838 kJ/g mol and the heat of evaporation is +44.012 kJ/g mol at 25°C and 1 atm, what is the standard heat of formation of $H_2O(g)$?

Solution

Basis: 1 g mol of H_2O

We shall proceed as in Example 4.23 to add the known chemical equations and the phase transitions to yield the desired chemical equation and carry out the same operations on the enthalpy changes. For reaction A, $\Delta \hat{H}^{\circ}_{\text{rxn}} = \sum \Delta \hat{H}^{\circ}_{f \text{ products}} - \sum \hat{H}^{\circ}_{f \text{ reactants}}$:

A: $H_2(g) + \frac{1}{2}O_2(g) \longrightarrow H_2O(l)$ $\Delta \hat{H}^{\circ}_{\text{rxn}} = -285.838$ kJ/g mol

B: $H_2O(l) \longrightarrow H_2O(g)$ $\Delta \hat{H}^{\circ}_{\text{vap}} = + 44.012$ kJ/g mol

$A + B$: $H_2(g) + \frac{1}{2}O_2(g) \longrightarrow H_2O(g)$

$\Delta \hat{H}^{\circ}_{\text{rxn } A} + \Delta \hat{H}^{\circ}_{\text{vap}} = \Delta \hat{H}^{\circ}_{\text{rxn } H_2O(g)} = \Delta \hat{H}^{\circ}_{f \ H_2O(g)} = -241.826$ kJ/g mol

---

You can see that any number of chemical equations can be treated by algebraic methods, and the corresponding heats of reaction can be added or subtracted in the same fashion as are the equations. By carefully following these rules of procedure, you will avoid most of the common errors in thermochemical calculations.

To simplify matters, the value cited for ΔH_{vap} of water in Example 4.25 was at 25°C and 1 atm. To calculate this value, if the final state for water is specified as $H_2O(g)$ at 25°C and 1 atm, the following enthalpy changes should be taken into account if you start with $H_2O(l)$ at 25°C and 1 atm:

$$\Delta H_{vap} \text{ at } 25°C \text{ and 1 atm} \begin{cases} H_2O(l) \text{ 25°C, 1 atm} \\ \quad\downarrow \Delta H_1 \\ H_2O(l) \text{ 25°C, vapor pressure at 25°C} \\ \quad\downarrow \Delta H_2 = \Delta H_{vap} \text{ at the vapor pressure of water} \\ H_2O(g) \text{ 25°C, vapor pressure at 25°C} \\ \quad\downarrow \Delta H_3 \\ H_2O(g) \text{ 25°C, 1 atm} \end{cases}$$

For practical purposes the value of ΔH_{vap} at 25°C and the vapor pressure of water of 43,911 J/g mol (10,495 cal/g mol) will be adequate for engineering calculations.

There are many sources of tabulated values for the standard heats of formation. A good source of extensive data is the *National Bureau of Standards Bulletin* 500, and its supplements by F. K. Rossini, or the serial publications of the Thermodynamics Research Center at Texas A & M University (API Research Project No. 44 and the TRC Data Project). A condensed set of values for the heats of formation may be found in Appendix F. If you cannot find a standard heat of formation for a particular compound in reference books or in the chemical literature, $\Delta \hat{H}_f^\circ$ may be estimated by the methods described in Verma and Doraiswamy[27] or by some of the authors listed as references in their article. Remember that the values for the standard heats of formation are negative for exothermic reactions.

Self-Assessment Test

1. Calculate the standard heat of formation of CH_4 given the following experimental results at 25°C and 1 atm:

$$H_2(g) + \tfrac{1}{2}O_2(g) \longrightarrow H_2O(l) \qquad \Delta H = -285.84 \text{ kJ/g mol } H_2$$
$$C \text{ (graphite)} + O_2(g) \longrightarrow CO_2(g) \qquad \Delta H = -393.51 \text{ kJ/g mol } C$$
$$CH_4(g) + 2O_2(g) \longrightarrow CO_2(g) + 2H_2O(l) \qquad \Delta H = -890.36 \text{ kJ/g mol } CH_4$$

Compare your answer with that found in the table of heats of formation listed in Appendix F.

2. Calculate the standard heat of reaction for the following reaction from the heats of formation:

$$C_6H_6(g) \longrightarrow 3C_2H_2(g)$$

[27]K. K. Verma and L. K. Doraiswamy, *Ind. Eng. Chem. Fundamentals*, v. 4, p. 389 (1965).

4.7-2 Standard Heat of Combustion

> ***Your objectives in studying this section are to be able to:***
>
> 1. Compute heats of combustion from experimental data for the enthalpy change (including phase changes) of a process with a reaction taking place.
> 2. Look up heats of combustion in reference tables for a given compound.
> 3. List the standard conventions and reference states used for reactions associated with the standard heat of combustion.
> 4. Calculate the standard heat of reaction from tabulated standard heats of combustion for a given reaction.
> 5. Calculate standard heats of formation from heats of combustion and/or standard heats of combustion from standard heats of formation.
> 6. Calculate the standard higher heating value from the lower heating value or the reverse.

Standard heats of combustion are the second method of recording thermochemical data. The standard heats of combustion do not have the same standard states as the standard heats of formation. The conventions used with the standard heats of combustion are

(a) The compound is oxidized with oxygen or some other substance to the products $CO_2(g)$, $H_2O(l)$, $HCl(aq)$,[28] and so on.

(b) The reference conditions are still 25°C and 1 atm.

(c) Zero values of $\Delta \hat{H}_c^\circ$ are assigned to certain of the oxidation products as, for example, $CO_2(g)$, $H_2O(l)$, and $HCl(aq)$, and to $O_2(g)$ itself.

(d) If other oxidizing substances are present, such as S or N_2, or if Cl_2 is present, it is necessary *to make sure that states of the products are carefully specified and are identical to (or can be transformed into) the final conditions which determine the standard state as shown in Appendix F.*

The standard heat of combustion of an unoxidized compound can never be positive but is always negative. A positive value would mean that the substance would not burn or oxidize.

In the search for better high-energy fuels, the heat of combustion of each element in terms of energy per unit mass is of interest. You can see from Fig. 4.15 that the best fuels will use H, Be or B as building blocks. With everything else held constant, the greater the heat of combustion per unit mass, the longer the range of an airplane or missile.

For a fuel such as coal or oil, the standard heat of combustion is known as the *heating value* of the fuel. To determine the heating value, a weighed sample is

---

[28] $HCl(aq)$ represents an infinitely dilute solution of HCl (see Sec. 4.8).

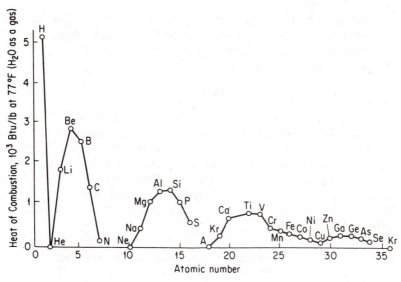

Fig. 4.15

burned in oxygen in a calorimeter bomb, and the energy given off is detected by measuring the temperature increase of the bomb and surrounding apparatus.

Because the water produced in the calorimeter is in the vicinity of room temperature, although the gas within the bomb is saturated with water vapor at this temperature, essentially all the water formed in the combustion process is condensed into liquid water. This is not like a real combustion process in a furnace, where the water remains as a vapor and passes up the stack. Consequently, we have two heating values for fuels containing hydrogen: (1) the *gross*, or *higher*, *heating value*, in which all the water formed is condensed into the liquid state, and (2) the *net*, or *lower*, *heating value*, in which all the water formed remains in the vapor state. The value you determine in the calorimeter is the gross heating value; this is the one reported along with the fuel analysis and should be presumed unless the data state specifically that the net heating value is being reported.

You can estimate the heating value of a coal within about 3% from the Dulong formula[29]:

The higher heating value (HHV) in British thermal units per pound

$$= 14,544C + 62,028\left(H - \frac{O}{8}\right) + 4050\,S$$

where   C = weight fraction carbon

         S = weight fraction net sulfur

$H - \dfrac{O}{8}$ = weight fraction net hydrogen

         = total weight fraction hydrogen − $\frac{1}{8}$ weight fraction oxygen

[29]H. H. Lowry, ed., *Chemistry of Coal Utilization*, Wiley, New York, 1945, Chap. 4.

The values of C, H, S, and O can be taken from the fuel or flue-gas analysis. If the heating value of a fuel is known and the C and S are known, the approximate value of the net H can be determined from the Dulong formula. A general relation between the gross heating and net heating values is

$$\text{net Btu/lb coal} = \text{gross Btu/lb coal} - 91 \, (\% \text{ total H by weight})$$

Urban refuse or municipal solid waste is produced at a rate of about 11 kg (5 lb) per person per day, and contains both organic as well as nonorganic materials. The breakdown of the components in a typical municipal refuse was listed in Problem 1.54. Combustion of organic waste represents a supplemental source of supply of energy. Table 4.8 lists the proximate and ultimate analysis of composite municipal refuse, and Table 4.9 lists the HHV for typical wastes. If you "watch calories," Table 4.10 shows the calories in various foods. Table 4.11 indicates some heating values of typical coals.

TABLE 4.8 ANALYSES OF A COMPOSITE MUNICIPAL REFUSE*

Proximate Analysis (%)	
Moisture	20.00
Volatile matter	52.70
Fixed carbon	7.30
Ash and metal	20.00
	100.00
Ultimate Analysis (%)	
Moisture	20.00
Carbon	29.83
Hydrogen	3.99
Oxygen	25.69
Nitrogen	0.37
Sulfur	0.12
Ash and metal	20.00
	100.00

*Btu/lb: 5260 + 182 = 5442 (the 182 figure is from 50% oxidation of metals).

TABLE 4.9 HIGHER HEATING VALUE OF MUNICIPAL REFUSE (kJ/kg)

Refuse Component	As Delivered	Dry Basis	Refuse Component	As Delivered	Dry Basis
Lawn grass	4,780	19,320	Average	—	20,050
Meat scraps, cooked	32,260	32,260	Mail	14,150	14,820
Newspaper	18,530	19,710	Cardboard	16,370	17,260
Shrub cuttings	6,290	20,300	Ripe tree leaves	18,550	20,610
Vegetable food waste	4,170	19,220	Magazines	12,210	12,730

TABLE 4.10   CALORIE COUNTS IN VARIOUS FOODS

Food	Portion	Kilo-calories	Food	Portion	Kilo-calories
Beer	12 oz	165	Orange juice	½ glass	56
Chicken, broiled	½	308	Soft drink	1 can	73
Coffee	1 cup	0	Toast	2 pieces	120
Martini	1 oz	168	Trout	1 lb	224

SOURCE: R. Eshleman and M. Winston, *The American Heart Association Cookbook*, David McKay, New York, 1975.

TABLE 4.11   HIGHER HEATING VALUES OF TYPICAL COALS

Class	kJ/kg	Btu/lb
Meta-anthracite	26,680	11,480
Anthracite	33,110	14,250
Low-volatile bituminous	35,370	15,220
High-volatile bituminous	27,790	11,960
Subituminous	23,330	10,040
Lignite	17,290	7,440

EXAMPLE 4.26   Heating Value of Coal

Coal gasification consists of the chemical transformation of solid coal into gas. The heating values of coal differ, but the higher the heating value, the higher the value of the gas produced (essentially methane, carbon monoxide, hydrogen, etc.). The following coal has a reported heating value of 29,770 kJ/kg as received. Assuming that this is the gross heating value, calculate the net heating value.

component	percent
C	71.0
H_2	5.6
N_2	1.6
Net S	2.7
Ash	6.1
O_2	13.0
Total	100.0

Solution

The corrected ultimate analysis shows 5.6% hydrogen on the as-received basis.

Basis: 100 kg of coal as received

The water formed on combustion is

$$\frac{5.6 \text{ kg H}_2}{} \left| \frac{1 \text{ kg mol H}_2}{2.02 \text{ kg H}_2} \right| \frac{1 \text{ kg mol H}_2\text{O}}{1 \text{ kg mol H}_2} \left| \frac{18 \text{ kg H}_2\text{O}}{1 \text{ kg mol H}_2\text{O}} \right. = 50 \text{ kg H}_2\text{O}$$

The energy required to evaporate the water is

$$\frac{50 \text{ kg H}_2\text{O}}{100 \text{ kg coal}} \left| \frac{2370 \text{ kJ}}{\text{kg H}_2\text{O}} \right. = \frac{1185 \text{ kJ}}{\text{kg coal}}$$

The net heating value is

$$29{,}770 - 1185 = 28{,}595 \text{ kJ/kg}$$

The value 2370 kJ/kg is not the latent heat of vaporization of water at 25°C (2440 kJ/kg) but includes the effect of a change from a heating value at constant volume to one at constant pressure (-70 kJ/kg) as described in a later section of this chapter.

Standard heats of reaction can be calculated from standard heats of combustion by an equation analogous to Eq. (4.34):

$$\Delta H_{\text{rxn}}^{\circ} = -\left(\sum \Delta H_{c\ \text{products}}^{\circ} - \sum H_{c\ \text{reactants}}^{\circ} \right)$$
$$= -\left(\sum_i n_{\text{prod } i}\, \Delta \hat{H}_{c\ \text{prod } i}^{\circ} - \sum_i n_{\text{react } i}\, \Delta \hat{H}_{c\ \text{react } i}^{\circ} \right) \qquad (4.35)$$

Note: The minus sign in front of the summation expression occurs because the choice of reference states is zero for the right-hand side of the standard equations.

EXAMPLE 4.27 Calculation of Heat of Reaction from Heat of Combustion Data

Compute the heat of reaction of the following reaction from standard heat of combustion data:

$$\text{C}_2\text{H}_5\text{OH(l)} + \text{CH}_3\text{COOH(l)} \longrightarrow \text{C}_2\text{H}_5\text{OOCCH}_3\text{(l)} + \text{H}_2\text{O(l)}$$
ethyl alcohol   **acetic acid**   **ethyl acetate**

Solution

Basis: 1 g mol of $\text{C}_2\text{H}_5\text{OH}$

tabulated data	$C_2H_5OH(l)$	$CH_3COOH(l)$	$C_2H_5OOCCH_3(l)$	$H_2O(l)$
$\Delta \hat{H}_c^{\circ}$ per mole at 25°C and 1 atm (kJ/g mol)	-759.18	-484.14	-1251.94	0

We use Eq. (4.35):

$$\Delta H_{\text{rxn}}^{\circ} = -[-1251.94 - (-759.18 - 484.14)] = +8.62 \text{ kJ/g mol}$$

EXAMPLE 4.28 Calculation of Heat of Formation from Heats of Combustion

Calculate the standard heat of formation of $C_2H_2(g)$ (acetylene) from the standard heat of combustion data.

Solution

The standard heat of formation for $C_2H_2(g)$ would be expressed in the following manner (basis: 1 g mol C_2H_2):

$$2C(\beta) + H_2(g) \longrightarrow C_2H_2(g) \qquad \Delta H_f^\circ = ?$$

How can we obtain this equation manipulating known equations and their associated standard heat of combustion data? The procedure is to take the chemical equation that gives the standard heat of combustion of $C_2H_2(g)$ (Eq. A below) and add or subtract other known combustion equations (B and C) so that the desired equation is finally obtained algebraically. The details are as follows:

$$\Delta \hat{H}_c^\circ$$
(kJ/g mol)

$$
\begin{array}{lll}
A\colon & C_2H_2(g) + 2\frac{1}{2}O_2(g) \longrightarrow 2CO_2(g) + H_2O(l) & -721.804 \\
B\colon & C(\beta) \quad + O_2(g) \quad \longrightarrow CO_2(g) & -218.557 \\
C\colon & H_2(g) \quad + \frac{1}{2}O_2(g) \quad \longrightarrow H_2O(l) & -158.743 \\
\end{array}
$$

$-A + 2B + C$:
$$2C(\beta) \quad + H_2(g) \quad \longrightarrow C_2H_2(g)$$

$$\Delta H_{net}^\circ = \Delta \hat{H}_f^\circ = -(-721.804) + 2(-218.557) + (-158.743) = +125.947 \; \frac{kJ}{g\;mol}$$

EXAMPLE 4.29 Combination of Heats of Reaction at 25°C

The following heats of reaction are known from experiments for the reactions below at 25°C in the standard thermochemical state:

$$\Delta \hat{H}_{rxn}^\circ$$

rxn	(*kcal/g mol*)
1. $C_3H_6(g) + H_2(g) \longrightarrow C_3H_8(g)$	−29.6
2. $C_3H_8(g) + 5O_2(g) \longrightarrow 3CO_2(g) + 4H_2O(l)$	−530.6
3. $H_2(g) + \frac{1}{2}O_2(g) \longrightarrow H_2O(l)$	−68.3
4. $H_2O(l) \longrightarrow H_2O(g)$	+10.5 ($\Delta \hat{H}_{vap}^\circ$)
5. C (diamond) + $O_2(g) \longrightarrow CO_2(g)$	−94.50
6. C (graphite) + $O_2(g) \longrightarrow CO_2(g)$	−94.05

Calculate the following:
  (a) The standard heat of formation of propylene (C_3H_6 gas)
  (b) The standard heat of combustion of propylene (C_3H_6 gas)
  (c) The net heating value of propylene in Btu/ft³ measured at 60°F and 30 in. Hg saturated with water vapor

Solution

(a) Reference state: 25°C, 1 atm; $C(\beta)$ (graphite) and $H_2(g)$ are assigned zero values.

$$\text{Desired}: 3C(\beta) + 3H_2(g) \longrightarrow C_3H_6(g)$$
$$\Delta H_f^\circ = ?$$

		kcal
(1) $C_3H_6(g) + H_2(g) \longrightarrow C_3H_8(g)$		$\Delta H_1 = -29.6$
(2) $C_3H_8(g) + 5O_2(g) \longrightarrow 3CO_2(g) + 4H_2O(l)$		$\Delta H_2 = -530.6$
(3) $-4[H_2(g) + \frac{1}{2}O_2(g) \longrightarrow H_2O(l)]$		$-4\Delta H_3 = +273.2$
(6) $-3[C(\beta) + O_2(g) \longrightarrow CO_2(g)]$		$-3\Delta H_6 = +282.2$

$(7) = (1) + (2) + (3) + (6)$:
$$C_3H_6(g) \longrightarrow 3C(\beta) + 3H_2(g) \qquad \Delta H_{rxn}^\circ = -4.8$$
$$\Delta H_f^\circ{}_{C_3H_6(g)} = +4.8 \text{ kcal/g mol}$$

(b) Reference state: 25°C, 1 atm; $CO_2(g)$ and $H_2O(l)$ have zero values.

$$\text{Desired}: C_3H_6(g) + \tfrac{9}{2}O_2(g) \longrightarrow 3CO_2(g) + 3H_2O(l)$$

		kcal
(1) $C_3H_6(g) + H_2(g) \longrightarrow C_3H_8(g)$		$\Delta H_1 = -29.6$
(2) $C_3H_8(g) + 5O_2(g) \longrightarrow 3CO_2(g) + 4H_2O(l)$		$\Delta H_2 = -530.6$
(3) $-[H_2(g) + \frac{1}{2}O_2(g) \longrightarrow H_2O(l)]$		$-\Delta H_3 = +68.3$

$(8) = (1) + (2) + (3)$:
$$C_3H_6(g) + \tfrac{9}{2}O_2(g) \longrightarrow 3CO_2(g) + 3H_2O(l) \qquad \Delta H_{rxn}^\circ = -491.9$$
$$\Delta H_c^\circ{}_{C_3H_6(g)} = -491.9 \text{ kcal/g mol}$$

(c) Net heating value

$$\text{Desired}: C_3H_6(g) + \tfrac{9}{2}O_2(g) \longrightarrow 3CO_2(g) + 3H_2O(g)$$

		kcal
(8) $C_3H_6(g) + \frac{9}{2}O_2(g) \longrightarrow 3CO_2(g) + 3H_2O(l)$		$\Delta H_8 = -491.9$
(4) $3H_2O(l) \longrightarrow 3H_2O(g)$		$3\Delta H_4 = +31.5$

$(9) = (8) + (4)$:
$$C_3H_6(g) + \tfrac{9}{2}O_2(g) \longrightarrow 3CO_2(g) + 3H_2O(g) \qquad \Delta H_{rxn}^\circ = -460.4$$
$$\Delta H_c' = -460,400 \text{ cal/g mol}$$
$$\Delta H_c' = (-460,400)(1.8) = -829,000 \text{ Btu/lb mol}$$

At 60°F, 30 in. Hg., saturated with water vapor,

$$\text{vapor pressure of } H_2O \text{ at } 60°F = 13.3 \text{ mm Hg}$$

$$\text{pressure on gas} = (760)\left(\frac{30.00}{29.92}\right) - 13.3$$
$$= 763 - 13.3 = 749.7 \simeq 750 \text{ mm Hg}$$

$$\text{molal volume} = \frac{359}{} \left| \frac{760}{750} \right| \frac{(460 + 60)}{(460 + 32)} = 384 \text{ ft}^3/\text{lb mol}$$

$$\text{heating value} = \frac{829,000}{384} = 2160 \text{ Btu/ft measured at } 60°F \text{ and } 30 \text{ in. Hg satd.}$$

At 60°F, 30 in. Hg, dry,

$$\text{molal volume} = 379 \text{ ft}^3/\text{lb mol}$$

$$\text{net heating value} = \frac{829,000}{379} = 2185 \text{ Btu/ft}^3$$

One of the common errors made in these thermochemical calculations is to forget that the standard state for heat of combustion calculations is liquid water, and that if gaseous water is present, a phase change (the heat of vaporization or heat of condensation) must be included in the calculations. Don't you forget!

With adequate data available you will find it simpler to use only the heats of combustion or only the heats of formation in the same algebraic calculation for a heat of reaction. Mixing these two sources of enthalpy change data, unless you take care, will only lead to error and confusion. Of course, when called upon, you should be able to calculate the heat of formation from the heat of combustion, or the reverse, being careful to remember and take into account the fact that the standard states for these two types of calculations are different. Since the heat of combustion values are so large, calculations made by subtracting these large values from each other can be a source of error. To avoid other major errors, carefully check all signs and make sure all equations are not only balanced but are written according to the proper conventions.

Self-Assessment Test

1. Calculate the standard heat of combustion for CH_4 given the following experimental data at 25°C and 1 atm:

$$H_2(g) + \tfrac{1}{2}O_2(g) \longrightarrow H_2O(g) \qquad\qquad \Delta H = -241.826 \text{ kJ/g mol}$$
$$C \text{ (graphite)} + O_2(g) \longrightarrow CO_2(g) \qquad\qquad \Delta H = -393.51 \text{ kJ/g mol}$$
$$CH_4(g) + 2O_2(g) \longrightarrow CO_2(g) + 2H_2O(g) \qquad \Delta H = -802.34 \text{ kJ/g mol}$$

2. Calculate the standard heat of reaction for the Sachse process (in which acetylene is made by partial combustion of LPG) from heat of combustion data:

$$C_3H_8(l) + 2O_2(g) \longrightarrow C_2H_2(g) + CO(g) + 3H_2O(l)$$

3. Can a standard heat of combustion ever be positive?

4. Calculate the standard heat of formation of C_3H_8 gas (propane) from standard heat of combustion data.

5. A synthetic gas analyzes 6.1% CO_2, 0.8% C_2H_4, 0.1% O_2, 26.4% CO, 30.2% H_2, 3.8% CH_4, and 32.6% N_2. What is the heating value of the gas measured at 60°F, saturated, when the barometer reads 30.0 in. Hg?

4.7-3 Heat of Reaction at Constant Pressure or Constant Volume

Your objectives in studying this section are to be able to:

1. Calculate the heat transfer, Q, from a bomb calorimeter (constant volume) or a steady flow calorimeter (constant pressure), Q_p, from theory or experimental data.
2. Calculate the heat of reaction for either process.

Consider the heat of reaction of a substance obtained in a bomb calorimeter, such as in a bomb in which the volume is constant but not the pressure. For such a process (the system is the material in the bomb), the general energy balance, Eq. (4.24), reduces to (with no work, mass flow, nor kinetic or potential energy effects)

$$\Delta U = U_{t_2} - U_{t_1} = Q_v \qquad (4.36)$$

Here Q_v designates the heat transferred from the bomb. Next, let us assume that the heat of reaction is determined in a steady-state flow calorimeter with $1 =$ entering fluid and $2 =$ exiting fluid, and with $W = 0$, $\Delta P = 0$, and $\Delta K = 0$. Then if the process takes place at constant pressure the general energy balance reduces to

$$\Delta H = Q_p$$

Here Q_p designates the heat transferred from the flow calorimeter.

If we subtract Q_v from Q_p and use $H = U + pV$, we find that

$$Q_p - Q_v = \Delta H_p - \Delta U_v = \Delta H_p - [\Delta(H - pV)]_v$$
$$= (H_2 - H_1) - [(H - pV)_{t_2} - (H - pV)_{t_1}] \qquad (4.37)$$

Suppose, furthermore, that the enthalpy change for the batch, constant-volume process is made identical to the enthalpy difference between the outlet and inlet in the flow process by a suitable adjustment of the temperature of the water baths surrounding the calorimeters. Then

$$H_{t_2} - H_{t_1} = H_2 - H_1$$

and

$$Q_p - Q_v = (pV)_{t_2} - (pV)_{t_1} \qquad (4.38)$$

To evaluate the terms on the right-hand side of Eq. (4.38), we can assume for solids and liquids that the $\Delta(pV)$ change is negligible and can be ignored. Therefore, the only change which must be taken into account is for gases present as products and/or reactants. If, for simplicity, the gases are assumed to be ideal, then at constant temperature

$$pV = nRT$$
$$\Delta(pV) = \Delta n\ RT$$

and

$$Q_p - Q_v = \Delta n\ (RT) \tag{4.39}$$

Equation (4.39) gives the difference between the heat of reaction for the constant-pressure experiment and the constant-volume experiment.

EXAMPLE 4.30 Difference between Heat of Reaction at Constant Pressure and at Constant Volume

Find the difference between the heat of reaction at constant pressure and at constant volume for the following reaction at 25°C (assuming that it could take place):

$$C(s) + \tfrac{1}{2}O_2(g) \longrightarrow CO(g)$$

Solution

Basis: 1 mole of C(s)

Examine Fig. E4.30. The system is the bomb; $Q = +$ when heat is absorbed by the bomb.

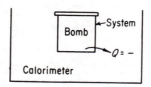

Fig. E4.30

From Eq. (4.39), $Q_p - Q_v = \Delta n(RT)$.

$$\Delta n = 1 - \tfrac{1}{2} = +\tfrac{1}{2}$$

since C is a solid.

$$Q_p - Q_v = \tfrac{1}{2}RT = 0.5[8.31\ \text{J/(g mol)(K)}](298\ \text{K}) = 1238\ \text{J/g mol}$$

If the measured heat evolved from the bomb was 111,759 J,

$$Q_p = Q_v + 1238 = -111,759 + 1238 = -110,521\ \text{J}$$

The size of the this correction is relatively insignificant compared to either the quantity Q_p or Q_v. In any case, the heat of reaction calculated from the bomb experiment is

$$\Delta \hat{H}_{\text{rxn}_v} = \frac{Q_v}{n} = \frac{-111{,}759 \text{ J}}{\text{g mol C(s)}}$$

Since reported heats of reaction are normally for constant-pressure processes, the value of $-111{,}759$ would not be reported, but instead you would report that

$$\Delta \hat{H}_{\text{rxn}_p} = \frac{Q_p}{n} = \frac{-110{,}521 \text{ J}}{\text{g mol C(s)}}$$

Self-Assessment Test

1. Is Q_p always bigger than Q_v? Explain.

2. The Claude ammonia synthesis gas reaction

$$CO(g) + 3H_2(g) \longrightarrow CH_4(g) + H_2O(l)$$

is carried out in a constant-volume bomb reactor. The molal ratio of H_2 to CO in the bomb at the start is 3 : 1. The reaction is complete. The initial and final temperature is 25°C. The initial pressure is 4 atm. The volume of the bomb is 1.50 liters.

     (a) Calculate the final pressure in the bomb reactor.

     (b) Calculate the heat transfer to or from the reactor—state which.

4.7-4 Incomplete Reactions

> *Your objectives in studying this section are to be able to:*
>
> 1. Calculate the heat of reaction at standard conditions for processes in which the reactants are not fed in stoichiometric quantities and the reaction may not go to completion.
> 2. Solve simple problems combining both material and energy balances.

If an incomplete reaction occurs, you should calculate the standard heat of reaction only for the products which are actually formed from the reactants that actually react. In other words, only the portion of the reactants that actually undergo some change and liberate or absorb some energy are to be considered in calculating the overall standard heat of the reaction. If some material passes through unchanged, you should not include it in the *standard heat of reaction* calculations (however, when the reactants or products are at conditions *other* than 25°C and 1 atm, whether they react or not, you must include them in the enthalpy calculations as explained in Sec. 4.7.5).

EXAMPLE 4.31 Incomplete Reactions

An iron pyrite ore containing 85.0% FeS_2 and 15.0% gangue (inert dirt, rock, etc.) is roasted with an amount of air equal to 200% excess air according to the reaction

$$4FeS_2 + 11O_2 \longrightarrow 2Fe_2O_3 + 8SO_2$$

in order to produce SO_2. All the gangue plus the Fe_2O_2 end up in the solid waste product (cinder), which analyzes 4.0% FeS_2. Determine the standard heat of reaction per kilogram of ore.

Solution

The material balance for the problem must be worked out prior to determining the standard heat of reaction.

Basis: 100 kg of ore

ore component	kg = percent	mol. wt.	kg mol
FeS_2	85.0	120	0.708
Gangue	15.0		
Total	100.0		

Required O_2: $\dfrac{0.708 \text{ kg mol } FeS_2}{} \left| \dfrac{11 \text{ kg mol } O_2}{4 \text{ kg mol } FeS_2} \right. = 1.95 \text{ kg mol } O_2$

Entering O_2: $(1 + 2)(1.95) = 5.84 \text{ kg mol } O_2$

Entering N_2: $5.84\left(\dfrac{79}{21}\right) = 22.0 \text{ kg mol } N_2$

The solid waste comprises Fe_2O_3, gangue, and unburned F_2S_3. We need to compute the amount of unburned FeS_2 on the basis of 100 kg of ore. Let x = kg of FeS_2 that does not burn. Then the cinder comprises

component	kg
FeS_2	x
Gangue	15
Fe_2O_3: $\dfrac{(85 - x) \text{ kg } FeS_2 \text{ burned}}{120 \dfrac{\text{kg } FeS_2}{\text{kg mol } FeS_2}} \left| \dfrac{2 \text{ mol } Fe_2O_3}{4 \text{ mol } FeS_2} \right| \dfrac{160 \text{ kg } Fe_2O_3}{\text{kg mol } Fe_2S_3} = \left(\dfrac{2}{3}\right)(85 - x)$	

Thus

$$0.040 = \frac{x}{x + 15 + (\frac{2}{3})(85 - x)}$$

$$x = 2.91 \text{ kg } FeS_2$$

$$\tfrac{2}{3}(85 - x) = 54.7 \text{ kg } Fe_2O_3$$

The composition of the solid waste is

solid waste component	kg	mol. wt.	kg mol
FeS_2	2.91	120	0.0242
Fe_2O_3	54.7	160	0.342
Gangue	15		

Finally, the FeS_2 that is oxidized is

$$\frac{(85 - 2.91) \text{ kg FeS}_2 \mid 1 \text{ kg mol FeS}_2}{120 \text{ kg FeS}_2} = 0.684 \text{ kg mol FeS}_2$$

Corresponding to 0.68 kg mol of FeS_2 oxidized:

$$\frac{0.684 \text{ kg mol FeS}_2 \mid 8 \text{ kg mol SO}_2}{4 \text{ kg mol FeS}_2} = 1.37 \text{ kg mol SO}_2 \text{ produced}$$

$$\frac{0.684 \text{ kg mol FeS}_2 \mid 11 \text{ kg mol O}_2}{4 \text{ kg mol FeS}_2} = 1.88 \text{ kg mol O}_2 \text{ used}$$

Tabulated heats of formation in kilocalories per gram mole are:

component:	$FeS_2(c)$	$O_2(g)$	$Fe_2O_3(c)$	$SO_2(g)$
$\Delta \hat{H}_f^\circ$	-42.520	0	-196.500	-70.960

$$\Delta H_{rxn}^\circ = (1.37)(-70.960) + (0.342)(-196.500) - (0.685)(-42.520)$$
$$= -135.292 \text{ kcal}$$
$$\Delta H_{rxn}^\circ = 1.353 \text{ kcal/kg ore}$$

Note that only the moles reacting and produced are used in computing the heat of reaction. The unoxidized FeS_2 is not included, nor is the N_2. Why does the O_2 used contribute a value of zero to the ΔH_{rxn}°? Because the ΔH_f° of the O_2 is zero.

Self-Assessment Test

1. A dry low-Btu gas with the analysis of CO: 20%, H_2: 20% and N_2: 60% is burned with 200% excess dry air which enters at 25°C. If the exit gases leave at 25°C, calculate the standard heat of reaction per unit volume of entering gas measured at standard conditions (25°C and 1 atm).

4.7-5 Energy Balance When the Products and Reactants Are Not at 25°C

Your objectives in studying this section are to be able to:

1. Calculate heats of reaction at other than standard temperature and pressure.
2. Calculate how much material must be introduced to provide a prespecified quantity of heat transfer.
3. Apply the general energy balance to processes involving reactions.

You no doubt realize that the standard state of 25°C for the heats of formation is only by accident the temperature at which the products enter and the reactants leave a process. In most instances the temperatures of the materials entering and leaving will be higher or lower than 25°C. However, since enthalpies (and hence heats of reaction) are point functions, you can use Eq. (4.24) together with Eq. (4.33), (4.34), or (4.35), to answer questions about the process being analyzed. Typical questions might be:

(a) What is the heat of reaction at a temperature other than 25°C, but still at 1 atm?

(b) What is the temperature of an incoming or exit stream?

(c) What is the temperature of the reaction?

(d) How much material must be introduced to provide a specified amount of heat transfer?

Consider the process illustrated in Fig. 4.16, for which the reaction is

$$aA + bB \longrightarrow cC + dD$$

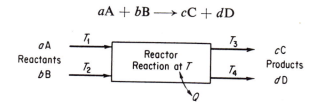

Fig. 4.16 Process with reaction.

In employing Eqs. (4.24) and (4.33), you should always **first choose a reference state** at which the heat of reaction is known. This usually turns out to be 25°C and 1 atm. (If no reaction takes place, the reference state can be an inlet or outlet stream temperature.) Then the next step is to calculate the enthalpy changes for each stream entering and leaving *relative to this reference state*. Finally, the enthalpy changes are summed up, including the heat of reaction calculated in the standard state by Eq. (4.33), (4.34), or (4.35) and the other terms in Eq. (4.24) are introduced if pertinent. Usually, it proves convenient in calculating the enthalpy change to compute the sensible heat changes and the heat of reaction separately. Methods you can use to determine the sensible heat ΔH values for the individual streams are:

(a) Obtain the enthalpy values from a set of published tables (e.g., the steam tables or tables such as in Appendix D).

(b) Use $\Delta H = C_{p_{m2}}(T_2 - T_{\text{ref}}) - C_{p_{m1}}(T_1 - T_{\text{ref}})$, as discussed in Sec. 4.2.

(c) Analytically, graphically, or numerically, find

$$\Delta H = \int_{T_1}^{T_2} C_p \, dT$$

for each component individually, using the respective heat capacity equations.

Any phase changes taking place among the reactants or products not accounted for in the thermochemical calculations must also be taken into account in applying the general energy balance.

Let us first demonstrate how to calculate the heat of reaction at a temperature other than 25°C. By this we mean that the reactants enter and the products leave at the same temperature—a temperature different from the standard state of 25°C. Figure 4.17 illustrates the information flow for the calculations corresponding to

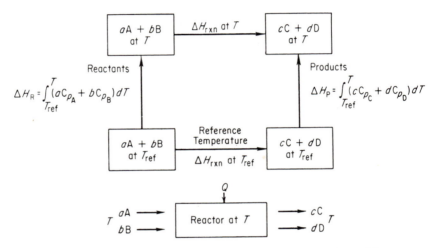

Fig. 4.17 Heat of reaction at a temperature other than standard conditions.

those associated with Eq. (4.33) assuming a steady-state process ($\Delta E = 0$), no kinetic or potential energy changes, and $W = 0$. The general energy balance, Eq. (4.24), reduces to

$$Q = \Delta H$$

or

$$Q = (\Delta H_P - \Delta H_R) + (\Delta \hat{H}_{rxn_{T_{ref}}}) \qquad (4.40)$$

in terms of the notation in Fig. 4.17. By definition, Q, as calculated by Eq. (4.40), is equal to the "heat of reaction at the temperature T," so that

$$\Delta H_{rxn_T} = \Delta H_{rxn_{T_{ref}}} + \Delta H_P - \Delta H_R \qquad (4.41)$$

To indicate a simplified way to calculate $\Delta H_P - \Delta H_R$, suppose that the heat capacity equations are expressed as

$$C_p = \alpha + \beta T + \gamma T^2 \tag{4.42}$$

Then, to obtain $\Delta H_P - \Delta H_R$, we add up the enthalpy changes for the products and subtract those for the reactants. Rather than integrate separately, let us consolidate like terms as follows. Each heat capacity equation is multiplied by the proper number of moles:

$$aC_{p_A} = a(\alpha_A + \beta_A T + \gamma_A T^2) \tag{4.43a}$$

$$bC_{p_B} = b(\alpha_B + \beta_B T + \gamma_B T^2) \tag{4.43b}$$

$$cC_{p_C} = c(\alpha_C + \beta_C T + \gamma_C T^2) \tag{4.43c}$$

$$dC_{p_D} = d(\alpha_D + \beta_D T + \gamma_D T^2) \tag{4.43d}$$

Then we define a new term ΔC_p which is equal to

original expression   equivalent
new term

$$cC_{p_C} + dC_{p_D} - (aC_{p_A} + bC_{p_B}) = \Delta C_p$$

and

$$[(c\alpha_C + d\alpha_D) - (a\alpha_A + b\alpha_B)] = \Delta\alpha$$

$$T[(c\beta_C + d\beta_D) - (a\beta_A + b\beta_B)] = T\,\Delta\beta \tag{4.44}$$

$$T^2[(c\gamma_C + d\gamma_D) - (a\gamma_A + b\gamma_B)] = T^2\,\Delta\gamma$$

Simplified, ΔC_p can be expressed as

$$\Delta C_p = \Delta\alpha + \Delta\beta T + \Delta\gamma T^2 \tag{4.45}$$

Furthermore,

$$\Delta H_P - \Delta H_R = \int_{T_R}^{T} \Delta C_p\, dT = \int_{T_R}^{T} (\Delta\alpha + \Delta\beta T + \Delta\gamma T^2)\, dT$$

$$= \Delta\alpha(T - T_R) + \frac{\Delta\beta}{2}(T^2 - T_R^2) + \frac{\Delta\gamma}{3}(T^3 - T_R^3) \tag{4.46}$$

where we let $T_R = T_{\text{ref}}$ for simplicity.

If the integration is carried out without definite limits,

$$\Delta H_P - \Delta H_R = \int (\Delta C_p)\, dT = \Delta\alpha T + \frac{\Delta\beta}{2}T^2 + \frac{\Delta\gamma}{3}T^3 + C \tag{4.47}$$

where C is the integration constant.

Finally, ΔH_{rxn} at the new temperature T is

$$\Delta H_{rxn_T} = \Delta H_{rxn_{T_R}} + \Delta\alpha(T - T_R) + \frac{\Delta\beta}{2}(T^2 - T_R^2) + \frac{\Delta\gamma}{3}(T^3 - T_R^3) \quad (4.48a)$$

or

$$\Delta H_{rxn_T} = \Delta H_{rxn_{T_R}} + \Delta\alpha T + \frac{\Delta\beta}{2}T^2 + \frac{\Delta\gamma}{3}T^3 + C \quad (4.48b)$$

as the case may be. Using Eq. (4.48a) and knowing ΔH_{rxn} at the reference temperature T_R, you can easily calculate ΔH_{rxn} at any other temperature.

Equation (4.48b) can be consolidated into

$$\Delta H_{rxn_T} = \Delta H_0 + \Delta\alpha T + \frac{\Delta\beta}{2}T^2 + \frac{\Delta\gamma}{3}T^3 \quad (4.49)$$

where

$$\Delta H_0 = \Delta H_{rxn_{T_R}} + C$$

Now, if ΔH_{rxn} is known at any temperature T, you can calculate ΔH_0 as follows:

$$\Delta H_0 = \Delta H_{rxn_T} - \Delta\alpha(T) - \frac{\Delta\beta}{2}(T^2) - \frac{\Delta\gamma}{3}T^3 \quad (4.50)$$

Then you can use this value of ΔH_0 to assist in the calculation of ΔH_{rxn} at any other temperature. Probably the easiest way to compute the necessary enthalpy changes is to use enthalpy data obtained directly from published tables. Do not forget to take into account phase changes, if they take place, in the enthalpy calculations. If there is a phase change in one or more of the streams entering or leaving the process, you can conceptually think of the enthalpy changes that take place as shown in Fig. 4.18.

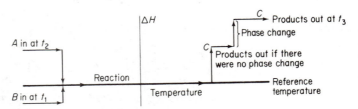

Fig. 4.18 Enthalpy changes in a reacting system with a phase change taking place.

EXAMPLE 4.32 Calculation of Heat of Reaction at a Temperature Different from Standard Conditions

An inventor thinks he has developed a new catalyst that can make the gas-phase reaction

$$CO_2 + 4H_2 \longrightarrow 2H_2O + CH_4$$

proceed with 100% conversion. Estimate the heat that must be provided or removed if the gases enter and leave at 500°C.

Solution

In effect we need to calculate heat of reaction at 500°C from Eq. (4.41) or Q in Eq. (4.40). For illustrative purposes we use the technique based on Eqs. (4.49) and (4.50).

$$\text{Basis: 1 g mol of } CO_2(g)$$

tabulated data	$CO_2(g) + 4H_2(g) \longrightarrow 2H_2O(g) + CH_2(g)$			
$-\Delta \hat{H}_f^\circ$ (J/g mol)	393,513	0	241,827	74,848

$$\Delta \hat{H}_{rxn_{298°K}} = [-74,848 - 2(241,827)] - [4(0) - 393,513]$$
$$= -164,989 \text{ J/g mol } CO_2$$

First we shall calculate ΔC_p; units are $\dfrac{\text{cal}}{(\text{g mol})(\text{K})}$

$$C_{p\,CO_2} = 6.393 + 10.100 \times 10^{-3}T - 3.405 \times 10^{-6}T^2 \qquad T \text{ in K}$$
$$C_{p\,H_2} = 6.424 + 1.039 \times 10^{-3}T - 0.078 \times 10^{-6}T^2 \qquad T \text{ in K}$$
$$C_{p\,H_2O} = 6.970 + 3.464 \times 10^{-3}T - 0.483 \times 10^{-6}T^2 \qquad T \text{ in K}$$
$$C_{p\,CH_4} = 3.204 + 18.41 \times 10^{-3}T - 4.48 \times 10^{-6}T^2 \qquad T \text{ in K}$$

$$\Delta\alpha = [1(3.204) + 2(6.970)] - [1(6.393) + 4(6.424)]$$
$$= -14.945$$
$$\Delta\beta = [1(18.41) + 2(3.464)](10^{-3}) - [1(10.100) + 4(1.039)](10^{-3})$$
$$= 11.087 \times 10^{-3}$$
$$\Delta\gamma = [1(-4.48) + 2(-0.483)](10^{-6}) - [1(-3.405) + 4(-0.078)](10^{-6})$$
$$= -1.729 \times 10^{-6}$$
$$\Delta C_p = -14.945 + 11.082 \times 10^{-3}T - 1.729 \times 10^{-6}T^2$$

Next we find ΔH_0, using as a reference temperature 298 K.

$$\Delta H_0 = \Delta H_{rxn_{298}} - \Delta\alpha T - \frac{\Delta\beta}{2}T^2 - \frac{\Delta\gamma}{3}T_3$$

$$= -164,989 - 4.184\left[(-14.945)(298) - \frac{11.082 \times 10^{-3}}{2}(298)^2\right.$$

$$\left. - \frac{-1.729 \times 10^{-6}}{3}(298)^3\right]$$

$$= -164,989 + 18,634 - 2059 + 64 = -148,350 \text{ J}$$

Then, with ΔH_0 known, the ΔH_{rxn} at 773 K can be determined.

$$\Delta H_{rxn_{773} \text{ K}} = \Delta H_0 + \Delta \alpha T + \frac{\Delta \beta}{2} T^2 + \frac{\Delta \gamma}{3} T_3$$

$$= -148{,}350 - 4.184 \left[14.945(773) + \frac{11.082 \times 10^{-3}}{2}(773)^2 \right.$$

$$\left. - \frac{1.729 \times 10^{-6}}{3}(773)^3 \right]$$

$$= -183{,}940 \text{ J}$$

or 183,940 J/g mol CO_2 must be removed.

EXAMPLE 4.33 Calculation of Heat of Reaction at a Temperature Different from Standard Conditions

Repeat the calculation of the previous example using enthalpy values from Table 4.4b and Appendix D.

Solution

The heat of reaction at 25°C and 1 atm from the previous example is

$$\Delta \hat{H}_{rxn_{298K}} = -164{,}989 \frac{\text{kJ}}{\text{kg mol}}$$

$\Delta \hat{H}$ (kJ/kg mol); reference is 0°C

temp. (°C)	CO_2	H_2	H_2O	CH_4
25	912	718	837	879
500	22,342	14,640	17,799	23,974

From Eq. (4.41),

$$\Delta H_{rxn_{500°C}} = \Delta H_{rxn_{25°C}} + \Delta H_{products} - \Delta H_{reactants}$$

$$= -164{,}989 + [(1)(23{,}974 - 879) + (2)(17{,}799 - 837)]$$

$$- [(1)(22{,}342 - 912) + (4)(14{,}640 - 718)]$$

$$= -185{,}088 \frac{\text{kJ}}{\text{kg mol } CO_2}$$

Note that the enthalpies of the products and of the reactants are both based on the reference temperature of 25°C. The answer is not quite the same as in Example 4.32 because the heat capacity data used there were not quite the same as those used in calculating the ΔH values in the tables.

EXAMPLE 4.34 Application of the Energy Balance to a Process in Which a Reaction Occurs

Carbon monoxide at 50°F is completely burned at 2 atm pressure with 50% excess air that is at 1000°F. The products of combustion leave the combustion chamber at 800°F. Calculate the heat evolved from the combustion chamber expressed as British thermal units per pound of CO entering.

Solution

Steps 1, 2, 3, and 4: Refer to Fig. E 4.34a. A material balance is needed before an energy balance can be made.

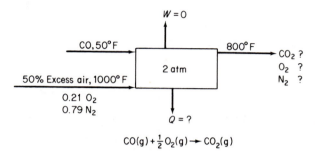

Fig. E4.34a

Steps 5 and 8: We have three elements, hence can make three independent balances. Because we have three unknown compositions, a unique solution exists, a solution that can be obtained by direct addition or subtraction.

Step 6:

Basis: 1 lb mol of CO (easier to use than 1 lb of CO)

Steps 7 and 9:

Amount of air entering:

$$\frac{1 \text{ lb mol CO}}{} \left| \frac{0.5 \text{ lb mol O}_2}{\text{lb mol CO}} \right| \frac{1.5 \text{ lb mol O}_2 \text{ used}}{1.0 \text{ lb O}_2 \text{ mol needed}} \left| \frac{1 \text{ lb mol air}}{0.21 \text{ lb mol O}_2} \right.$$

$$= 3.57 \text{ lb mol air}$$

Amount of O_2 leaving, unreacted:

$$0.21(3.57) - 0.5 = 0.25 \text{ lb mol O}_2$$

Amount of N_2 leaving:

$$0.79(3.57) = 2.82 \text{ lb mol N}_2$$

Amount of CO_2 leaving:

$$1 \text{ lb mol}$$

Steps 3a and 3b: The general energy balance

$$\Delta E = -\Delta[(\hat{H} + \hat{P} + \hat{K})m] + Q - W$$

reduces to $Q = \Delta H$. Why? Only Q is unknown.

Step 9 Continued: Enthalpy data have been taken from Table 4.4a and Appendix D. The heat of reaction at 25°C (77°F) and 1 atm from Example 4.23 after conversion of units is −67,636 cal/g mol of CO or −121,745 Btu/lb mol of CO. We can assume that the slightly higher pressure of 2 atm has no effect on the heat of reaction or the enthalpy values. Figure E4.34b shows the sensible heat (enthalpy) values for the entering and exiting materials.

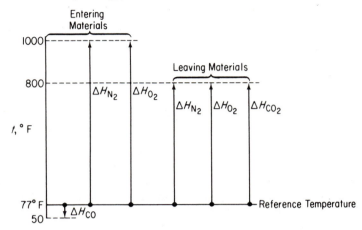

Fig. E4.34b

ΔH (Btu/lb mol; reference is 32°F)

temp. (°F)	CO	air	O_2	N_2	CO_2
50	125.2	—	—	—	—
77	313.3	312.7	315.1	313.2	392.2
800	—	—	5690	5443	8026
1000	—	6984	—	—	—

$$Q = \Delta H_{rxn\,77°F} + \Delta H_{products} - \Delta H_{reactants}$$

(a)
$$\Delta \hat{H}_{rxn\,77°F} = -121,745 \text{ Btu/lb mol}$$

(b)
$$\Delta H_{products} = \Delta H_{800°F} - \Delta H_{77°F}$$

$$= \overbrace{(1)(8026 - 392.2)}^{CO_2} + \overbrace{(2.82)(5443 - 313.2)}^{N_2}$$

$$+ \overbrace{(0.25)(5690 - 315.1)}^{O_2}$$

$$= 7634 + 14,469 + 1344$$

$$= 23,447 \text{ Btu/lb mol}$$

(c)
$$\Delta H_{reactants} = \overbrace{\Delta H_{1000°F} - \Delta H_{77°F}}^{air} + \overbrace{\Delta H_{50°F} - \Delta H_{77°F}}^{CO}$$

$$= (3.57)(6984 - 312.7) + (1)(125.2 - 313.3)$$

$$= 23,817 - 188.1 = 23,629 \text{ Btu/lb mol}$$

(d)
$$Q = -121,745 + 23,447 - 23,629$$
$$= -121,927 \text{ Btu/lb mol}$$

$$\frac{-121,927 \text{ Btu}}{\text{lb mol CO}} \left| \frac{1 \text{ lb mol CO}}{28 \text{ lb CO}} \right. = -4355 \text{ Btu/lb CO}$$

Self-Assessment Test

1. Calculate the heat of reaction at 1000 K and 1 atm for the reaction

$$H_2(g) + \tfrac{1}{2}O_2(g) \longrightarrow H_2O(g)$$

2. Methane is burned in a furnace with 100% excess dry air to generate steam in a boiler. Both the air and the methane enter the combustion chamber at 500°F and 1 atm, and the products leave the furnace at 2000°F and 1 atm. If the effluent gases contain only CO_2, H_2O, O_2, and N_2, calculate the amount of heat absorbed by the water to make steam per pound of methane burned.

3. A mixture of metallic aluminum and Fe_2O_3 can be used for high-temperature welding. Two pieces of steel are placed end to end and the powdered mixture is applied and ignited. If the temperature desired is 3000°F and the heat loss is 20% of $(\Delta H_{rxn_{25°C}} + \Delta H_{prod} - \Delta H_{react})$ by radiation, what weight in pounds of the mixture (used in the molecular proportions of $2Al + 1Fe_2O_3$) must be used to produce this temperature in 1 lb of steel to be welded? Assume that the starting temperature is 65°F.

$$2Al + Fe_2O_3 \longrightarrow Al_2O_3 + 2Fe$$

Data	C_p, Solid [Btu/(lb)(°F)]	Heat of Fusion (Btu/lb)	Melting Point (°F)	C_p, Liquid [Btu/(lb)(°F)]
Fe and steel	0.12	86.5	2800	0.25
Al_2O_3	0.20	—	—	—

4.7-6 Temperature of a Reaction

> **Your objectives in studying this section are to be able to:**
>
> 1. Calculate the temperature of an entering or exiting steam of a process given all the other necessary information to obtain a unique solution.
> 2. Calculate the adiabatic reaction temperature.

We are now equipped to determine what is called the *adiabatic reaction temperature*. This is the temperature obtained inside the process when (1) the reaction is carried out under adiabatic conditions, that is, there is no heat interchange between the container in which the reaction is taking place and the surroundings; and (2) when there are no other effects present, such as electrical effects, work, ionization, free radical formation, and so on. In calculations of flame temperatures for combustion reactions, the adiabatic reaction temperature *assumes complete combustion*. Equilibrium considerations may dictate less than complete combustion for an actual case. For example, the adiabatic flame temperature for the combustion of CH_4 with theoretical air has been calculated to be 2010°C; allowing for incomplete combustion, it would be 1920°C. The actual temperature when measured is 1885°C.

The adiabatic reaction temperature tells us the temperature ceiling of a process. We can do no better, but of course the actual temperature may be less. The adiabatic reaction temperature helps us select the types of materials that must be specified for the container in which the reaction is taking place. Chemical combustion with air produces gases at a maximum temperature of 2500 K which can be increased to 3000 K with the use of oxygen and more exotic oxidants, and even this value can be exceeded although handling and safety problems are severe. Applications of such hot gases lie in the preparation of new materials, micromachining, welding using laser beams, and the direct generation of electricity using ionized gases as the driving fluid.

To calculate the adiabatic reaction temperature, you assume that all the energy liberated from the reaction at the reference temperature plus that brought in by the entering stream (relative to the same base temperature) is available to raise the temperature of the products. We assume that the products leave at the temperature of the reaction, and thus if you know the temperature of the products, you automatically know the temperature of the reaction. In effect, for this adiabatic process we can apply Eq. (4.40).

Since no heat escapes and the energy liberated is allowed only to increase the enthalpy of the products of the reaction, we have

$$\Delta H_{products} = \boxed{\Delta H_{reactants} - \Delta H_{rxn}^{\circ}} \qquad (4.51)$$

"energy pool"

Any phase changes that take place and are not accounted for by the chemical equation must be incorporated into the "energy pool." Because of the character of the information available, the determination of the adiabatic reaction temperature or flame temperature may involve a trial-and-error solution; hence the iterative solution of adiabatic flame temperature problems is often carried out on the computer.

EXAMPLE 4.35 Adiabatic Flame Temperature

Calculate the theoretical flame temperature for CO burned at constant pressure with 100% excess air, when the reactants enter at 100°C.

Solution

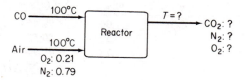

Fig. E4.35

$$CO(g) + \tfrac{1}{2}O_2(g) \longrightarrow CO_2(g)$$

Basis: 1 g mol of CO(g); ref. temp. 25°C

Material balance:

entering reactants			*exit products*	
component	*moles*		*component*	*moles*
CO(g)	1.00		CO_2(g)	1.00
O_2(req.)	0.50		O_2(g)	0.50
O_2(xs)	0.50		N_2(g)	3.76
O_2(total)	1.00			
N_2	3.76			
Air	4.76			

From Example 4.23, ΔH_{rxn} at 25°C $= -282{,}990$ J. Then, for Eq. (4.51), $\Delta H_{reactants}$:

reactants (100°C, 373 K)

component	moles	$\Delta \hat{H}$ (J/g mol)	ΔH (J)
CO(g)	1.00	2167	2,167
Air	4.76	2196	10,443
		$\Sigma \Delta H_R =$	12,610

$$\Delta H_{products} = 12{,}610 + 282{,}990 = 295{,}600 \text{ J}$$

To find the temperature which yields a $\Delta H_{products}$ of 295,600 J, the simplest procedure is to assume various values of the exit temperature of the products until the $\Sigma \Delta H_P = 295{,}600$. Assume that TFT (theoretical flame temperature) = 2000 K.

component	moles	$\Delta \hat{H}$ (J/g mol)	ΔH (J)
CO_2	1.00	92,470	92,470
O_2	0.50	59,915	29,960
N_2	3.76	56,900	213,950
		$\Sigma \Delta H_P =$	336,380

Assume that TFT = 1750 K.

component	moles	$\Delta \hat{H}$ (J/g mol)	ΔH (J)
CO_2	1.00	77,450	77,450
O_2	0.50	50,555	25,280
N_2	3.76	47,940	180,260

$$\sum \Delta H_P = 282,990$$

Make a linear interpolation:

$$\text{TFT} = 1750 + \frac{295,600 - 282,990}{336,380 - 282,990}(250) = 1750 + 59 = 1809$$

$$= 1809 \text{ K} \Leftrightarrow 1536°C$$

Self-Assessment Test

1. Calculate the theoretical flame temperature when hydrogen burns with 400% excess dry air at 1 atm. The reactants enter at 100°C.

Section 4.8 Heats of Solution and Mixing

Your objectives in studying this section are to be able to:

1. Distinguish between ideal solutions and real solutions.
2. Calculate the heat of mixing, or the heat of dissolution, at standard conditions given the moles of the materials forming the mixture.
3. Calculate the standard integral heat of solution.
4. Define the standard integral heat of solution at infinite dilution.
5. Apply the energy balance to problems in which the heat of mixing is significant.

So far in our energy calculations, we have been considering each substance to be a completely pure and separate material. The physical properties of an ideal solution or mixture may be computed from the sum of the properties in question for the individual components. For gases, mole fractions can be used as weighting values, or, alternatively, each component can be considered to be independent of the others. In most instances so far in this book we have used the latter procedure. Using the former technique, as we did in a few instances, we could write down, for the heat capacity of an ideal mixture,

$$C_{p \text{ mixture}} = x_A C_{p_A} + x_B C_{p_B} + x_C C_{p_C} + \cdots \qquad (4.52)$$

or, for the enthalpy,

$$\Delta \hat{H}_{\text{mixture}} = x_A \, \Delta \hat{H}_A + x_B \, \Delta \hat{H}_B + x_C \, \Delta \hat{H}_C + \cdots \tag{4.53}$$

These equations are applicable to ideal mixtures only.

When two or more pure substances are mixed to form a gas or liquid solution, we frequently find heat is absorbed or evolved upon mixing. Such a solution would be called a "real" solution. The total heat of mixing (ΔH_{mixing}) has to be determined experimentally, but can be retrieved from tabulated experimental (smoothed) results, once such data are available. This type of energy change has been given the formal name *heat of solution* when one substance dissoves in another; and there is also the negative of the heat of solution, the *heat of dissolution*, for a substance that separates from a solution. Tabulated data for heats of solution appear in Table 4.12 in terms

TABLE 4.12   HEAT OF SOLUTION OF HCL*
(AT 25°C AND 1 ATM)

Composition	Total Moles H$_2$O Added to 1 mole HCl	$-\Delta \hat{H}°$ for Each Incremental Step (J/g mol)	Integral Heat of Solution: Cumulative $-\Delta \hat{H}°$ (J/g mol)
HCl(g)	0		
HCl·1J$_2$O	1	26,225	26,225
HCl·2H$_2$O	2	22,593	48,818
HCl·3H$_2$O	3	8,033	56,851
HCl·4H$_2$O	4	4,351	61,202
HCl·5H$_2$O	5	2,845	64,047
HCl·8H$_2$O	8	4,184	68,231
HCl·10H$_2$O	10	1,255	69,486
HCl·15H$_2$O	15	1,503	70,989
HCl·25H$_2$O	25	1,276	72,265
HCl·50H$_2$O	50	1,013	73,278
HCl·100H$_2$O	100	569	73,847
HCl·200H$_2$O	200	356	74,203
HCl·500H$_2$O	500	318	74,521
HCl·1000H$_2$O	1,000	163	74,684
HCl·5000H$_2$O	5,000	247	74,931
HCl·50,000H$_2$O	50,000	146	75,077
HCl·∞H$_2$O		67	75,144

SOURCE: *National Bureau of Standards Circular 500* (see Reference 21 in Table 4.5).

*To convert to cal/g mol multiply by 0.2390.

of energy per mole for consecutively added quantities of solvent to solute; the gram mole refers to the gram mole of solute. Heats of solution are somewhat similar to heats of reaction in that an energy change takes place because of differences in the forces of attraction of the solvent and solute molecules. Of course, these energy changes are much smaller than those we find accompanying the breaking and combining of chemical bonds. Heats of solution are conveniently treated in exactly the same way as are the heats of reaction in the energy balance.

The solution process can be represented by an equation such as the following:

$$HCl(g) + 5H_2O \longrightarrow HCl \cdot 5H_2O$$

or

$$HCl(g) + 5H_2O \longrightarrow HCl(5H_2O)$$
$$\Delta H^{\circ}_{soln} = -64,047 \text{ J/g mol HCl(g)}$$

The expression $HCl(5H_2O)$ means that 1 mole of HCl has been dissolved in 5 moles of water, and the enthalpy change for the process is $-64,047$ J/g mol of HCl. Table 4.12 shows the heat of solution for various cumulative numbers of moles of water added to 1 mole of HCl.

The *standard integral heat of solution* is the cumulative ΔH°_{soln} as shown in the last column for the indicated number of molecules of water. As successive increments of water are added to the mole of HCl, the cumulative heat of solution (the integral heat of solution) increases, but the incremental enthalpy change decreases as shown in Table 4.12. Note that both the reactants and products have to be at standard conditions. The heat of dissolution would be just the negative of these values. The integral heat of the solution is plotted in Fig. 4.19, and you can see that an asymptotic

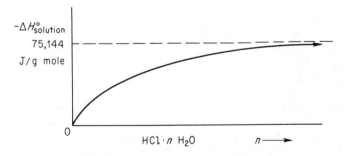

Fig. 4.19 Integral heat of solution of HCl in water.

value is approached as the solution becomes more and more dilute. At infinite dilution this value is called the *standard integral heat of solution at infinite dilution* and is $-75,144$ J/g mol of HCl. What can you conclude about the reference state for pure HCl from Fig. 4.19? In Appendix H are other tables presenting standard integral heat of solution data. Since the energy changes for heats of solution are point functions, you can easily look up any two concentrations of HCl and find the energy change caused by adding or subtracting water.

EXAMPLE 4.36 Heat of Solution

Calculate the standard heat of formation of HCl in 5 g mol of water.

Solution

We treat the solution process in an identical fashion to a chemical reaction:

$$J/g\ mol$$

$\frac{1}{2}H_2(g) + \frac{1}{2}Cl_2(g)$	$= HCl(g)$	$A:$	$\Delta\hat{H}_f^\circ = -92{,}312$
$HCl(g) + 5H_2O$	$= HCl(5H_2O)$	$B:$	$\Delta\hat{H}_{soln}^\circ = -64{,}048$
$\frac{1}{2}H_2(g) + \frac{1}{2}Cl_2(g) + 5H_2O = HCl(5H_2O)$		$A+B:$	$\Delta\hat{H}_f^\circ = -156{,}360$

It is important to remember that the heat of formation of H_2O itself does not enter into the calculation. The heat of formation of HCl in an infinitely dilute solution is

$$\Delta H_f^\circ = -92.312 - 75.144 = -167.456\ kJ/g\ mol$$

Another type of heat of solution which is occasionally encountered is the partial molal heat of solution. Information about this thermodynamic property can be found in most standard thermodynamic texts or in books on thermochemistry, but we do not have the space to discuss it here.

One point of special importance concerns the formation of water in a chemical reaction. When water participates in a chemical reaction in solution as a reactant or product of the reaction, you must include the heat of formation of the water as well as the heat of solution in the energy balance. Thus, if HCl reacts with sodium hydroxide to form water and the reaction is carried out in only 2 moles of water to start with, it is apparent that you will have 3 moles of water at the end of the process. Not only do you have to take into account the heat of reaction when the water is formed, but there is also a heat of solution contribution. If gaseous HCl reacts with crystalline sodium hydroxide and the product is 1 mole of gaseous water vapor, you could employ the energy balance without worrying about the heat of solution effect.

If an enthalpy–concentration diagram is available for the system in which you are interested, the enthalpy changes can be obtained directly from the diagram by convenient graphical methods. See Sec. 5.2 for illustrations of this technique.

EXAMPLE 4.37 Application of Heat of Solution Data

An ammonium hydroxide solution is to be prepared at 77°F by dissolving gaseous NH_3 in water. Prepare charts showing:

    (a) The amount of cooling needed (in Btu) to prepare a solution containing 1 lb mol of NH_3 at any concentration desired

(b) The amount of cooling needed (in Btu) to prepare 100 gal of a solution of any concentration up to 35% NH_3

(c) If a 10.5% NH_3 solution is made up without cooling, at what temperature will the solution be after mixing?

Solution

Heat of solution data have been taken from *NBS Circular* 500.

(a)   Basis: 17 lb of NH_3 = 1 lb mol of NH_3

Reference temperature = $77°F \leftrightarrow 25°C$

To convert from kilocalories per gram mole to British thermal units per pound mole, multiply by 1800.

Description	State	$-\Delta \hat{H}_f^\circ$ (kcal/g mol)	$-\Delta \hat{H}_f^\circ$ (Btu/lb mol)	$-\Delta \hat{H}_{soln}^\circ$ (Btu/lb mol)	Weight $\%$ HN$_3$
$1H_2O$	g	11.04	19,900	0	100
$1H_2O$	aq	18.1	32,600	12,700	48.5
$2H_2O$	aq	18.7	33,600	13,700	32.0
$3H_2O$	aq	18.87	34,000	14,100	23.9
$4H_2O$	aq	18.99	34,200	14,300	19.1
$5H_2O$	aq	19.07	34,350	14,450	15.9
$10H_2O$	aq	19.23	34,600	14,700	8.63
$20H_2O$	aq	19.27	34,700	14,800	4.51
$30H_2O$	aq	19.28	34,700	14,800	3.05
$40H_2O$	aq	19.28	34,700	14,800	2.30
$50H_2O$	aq	19.29	34,750	14,850	1.85
$100H_2O$	aq	19.30	34,750	14,850	0.94
$200H_2O$	aq	19.32	34,800	14,900	0.47
∞H_2O	aq	19.32	34,800	14,900	0.0

Standard heats of solution have been calculated from the cumulative data as follows:

$$\Delta \hat{H}_{soln}^\circ = \Delta \hat{H}_f^\circ - \Delta \hat{H}_{f\ \text{gas}}^\circ$$

For example, for $NH_3 \cdot 1H_2O$:

$$\Delta \hat{H}_{soln}^\circ = -32,600 - (-19,900)$$
$$= -12,700 \text{ Btu/lb mol } NH_3$$

Weight percents have been computed as follows:

$$\text{wt }\% \text{ NH}_3 = \frac{\text{lb NH}_3(100)}{\text{lb H}_2\text{O} + \text{lb NH}_3}$$

$$\text{wt}\% \text{ NH}_3 \text{ for } 1H_2O: = \frac{17(100)}{18(1) + 17} = 48.5\%$$

The heat of solution values shown in Fig. E4.37a are equivalent to the cooling duty required.

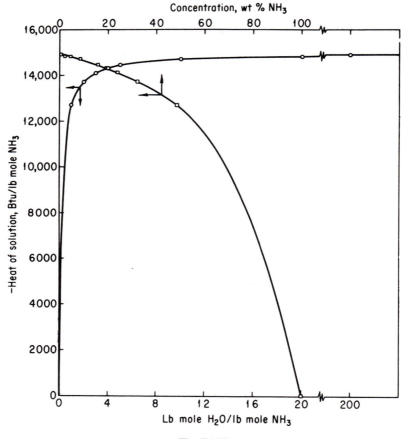

Fig. E4.37a

(b) This part of the problem requires that a new basis be selected, 100 gal of solution. Additional data concerning densities of NH_4OH are shown in the table below.

Sample calculations are as follows:

$$\text{density (lb/100 gal)} = \frac{(\text{sp gr soln})(62.4 \text{ lb/ft}^3)(1.003)(100)}{7.48 \text{ gal/ft}^3}$$

Note: 1.003 is the specific gravity of water at 77°F.

$$\text{density at } 32.0\% \text{ NH}_3 = \frac{0.889(62.4)(1.003)(100)}{7.48} = 741 \text{ lb/100 gal}$$

$$\left.\begin{array}{c}\text{cooling req'd} \\ \text{Btu/100 gal}\end{array}\right\} = \left(\frac{\text{lb mol NH}_3}{100 \text{ gal}}\right)\left(-\Delta H^\circ_{\text{soln}} \frac{\text{Btu}}{\text{lb mol NH}_3}\right)$$

$$\left.\begin{array}{c}\text{cooling req'd}\\\text{for 100 gal}\\32.0\%\ NH_3\ \text{soln}\end{array}\right\} = 13.94(13,700) = 191,000\ \text{Btu/100 gal soln}$$

These data are portrayed in Fig. E4.37b.

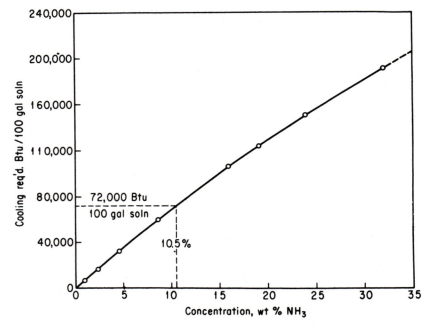

Fig. E4.37b

Basis: 100 gal of solution at concentrations and densities shown

% HN₃	Sp gr* at 4°C	Density (lb/100 gal)	lb HN₃ / 100 gal	lb mol NH₃ / 100 gal	$-\Delta\hat{H}^\circ_{\text{soln}}$ ($\frac{\text{Btu}}{\text{lb mol HN}_3}$)	Cooling Req'd per 100 gal (Btu)
32.0	0.889	741	237	13.94	13,700	191,000
23.9	0.914	761	182	10.70	14,100	151,000
19.1	0.929	774	148	8.70	14,300	124,000
15.9	0.940	784	124.7	7.32	14,450	106,000
8.63	0.965	805	69.4	4.08	14,700	60,000
4.51	0.981	819	37.0	2.18	14,800	32,200
3.05	0.986					
2.30	0.990	826	19.0	1.12	14,800	16,600
1.85	0.992					
0.94	0.995	830	7.8	0.46	14,850	6,800
0.47	0.998					
0.0	1.000	834	0	0	14,900	0

*SOURCE: N. A. Lange, *Handbook of Chemistry*, 8th ed., Handbook Publishers, Sandusky, Ohio, 1958.

(c)   Basis: 100 gal of solution at 10.5% NH_3

$$\Delta E = \Delta U = mC_v\,\Delta T \simeq mC_p\,\Delta T = 72{,}000 \text{ Btu/100 gal} \qquad [\text{see Fig. E4.37(b)}]$$

$$\text{sp gr soln} = 0.955$$

$$m = \frac{0.955(6.24)(1.003)(100)}{7.48} = 800 \text{ lb}$$

$$C_p\ 10.5\%\ NH_3\ \text{soln} = 4.261 \text{ J/(g)(°C)}$$

$$= 1.02 \text{ Btu/(lb)(°F)}$$

$$\Delta T = \frac{\Delta H}{mC_p} = \frac{72{,}000}{(800)(1.02)} = 88°F$$

$$T_{\text{final}} = 77 + 88 = 165°F$$

Self-Assessment Test

1. Is a gas mixture an ideal solution?

2. Give (*a*) two examples of exothermic mixing of two liquids and (*b*) two examples of endothermic mixing based on your experience.

3. (*a*) What is the reference state for H_2O in the table for the heat of solution of HCl? (*b*) What is the value of the enthalpy of H_2O in the reference state?

4. Use the heat of solution data in Appendix H to determine the heat transferred per mole of entering solution into or out of (state which) a process in which 2 g mol of a 50 *mole* % solution of sulfuric acid at 25°C is mixed with water at 25°C to produce a solution at 25°C containing a mole ratio of $10H_2O$ to $1H_2SO_4$.

5. Calculate the heat that must be added per ton of 50 wt % H_2SO_4 produced by the process shown.

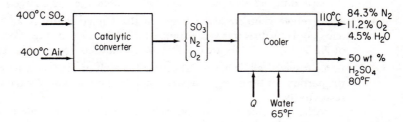

SUPPLEMENTARY REFERENCES

1. Biskup, B., and P. Tausk, *Calculation of Properties Using Corresponding State Methods,* Elsevier, Amsterdam, 1979.

2. Canjar, L., and F. Manning, *Thermodynamic Properties and Reduced Correlations for Gases,* Gulf Publishling Co., Houston, Tex., 1967.

3. Edmister, W. C., *Applied Hydrocarbon Thermodynamics*, Gulf Publishing Co., Houston, Tex., 1961.

4. Hougen, O. A., K. M. Watson, and R. A. Ragatz, *Chemical Process Principles*, Part II, 2nd ed., Wiley, New York, 1959.

5. Leland, T. W., and P. S. Chappelear, "The Corresponding States Principle," *Ind. Eng. Chem.*, v. 60, no. 7, p. 15 (1968).

6. Mott-Smith, M., *The Concept of Energy Simply Explained*, Dover, New York, 1964.

7. National Academy of Sciences, *Energy in Transition*, W. H. Freeman, San Francisco, 1980.

8. Smith, J. M., and H. C. Van Ness, *Introduction to Chemical Engineering Thermodynamics*, 2nd ed., McGraw-Hill, New York, 1959.

9. Stobaugh, R., and D. Yergin, *Energy Future*, Random House, New York, 1979.

PROBLEMS[30]

Section 4.1

4.1. The energy from the sun incident on the surface of the earth averages 2.0 cal/(min)(cm²). It has been proposed to use space stations in synchronous orbits 36,000 km from earth to collect solar energy. How large a collection surface is needed (in m²) to obtain 10^{11} watts of electricity? Assume that 10% of the collected energy is converted to electricity. Is this a reasonable size?

4.2. The world's largest plant that obtains energy from tidal changes is at Saint Malo, France. The plant uses both the rising and falling cycle (one period in and out is 6 hr 10 min in duration). The tidal range from low to high is 14 m, and the tidal estuary (the LaRance River) is 21 km long with an area of 23 km². Assume that the efficiency of the plant in converting potential to electrical energy is 85%, and estimate the average power produced by the plant. (*Note:* Also assume that after high tide, the plant does not release water until the sea level drops 7 m, and after a low tide does not permit water to enter the basin until the level outside the basin rises 7 m, and the level differential is maintained during discharge and charge.)

4.3. The average geothermal heat flux in the United States is about 1.4×10^{-6} cal/(cm)²(s). How does this value compare with the 1980 average electrical consumption of 2.2×10^{8} kW? The area of the United States is about 3.1×10^{6} mi².

4.4. Explain specifically what the system is for each of the following processes; indicate the portions of the energy transfer that are heat and work by the symbols Q and W, respectively.
  (a) A liquid inside a metal can, well insulated on the outside of the can, is shaken very rapidly in a vibrating shaker.
  (b) Hydrogen is exploded in a calometric bomb and the water layer outside the bomb rises in temperature by 1°C.
  (c) A motor boat is driven by an outboard-motor propeller.
  (d) Water flows through a pipe at 1.0 m/min, and the temperature of the water and the air surrounding the pipe are the same.

---

[30]An asterisk designates problems appropriate for computer solution. Refer also to the computer problems at the end of the chapter.

4.5.   Draw a simple sketch of each of the following processes, and, in each, label the system boundary, the system, the surroundings, and the streams of material and energy that cross the system boundary.
  (a) Water enters a boiler, is vaporized, and leaves as steam. The energy for vaporization is obtained by combustion of a fuel gas with air outside the boiler surface.
  (b) The steam enters a rotary steam turbine and turns a shaft connected to an electric generator. The steam is exhausted at a low pressure from the turbine.
  (c) A battery is charged by connecting it to a source of current.
  (d) A tree obtains water and minerals through its roots and gives off carbon dioxide from its leaves. The tree, of course, gives off oxygen at night.

4.6.   Draw a simple sketch of the following processes; indicate the system boundary; and classify the system as open or closed.
  (a) Automobile engine   **(b)** Water wheel
  (c) Pressure cooker   **(d)** Human being
  (e) River   **(f)** The earth and its atmosphere
  (g) Air compressor   **(h)** Coffee pot

4.7.   Are the following variables intensive or extensive variables? Explain for each.
  (a) Pressure   **(b)** Volume
  (c) Specific volume   **(d)** Refractive index
  (e) Surface tension

4.8.   Classify the following measurable physical characteristics of a gaseous mixture of two components as (1) an intensive property, (2) an extensive property, (3) both, or (4) neither:
  (a) Temperature   **(b)** Composition
  (c) Pressure   **(d)** Mass

4.9.   Convert 45.0 Btu/lb_m to the following:
  (a) cal/kg   **(b)** J/kg
  (c) kWh/kg   **(d)** $(ft)(lb_f)/lb_m$

4.10.   A sedan and a semi-trailer truck collide head-on on a freeway. The truck weighs 28,000 lb and the sedan weighs 5000 lb. At the moment of impact, both the truck and the sedan are going at 60.0 mi/hr.
  (a) What was the car's kinetic energy in $(ft)lb_f$? in joules?
  (b) What was the truck's kinetic energy in $(ft)(lb_f)$? in joules?
  (c) How much kinetic energy is changed into other forms of energy in the collision by the time the two vehicles come to rest?

4.11.   Before it lands, a vehicle returning from space must convert its enormous kinetic energy to heat. To get some idea of what is involved, a vehicle returning from the moon at 25,000 mi/hr can, in converting its kinetic energy, increase the internal energy of the vehicle sufficiently to vaporize it. Obviously, a large part of the total kinetic energy must be transferred from the vehicle. How much kinetic energy does the vehicle have (in Btu)? How much energy must be transferred by heat if the vehicle is to heat up only 20°F/lb?

4.12.   Suppose that a constant force of 40.0 N is exerted to move an object for 6.00 m. What is the work accomplished (on an ideal system) expressed in the following:
  (a) joules   **(b)** $(ft)(lb_f)$
  (c) cal   **(d)** Btu

4.13. The energy content of 1 kg of gasoline is 12,500 watt-hours. A similar weight of lead-acid storage batteries can contain but 30 watt-hours, and it is this 400 : 1 difference which has for 50 years or more frustrated those those who would replace conventional automobiles with a fleet of electric cars.

The difference can be reduced by regenerative braking, that is, using the electric motor as a generator to transfer the car's kinetic to electrical energy. For a 3500-lb car cruising at 50 mi/h, how much energy (in watt-hours) is delivered to the battery by regenerative braking if the conversion efficiency is 70%?

4.14. Steam is used to cool a polymer reaction. The steam in the steam chest of the apparatus is found to be at 250°C and 4000 kPa absolute during a routine measurement at the beginning of the day. At the end of the day the measurement showed that the temperature was 650°C and the pressure 10,000 kPa absolute. What was the internal energy change of 1 kg of steam in the chest during the day? Obtain your data from the steam tables.

4.15. In a computer program you find that the relation for the specific enthalpy (J/g mol) of a gas is given by

$$\hat{H} = 32.97t + 6.694 \times 10^{-3}t^2 - 6.745 \times 10^{-7}t^3 + 5.481 \times 10^{-5}p$$

where t is in °C and p is in kPa, and the reference state is unspecified.
(a) Determine the value of the heat capacity C_p of the gas in J/(kg mol) (°C) at 100°C and 100 kPa.
(b) What is a function that expresses the specific internal energy (for the same reference state) assuming that the gas acts as an ideal gas.
(c) Determine the value of the heat capacity C_v of the gas in J/(kg mol) (°C) at 100°C and 100 kPa.

4.16. (a) Ten pound moles of an ideal gas are originally in a tank at 100 atm and 40°F. The gas is heated to 440°F. The specific molal enthalpy, $\Delta\hat{H}$, of the ideal gas is given by the equation

$$\Delta\hat{H} = 300 + 8.00t$$

where $\Delta\hat{H}$ is in Btu/lb mol and t is the temperature in °F.
(1) Compute the volume of the container (ft³).
(2) Compute the final pressure of the gas (atm).
(3) Compute the enthalpy change of the gas.
(b) Use the equation above to develop an equation giving the molal internal energy, ΔU, in cal/g mol as a function of temperature, t, in °C.

4.17. You have calculated that specific enthalpy of 1 kg mol of an ideal gas at 300 kN/m² and 100°C is 6.05×10^5 J/kg mol (with reference to 0°C and 100 kN/m²). What is the specific internal energy of the gas?

4.18. To exploit geothermal energy, it is proposed to fracture the earth's crust at a depth of 2 to 6 km with water at 5000 kPa. In the projected heat exchange 143 kg/s of water will be heated from 65°C to 280°C. What will be the enthalpy change of the water in joules per second?

4.19. One hundred and fifty pounds of CO_2 has a specific enthalpy of 45,000 (ft)(lb$_f$)/lb$_m$

with reference to 0°F and 1 psia. What is the enthalpy of this CO_2 in Btu/lb mole? What is the total internal energy if the pressure is 6 atm and the temperature is 50°F?

4.20. One kilogram mole of steam at 150°C and 150 kPa is expanded to 210°C and 100 kPa. Thereafter it is condensed to water at 25°C and 100 kPa, and finally heated back to 150°C and 150 kPa. What is the overall **(a)** enthalpy change and **(b)** internal energy change per kilogram mole of steam for the process?

4.21. State which of the following variables are point, or state, variables, and which are not; explain your decision in one sentence for each one:
(a) Pressure   **(b)** Density
(c) Molecular weight   **(d)** Heat capacity
(e) Internal energy   **(f)** Ionization constant

Section 4.2

4.22.* Experimental values calculated for the heat capacity of ammonia from -40 to 1200°C are:

t (°C)	C_p° [cal/(g mol)(°C)]	t (°C)	C_p° [cal/(g mol)(°C)]
-40	8.180	500	12.045
-20	8.268	600	12.700
0	8.371	700	13.310
18	8.472	800	13.876
25	8.514	900	14.397
100	9.035	1000	14.874
200	9.824	1100	15.306
300	10.606	1200	15.694
400	11.347		

Fit the data by least squares for the following two functions:

$$\left.\begin{array}{l} C_p^\circ = a + bt + ct^2 \\ C_p^\circ = a + bt + ct^2 + dt^3 \end{array}\right\} \quad \text{where } t \text{ is in °C}$$

4.23.* Experimental values of the heat capacity C_p have been determined in the laboratory as follows; fit a second-order polynomial in temperature to the data ($C_p = a + bT + cT^2$):

T (°C)	C_p (cal/(g mol)(°C)]
100	9.69
200	10.47
300	11.23
400	11.79
500	12.25
600	12.63
700	12.94

(The data are for carbon dioxide.)

4.24. The heat capacity for gaseous *n*-pentane is given by $C_p^\circ = 27.50 + 8.148 \times 10^{-2}t - 4.538 \times 10^{-5}t^2 + 10.10 \times 10^{-9}t^3$, where C_p° is in cal/(g mol) (°C) and t is in °C. Modify the relation for C_p° so that t can be expressed in °F and the units of C_p are Btu/(lb mol) (°F).

4.25. Your assistant has developed the following equation to represent the heat capacity of air (with C_p in cal/(g mol) (K) and T in K):

$$C_p = 6.39 + 1.76 \times 10^{-3}T - 0.26 \times 10^{-6}T^2$$

(a) Derive an equation giving C_p but with the temperature expressed in °C.
(b) Derive an equation giving C_p in terms of Btu per pound per degree Fahrenheit with the temperature being expressed in degrees Fahrenheit.

4.26. Plot the heat capacity (C_p) of benzene from −250 to 300°C as a function of the temperature in °C. Mark on the axis the melting point (and temperature) and the boiling point (and temperature). Take the data from a reference handbook(s).

4.27. Estimate the heat capacity of sand (SiO_2) in the vicinity of 25°C. Use Kopp's rule, and compare with the best estimate you can find in a reference book as well as the best estimate you can find for SiO_2 as a pure compound.

4.28. Repeat problem 4.27 but for the following liquids at 25°C: acetonitrile (CH_3CN), ethyl ether [$(C_2H_5)_2O$], and arsenic trichloride ($AsCl_3$).

4.29. One hundred pounds of a 35°API distillate with an average boiling point of 250°F is cooled from 400°F to 300°F. Estimate the heat capacity of the distillate at each temperature using the equation in the text and the charts in Appendix K.

4.30. Estimate the heat capacity of gaseous isobutane at 1000 K and 200 mm Hg by using the Kothari–Doraiswamy relation

$$C_p = A + B \log_{10} T_r$$

from the following experimental data at 200 mm Hg:

C_p [cal/(K)(g mol)]	23.25	35.62
Temp. (K)	300	500

The experimental value is 54.40; what is the percentage error in the estimate?

Section 4.3

4.31. One gram mole of air is heated from 400°C to 1000°C. Calculate ΔH by integrating the heat capacity equation. Also calculate ΔU.

4.32. Calculate the enthalpy change (in J/kg mol) that takes place in raising the temperature of 1 kg mol of the following gas mixture from 50°C to 550°C.

Component	Mole %
CH_4	80
C_2H_6	20

4.33. What is the change in enthalpy of 10 lb of CO_2 when heated from 140°F to 300°F at 1 atm? Use the heat capacity equation for CO_2 from Appendix E.

4.34. Use the steam tables to calculate the enthalpy change (in joules) of 2 kg mol of steam when heated from 400 K and 100 kPa to 900 K and 100 kPa. Repeat using the table in the text for the enthalpies of combustion gases. Repeat using the heat capacity for steam. Compare your answers. Which is most accurate?

4.35.* The heat capacity of chlorine has been determined experimentally as follows:

T (°C)	C_p [cal/(g mol)(K)]
0	8.00
18	8.08
25	8.11
100	8.36
200	8.58
300	8.71
400	8.81
500	8.88
600	8.92
700	8.95
800	8.98
900	9.01
1000	9.03
1100	9.05
1200	9.07

Calculate the enthalpy change required to raise 1 g mol of chlorine from 0°C to 1200°C, and show the method of calculation.

4.36. One pound mole of a gas of composition 20% He, 20% CH_4, 40% CO_2, 20% N_2 is heated from 100°F to 300°F at a constant pressure of 1 atm. Find ΔH and ΔU for the process.

4.37. One kilogram mole of NO is heated at constant pressure from 10°C to 600°C. What is the enthalpy change? Compare with that for NO as an *ideal* gas.

4.38. Determine the mean heat capacity between T_1 and T_2 if C_p is expressed as
(a) $C_p = a + bT$ (T in degrees absolute)
(b) $C_p = a + bT - cT^{-2}$ (T in degrees absolute)
(c) $C_p = a - bT - cT^{1/2}$ (T in °C)

4.39. Use the tables for the mean heat capacity of the combustion gases to compute the enthalpy change (in Btu) that takes place when a mixture of 6.00 lb mol of gaseous H_2O and 4.00 lb mol of CO_2 is heated from 60°F to 600°F.

4.40. Use the heat capacity data for carbon dioxide to compute the mean heat capacity for CO_2 between 100 and 200°C. Repeat using the table for the combustion gases in the text. Compare both with the mean heat capacity calculated from the table in the text.

4.41. Calculate ΔH (in J/kg mol) for 2 kg of CO_2 that is cooled from 80°C and 2700 kPa to 0°C and 200 kPa by use of:

(a) The CO_2 chart in the appendix

(b) The table of enthalpies of the combustion gases

Section 4.4

4.42. Water can exist in various states, depending on its temperature, pressure, and so on. From what you have learned in Sec. 3.7, what is the state (gaseous, liquid, solid, or combinations thereof) for water at the following conditions:

(a) 250°F and 1 atm

(b) −10°C and 720 mm Hg

(c) 32°F and 100 psia

4.43. Calculate the enthalpy change (in joules) that occurs when 1 kg of benzene vapor at 150°C and 100 kPa condenses to a solid at −20.0°C and 100 kPa.

4.44. The vapor pressure of phenylhydrazine has been found to be (from 25 to 240°C)

$$\log_{10} p^* = 7.9046 - \frac{2366.4}{T + 230}$$

where p^* is in atm and T is in °C. What is the heat of vaporization of phenylhydrazine (in Btu/lb) at 200°F?

4.45. Check the precision of the empirical relation proposed by Watson to calculate heats of vaporization. Take water as a test example. With boiling water at 100°C as the reference temperature, estimate the heat of vaporization of water at 300°C from Watson's relation. Compare your result with the actual value from the steam tables.

4.46.* Make an Othmer plot for *n*-butane, and calculate the latent heat of vaporization at 280°F; compare this value with the one calculated from the following experimental data:

T (°F)	p^* (atm)	$V_{(l)}$ (ft³/lb)	$V_{(g)}$ (ft³/lb)
260	24.662	0.0393	0.222
270	27.134	0.0408	0.192
280	29.785	0.0429	0.165
290	32.624	0.0458	0.138
305.56 (T_c)	37.47	0.0712	0.0712

SOURCE: Data from H. W. Prengle, L. R. Greenhaus, and R. York, *Chem. Eng. Progr.*, v. 44, p. 863 (1948). The reported value of ΔH is 67.3 Btu/lb.

4.47. Make an Othmer plot to determine the heat of vaporization of _____ at _____ °C. Compare with an experimental value if one can be found in a handbook.

4.48.* (a) Using the following data on the vapor pressure of Cl_2, plot $\log p^*$ against $1/T$, and calculate the latent heat of vaporization of Cl_2 as a function of temperature:

T (°F)	p^* (psia)
−22	17.8
32	53.5
86	126.4
140	258.5
194	463.5
248	771
284	1050

How satisfactory are these data?

(b) The vapor pressure of Cl_2 is given by Lange[31] by the equation

$$\log p^* = A - \frac{B}{C + T}$$

where p^* is the vapor pressure in mm Hg, T is in °C, and for chlorine, $A =$ 6.86773, $B = 821.107$, and $C = 240$. Calculate the latent heat of vaporization of chlorine at the normal boiling point (in Btu/lb). Compare with the experimental value.

4.49. Use of the steam tables and chart:

(a) What is the enthalpy change needed to change 3 lb of liquid water at 32°F to steam at 1 atm and 300°F?

(b) What is the enthalpy change needed to heat 3 lb of water from 60 psia and 32°F to steam at 1 atm and 300°F?

(c) What is the enthalpy change needed to heat 1 lb of water at 60 psia and 40°F to steam at 300°F and 60 psia?

(d) What is the enthalpy change needed to change 1 lb of a water–steam mixture of 60% quality to one of 80% quality if the mixture is at 300°F?

(e) Calculate the ΔH value for an isobaric (constant pressure) change of steam from 120 psia and 500°F to saturated liquid.

(f) Repeat part (e) for an isothermal change to saturated liquid.

(g) Does an enthalpy change from saturated vapor at 450°F to 210°F and 7 psia represent an enthalpy increase or decrease? a volume increase or decrease?

(h) In what state is water at 40 psia and 267.24°F? At 70 psia and 302°F? At 70 psia and 304°F?

(i) A 2.5-ft^3 tank of water at 160 psia and 363.5°F has how many cubic feet of liquid water in it? Assume that you start with 1 lb of H_2O. Could it contain 5 lb of H_2O under these conditions?

(j) What is the volume change when 2 lb of H_2O at 1000 psia and 200°F expands to 245 psia and 460°F?

(k) Ten pounds of wet steam at 100 psia has an enthalpy of 9000 Btu. Find the quality of the wet steam.

4.50. A government agency has proposed the use of a heat pump as an energy-saving device in homes. The heart of the system would be a $7\frac{1}{2}$-ft-deep insulated tank containing

[31] N. A. Lange, *Handbook of Chemistry*, 9th ed., Handbook Publishers, Sandusky, Ohio, 1956, p. 1426.

roughly 3000 ft³ of water. The tank would occupy about one-third of the house basement or be installed beneath it.

In winter, the heat pump would draw energy from the water for heating. Removal of energy from the water would turn it to ice over several months, it is said. In the summer, the ice would be used to chill air that would be blown through the house to provide air conditioning. The ice would gradually melt and the water would warm up to store heat for winter use.

What is the approximate maximum number of Btu (i.e., the enthalpy change) that can be utilized in the water tank for air conditioning in the summer if it is initially 80% ice, and the air-conditioner intake cannot rise above 50°F? Comment on the feasibility of the proposed system.

4.51. Use the steam tables to determine the answers to the following questions:

(a) Liquid water exists at a pressure such that its boiling point is 575°F. How many Btu is required to change 1 lb of liquid water at 575°F to 1 lb of dry saturated water vapor at 575°F? What is the pressure in psia at which water boils at 575°F?

(b) Water vapor exists at 240.0 psia and a temperature of 423.86°F. Is this vapor superheated? If so, how many degrees of superheat does it contain?

(c) Water at 50°F is heated to 495°F, at which temperature it boils. What is the mean heat capacity [in Btu/(lb)(°F)] for liquid water over this range?

(d) Wet steam at 315°F has an enthalpy of 1150.2 Btu/lb. What is the quality of this steam?

4.52. For an orbital vehicle composed of three basic elements—spacecraft, propellant tanks, and rocket engines—it is generally agreed that the spacecraft and rocket engines should be recovered for reuse. The main point of controversy concerns what portion of the propellant tanks should be made reusable. Propellant containers have been the most weighty and bulky elements of large rocket vehicles. On a cost-per-pound basis, these containers cost the least among major vehicle elements, but still, they are not cheap. The ideal solution would be to eliminate the tanks entirely. But how would the propellants be contained and supported? One answer is to freeze them. Most substances can be frozen, and propellants are no exception. Estimate the sum of the heat of melting of solid oxygen and the heat of vaporization of liquid oxygen per unit mass at low pressure to determine the amount of energy that will be removed from the rocket exhaust to vaporize the solid oxygen. What fraction of the total enthalpy change from 54.4 K (melting point) to 90.2 K (boiling point) would you estimate is provided by the latent heats (enthalpy change of the phase transitions)?

Section 4.5

4.53. Five kilograms of steam in an engine at 800 kPa and 170°C is allowed to expand isothermally into surrounding steam at 170°C and at a constant pressure of 200 kPa. Calculate Q, W, ΔU, and ΔH for the 5 kg for this process.

4.54. One gram mole of an *ideal* gas whose C_p is 300 J/(g mol)(K) is confined in a reservoir with a floating top such that the pressure on the gas is 400 kPa no matter what its volume is. In the morning the gas is at 10°C, but late in the afternoon its temperature rises to 32°C.

(a) Determine how much heat has been transferred into the tank, the work done by the gas, and the internal energy change for the gas.

(b) If the system returns to its original state in the evening, again find Q, W, and the internal energy and enthalpy changes for the cooling process.

(c) What are Q, W, and the internal energy and enthalpy changes for the overall process of heating and cooling?

4.55. **(a)** A large piston does 12,500 (ft)(lb$_f$) of work in compressing 3 ft^3 of air to 25 psia. Five pounds of water circulates into and out of the water jacket around the piston and increases in temperature 2.3°F by the end of the process. What is the change of internal energy of the air?

(b) Suppose that the water does not circulate but is stationary and increases in temperature the same amount. What is the change now in the internal energy of the air?

(c) Suppose that the piston after compressing the gas returns to its original position, and all the 12,500 (ft)(lb$_f$) of work is recovered. What will the water temperature be in part (b)?

4.56. Calculate the work done when 1 kg mol of water evaporates completely at 100°C. Express your results in joules.

4.57. A household freezer is placed inside an insulated sealed room. If the freezer door is left open with the freezer operating, will the temperature of the room increase or decrease? Explain your answer.

4.58. A car weighing 3000 lb, which is initally at rest, is pushed up an inclined hill for a linear distance of 50 ft, at which time it is 5 ft higher than when it started and is moving at the speed of 10 mi/hr. The average force exerted was 600 lb.

(a) What was the total work done on the car?

(b) What was the potential energy, kinetic energy, and total energy acquired by the car?

(c) Why are the calculated values obtained in parts (a) and (b) not the same?

4.59. A system consists of 25 lb of water vapor at the dew point. The system is compressed isothermally at 400°F, and 988 Btu of work is done on the system by the surroundings. What volume of liquid was present in the system before and after compression?

4.60. Calculate the required quantities listed below using data from the steam tables for the properties of water.

(a) What is W, ΔU, and ΔH (in joules) for 1 kg of water as it is vaporized at a constant temperature of 100°C and a constant pressure of 1 atm? For this change, how many joules of heat have to be added to the water?

(b) A stream of 200 kg/s of water at 80°C is mixed with a stream of 300 kg/s of water at 20°C. Calculate the mass flow rate and the enthalpy (in J/kg) of the resulting mixture.

(c) Water at 26°C is flowing in a straight horizontal insulated pipe of 2.0 cm inside diameter at a velocity of 7 m/s. The water flows into a section where the diameter suddenly increases to 8.0 cm. What is the change in enthalpy of the water (in J/kg)? What is the exit mass flow rate (in kg/s)?

4.61. Steam fills tank 1 at 1000 psia and 700°F. The volume of tank 1 equals the volume of tank 2. A vacuum exists in tank 2 initially. See Fig. P4.61. The valve connecting the two tanks is opened, and isothermal expansion of the steam occurs from tank 1 to tank 2. Find the enthalpy change per pound of steam for the entire process.

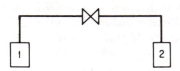

Fig. P4.61

4.62. One pound of water at 70°F is put into a closed tank of constant volume. If the temperature is brought to 80°F by heating, find Q, W, ΔU, and ΔH for the process (in Btu). If the same temperature change is accomplished by a stirrer that agitates the water, find Q, W, ΔE, and ΔH. What assumptions are needed?

4.63. A 55-gal drum of fuel oil (15°API) is to be heated from 70°F to 180°F by means of an immersed steam coil. Steam is available at 220°F. How much steam is required if it leaves the coil as liquid water at 1 atm pressure?

4.64. Oil of average C_p 0.8 Btu/(lb)(°F) flows at 2000 lb/min from an open reservoir standing on a hill 1000 ft high into another open reservoir at the bottom. To ensure rapid flow, heat is put into the pipe at the rate of 100,000 Btu/hr. What is the enthalpy change in the oil per pound? Suppose that a 1-hp pump (50% efficient) is added to the pipeline to assist in moving the oil. What is the enthalpy change per pound in the oil now?

4.65. Write the simplified energy balances for the following changes:
(a) A fluid flows steadily through a poorly designed coil in which it is heated from 70°F to 250°F. The pressure at the coil inlet is 120 psia, and at the coil outlet is 70 psia. The coil is of uniform cross section, and the fluid enters with a velocity of 2 ft/sec.
(b) A fluid is expanded through a well-designed adiabatic nozzle from a pressure of 200 psia and a temperature of 650°F to a pressure of 40 psia and a temperature of 350°F. The fluid enters the nozzle with a velocity of 25 ft/sec.
(c) A turbine directly connected to an electric generator operates adiabatically. The working fluid enters the turbine at 1400 kPa absolute and 340°C. It leaves the turbine at 275 kPa absolute and at a temperature of 180°C. Entrance and exit velocities are negligible.
(d) The fluid leaving the nozzle of part (b) is brought to rest by passing through the blades of an adiabatic turbine rotor and leaves the blades at 40 psia and at 400°F.
(e) A fluid is allowed to flow through a cracked (slightly opened) valve from a region where its pressure is 1400 kPa absolute and 350°C to a region where its pressure is 275 kPa absolute, the whole operation being adiabatic.

4.66. A steam turbine receives 10,000 kg/hr of steam and delivers 1 MW (1000 kW). Neglect any heat losses.
(a) Neglecting velocity changes, find the change in enthalpy across the turbine.
(b) Assume that the entrance velocity is 15 m/s and the exit velocity 300 m/s. Find ΔH.
(c) Assume that the inlet steam pipe is 3 m higher than the exhaust. Find ΔH.

4.67. A desalted crude oil (40°API) is being heated in a parallel-flow heat exchanger from 70°F to 200°F with a straight-run gasoline vapor available at 280°F. If 400 gal/hr of gasoline is available (measured at 60°F, 63°API), how much crude can be heated per hour? Would countercurrent flow of the gasoline to the crude change your answer to this problem? Assume that $K = 11$ for the oil and 12.2 for the gasoline. C_p of the vapor is 0.95 Btu/(lb)(°F).

4.68. A compressor station delivers gas at 290°F and 1200 psia from a supply at 200°F and 200 psia. If 3000 Btu/lb mole of gas that is compressed is removed from the compressor by intercooling, find the work done (in Btu) by the compressor per pound mole of gas delivered. The gas is (a) air, (b) propane, (c) methane, (d) CO_2, and (e) N_2. If the compression were adiabatic, what would your answer be? If the compression were isothermal at 200°F (with the same heat interchange), what would your answer be?

4.69. The process of throttling (i.e., expansion of a gas through a small orifice) is an important part of refrigeration. If wet steam (fraction vapor is 0.975) at 700 kPa absolute is throttled to 1 atm, find the following for the process:
(a) Heat transferred to or from steam
(b) Work done
(c) Enthalpy change
(d) Internal energy change
(e) Exit temperature

4.70. In one stage of a process for the manufacture of liquid air, air as a gas at 4 atm absolute and 250 K is passed through a long, insulated 3-in.-ID pipe in which the pressure drops 3 psi because of frictional resistance to flow. Near the end of the line, the air is expanded through a valve to 2 atm absolute. State all assumptions.
(a) Compute the temperature of the air just downstream of the valve.
(b) If the air enters the pipe at the rate of 100 lb/hr, compute the velocity just downstream of the valve.

4.71. Solid CO_2 is being produced by adiabatically expanding high-pressure CO_2 at 70°F through a valve to a pressure of 20.0 psia.
(a) If 30% conversion of the initial CO_2 to the solid is desired, at what pressure must the CO_2 be before expansion?
(b) Is the high-pressure CO_2 gas, liquid, or a mixture of each?
(c) What is the approximate volume change per pound of CO_2 during the expansion?

4.72. Examine the simplified refinery power plant flow sheet in Fig. P4.72 and calculate the energy loss (in Btu/hr) for the plant. Also calculate the energy loss as a percentage of the energy input.

4.73. Frozen-food quality is taking a big step forward with the introduction of ultrafast freezing as a commercial reality. The key to this process is the use of −320°F liquid nitrogen to do the freezing. For example, in the freezing of fish fillets, after precooling by vaporized nitrogen, the fillets enter a freezing section, where they pass through a liquid nitrogen spray which vaporizes into −320°F gas. The resulting cold gas is used to remove sensible heat from the entering fish and exhausts to the atmosphere at −10°F or warmer. A well-insulated tunnel, 6 ft wide by 32 ft long, has a capacity of 1000 lb/hr of fish fillets. Estimate the gallons of liquid N_2 used per hour. (The actual consumption was reported to be 130 to 150 gal/hr.)

4.74. Start with the general energy balance, and simplify it for each of the processes listed below to obtain an energy balance that represents the process. Label each individual

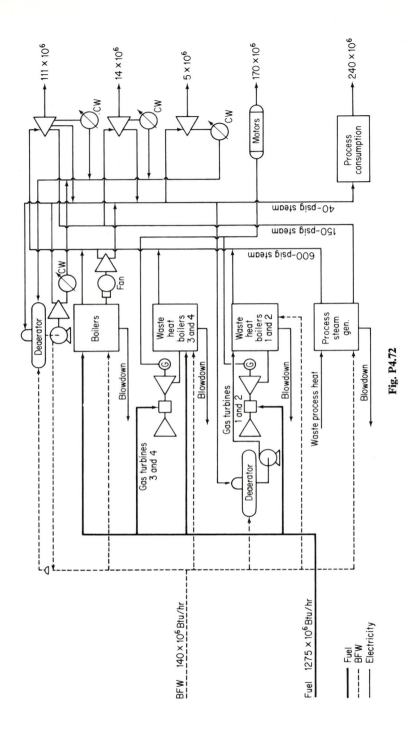

Fig. P4.72

term in the general energy balance by number, and list by their numbers the terms retained or deleted followed by your explanation. (You do not have to calculate any quantities in this problem.)

(a) One hundred kilograms per second of water is cooled from 100°C to 50°C in a heat exhanger. The heat is used to heat up 250 kg/s of benzene entering at 20°C. Calculate the exit temperature of benzene.

(b) A feedwater pump in a power generation cycle pumps water at the rate of 500 kg/min from the turbine condensers to the steam generation plant, raising the pressure of the water from 6.5 kPa to 2800 kPa psia. If the pump operates adiabatically with an overall mechanical efficiency of 50% (including both pump and its drive motor), calculate the electric power requirement of the pump motor (in kW). The inlet and outlet lines to the pump are of the same diameter. Neglect any rise in temperature across the pump due to friction (i.e., the pump may be considered to operate isothermally).

(c) A caustic soda solution is placed in a mixer together with water to dilute it from 20% NaOH in water to 10%. What is the final temperature of the mixture if the materials initially are at 25°C and the process is adiabatic?

4.75. Determine the steam rate required in the diagrammed still in Fig. P4.75. Assume that the K of the exit streams $\simeq 11$.

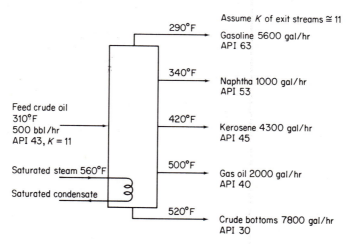

Fig. P4.75

4.76. The large increase in the emission of SO_2 has been caused by the large increase in fuel consumption for electricity and other industrial plants. Both the fuel oil and coal used in these plants contain sulfur that forms SO_2 (and a little SO_3) in the exhaust gases. Recovery of the sulfur in the exhaust gases competes with desulfurization of the fuel as a means of preventing air pollution. In a plant, 10,000 ft³/hr of a combustion gas at 1000°F and 1 atm is produced from a high-sulfur fuel. The gas, which has the following analysis:

$$
\begin{array}{ll}
SO_2 & 1.1\% \\
CO_2 & 14.5\% \\
O_2 & 3.4\% \\
N_2 & \underline{81.0\%} \\
& 100.0\%
\end{array}
$$

is to be cooled from 1000°F to 400°F by mixing with *dry* air whose initial temperature is 65°F and pressure is 760 mm Hg. See Fig. P4.76. What volume of air per hour measured at 65°F and 760 mm Hg pressure is required to produce an air–gas mixture having a temperature of 400°F?

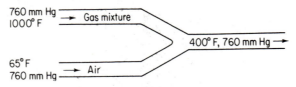

Fig. P4.76

4.77. A distillation process has been set up to separate an ethylene-ethane mixture as shown in the figure below. The product stream will consist of 98% ethylene and it is desired to recover 97% of the ethylene in the feed. The feed, 1000 lb per hour of 35% ethylene, enters the Pre-heater as a subcooled liquid (Temperature = −100°F, Pressure = 250 psia). The feed experiences a 20°F temperature rise before it enters the still. The heat capacity of liquid ethane may be considered to be constant and equal to 0.65 Btu/(lb)(°F) and the heat capacity of ethylene may be considered to be constant and equal to 0.55 Btu/(lb)(°F). Heat capacities and saturation temperatures of mixtures may be determined on a weight fraction basis. An optimum reflux ratio of 6.1 lb reflux/lb product has been previously determined and will be used. Operating pressure in the still will be 250 psia. Additional data is as follows:

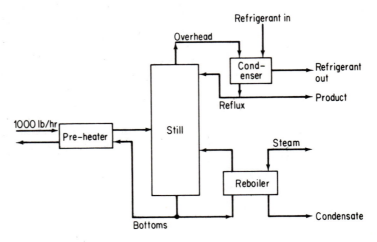

Fig. P4.77

Pressure = 250 Psia

Component	Temp. Sat., °F	Heat of Vaparization, Btu/lb
C_2H_6	10°	140
C_2H_4	−30°	135

Determine the

(1) Pounds of 30 psig steam required in the reboiler per pound of feed.

(2) Gallons of refrigerant required in the condenser per hour assuming a 25°F rise in the temperature of the refrigerant. (Heat capacity is approximately 1.0 Btu/(lb)(°F) and Density = 50 lb/ft³)

(3) The temperature of the bottoms as it leaves the Pre-heater.

4.78. In a letter to the editor of a journal, the following was written:

In his discussions of the conservation of energy, the author said that teachers should refrain from any suggestions that energy and mass are interconvertible. He states (1) that Einstein's relation $E = mc^2$ is an equality and not a description of a transformation and (2) that it is in no way proper to infer that, *for any change*, either mass or energy is not each separately conserved. From the viewpoint of relativity theory, the total mass of the products of an exothermic chemical reaction will be less than the mass of the reactants by $\Delta m = \Delta E/c^2$, where ΔE is the energy evolved. For example, if $\Delta E = 100$ kcal/mol $= 4.2 \times 10^5$ J/mol, then $\Delta m = 4.7 \times 10^{-9}$ g/mol. One would be hard pressed to measure such a change, yet it is of importance *conceptually*. There may be some semantic problems, but I would say that mass is not conserved—separately from the energy changes—in such reactions.

Comment on this letter. Is mass conserved or not?

4.79. Ammonia at 10°C and 900 kPa pressure is supplied to a heat exchanger to cool a liquid hydrocarbon solution. The NH_3 leaves the exchanger at −18°C and 120 kPa pressure. The heat duty is 2800 J/s. How much NH_3 is required per hour? NH_3 data are available in most reference books and the *Chemical Engineers' Handbook*.

4.80. The feed to a chemical reactor is 1000 kg mol/hr with an enthalpy of 2.00×10^6 J/kg mol. Chemical reaction generates 4.63×10^9 J of energy. The products of the reactions (the exit stream of reactor) have a flow rate of 500 kg mol/hr with an enthalpy of 8.00×10^6 J/kg mol. What is the heat loss to the surrounding from the reactor?

4.81. Your company produces small power plants that generate electricity by expanding waste process steam in a turbine. You are asked to study the turbine to determine if it is operating as efficiently as possible. One way to ensure good efficiency is to have the turbine operate adiabatically. Measurements show that for steam at 500°F and 250 psia

(1) The work output of the turbine is 86.5 hp.

(2) The rate of steam usage is 1000 lb/hr.

(3) The steam leaves the turbine at 14.7 psia and consists of 15% moisture (i.e., liquid H_2O).

Is the turbine operating adiabatically? Support your answer with calculations.

4.82. What is T_4 for two perfect gas streams flowing at the same (mole) rate in a counter-current heat exchanger? See Fig. P4.82. Assume that $C_{p_A} = C_{p_B}$ and are constant. Neglect any heat loss from the exchanger.

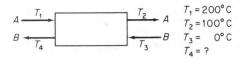

$$T_1 = 200°C$$
$$T_2 = 100°C$$
$$T_3 = 0°C$$
$$T_4 = ?$$

Fig. P4.82

4.83. Solvent emissions are likely to be specifically regulated in areas of the country with present or putative smog problems. These regulations may limit total hydrocarbon emissions. To reduce solvent vapor losses and prevent explosions, a mixture of CO_2 (60.0%) and benzene vapor (40.0%) at 93°C and 100 kPa is cooled to 4°C and 100 kPa. Compute the heat removal required per kg mol of the original gas, and the fraction of the benzene recovered as liquid.

4.84. Humid air at 1 atm and 200°F, containing 0.0645 lb of H_2O/lb dry air, enters a cooler at the rate of 1000 lb of dry air per hour (plus the accompanying water vapor). The air leaves the cooler at 100°F, saturated with water vapor (0.0434 lb of H_2O/lb of dry air).

  (a) How many pounds of water are condensed per pound of dry air entering the process?

  (b) How much heat is given up by the process per hour?

4.85. A dryer with a rating of 5.0×10^5 J/s is available to dry the products leaving the separator of a plant, as shown in Fig. P4.85. Is the heat input to the dryer large enough to do the job?

	C_p [J/(kg)(°C)]	H_{vap} (J/kg)
Urea	2500	
Solvent	2800	4.50×10^5

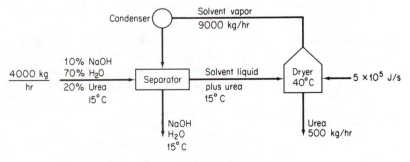

Fig. P4.85

Section 4.6

4.86. Solid carbon dioxide (dry ice) has innumerable uses in industry and in research. Because it is easy to manufacture, the competition is severe, and it is necessary to make dry ice very cheaply to be successful in selling it. In a proposed plant to make dry ice, the gaseous CO_2 is compressed isothermally and essentially reversibly from 6 psia and 40°F to a specific volume of 0.05 ft³/lb$_m$.

(a) What is the final state of the compressed CO_2?

(b) Compute the work of compression.

(c) What is the heat removed?

(d) If the actual efficiency of the compressor (relative to a reversible compressor) is 85% and the electricity to run the compressor motor costs $0.02/kWh, what is the cost of compression of the solid CO_2 in dollars per pound of dry ice?

4.87. An ideal gas with a heat capacity of $C_p = 27 + 11T \times 10^{-3}$ [where T is in K and C_p is in J/(g mol)(K)] passes through a compressor at the rate of 50 kg mol/min. The entering conditions of the gas are 100 kPa and 21°C. The exit conditions are 1500 kPa and 650°C. Ignore potential and kinetic energy effects, and assume that the process is adiabatic. Is it a reversible process? Show how you can prove by equations whether it is or is not.

4.88. A power plant is as shown in Fig. P4.88. If the pump moves 100 gal/min into the boiler with an overall efficiency of 40%, find the horsepower required for the pump. List all additional assumptions required.

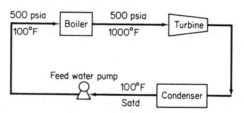

Fig. P4.88

4.89. A firefighter drives a pumper up to a river and pumps water through the hose to a fire 100 ft higher than the river. The water comes out of the fire hose at 100 ft/sec. Ignoring friction, what horsepower engine is required to deliver 5000 gal of water per minute? The river flows at 2 mi/hr.

4.90. A phase change (condensation, melting, etc.) of a pure component is an example of a reversible process because the temperature and pressure remain constant during the change. Use the definition of work to calculate the work done by butane when 1 kg of saturated liquid butane at 70 kPa vaporizes completely. Can you calculate the work done by the butane from the energy balance alone?

4.91. Calculate the work done when 1 lb mol of water in a open vessel evaporates completely at 212°F. Express your results in Btu.

4.92. One kilogram of steam goes through the following reversible process. In its initial state (state 1) it is at 2700 kPa and 540°C. It is then expanded isothermally to state 2, which is at 700 kPa. Then it is cooled at constant volume to 400 kPa (state 3). Next it is cooled at constant pressure to a volume of 0.4625 m³/kg (state 4). Then it is compressed adiabatically to 2700 kPa, and 425°C (state 5), and finally it is heated at constant pressure back to the original state.

(a) Sketch the path of each step in a *p–V* diagram.

(b) Compute ΔU and ΔH for each step and for the entire process.

(c) Compute Q and W whenever possible for each step of the process.

4.93. An office building requires water at two different floors. A large pipe brings the city water supply into the building in the basement level, where a booster pump is located. The water leaving the pump is transported by smaller insulated pipes to the second and fourth floors, where the water is needed. Calculate the *minimum* amount of work per unit time (in horsepower) that the pump must do in order to deliver the necessary water, as indicated in Fig. P4.93. (*Minimum* refers to the fact that you should neglect the friction and pump energy losses in your calculations.) The water does not change temperature.

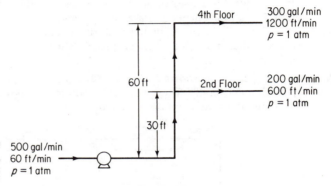

Fig. P4.93

4.94. Water at 20°C is being pumped from a constant-head tank open to the atmosphere to an elevated tank kept at a constant pressure of 1150 kPa in an experiment as shown in Fig. P4.94. If water is flowing in the 5.0-cm line at a rate of 0.40 m³/min, find

(a) The rating of the pump in joules per kilogram

(b) The rating of the pump in joules per minute

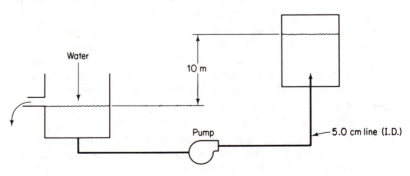

Fig. P4.94

The pump and motor have an overall efficiency of 70% and the energy loss in the line can be determined to be 60.0 J/kg.

Section 4.7

4.95. Compare prices of five fuels:
(a) Natural gas (all CH_4) at $2.20 per 10^3 ft^3
(b) No. 2 fuel oil (33°API) at $0.85 per gallon
(c) Dry yellow pine at $95 per cord
(d) High-volatile A bituminous coal at $55.00 per ton
(e) Electricity at $0.032 per kilowatt
List the cost of each in descending order per 10^6 Btu. As far as energy consumption is concerned, would you be better off with a gas dryer or an electric dryer? Should you heat your house with wood, No. 2 fuel oil, or coal? (*Note:* A cord of wood is a pile 8 ft long, 4 ft high, and 4 ft wide.)

4.96. How would you determine the heat of formation of gaseous fluorine at 25°C and 1 atm?

4.97. Given the data in the table, determine:
(a) ΔH_f° for FeO(s)
(b) ΔH_{rxn}° for FeO(s) $+ 2H^+ \longrightarrow H_2O(l) + Fe^{2+}$

Reaction	ΔH_{rxn}° (J)
Fe(s) $+ 2H^+ \longrightarrow H_2(g) + Fe^{2+}$	$-86{,}200$
2Fe(s) $+ \frac{3}{2}O_2(g) \longrightarrow Fe_2O_3(s)$	$-822{,}200$
2FeO(s) $+ \frac{1}{2}O_2(g) \longrightarrow Fe_2O_3(s)$	$-284{,}000$

4.98. What is the heat of formation of 1 g mol of S gas at 717.76 K?

4.99. Calculate the heat of reaction at 77°F and 1 atm for the following reactions [these are all methods of manufacturing $H_2(g)$]:
(a) $CH_4(g) + H_2O(g) \longrightarrow CO(g) + 3H_2(g)$
(b) $CO(g) + H_2O(g) \longrightarrow CO_2(g) + H_2(g)$
(c) $CH_4(g) \longrightarrow C(s) + 2H_2(g)$
(d) $3Fe(s) + 4H_2O(g) \longrightarrow Fe_3O_4(s) + 4H_2(g)$
(e) $2H_2O(l) \longrightarrow 2H_2(g) + O_2(g)$
(f) $2NH_3(g) \longrightarrow N_2(g) + 3H_2(g)$

4.100. What is the heat of reaction in J at 25°C and 1 atm for the following reactions?
(a) $NH_3(g) \longrightarrow \frac{1}{2}N_2(g) + \frac{3}{2}H_2(g)$
(b) $2H_2O(g) \longrightarrow 2H_2(g) + O_2(g)$
(c) $H_2O(l) \longrightarrow H_2(g) + \frac{1}{2}O_2(g)$

4.101. Calculate the heat of reaction at the standard reference state for the following reactions:
(a) $CO_2(g) + H_2(g) \longrightarrow CO(g) + H_2O(l)$
(b) $2CaO(s) + 2MgO(s) + 4H_2O(l) \longrightarrow 2Ca(OH)_2(s) + 2Mg(OH)_2(s)$
(c) $Na_2SO_4(s) + C(s) \longrightarrow Na_2SO_3(s) + CO(g)$
(d) $NaCl(s) + H_2SO_4(l) \longrightarrow NaHSO_4(s) + HCl(g)$

(e) $NaCl(s) + 2SO_2(g) + 2H_2O(l) + O_2(g) \longrightarrow 2Na_2SO_4(s) + 4HCl(g)$

(f) $SO_2(g) + \frac{1}{2}O_2(g) + H_2O(l) \longrightarrow H_2SO_4(l)$

(g) $N_2(g) + O_2(g) \longrightarrow 2NO(g)$

(h) $Na_2CO_3(s) + 2Na_2S(s) + 4SO_2(g) \longrightarrow 3Na_2S_2O_3(s) + CO_2(g)$

(i) $NaNO_3(s) + Pb(s) \longrightarrow NaNO_2(s) + PbO(s)$

(j) $Na_2CO_3(s) + 2NH_4Cl(s) \longrightarrow 2NaCl(s) + 2NH_3(g) + H_2O(l) + CO_2(g)$

(k) $Ca_3(PO_4)_2(s) + 3SiO_2(s) + 5C \longrightarrow 3CaSiO_3(s) + 2P(s) + 5CO(g)$

(l) $P_2O_5(s) + 3H_2O(l) \longrightarrow 2H_3PO_4(l)$

(m) $CaCN_2(s) + 2NaCl(s) + C(s) \longrightarrow CaCl_2(s) + 2NaCN(s)$

(n) $NH_3(g) + HNO_3(aq) \longrightarrow NH_4NO_3(aq)$

(o) $2NH_3(g) + CO_2(g) + H_2O(l) \longrightarrow (NH_4)_3CO_2(aq)$

(p) $(NH_4)_2CO_3(aq) + CaSO_4 \cdot 2H_2O(s) \longrightarrow CaCO_3(s) + 2H_2O(l)$
    $+ (NH_4)_2SO_4(aq)$

(q) $CuSO_4(aq) + Zn(s) \longrightarrow ZnSO_4(aq) + Cu(s)$

(r) $\underset{\text{benzene}}{C_6H_6(l)} + Cl_2(g) \longrightarrow \underset{\text{chlorobenzene}}{C_6H_5Cl(l)} + HCl(g)$

(s) $CS_2(l) + Cl_2(g) \longrightarrow S_2Cl_2(l) + CCl_4(l)$

(t) $\underset{\text{ethylene}}{C_2H_4(g)} + HCl(g) \longrightarrow \underset{\text{ethylchloride}}{CH_3CH_2Cl(g)}$

(u) $\underset{\text{methyl alcohol}}{CH_3OH(g)} + \frac{1}{2}O_2(g) \longrightarrow \underset{\text{formaldehyde}}{H_2CO(g)} + H_2O(g)$

(v) $\underset{\text{acetylene}}{C_2H_2(g)} + H_2O(l) \longrightarrow \underset{\text{acetaldehyde}}{CH_3CHO(l)}$

(w) $\underset{\text{n-butane}}{n\text{-}C_4H_{10}(g)} \longrightarrow \underset{\text{ethylene}}{C_2H_4(g)} + \underset{\text{ethane}}{C_2H_6(g)}$

(x) $\underset{\text{ethylene}}{C_2H_4(g)} + \underset{\text{benzene}}{C_6H_6(l)} \longrightarrow \text{ethyl benzene(l)}$

(y) $2\underset{\text{ethylene}}{C_2H_4(g)} \longrightarrow \underset{\text{1-butane}}{C_4H_8(l)}$

(z) $\underset{\text{propene}}{C_3H_6(g)} + \underset{\text{benzene}}{C_6H_6(l)} \longrightarrow \underset{\text{isopropyl benzene}}{\text{cumene(l)}}$

4.102. The process of converting SO_2 to sulfuric acid with the aid of nitrogen oxides can be expressed on an overall basis as follows:

$$SO_2(g) + NO_2(g) + H_2O(l) \longrightarrow H_2SO_4(l) + NO(g)$$

This process has been used by the chemical industry for over a century in what is known as the chamber process. With the high dilution of N_2 that accompanies the SO_2, the raw gases have to pass at low velocities through large lead chambers in order to react sufficiently. If 10,000 ft³/hr of NO (measured at S.C.) and 2750 lb/hr of H_2SO_4 are produced, what is the cooling required if the reaction is to proceed isothermally at 77°F?

4.103. To take care of peak loads in the winter, the local gas company wishes to supplement its limited supply of natural gas (CH_4) by catalytically hydrogenating a heavy liquid petroleum oil that it can store in tanks. The supplier says that the oil has a U.O.P. Characterization Factor of 12.0 and an API gravity of 30°API. To minimize carbon deposition on the catalyst, twice the required amount of hydrogen is introduced with the oil. The reactants enter at 70°F and after heat exchange the products leave at 70°F. Calculate the following, per 1000 ft³ of CH_4 at the standard conditions in the gas industry:

(a) The pounds oil required

(b) The heat input or output to the catalytic reactor

4.104. A technician determined the following standard heats of combustion at 25°C and 1 atm for benzene:

$$C_6H_6(g): \quad \Delta H_c = -3{,}301.5 \text{ kJ/g mol}$$

$$C_6H_6(l): \quad \Delta H_c = -3{,}267.6 \text{ kJ/g mol}$$

What is the heat of combustion for both liquid and gaseous C_6H_6 if the final product is $H_2O(g)$?

4.105. Calculate the heat of combustion for neopentane gas (2, 2-dimethyl propane) given that the heat of formation of gaseous neopentane is -1.6598×10^5 J/g mol.

4.106. Calculate the heat of formation of isopentane gas (2-methylbutane) at 25°C and 1 atm given that the heat of combustion of gaseous isopentane is -3.5281×10^6 J/g mole based on H_2O as a liquid.

4.107. The standard heats of combustion of various compounds are listed below. Calculate the standard heat of formation of each.

Compound	$-\Delta H_c$ (kcal/g mol)
(a) cyclopentane(g), C_5H_{10}	793.39
(b) cyclopentane(l), C_5H_{10}	786.54
(c) n-propylbenzene(l), C_9H_{12}	1247.19
(d) 3-ethylhexane(g), C_8H_{18}	1316.87
(e) n-hexane(g), C_6H_{14}	1002.57

4.108. Determine the heat of reaction at 25°C and 1 atm data for the following reactions from heat of combustion data:
(a) $H_2S(g) + \frac{3}{2}O_2(g) \longrightarrow SO_2(g) + H_2O(g)$
(b) $N_2(g) + O_2(g) \longrightarrow 2NO(g)$
(c) $NO(g) + \frac{5}{2}H_2(g) \longrightarrow NH_3(g) + H_2O(g)$

4.109. A mixture of gaseous methane (CH_4) and hydrogen (H_2) was found to have a heat of combustion at 25°C of -150.0 kcal/g mol; however, this value was determined under conditions such that the water produced left as a vapor rather than as a liquid as required for the standard values. Calculate the composition of the mixture.

4.110. What is the higher heating value of 1 m³ of n-propylbenzene measured at 25°C and 1 atm and with a relative humidity of 40%?

4.111. An off-gas from a crude oil topping plant has the following composition:

Component	Vol %
methane	88
ethane	6
propane	4
butane	2

(a) Calculate the higher heating value on the following bases: **(1)** Btu per pound, **(2)**

Btu per mole, and (3) Btu per cubic foot of off-gas measured at 60°F and 760 mm Hg.

(b) Calculate the lower heating value on the same three bases indicated above.

4.112. The chemist for a gas company finds a gas analyzes 9.2% CO_2, 0.4% C_2H_4, 20.9% CO, 15.6% H_2, 1.9% CH_4, and 52.0% N_2. What should the chemist report as the gross heating value of the gas?

4.113. Calculate the heat of reaction at 25°C for the following reaction if it takes place at constant volume.

$$C_2H_4(g) + 2H_2(g) \longrightarrow 2CH_4(g)$$

4.114. If 1 lb mol of Cu and 1 lb mol of H_2SO_4(100%) react together completely in a bomb calorimeter, how many Btu is absorbed (or evolved)? Assume that the products are H_2(g) and solid $CuSO_4$. The initial and final temperatures in the bomb are 25°C.

4.115. Acetylene gas at 25°C is burned at 1 atm with 20% excess air at 35°C and containing no moisture. What is the temperature of the gas leaving the burner if the burner is sufficiently insulated so that no heat is transferred to the surroundings during the combustion? Assume 100% completion of the reaction.

4.116. An older way to make HCl gas is by the following reaction:

$$NaHSO_4(c) + NaCl(c) = Na_2SO_4(c) + HCl(g)$$

Calculate the standard heat of reaction per kilogram of $NaHSO_4$ for this reaction if 40.0 kg of $NaHSO_4$ is mixed with 25.0 kg of NaCl, and the percent conversion in the reaction is 80%.

4.117. Calculate the heat of reaction of the following reactions at the stated temperature:

(a) $MgO(s) + C(s) + Cl_2(g) \xrightarrow{850°C} MgCl_2(g) + CO(g)$

(b) $CH_3OH(g) + \frac{1}{2}O_2(g) \xrightarrow{200°C} H_2CO(g) + H_2O(g)$
    methyl alcohol   formaldehyde

(c) $SO_2(g) + \frac{1}{2}O_2(g) \xrightarrow{300°C} SO_3(g)$

4.118. Sulfur can be recovered from the H_2S in natural gas by the following reaction:

$$3H_2S(g) + 1\frac{1}{2}O_2(g) \longrightarrow 3S(s) + 3H_2O(g)$$

For the materials as shown, with H_2S and O_2 entering the process at 300°F and the products leaving the reactor at 100°F, calculate the heat of reaction in British thermal units per pound of sulfur formed. Assume that the reaction is complete. If it is only 75% complete, will this change your answer, and if so, how much?

4.119. Calculate the net heating value of hydrogen at 200°C.

4.120. What is the gross heating value of H_2 at 0°C? Gross heating value means the heat of reaction with the product H_2O as a liquid at 1 atm.

$$H_2(g) + \frac{1}{2}O_2(g) \longrightarrow H_2O(l)$$

$$\Delta H_{rxn} \text{ at } 25°C = -285,810 \text{ J/g mol } H_2$$

$$C_p \text{ of } O_2 \text{ and } N_2 = 2.79 \text{ J/(g mol)(K)}$$

What is the net heating value (i.e., heat of reaction) with H_2O as a gas?

4.121. The net heating value of CH_4 at 50°C is 191,870 cal/g mol. What is the gross heating value at 0°C?

C_p for $CH_4 = 8.52 + 0.0096\,T$
$C_p = $ cal/(g mol)(°C)
$T = $ °C

T (°C)	p^*_{mm}	$H_2O\ \Delta H_{vap}$ (cal/g)
0	4.6	597
18	15.5	586
25	23.7	582
50	92.3	568
100	760.0	539

4.122. Propane, butane, or liquefied petroleum gas (LPG) has seen practical service in passenger automobiles for 30 years or more. Because it is used in the vapor phase, it pollutes less than gasoline but more than natural gas. A number of cars in the Clean Air Car Race ran on LPG. The table below lists their results and those for natural gas. It must be kept in mind that these vehicles were generally equipped with platinum catalyst reactors and with exhaust-gas recycle. Therefore, the gains in emission control did not come entirely from the fuels.

	Natural Gas, Avg. 6 Cars	LPG, Avg. 13 Cars	1975 Fed. Std.
HC (g/mi)	1.3	0.49	0.22
CO (g/mi)	3.7	4.55	2.3
NO_x (g/mi)	0.55	1.26	0.6

Suppose that in a test butane gas at 100°F is burned completely with the stoichiometric amount of heated air at 400°F and a dew point of 77°F in an engine. To cool the engine, there are 12.5 lb of steam at 100 psia and 95% quality generated from water at 77°F per pound of butane burned. It may be assumed that 7% of the gross heating value of the butane is lost as radiation from the engine. Will the exhaust gases leaving the engine exceed the temperature limit of the catalyst of 1500°F?

4.123. The Smog Abatement Co. is authorized to market and install conversion kits so that automobiles can use natural gas. It requires about 4 hr to install the conversion equipment in each car and costs about $400 to $500 per car. Customers can recover the expense of conversion because natural gas can be sold to the public more cheaply than gasoline for the equivalent amount of energy, according to the company. To provide some specific numbers to promote the conversion service, you are asked to make the following computations. Assume that the automobile engine will burn ethane, C_2H_6.
(a) For theoretical complete combustion with dry air, what weight of air is used per hour when the motor uses 480 ft³ of fuel per hour measured at 32°F and 760 mm Hg pressure?
(b) Assume that the fuel enters the engine at 150°F and the air at 250°F because of

heat exchange with the exhaust gases after they leave the engine. Assume that the exhaust gases leave the engine at 2000°F and that 15% of the net heating value of the fuel is lost by radiation from the engine. Calculate the amount of energy that is available to operate the engine.

4.124. In the production of ethylene from ethane and propane, the following yields were found:

Reaction	Percent of Reactant Proceeding by Process Shown
$C_2H_6 \longrightarrow C_2H_4 + H_2$	100
$C_3H_8 \longrightarrow C_2H_4 + CH_4$	60
$C_3H_8 \longrightarrow C_3H_6 + H_2$	30
$2C_3H_8 \longrightarrow 2C_2H_6 + C_2H_4$	10

If a mixture of 50% ethane (C_2H_6) and 50% propane (C_3H_8) is passed through the reactor, what is the standard heat of reaction per mole of feed?

4.125. At 25°C in a bomb calorimeter it is found that Mg unites with $\frac{1}{2}O_2$ to form MgO with the evolution of 146,000 cal/g mol of MgO. What is the heat of reaction at constant pressure at 2800°C?

MgO (melting point 2800°C):

t (°C)	0	50	100	200	400	800	1000	1500	2000	2500
C_p	0.21	0.24	0.25	0.26	0.28	0.31	0.32	0.36	0.39	0.42

cal/(g)(°C)

Mg (melting point 651°C): ΔH fusion $= 78$ cal/g

C_p solid [cal/(g mol)(°C)] $= 4.85 + 3.76 \times 10^{-3}T - 7.47 \times 10^{-7}T^2$ (T in K)

C_p liquid [cal/(g mol)(°C)] $= 8.76$

4.126. The reaction $SO_2 + \frac{1}{2}O_2 \longrightarrow SO_3$ would seem to offer a simple method of making sulfur trioxide. However, both the reaction rate and equilibrium considerations are unfavorable for this reaction. At 600°C the reaction is only about 70% complete. If the products leave the reactor at 600°C and the reactants enter at 25°C, what is the heat evolved or absorbed from the system under these conditions?

4.127. Sulfur dioxide gas is oxidized in 100% excess air with 80% conversion to SO_3. The gases enter the converter at 400°C and leave at 450°C. How many joules are absorbed in the heat interchanger of the converter per kilogram mole of SO_2 introduced?

4.128. *n*-Heptane is dehydrocyclicized in the hydroforming process to toluene by means of catalysts as follows:

$$C_7H_{16} \longrightarrow C_6H_5CH_3 + 4H_2$$

Assuming that a yield of 35% of the theoretical is obtained under the conditions

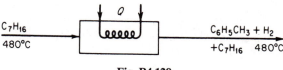

Fig. P4.128

shown in Fig. P4.128, how much heat is required in the process per 1000 kg of toluene produced? *Note:*

$$C_{p_m}(C_7H_{16})_{25-480°C} = 2.34 \text{ kJ/(kg)(K)}$$
$$C_{p_m}(C_6H_5CH_3)_{25-480°C} = 2.59 \text{ kJ/(kg)(K)}$$

4.129. Ammonia oxidation furnishes a large amount of nitric acid for industry. When NH_3 is oxidized using either air or oxygen in the presence of a suitable catalyst, at least two reactions are possible:

$$4NH_3 + 5O_2 \longrightarrow 4NO + 6H_2O(g) \tag{1}$$
$$4NH_3 + 3O_2 \longrightarrow 2N_2 + 6H_2O(g) \tag{2}$$

The first reaction yields NO which can be converted into HNO_3, while the second results in a loss of fixed nitrogen. At 1000°C, 92% of the NH_3 can be converted into NO while the remainder goes by reaction (2). What is the heat of reaction at 1000°C per kilogram of NH_3 supplied to the reactor if the catalytic burning is with the stoichiometric amount of pure oxygen? What would be the change in the heat of reaction if 10% excess air at 1000°C [based on reaction (1)] were used?

4.130. In the manufacture of benzaldehyde, a mixture of toluene and air is passed through a catalyst bed where it reacts to form benzaldehyde:

$$C_6H_5CH_3 + O_2 \longrightarrow C_6H_5CHO + H_2O$$

Dry air and toluene gas are fed to the converter at a temperature of 350°F and at atmospheric pressure. To maintain a high yield, air is supplied in 100% excess over that required for complete conversion of the toluene charged. The degree of completion of the reaction, however, is only 13%. Owing to the high temperature in the catalyst bed, 0.5% of the toluene charged burns to form CO_2 and H_2O:

$$C_6H_5CH_3 + 9O_2 \longrightarrow 7CO_2 + 4H_2O$$

Cooling water is circulated through a jacket on the converter, entering at 80°F and leaving at 105°F. The hot gases leave the converter at 379°F. During a 4-hr test run, a water layer amounting to 29.3 lb was collected after the exit gases had been cooled. Calculate the following:

(a) The composition of the inlet stream to the converter
(b) The composition of the exhaust gas from the converter
(c) The gallons per hour of cooling water required in the converter jacket
Additional data: The heat capacities of both toluene and benzaldehyde *gas* may be taken as 31 Btu/(lb mol)(°F); C_p for benzaldehyde liquid is 0.43 cal/(g)(°C).

4.131. Calculate the theoretical flame temperature for CO burned at 110 kPa absolute pressure with 100% excess air, when the reactants enter at 93°C.

4.132. If CO at constant pressure is burned with excess air and the theoretical flame temperature is 1255 K, what was the percentage of excess air used? The reactants enter at 93°C.

4.133. Which substance will give the higher theoretical flame temperature if the inlet percentage of excess air and temperature conditions are identical: (1) CH_4, (2) C_2H_6 or (3) C_4H_8?

4.134. A mixed coke oven and synthetic gas analysis is:

CO_2	3.4%
O_2	0.3%
N_2	12.0%
CO	17.4%
H_2	36.8%
CH_4	24.9%
C_2H_4	3.7%
C_6H_6	1.5%

(a) Calculate the gross and net heating values of this gas in Btu/ft³ (S.C. gas industry).

(b) When this gas at 65°F is burned with the theoretical amount of dry air at 65°F, what will be the theoretical flame temperature?

(c) If the air is in 50% excess and is preheated to 400°F, what will be the theoretical flame temperature?

4.135. A power plant burns natural gas (90% CH_4, 10% C_2H_6) at 25°C and 1 atm with 70% excess air at the same conditions. Calculate the theoretical maximum temperature in the boiler if all products are in the gaseous state.

4.136. In Problem 4.135, if the air is preheated to 260°C, what will be the maximum temperature?

Section 4.8

4.137. (a) In a petrochemical plant 1 ton/hr of 50% by weight H_2SO_4 is being produced at 250°F by concentrating 30% H_2SO_4 supplied at 100°F. From the data provided, find the heat that has to be supplied per hour.

Btu Evolved at 100°F per lb mol H_2SO_4	Moles of H_2O per mol of H_2SO_4
18,000	2
21,000	3
23,000	4
24,300	5
27,800	10
31,200	22

C_p [Btu/(lb)(°F)]	30% H_2SO_4	50% H_2SO_4
100	0.80	0.70
250	0.75	0.60

The C_p of water vapor is about 0.50, and for water use, 1.0.

(b) Using the same data as in part (a), what is the temperature of a solution of 30% by weight H_2SO_4 if it is made from 50% H_2SO_4 at 100°F and pure water at 200°F if the process is adiabatic?

4.138.* (a) From the data plot the enthalpy of 1 *mole of solution* at 27°C as a function of the weight percent HNO_3. Use as reference states liquid water at 0°C and liquid HNO_3 at 0°C. You can assume that C_p for H_2O is 75 J/(g mol)(°C) and for HNO_3, 125 J/(g mol)(°C).

$-\Delta H_{soln}$ at 27°C (J/g mol HNO_3)	Moles H_2O Added to 1 mole HNO_3
0	0.0
3,350	0.1
5,440	0.2
6,900	0.3
8,370	0.5
10,880	0.67
14,230	1.0
17,150	1.5
20,290	2.0
24,060	3.0
25,940	4.0
27,820	5.0
30,540	10.0
31,170	20.0

(b) Compute the energy absorbed or evolved at 27°C on making a solution of 4 moles of HNO_3, and 4 moles of water by mixing a solution of $33\frac{1}{3}$ mole % acid with one of 60 mole % acid.

4.139.* *National Bureau of Standards Circular 500* gives the following data for calcium chloride (mol. wt. 111) and water:

Formula	State	$-\Delta H_f$ at 25°C (kcal/g mol)
H_2O	Liquid	68.317
	Gas	57.798
$CaCl_2$	Crystal	190.0
	in 25 moles of H_2O	208.51
	50	208.86

Note: Table continues on following page.

Formula	State	$-\Delta H_f$ at 25°C (kcal/g mol)
	100	209.06
	200	209.20
	500	209.30
	1000	209.41
	5000	209.60
	∞	209.82
$CaCl_2 \cdot H_2O$	Crystal	265.1
$CaCl_2 \cdot 2H_2O$	Crystal	335.5
$CaCl_2 \cdot 4H_2O$	Crystal	480.2
$CaCl_2 \cdot 6H_2O$	Crystal	623.15

Calculate the following:
(a) The energy evolved when 1 lb mol of $CaCl_2$ is made into a 20% solution at 77°F
(b) The heat of hydration of the dihydrate to the hexahydrate
(c) The energy evolved when a 20% solution containing 1 lb mol of $CaCl_2$ is diluted with water to 5% at 77°F

4.140. A vessel contains 100 g of an NH_4OH–H_2O liquid mixture at 25°C and 1 atm that is 15.0% by weight NH_4OH. Just enough aqueous H_2SO_4 is added to the vessel from an H_2SO_4 liquid mixture at 25°C and 1 atm (i.e., 25.0 mole % H_2SO_4) so that the reaction to $(NH_4)_2SO_4$ is complete. After the reaction, the products are again at 25°C and 1 atm. How much heat (J) is absorbed or evolved by this process? It may be assumed that the final volume of the products is equal to the sum of the volumes of the two initial mixtures.

PROBLEMS TO PROGRAM ON THE COMPUTER

4.1. The enthalpy of a mixture of constant composition at any constant temperature T is

$$\Delta H_{\text{mixture}}(T, p) = \sum_{i=1}^{n} x_i \sum H_i^0(T, p) - \Delta H_{\text{mixture}}^D(T, p)$$

where the deviation, ΔH^D, is a correction because of high pressure. Use of the Benedict–Webb–Rubin (BWR) equation (see Table 3.2) gives the following for ΔH^D:

$$\Delta H^D = \left(B_0 RT - 2A_0 - 4\frac{C_0}{T^2}\right)\rho + \frac{1}{2}(2bRT - 3a)\rho^2 + \frac{6}{5}(a\alpha\rho^5)$$
$$+ \frac{c\rho}{T^2}\left[\frac{3[1 - \exp(-\gamma\rho^2)]}{\gamma\rho^2} - \left(\frac{1}{2} - \gamma\rho^2\right)\exp(-\gamma\rho^2)\right]$$

Use the density values calculated in computer problem 3.3 for ρ, and prepare an additional subroutine to give $\Delta H_{\text{mixture}}^D(T, p)$ for selected T and p pairs. Refer to problem 3.3 for other pertinent information.

4.2. Prepare a computer program to find the adiabatic flame temperature of a gas such as in problem 4.111. Assume that the principal reactant is completely burned.

4.3.   Prepare a more general program than the one in problem 4.2 to compute the adiabatic flame temperature. Assume that several hydrocarbons of the form CH_n can be burned with various amounts of excess air. The percentage of excess air is to be read into the program as well as the type of fuel gas or solid. The only combustion products that need to be considered are carbon dioxide, carbon monoxide, water, hydrogen, oxygen, and nitrogen. Make provision for the following additional parameters:

(a) Pressure in the burner
(b) Humidity of the incoming air
(c) Number of moles of each hydrocarbon and of nitrogen in the fuel stream
(d) Standard heats of formation for all the hydrocarbon gases burned and the products (Assume that gaseous water is formed in the reaction.)
(e) Heat capacities of all components
(f) Inlet reactant temperatures, which may be different for each reactant.

The gases can be treated as an ideal mixture.

INDUSTRIAL CHEMICAL DATA

HOW MADE	MAJOR END USES	MAJOR DERIVATIVES	ANNUAL PRODUCTION, PRICES (USA 1981)
	Soda ash	Na_2CO_3	
Recovered from trona ore or brine, synthesized from salt and limestone using ammonia	Glass 55%, chemicals manufacture 25%, pulp and paper 5%, soap and detergents 5%		8.4×10^6 tons $40–$86/ton

Urea   H$_2$N—C—NH$_2$ (with O double-bonded to C)

HOW MADE	MAJOR END USES	MAJOR DERIVATIVES	ANNUAL PRODUCTION, PRICES (USA 1981)
Reaction of ammonia and carbon dioxide under pressure	Fertilizer 75%, adhesives and plastics 15%, animal feed 5%	Urea-formaldehyde resins 10%, melamine resins 5%	6.9×10^6 tons $160–$190/ton

Benzene

HOW MADE	MAJOR END USES	MAJOR DERIVATIVES	ANNUAL PRODUCTION, PRICES (USA 1981)
Extracted from catalytic reformate made in oil refineries, from pyrolysis gasoline made in steam-cracking olefins plants, and from light oils in coking coal; dealkylation of toluene	Polystyrene 25%, nylon 20%, other styrenic polymers 10%, rubber 5%	Ethylbenzene 50%, cumene 15%, cyclohexane 15%, aniline 5%	1.6×10^9 gal $1.65/gal

Adapted from Chemical and Engineering News by permission of the American Chemical Society.

5

APPLICATIONS
OF COMBINED MATERIAL
AND ENERGY BALANCES

Now that you have accumulated some experience in making energy balances and in the principles of thermochemistry, it is time to apply this knowledge to situations and problems involving both material and energy balances. You have already encountered some simple examples of combined material and energy balances, as, for example, in the calculation of the adiabatic reaction temperature, where a material balance provides the groundwork for the writing of an energy balance. In fact, in all energy balance problems, however trifling they may seem, you must know the amount of material entering and leaving the process if you are to apply successfully the appropriate energy balance equation(s).

In this chapter we consider both those problems that require you to make a preliminary material balance prior to making an energy balance and problems that require you to solve simultaneous material and energy balances. In line with this discussion we take up two special types of charts, enthalpy–concentration charts and humidity charts, which are prepared by combining material and energy balances and which are useful in a wide variety of practical problems. Figure 5.0 shows the relationships among the topics to be discussed in this chapter and the relationships with topics in previous chapters.

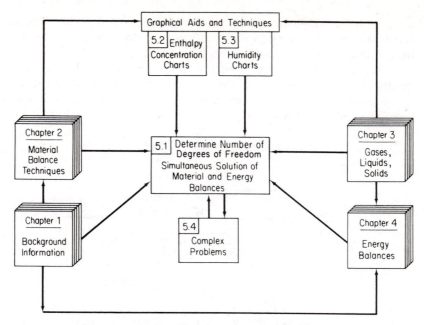

Fig. 5.0 Hierarchy of topics to be studied in this chapter (section numbers are in the upper left-hand corner of the boxes).

Section 5.1   Simultaneous Material and Energy Balances

Your objectives in studying this section are to be able to:

1. Calculate the number of degrees of freedom (decision variables) for single units and simple combination of units.
2. Solve combined material and energy balance problems with and without chemical reaction that are more complex than those of Chap. 4 by using the 10-step strategy.

For any specified system or piece of equipment you can write:

(a) A total material balance

(b) A material balance for each chemical component (or each atomic species, if preferred)

You can use the energy balance to add one additional independent overall equation to your arsenal, but you cannot make energy balances for each individual chemical component in a mixture or solution. As we mentioned previously, you need one independent equation for each unknown in a given problem. The energy balance often provides the extra piece of information to help you resolve an apparently insuperable calculation composed solely of material balances.

We can illustrate any given system or piece of equipment *in the steady state* by drawing a diagram such as Fig. 5.1. Figure 5.1 could represent a boiler, a distillation column, a drier, a human being, or a city, and takes into account processes involving a chemical reaction.

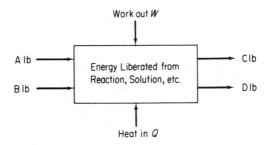

Fig. 5.1 Sketch of generalized flow process with chemical reaction.

Let x_i be the weight fraction of any component and $\Delta \hat{H}_i$ be the enthalpy per unit mass of any component with respect to some reference temperature. If the subscript numbers 1, 2, 3, and so on, represent the components in each stream, you can write the following material balances in algebraic form:

balance	*in*	=	*out*
Total	$A + B$	=	$C + D$
Component 1	$Ax_{A_1} + Bx_{B_1}$	=	$Cx_{C_1} + Dx_{D_1}$
Component 2	$Ax_{A_2} + Bx_{B_2}$	=	$Cx_{C_2} + Dx_{D_2}$
etc.			

As previously discussed, the number of independent equations is normally one less than the total number of equations.

In addition, an overall energy balance can be written as (ignoring kinetic and potential energy changes)

$$Q - W = (C\Delta \hat{H}_C + D\Delta \hat{H}_D) - (A\Delta \hat{H}_A + B\Delta \hat{H}_B)$$

For a more complex situation, examine the interrelated pieces of equipment in Fig. 5.2.

Again the streams are labeled, and the components are 1, 2, 3, and so on. How should you attack a problem of this nature? From an overall viewpoint, the problem reduces to that shown in Fig. 5.1. In addition, a series of material balances and an

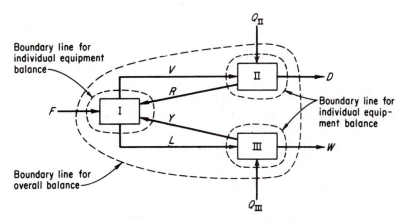

Fig. 5.2 Sketch for interrelated processes and streams.

balance	in	out
Entire process		
Total	$F = D + W$	
Component	$Fx_{F_i} = Dx_{D_i} + Wx_{W_i}$	
Energy	$Q_{\mathrm{II}} + Q_{\mathrm{III}} + F\,\Delta\hat{H}_F = D\,\Delta\hat{H}_D + W\,\Delta\hat{H}_W$	
Process I		
Total	$F + R + Y = V + L$	
Component	$Fx_{F_i} + Rx_{R_i} + Yx_{Y_i} = Vx_{V_i} + Lx_{L_i}$	
Energy	$F\,\Delta\hat{H}_F + R\,\Delta\hat{H}_R + Y\,\Delta\hat{H}_Y = V\,\Delta\hat{H}_V + L\,\Delta\hat{H}_L$	
Process II		
Total	$V = R + D$	
Component	$Vx_{V_i} = Rx_{R_i} + Dx_{D_i}$	
Energy	$Q_{\mathrm{II}} + V\,\Delta\hat{H}_V = R\,\Delta\hat{H}_R + D\,\Delta\hat{H}_D$	
Process III		
Total	$L = Y + W$	
Component	$Lx_{L_i} = Yx_{Y_i} + Wx_{W_i}$	
Energy	$Q_{\mathrm{III}} + L\,\Delta\hat{H}_L = Y\,\Delta\hat{H}_Y + W\,\Delta\hat{H}_W$	

energy balance can be made around each piece of equipment I, II, and III. Following Fig. 5.2 is a list of the material and energy balances that you can write for it. Naturally, all these equations are not independent, because, if you add up the total material balances around boxes I, II, and III, you simply have the grand overall material balance around the whole process. Similarly, if you add up the component balances for each box, you have a grand component balance for the entire process, and the

f the energy balances around boxes I, II, and III gives the overall energy balance.
the symmetry among these equations. The methods of writing and applying
balances are now illustrated by an example.

EXAMPLE 5.1 Simultaneous Material and Energy Balances

A distillation column separates 10,000 lb/hr of a 40% benzene–60% chlorobenzene liquid
solution which is at 70°F. The liquid product from the top of the column is 99.5% benzene,
while the bottoms (stream from the reboiler) contains 1% benzene. The condenser uses water
that enters at 60°F and leaves at 140°F, while the reboiler uses saturated steam at 280°F.
The reflux ratio (the ratio of the liquid overhead returned to the column to the liquid overhead
product removed) is 6 to 1. Assume that both the condenser and reboiler operate at 1 atm
pressure, that the temperature calculated for the condenser is 178°F and for the reboiler
268°F, and that the calculated fraction benzene in the vapor from the reboiler is 3.9 weight
% (5.5 mole %). Calculate the following:

(a) The pounds of overhead product (distillate) and bottoms per hour

(b) The pounds of reflux per hour

(c) The pounds of liquid entering the reboiler and the reboiler vapor per hour

(d) The pounds of steam and cooling water used per hour

Solution

Steps 1, 4, and 5: Figure E5.1 will help visualize the process and assist in pointing out what
additional data have to be determined. We have four unknowns, P, B, Q_C, and Q_S, but P and

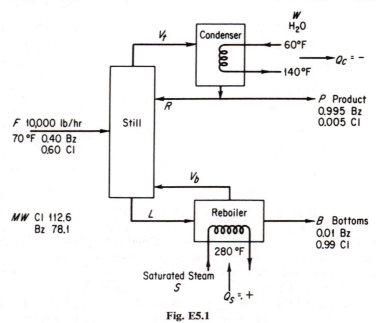

Fig. E5.1

B can be solved for separately from the two possible independent material balances. Do you know what they are? (*Hint:* The balances are for a system composed of all the units.) Then V_b can be calculated. Finally, the energy balance can be solved together with the material balances for the reboiler.

First we have to get some pertinent enthalpy or heat capacity data. Then we can make our material and energy balances. (The exit streams have the same composition as the solutions in the condenser or reboiler.)

Steps 2 and 3: Convert the reboiler analysis into mole fractions.

Basis: 100 lb of B

component	lb	mol. wt.	lb mol	mole fraction
Benzene	1	78.1	0.0128	0.014
Chlorobenzene	99	112.6	0.88	0.986
Total			0.8928	1.000

The heat capacity data for liquid benzene (Bz) and chlorobenzene (Cl) will be assumed to be as follows[1] (no enthalpy tables are available):

Temp. (°F)	C_p [Btu/(lb)(°F)]		$\Delta \hat{H}_{\text{vaporization}}$ (Btu/lb)	
	Cl	Bz	Cl	Bz
70	0.31	0.405		
90	0.32	0.415		
120	0.335	0.43		
150	0.345	0.45		
180	0.36	0.47	140	170
210	0.375	0.485	135	166
240	0.39	0.50	130	160
270	0.40	0.52	126	154

Step 6:   Basis: 10,000 lb of feed/hr

Step 4: Overall material balances:

Overall total material balance:

$$F = P + B$$
$$10,000 = P + B$$

Overall benzene balance:

$$Fx_F = Px_P + Bx_B$$
$$10,000(0.40) = P(0.995) + B(0.01)$$
$$10,000(0.40) = P(0.995) + (10,000 - P)(0.01)$$

[1] Data estimated from tabulations in D. Q. Kern, *Process Heat Transfer*, McGraw-Hill, New York, 1950, Appendix.

Step 9:
$$P = 3960 \text{ lb/hr} \longleftarrow \text{(a}_1\text{)}$$
$$B = 6040 \text{ lb/hr} \longleftarrow \text{(a}_2\text{)}$$

Step 4: Material balances around the condenser:

$$\frac{R}{P} = 6 \quad \text{or} \quad R = 6P = 6(3960) = 23{,}760 \text{ lb/hr} \longleftarrow \text{(b)}$$

Step 9:
$$V_t = R + P = 23{,}760 + 3960 = 27{,}720 \text{ lb/hr}$$

Step 4: Material balances around the reboiler:
Total:

$$L = B + V_b$$

Benzene:

$$Lx_L = Bx_B + V_b x_{V_b}$$
$$L = 6040 + V_b$$
$$Lx_L = 6040(0.01) + V_b(0.039)$$

Step 5: We have three unknowns and only two *independent* equations. We can write additional equations around the still, but these will not resolve the problem since we would still be left with one unknown more than the number of *independent* equations. This is the stage at which energy balances can be used effectively.

Steps 3, 4, 8, and 9: Overall energy balance: Let the reference temperature be 70°F; this will eliminate the feed from the enthalpy calculations. Assume that the solutions are ideal so that the thermodynamic properties (enthalpies and heat capacities) are additive. No work or potential or kinetic energy is involved in this problem. Thus

$$Q_{\text{steam}} + Q_{\text{condenser}} = \underbrace{P \int_{70}^{178} C_{P_P} \, dt}_{\Delta H_P} + \underbrace{B \int_{70}^{268} C_{P_B} \, dt}_{\Delta H_B} - \underbrace{F \int_{70}^{70} C_{P_F} \, dt}_{\Delta H_F = 0}$$

We know all the terms except Q_{steam} and $Q_{\text{condenser}}$ and still need more equations.

Energy balance on the condenser: This time we shall let the reference temperature equal 178°F; this choice simplifies the calculation, because then neither the R nor P streams have to be included. Assume that the product leaves at the saturation temperature in the condenser of 178°F. With the condenser as the system and the water as the surroundings, we have

system (condenser)	surroundings (water)
$\Delta H_{\text{condenser}} = Q_{\text{condenser}}$	$\Delta H_{\text{water}} = Q_{\text{water}}$

and, since

$$Q_{\text{system}} = -Q_{\text{surroundings}}, \qquad \Delta H_{\text{condenser}} = -\Delta H_{\text{water}}$$

$$V_t(-\Delta \hat{H}_{\text{vaporization}}) = -WC_{P_{H_2O}}(t_2 - t_1)$$
$$27{,}720[170(0.995) + 140(0.005)] = W(1)(140 - 60) = Q_{\text{water}} = -Q_{\text{condenser}}$$

Solving, $Q_C = -4.71 \times 10^6$ Btu/hr (heat evolved), and the cooling water used is

$$W = 5.89 \times 10^4 \text{ lb H}_2\text{O/hr} \longleftarrow \text{(d}_1\text{)}$$

Amount of steam used: We shall use the overall energy balance.

$$Q_{\text{steam}} = 3960 \frac{\text{lb}}{\text{hr}}\left(46.9 \frac{\text{Btu}}{\text{lb}}\right) + 6040 \frac{\text{lb}}{\text{hr}}\left(68.3 \frac{\text{Btu}}{\text{lb}}\right) + 4.71 \times 10^6 \frac{\text{Btu}}{\text{hr}}$$

In the preceding equation the $\int C_p \, dt$ was determined by first numerically integrating $\int C_p \, dt$ to get $\Delta\hat{H}$ for each component and then weighting the $\Delta\hat{H}$ values by their respective weight fractions:

$$\Delta\hat{H}_P = \int_{70}^{178} C_{P_P} \, dt \text{ Btu/lb} \qquad \Delta\hat{H}_B = \int_{70}^{268} C_{P_B} \, dt \text{ Btu/lb}$$

Bz	Cl	Avg.
47.0	36.2	46.9

Bz	Cl	Avg.
88.1	68.0	68.3

An assumption that the P stream was pure benzene and the B stream was pure chlorobenzene would be quite satisfactory since the value of Q_C is one order of magnitude larger than the sensible heat terms

$$Q_{\text{steam}} = 5.31 \times 10^6 \text{ Btu/hr}$$

From the steam tables, $\Delta\hat{H}_{\text{vap}}$ at 280°F is 923 Btu/lb. Assume that the steam leaves at its saturation temperature and is not subcooled. Then

$$\text{lb steam used/hr} = \frac{5.31 \times 10^6 \text{ Btu/hr}}{923 \text{ Btu/lb}} = 5760 \text{ lb/hr} \longleftarrow \text{(d}_2\text{)}$$

Energy balance around the reboiler:

$$Q_{\text{steam}} + L(\Delta\hat{H}_L) + V_b(\Delta\hat{H}_{V_b}) + B(\Delta\hat{H}_B)$$

Step 5: We know that $Q_{\text{steam}} = 5.31 \times 10^6$ Btu/hr. We do not know L or the value of $\Delta\hat{H}_L$. However, even if $\Delta\hat{H}_L$ is not known, the temperature of stream L entering the reboiler would not be more than 20°F lower than 268°F at the very most, and the heat capacity would be about that of chlorobenzene. We shall assume a difference of 20°F, but you can note in the calculations below that the effect of including $\Delta\hat{H}_L$, as opposed to ignoring it, is minor.

The value of V_b is still unknown, but all the other values ($\Delta\hat{H}_{V_b}$, B, $\Delta\hat{H}_B$) are known or can be calculated. Thus, if we combine the energy balance around the reboiler with the overall material balance around the reboiler, we can find both L and V_b.

Step 3, 4, 8 and 9:   Reference temperature: 268°F

Energy balance:

$$5.31 \times 10^6 \frac{\text{Btu}}{\text{hr}} + (L \text{ lb})\left[0.39 \frac{\text{Btu}}{\text{(lb)(°F)}}\right](-20°\text{F}) = V_b(0.99)(126) + (0.01)(154)] + B(0)$$

Material balance:

$$L = 6040 + V_b$$

$$5.31 + 10^6 - (6040 + V_b)(7.8) = 126.3 V_b$$

$$5.31 \times 10^6 - 0.047 \times 10^6 = 126.3 V_b + 7.8 V_b$$

$$V_b = \frac{5.26 \times 10^6}{134} = 39,300 \text{ lb/hr} \longleftarrow \text{\textcircled{c}}_1$$

$$L = V_b + B = 39,300 + 6040 = 45,340 \text{ lb/hr} \longleftarrow \text{\textcircled{c}}_2$$

If we had ignored the enthalpy of the L stream, then V_b would have been

$$V_b = \frac{5.31 \times 10^6}{126.3} = 42,100 \text{ lb/hr}$$

a difference of about 7.1%.

In practice, the solution of steady-state material and energy balances is quite tedious, a situation that has led engineers to the use of digital computer routines. Certain special problems are encountered in the solution of large-scale sets of equations, problems of sufficient complexity to be beyond our scope here.

An important aspect of combined material and energy balance problems is how to ensure that the process equations are determinate (i.e., have at least one solution). We can use the phase rule (discussed in Sec. 3.7.1) plus some common sense to provide a guide as to the number of variables that must be specified and the number of variables that can be unknown in any problem. The phase rule gives the relationship between the number of degrees of freedom and the *intensive* variables such as temperature, pressure, specific volume, and composition that are associated with each stream. According to Eq. (3.57), the number of degrees of freedom, F, for a single phase at equilibrium that contains N_{sp} chemical species is equal to the following count of intensive variables:

$$F = (N_{sp} - 1) + 2 = N_{sp} + 1$$

We have counted $(N_{sp} - 1)$ for the chemical species—do you remember that specifying N_{sp} compositions would be redundant because the summation of the mass (or mole) fractions is unity? The two other intensive variables might be temperature and pressure.

To completely specify a stream we need to know how much material is flowing (an extensive variable), so that N_v, the total number of variables for one stream is

$$N_v = (N_{sp} + 1) + 1 = N_{sp} + 2$$

The total number of variables for all streams entering or leaving the boundaries of the system would be

$$N_v = N_s(N_{sp} + 2) + N_Q + N_W + N_p$$

where N_s = total number of streams
N_Q = number of heat transfer variables
N_W = number of work variables
N_P = number of equipment parameters, such as kinetic rate constants, equilibrium constants, recycle ratio, and so on

To obtain the number of degrees of freedom, N_d (i.e., the number of decision or design variables to be specified), you reduce N_v by the number of *independent* constraints, N_r, involved in the process (including variables whose values are prespecified):

$$N_d = N_v - N_r$$

Recall that if a chemical reaction takes place in the process, the number of independent material balances is equal to the number of atomic species, or fewer in some instances (refer to Appendix L). What types of constraints do you have to include in the count for N_r? Here is a list:

(a) Material balances

(b) One energy balance

(c) Equipment constraints involving the equipment parameters

(d) Variables with prespecified values—one constraint for each specification (as long as the material and energy balances are still independent)

Note that if the energy balance is not involved in the process, the energy balance as a constraint will not be independent. It will be the same as an overall material balance.

EXAMPLE 5.2 Determining the Degrees of Freedom in a Process

We shall consider five typical processes, as depicted by the respective figures below, and for each ask the question: How many variables have to be specified [i.e., what are the degrees of freedom (N_d)] to make the problem of solving the combined material and energy balances determinate? All the processes will be steady-state ones, and the entering and exit streams single-phase streams.

(a) *Stream splitter* (Fig. E5.2a): We assume that $Q = W = 0$, and that the energy balance is not involved in the process. By implication of a splitter the temperatures, pressures,

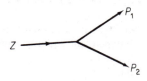

Fig. E5.2a

and compositions of the inlet and outlet streams are identical, and $N_p = 0$. The count of the total number of variables, total number of constraints, and degrees of freedom is as follows.

Total number of variables		
$N_v = N_s(N_{sp} + 2) =$		$3(N_{sp} + 2)$
Number of independent equality constraints		
Material balances	1	
Compositions of Z, P_1, and P_2		
are the same	$2(N_{sp} - 1)$	
$T_{P_1} = T_{P_2} = T_Z$	2	
$p_{P_1} = p_{P_2} = p_Z$	2	$2N_{sp} + 3$
Total no. d.f.		$N_{sp} + 3$

Did you note that we did not count N_{sp} material balances but only one? Why? Let us look at some of the balances:

$$\text{component 1:} \quad Zx_{1Z} = P_1 x_{1P_1} + P_2 x_{1P_2}$$

$$\text{component 2:} \quad Zx_{2Z} = P_1 x_{2P_1} + P_2 x_{2P_2}$$

$$\text{etc.}$$

If $x_{1Z} = x_{1P_1} = x_{1P_2}$ and the same is true for x_2, and so on, the only independent material balance is the total balance,

$$Z = P_1 + P_2$$

Do you understand the counts resulting from making the compositions equal in Z, P_1, and P_2? Write down the expressions for each component: $x_{iZ} = x_{iP_1} = x_{iP_2}$. Each set represents $2N_{sp}$ constraints. But you cannot specify every x_i in a stream, only $N_{sp} - 1$ of them. Do you remember why?

To make the problem determinate we might specify the values of the following decision variables:

Flow rate Z	1
Composition of Z	$N_{sp} - 1$
T_Z	1
p_Z	1
Ratio of split $\alpha = P_1/P_2$	1
Total no. d.f.	$N_{sp} + 3$

(b) *Mixer* (Fig. E5.2b): For this process we assume that $W = 0$, but Q is not.

Total number of variables (3 streams + Q)	$3(N_{sp} + 2) + 1$
Number of independent equality constraints	
Material balances	N_{sp}
Energy balance	1
Total no. d.f. $= 3(N_{sp} + 2) + 1 - N_{sp} - 1 = 2N_{sp} + 6$	

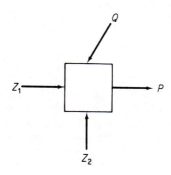

Fig. E5.2b

(c) *Heat exchanger* (Fig. E5.2c): For this process we assume that $W = 0$ (but not Q) and $N_p = 0$.

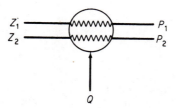

Fig. E5.2c

Total number of variables
$$N_v = N_s(N_{sp} + 2) + 1 = \qquad\qquad 4N_{sp} + 9$$
Number of independent equality constraints
Material balances	2	
Energy balance	1	
Composition of inlet and outlet streams the same	$2(N_{sp} - 1)$	$2N_{sp} + 1$
Total no. d.f.		$2N_{sp} + 8$

One might specify four temperatures, four pressures, $2(N_{sp} - 1)$ compositions in Z_1 and Z_2, and Z_1 and Z_2 themselves to use up $2N_{sp} + 8$ degrees of freedom.

(d) *Pump* (Fig. E5.2d): Here $Q = 0$ but W is not; $N_p = 0$.

Total number of variables
$$N_v = N_s(N_{sp} + 2) + 1 \qquad\qquad 2N_{sp} + 5$$
Number of independent equality constraints
Material balance	1	
Composition of inlet and outlet streams the same	$N_{sp} - 1$	
Energy balance	1	$N_{sp} + 1$
Total no. d.f.		$N_{sp} + 4$

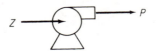

Fig. E5.2d

(e) *Two-phase well-mixed tank (stage) at equilibrium* (Fig. E5.2e) (L, liquid phase; V, vapor phase): Here $W = 0$. Although two phases exist (V_1 and L_1) at equilibrium inside the system, the streams entering and leaving are single-phase streams. By equilibrium we mean that both phases are at the same temperature and pressure and that an equation is known that relates the composition in one phase to that in the other for each component.

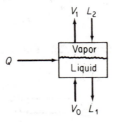

Fig. E5.2e

Total number of variables (4 streams $+ Q$)	$4(N_{sp} + 2) + 1$
Number of independent equality constraints	
Material balances	N_{sp}
Energy balance	1
Composition relations at equilibrium	N_{sp}
Temperatures in the two phases equal	1
Pressures in the two phases equal	1

$$\text{Total no. d.f.} = 4(N_{sp} + 2) + 1 - 2N_{sp} - 3 = 2N_{sp} + 6$$

In general you might specify the following variables to make the problem determinate:

Input stream L_2	$N_{sp} + 2$
Input stream V_0	$N_{sp} + 2$
Pressure	1
Q	1
Total	$2N_{sp} + 6$

Other choices are of course possible.

You can compute the degrees of freedom for combinations of like or different simple processes by proper combination of their individual degrees of freedom. In adding the degrees of freedom for units, you must eliminate any double counting either for variables or constraints and take proper account of interconnecting streams whose characteristics are often fixed only by implication.

EXAMPLE 5.3 Degrees of Freedom in Combined Units

Consider the mixer-separator shown in Fig. E5.3. Superficially, it would seem that we could carry out an analysis as follows to combine the units. For the mixer considered as a separate unit, from Example 5.2, d.f. $= 2N_{sp} + 6$. For the separator, an equilibrium unit, we find from Example 5.2, that d.f. $= N_{sp} + 4$. Hence the total d.f. $= 2N_{sp} + 6 + N_{sp} + 4 = 3N_{sp} + 10$. However, we have to eliminate certain redundant counts of variables and constraints for the combination:

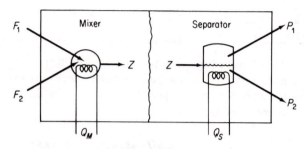

Fig. E5.3 Degrees of freedom in combined units.

Eliminate variable Z	$N_{sp} + 2$
Eliminate Q_s	1
Total	$N_{sp} + 3$
Eliminate one energy balance	1
Net change	$N_{sp} + 2$

The net d.f. for the entire unit would be $2N_{sp} + 8$, but this count is incorrect because you can see from Example 5.2 that the correct number of d.f. $= 2N_{sp} + 6$.

We can locate the source of this discrepancy if we note for the mixer that stream Z is actually a *two-phase stream*; hence $N_v = N_{sp} + 1$, not $N_{sp} + 2$. The proper analysis is as follows:

For the mixer:

	single-phase streams	two-phase stream	Q_M
Total no. variables	$2(N_{sp} + 2) + 1(N_{sp} + 1) + 1 = 3N_{sp} + 6$		
Constraints (as before)			$N_{sp} + 1$
Net			$2N_{sp} + 5$

For the separator:

Total no. variables	$2(N_{sp} + 2) + 1(N_{sp} + 1) + 1 = 3N_{sp} + 6$		
Constraints (as before)			$2N_{sp} + 3$
Net			$N_{sp} + 3$

Eliminate redundant counting:

Drop Q_S
Eliminate Z $\qquad\qquad\qquad\qquad\qquad \dfrac{1}{N_{sp}+1}$
$\qquad$ Together $\qquad\qquad\qquad\qquad\qquad\qquad\qquad \overline{N_{sp}+2}$

Consequently, the total degrees of freedom are

$$(2N_{sp}+5)+(N_{sp}+3)-(N_{sp}+2)=2N_{sp}+6$$

as in Example 5.2.

We do not have the space to illustrate additional combinations of simple units to form more complex units, but Kwauk[2] prepared several excellent tables summarizing the variables and degrees of freedom for distillation columns, absorbers, heat exchangers, and the like.

Self-Assessment Test

1. Is there any difference between the number of species present in a process and the number of components in the process?

2. Why are there $N_{sp}+2$ variables associated with each stream?

3. Determine the number of degrees of freedom for a still.

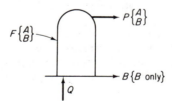

4. Determine the number of degrees of freedom in the following process:

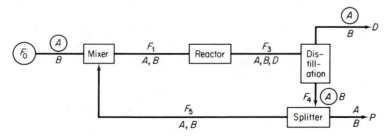

The encircled variables have known values. The reaction parameters in the reactor are

[2]M. Kwauk, *AIChE J.*, v. 2, p. 240 (1956).

known as is the fraction split at the splitter between F_4 and F_5. Each stream is a single phase.

5. The accompanying figure represents the schematic flow sheet of a distillation tower used to recover gasoline from the products of catalytic cracker.

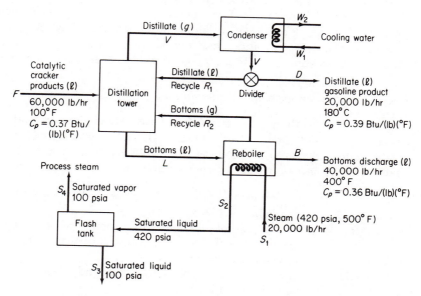

(a) How much heat must be transferred to the cooling water in the condenser? (Assume $\Delta \hat{H} = 0$ for oil components at $T = 32°F$.)

(b) It is desired to make 100 psia (saturated vapor) process steam by "flashing" the 420 psia saturated liquid exiting from the reboiler steam coil. Thus the pressure of stream S_2 is suddenly reduced to 100 psia in the flash tank. How many pounds per hour of S_4 can be produced?

Section 5.2   Enthalpy–Concentration Charts

Your objectives in studying this section are to be able to:

1. Prepare an enthalpy–concentration chart for a binary mixture given the enthalpy or heat capacity data for the pure components, the heat capacity data for mixtures at various concentrations, and the necessary heats of mixing at various concentrations.

2. Use the enthalpy–concentration chart in solving material and energy balances.

3. Employ graphical techniques to solve material and energy balance problems for adiabatic and nonadiabatic processes.

4. Apply the inverse lever arm principle.

An enthalpy–concentration chart is a convenient graphical method of representing enthalpy data for a binary mixture. If available,[3] such charts are useful in making combined material and energy balances calculations in distillation, crystallization, and all sorts of mixing and separation problems. You will find a few examples of enthalpy–concentration charts in Appendix I.

At some time in your career you may find you have to make numerous repetitive material and energy balance calculations on a given binary system and would like to use an enthalpy–concentration chart for the system, but you cannot find one in the literature or in your files. How do you go about constructing such a chart?

5.2-1   Construction of an Enthalpy–Concentration Chart

As usual, the first thing to do is choose a basis—some given amount of the mixture, usually 1 lb or 1 lb mole. Then choose a reference temperature ($\Delta H_0 = 0$ at t_0) for the enthalpy calculations. Assuming that 1 lb is the basis, you then write an energy balance for solutions of various compositions at various temperatures:

$$\Delta \hat{H}_{\text{mixture}} = x_A \Delta \hat{H}_A + x_B \Delta \hat{H}_B + \Delta \hat{H}_{\text{mixing}} \tag{5.1}$$

where $\Delta \hat{H}_{\text{mixture}}$ = enthalpy of 1 lb of the mixture

$\Delta \hat{H}_A, \Delta \hat{H}_B$ = enthalpies of the pure components per pound relative to the reference temperature (the reference temperature does not have to be the same for A as for B)

$\Delta \hat{H}_{\text{mixing}}$ = heat of mixing (solution) per pound at the temperature of the given calculation

In the common case where the $\Delta \hat{H}_{\text{mixing}}$ is known only at one temperature (usually 77°F), a modified procedure as discussed below would have to be followed to calculate the enthalpy of the mixture.

The choice of the reference temperatures for A and B locates the zero enthalpy datum on each side of the diagram (see Fig. 5.3). From enthalpy tables such as the steam tables, or by finding

$$(\Delta \hat{H}_A - \Delta \hat{H}_{A_0}) = \int_{t_0}^{t=77°F} C_{p_A} \, dt$$

you can find $\Delta \hat{H}_A$ at 77°F and similarly obtain $\Delta \hat{H}_B$ at 77°F. Now you have located points A and B. If one of the components is water, it is advisable to choose 32°F as t_0 because then you can use the steam tables as a source of data. If both compo-

---

[3] For a literature survey as of 1957, see Robert Lemlich, Chad Gottschlich, and Ronald Hoke, *Chem. Eng. Data Series*, v. 2, p. 32 (1957). Additional references: for CCl[4], see M. M. Krishnaiah et al., *J. Chem. Eng. Data*, v. 10, p. 117 (1965); for EtOH–EtAc, see Robert Lemlich, Chad Gottschlich, and Ronald Hoke, *Brit. Chem. Eng.*, v. 10, p. 703 (1965); for methanol–toluene, see C. A. Plank and D. E. Burke, *Hydrocarbon Processing*, v. 45, no. 8, p. 167 (1966); for acetone–isopropanol, see S. N. Balasubramanian, *Brit. Chem. Eng.*, v. 11, p. 1540 (1966); for alcohol–aromatic systems, see C. C. Reddy and P. S. Murti, *Brit. Chem. Eng.*, v. 12, p. 1231 (1967); for acetonitrile–water–ethanol, see Reddy and Murti, ibid., v. 13, p. 1443 (1968); for alcohol–aliphatics, see Reddy and Murti, ibid., v. 16, p. 1036 (1971); and for H_2SO_4, see D. D. Huxtable and D. R. Poole, *Proc. Intl. Solar Energy Soc.*, Winnipeg, August 15, 1976, v. 8, p. 178 (1977).

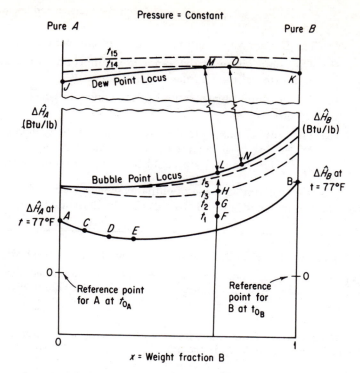

Fig. 5.3 Enthalpy–concentration diagram.

nents have the same reference temperature, the temperature isotherm will intersect on both sides of the diagram at 0, but there is no particular advantage in this.

Now that you have $\Delta \hat{H}_A$ and $\Delta \hat{H}_B$ at 77°F (or any other temperature at which $\Delta \hat{H}_{\text{mixing}}$ is known), you can calculate $\Delta \hat{H}_{\text{mixture}}$ at 77°F by Eq. (5.1). You can review the details about how to get $\Delta \hat{H}_{\text{mixing}}$ in Chap. 4. For various mixtures such as 10% A and 90% B, 20% A and 80% B, and so on, plot the calculated $\Delta \hat{H}_{\text{mixture}}$ as shown by the points C, D, E, and so on, in Fig. 5.3, and connect these points with a continuous line.

You can calculate $\Delta \hat{H}_{\text{mixture}}$ at any other temperature t, once this 77°F isotherm has been constructed, by again making an enthalpy balance as follows:

$$\Delta \hat{H}_{\text{mixture at any } t} = \Delta \hat{H}_{\text{mixture at 77°F}} + \int_{77°F}^{T} C_{p_x}\, dt \qquad (5.2)$$

where C_{p_x} is the heat capacity of the solution at concentration x. These heat capacities must be determined experimentally, although in a pinch they might be estimated. Points F, G, H, and so on, can be determined in this way for a given composition, and then additional like calculations at other fixed concentrations will give you enough points so that all isotherms can be drawn up to the bubble-point line for the mixture (1 atm).

To include the vapor region on the enthalpy concentration chart, you need to know:

(a) The heat of vaporization of the pure components (to get points J and K)

(b) The composition of the vapor in equilibrium with a given composition of liquid (to get the tie lines L–M, N–O, etc.)

(c) The dew-point temperatures of the vapor (to get the isotherms in the vapor region)

(d) The heats of mixing in the vapor region, usually negligible [to fix the points M, O, etc., by means of Eq. (5.1)] (Alternatively, the heat of vaporization of a given composition could be used to fix points M, O, etc.)

The construction and use of enthalpy–concentration diagrams for combination material–energy balance problems will now be illustrated.

EXAMPLE 5.4 Preparation of an Enthalpy–Concentration Chart for Ideal Mixtures

Prepare an enthalpy–concentration chart from the data listed below for the *n*-butane–*n*-heptane system. Assume that heats of mixing (solution) are negligible.

Solution

The compositions of the vapor and liquid phases as a function of temperature are also essential data and are listed in the first columns of the calculations.

100 PSIA, n-BUTANE–n-HEPTANE SYSTEM
REFERENCE STATE: 32°F AND 14.7 PSIA.

Temperature (°F)	$\Delta\hat{H}_{C_4}$ (*n*-butane) Btu/lb mol	$\Delta\hat{H}_{C_7}$ (*n*-heptane) Btu/lb mol
\multicolumn{3}{l}{Enthalpy of Saturated Hydrocarbon Liquids at 100 psia}		
140	3,800	7,100
160	4,500	7,700
200	6,050	9,100
240	7,700	11,400
280	9,400	14,000
320	11,170	16,700
360	13,350	19,400
\multicolumn{3}{l}{Enthalpy of Saturated Hydrocarbon Vapors at 100 psia}		
140	11,300	19,300
160	11,800	20,100
200	12,820	21,600
240	13,900	23,250
280	15,030	25,000
320	16,200	26,900
360	17,550	29,300

SOURCE: Data from E. G. Scheibel, *Petrol. Refiner*, v. 26, p. 116 (1947).

Intermediate values of enthalpy were interpolated from a plot of enthalpy vs. temperature prepared from the data given in the tables.

Basis: 1 lb mol of mixture

Reference conditions: 32°F and 14.7 psia

Step 1: Plot the boiling points of pure C_4 (146°F) and pure C_7 (358°F). Fix the enthalpy scale from 0 to 30,000 Btu/lb mole mixture. By plotting the temperature–enthalpy data on separate graphs you can find that the enthalpy of the pure C_4 at 146°F is 3970 Btu/lb mol and that of C_7 at 358°F is 19,250 Btu/lb mol. These are marked in Fig. E5.4 as points A and B, respectively.

Step 2: Now choose various other compositions of liquid between A and B (the corresponding temperatures determined from experimental measurements are shown in the table), and multiply the enthalpy of the pure component per mole by the mole fraction present:

$$\Delta \hat{H}_{\text{liquid mixture}} = \Delta \hat{H}_{C_4} x_{C_4} + \Delta \hat{H}_{C_7} x_{C_7}$$

No heat of mixing is present since ideality was assumed. In this way, the bubble-point line between A and B can be established.

ΔH–X CALCULATIONS FOR C_4–C_7 LIQUID (SATURATED) AT 100 PSIA

Temp. (°F)	Mole Fraction		Enthalpy (Btu/lb mol)				
	x_{C_4}	x_{C_7}	$\Delta \hat{H}_{C_4}$	$\Delta \hat{H}_{C_7}$	$\Delta \hat{H}_{C_4} x_{C_7}$	$\Delta \hat{H}_{C_7} x_{C_7}$	$\Delta \hat{H}_L = \Delta \hat{H}_{C_4} x_{C_4} + \Delta \hat{H}_{C_7} x_{C_7}$
358	0	1.0	13,200	19,250	0	19,250	19,250
349	0.02	0.98	12,700	18,650	254	18,250	18,504
340.5	0.04	0.96	12,220	18,000	490	17,280	17,770
331.5	0.06	0.94	11,750	17,380	710	16,300	17,010
315	0.10	0.90	11,000	16,200	1,100	14,600	15,700
296	0.15	0.85	10,100	14,900	1,515	12,650	14,165
278.5	0.20	0.80	9,350	13,750	1,870	11,000	12,870
248	0.30	0.70	8,050	11,850	2,420	8,300	10,720
223.5	0.40	0.60	7,000	10,400	2,800	6,240	9,040
204	0.50	0.50	6,200	9,400	3,100	4,700	7,800
188	0.60	0.40	5,600	8,700	3,360	3,480	6,840
173.8	0.70	0.30	5,000	8,120	3,500	2,400	5,940
160.5	0.80	0.20	4,500	7,700	3,600	1,540	5,140
151.0	0.90	0.10	4,150	7,400	3,740	740	4,480
147.8	0.95	0.05	4,000	7,310	3,800	366	4,166
146.0	1.0	0	3,970	7,280	3,970	0	3,970

Step 3: Plot the enthalpy values for the saturated pure vapor (dew point), points C and D. A supplementary graph again is required to assist in interpolating enthalpy values.

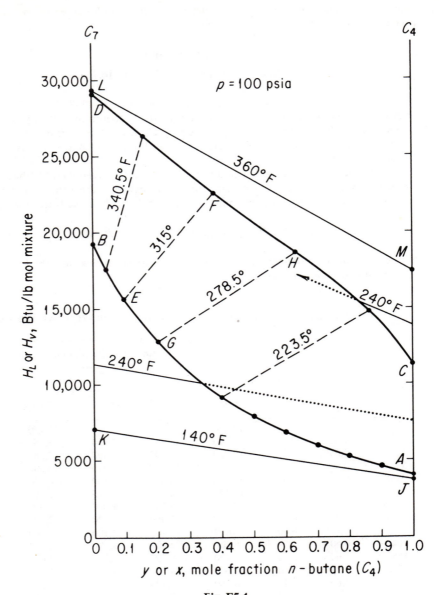

Fig. E5.4

Step 4: Fill in the saturated-vapor curve between C and D by the following calculations:

$$\Delta \hat{H}_V = \Delta \hat{H}_{C_4} y_{C_4} + \Delta \hat{H}_{C_7} y_{C_7}$$

ΔH–Y CALCULATIONS FOR C₄–C₇ VAPOR (SATURATED) AT 100 PSIA

Temp. (°F)	Mole Fraction		Enthalpy (Btu/lb mol)				$\Delta \hat{H}_V =$ $\Delta \hat{H}_{C_4} y_{C_4}$ $+ \Delta \hat{H}_{C_7} y_{C_7}$
	y_{C_4}	y_{C_7}	$\Delta \hat{H}_{C_4}$	$\Delta \hat{H}_{C_7}$	$\Delta \hat{H}_{C_4} y_{C_4}$	$\Delta \hat{H}_{C_7} y_{C_7}$	
358	0	1.0	17,500	29,250	0	29,250	29,250
349	0.084	0.914	17,150	28,700	1,400	26,200	27,640
340.5	0.160	0.84	16,850	28,100	2,700	23,600	26,300
331.5	0.241	0.759	16,500	27,550	3,970	20,860	24,830
315	0.378	0.622	16,050	26,700	6,060	16,600	22,660
296	0.518	0.482	15,800	25,800	8,050	12,420	20,470
278.5	0.630	0.370	14,980	24,900	9,425	9,210	18,635
248	0.776	0.224	14,100	23,600	10,950	5,290	16,240
223.5	0.857	0.143	13,420	22,550	11,500	3,220	14,720
204	0.907	0.093	12,950	21,700	11,750	2,020	13,770
188	0.942	0.058	12,500	21,100	11,780	1,225	13,005
173.8	0.968	0.032	12,100	20,600	11,700	659	12,359
160.5	0.986	0.014	11,800	26,100	11,620	282	11,902
151.0	0.996	0.004	11,500	19,720	11,450	79	11,529
147.8	0.999	0.001	11,450	19,600	11,430	20	11,450
146.0	1.000	0	11,400	19,540	11,400	0	11,400

Step 5: Draw tie lines (the dashed lines) between the bubble-point and dew-point lines (lines *E–F, G–H,* etc.). These lines of constant temperature show the equilibrium concentrations in the vapor and liquid phases. The experimental data were selected so that identical temperatures could be used in the calculations above for each phase; if such data were not available, you would have to prepare additional graphs or tables to use in interpolating the isobaric data (data at 100 psia) with respect to the mole fraction y or x. If you wanted to draw the lines at even temperature intervals, such as 140°F, 160°F, 180°F, and so on, you would also have to interpolate.

Step 6: Draw isothermal lines at 140°F, 160°F, and so on, in the liquid and vapor regions. Presumably these will be straight lines since there is no heat of mixing. Since

$$x_{C_7} + x_{C_4} = 1 \qquad \text{(a)}$$

$$\Delta \hat{H}_L = \Delta \hat{H}_{\text{mixture}} = x_{C_4} \Delta \hat{H}_{C_4} + x_{C_7} \Delta \hat{H}_{C_7} = x_{C_4} \Delta \hat{H}_{C_4} + (1 - x_{C_4}) \Delta \hat{H}_{C_7}$$

$$= (\Delta \hat{H}_{C_4} - \Delta \hat{H}_{C_7}) x_{C_4} + \Delta \hat{H}_{C_7} \qquad \text{(b)}$$

Because $\Delta \hat{H}_{C_4}$ and $\Delta \hat{H}_{C_7}$ are constant at any given temperature, $\Delta \hat{H}_{C_4} - \Delta \hat{H}_{C_7}$ is constant, and Eq. (b) is the equation of a straight line on a plot of $\Delta \hat{H}_{\text{mixture}}$ vs. x_{C_4}.

The 140°F isotherm is fixed by the points *J* and *K* in the liquid region, and the 360°F isotherm is fixed by points *L* and *M* in the vapor region:

	140°F, *liquid* $\Delta \hat{H}_L$ *(Btu/mol)*		360°F, *vapor* $\Delta \hat{H}_V$ *(Btu/mol)*
J (C₄)	3800	(C₄)	17,550
K (C₇)	7100	(C₇)	29,300

The 240°F isotherm (and others higher than 140°F) in the liquid region is real only up to the saturated-liquid line—the dotted portion is fictitious. Portions of isotherms in the vapor region lower than 360°F would also be fictitious, as, for example, at 240°F.

EXAMPLE 5.5 Application of the Enthalpy–Concentration Chart

One hundred pounds of a 73% NaOH solution at 350°F is to be diluted to give a 10% solution at 80°F. How many pounds of water at 80°F and ice at 32°F are required if there is no external source of cooling available? See Fig. E5.5. Use the steam tables and the NaOH–H$_2$O enthalpy–concentration chart in Appendix I as your source of data. (The reference conditions for the latter chart are $\Delta H = 0$ at 32°F for liquid water and $\Delta H = 0$ for an infinitely dilute solution of NaOH, with pure caustic having an enthalpy at 68°F of 455 Btu/lb above this datum.)

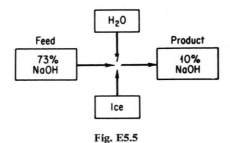

Fig. E5.5

Solution

The data required to make a material and an energy balance are as follows:

	From Appendix I		*From the steam tables*
NaOH			$\Delta \hat{H}$ of liquid H$_2$O
conc.	*temp.* (°F)	$\Delta \hat{H}$ (*Btu/lb*)	= 48 Btu/lb
73	350	468	$\Delta \hat{H}$ of ice $= -143$ Btu/lb
10	80	42	(or minus the heat of fusion)

We can make a material balance first.

Basis: 100 lb of 73% at 350°F

The tie element in the material balance is the NaOH, and using it we can find the total H$_2$O added (ice plus H$_2$O):

$$\frac{100 \text{ lb product}}{10 \text{ lb NaOH}} \Bigg| \frac{73 \text{ lb NaOH}}{100 \text{ lb feed}} = 730 \text{ lb product/100 lb feed}$$

less 100 lb feed

gives 630 lb H$_2$O added/100 lb feed

Next we make an energy (enthalpy) balance:

$$\Delta H_{\text{overall}} = 0 \qquad \text{or} \qquad \Delta H_{\text{in}} = \Delta H_{\text{out}}$$

To differentiate between the ice and the liquid water added, let us designate the ice as x lb and then the H_2O becomes $(630 - x)$ lb:

in			out
73% NaOH solution	H_2O	ice	10% NaOH solution

$$\frac{100 \text{ lb}}{} \left| \frac{468 \text{ Btu}}{\text{lb}} + \frac{(630 - x) \text{ lb}}{} \right| \frac{48 \text{ Btu}}{\text{lb}} + \frac{x \text{ lb}}{} \left| \frac{-143 \text{ Btu}}{\text{lb}} = \frac{730 \text{ lb}}{} \right| \frac{42 \text{ Btu}}{\text{lb}}$$

$$46{,}800 + 30{,}240 - 191x = 30{,}660$$

$$x = 243 \text{ lb ice at } 32°F$$

$$\text{liquid } H_2O \text{ at } 80°F = 387 \text{ lb}$$

EXAMPLE 5.6 Application of the Enthalpy–Concentration Chart

Six hundred pounds of 10% NaOH at 200°F is added to 400 lb of 50% NaOH at the boiling point. Calculate the following:

(a) The final temperature of the solution
(b) The final concentration of the solution
(c) The pounds of water evaporated during the mixing process

Solution

Basis: 1000 lb of final solution

Use the same NaOH–H_2O enthalpy–concentration chart as in Example 5.5 to obtain the enthalpy data. We can write the following material balance:

component	10% solution	+	50% solution	=	final solution	wt %
NaOH	60		200		260	26
H_2O	540		200		740	74
Total	600		400		1000	100

Next, the energy (enthalpy) balance (in Btu) is

$$
\begin{array}{ccccc}
10\% \text{ solution} & & 50\% \text{ solution} & & \text{final solution} \\
600(152) & + & 400(290) & = & \Delta H \\
91{,}200 & + & 116{,}000 & = & 207{,}200
\end{array}
$$

Note that the enthalpy of 50% solution at its boiling point is taken from the bubble-point curve at $x = 0.50$. The enthalpy per pound is

$$\frac{207{,}200 \text{ Btu}}{1000 \text{ lb}} = 207 \text{ Btu/lb}$$

On the enthalpy–concentration chart for $NaOH$–H_2O, for a 26% NaOH solution with an enthalpy of 207 Btu/lb, you would find that only a two-phase mixture of (1) saturated H_2O vapor and (2) $NaOH$–H_2O solution at the boiling point could exist. To get the fraction H_2O vapor, we have to make an additional energy (enthalpy) balance. By interpolation, draw the tie line through the point $x = 0.26$, $H = 207$ (make it parallel to the 220° and 250°F tie lines). The final temperature appears from Fig. E5.6 to be 232°F; the enthalpy of the liquid at the

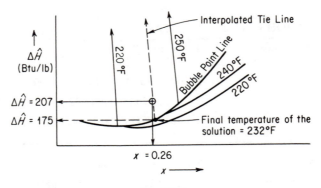

Fig. E5.6

bubble point is about 175 Btu/lb. The enthalpy of the saturated water vapor (no NaOH is in the vapor phase) from the steam tables at 232°F is 1158 Btu/lb. Let $x =$ lb of H_2O evaporated.

Basis: 1000 lb of final solution

$$x(1158) + (1000 - x)175 = 1000(207)$$

$$983x = 32,000$$

$$x = 32.6 \text{ lb of } H_2O \text{ evaporated}$$

5.2-2 Graphical Solutions on an Enthalpy–Concentration Chart

One of the major advantages of an enthalpy–concentration chart is that the same problems we have just used as examples, and a wide variety of other problems, can be solved graphically directly on the chart. We shall just indicate the scope of this technique, which in its complete form is usually called the Ponchon–Savaritt method; you can refer to some of the references for enthalpy–concentration charts at the end of the chapter for additional details.

Any steady-state process with three streams and a net heat interchange Q with the surroundings can be represented by a simple diagram as in Fig. 5.4. A, B, and C are in mass or moles. An overall material balance gives us

$$A + B = C \tag{5.3}$$

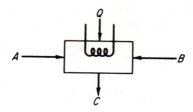

Fig. 5.4 Typical process with mass and heat interchange.

and a component material balance gives

$$Ax_A + Bx_B = Cx_C \tag{5.4}$$

An energy balance for a flow process (neglecting the work and the kinetic and potential energy effects) gives us[4]

$$Q = \Delta H$$

or

$$Q + AH_A + BH_B = CH_C \tag{5.5}$$

First we must choose some basis for carrying out the calculations, and then Q is dependent on this basis. Thus, if A is chosen as the basis, Q_A will be Q per unit amount of A, or

$$Q_A = \frac{Q}{A} \tag{5.6}$$

Similarly, if B is the basis,

$$Q_B = \frac{Q}{B} \tag{5.7}$$

Thus

$$Q = AQ_A = BQ_B = CQ_C \tag{5.8}$$

Let us choose A as the basis for working the problem, and then Eq. (5.5) becomes

$$AQ_A + AH_A + BH_B = CH_C \tag{5.9}$$

or

$$A(H_A + Q_A) + BH_B = CH_C \tag{5.10}$$

Similar equations can be written if B or C is chosen as the basis.

[4]In what follows H is always the specific enthalpy relative to some reference value, and we suppress the caret (⌃) and the symbol Δ to make the notation more compact.

Adiabatic processes. As a special case, consider the adiabatic process ($Q = 0$). Then (with A as the basis) the energy balance becomes

$$AH_A + BH_B = CH_C = (A + B)H_C = AH_C + BH_C$$

or

$$A(H_A - H_C) = B(H_C - H_B)$$

or

$$\frac{A}{B} = \frac{H_C - H_B}{H_A - H_C} \tag{5.11}$$

Combining material balances yields

$$Ax_A + Bx_B = Cx_C = (A + B)x_C = Ax_C + Bx_C$$

or

$$A(x_A - x_C) = B(x_C - x_B)$$

or

$$\frac{A}{B} = \frac{x_C - x_B}{x_A - x_C} \tag{5.12}$$

Obviously, then,

$$\frac{A}{B} = \frac{x_C - x_B}{x_A - x_C} = \frac{H_C - H_B}{H_A - H_C} \tag{5.13}$$

Similarly, solving the energy and material balances for A and C instead of A and B, we can get

$$\frac{A}{C} = \frac{x_C - x_B}{x_A - x_B} = \frac{H_C - H_B}{H_A - H_B} \tag{5.14}$$

and in terms of B and C we could get a like relation. Rearranging the two right-hand members of Eq. (5.13), we find that

$$\frac{H_C - H_B}{x_C - x_B} = \frac{H_A - H_C}{x_A - x_C} \tag{5.13a}$$

Also, from Eq. (5.14),

$$\frac{H_C - H_B}{x_C - x_B} = \frac{H_A - H_B}{x_A - x_B} \tag{5.14a}$$

On an enthalpy–concentration diagram (an H–x plot) the distances H_C–H_B, H_A–H_B, and H_A–H_C are projections on the H axis of a curve, while the distances x_C–x_B, x_A–x_B, x_A–x_C are the corresponding projections on the x axis (see Fig. 5.5). The form of Eqs. (5.13a) and (5.14a) is such that they tell us that they are each equations of a straight line on an H–x plot and that these lines pass through the points A, B, and C, which are represented on an H–x diagram by the coordinates $A(H_A, x_A)$, $B(H_B, x_B)$, and $C(H_C, x_C)$, respectively.

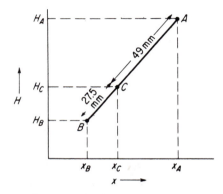

Fig. 5.5 Projections of the line BCA on the H and x axes.

Since Eqs. (5.13a) and (5.14a) both pass through point C, a common point, and have the same slope $(H_C - H_B)/(x_C - x_B)$, mathematical logic tells us they must be equations of the same line; that is, the lines are actually a single line passing through B, C, and A, as shown in Fig. 5.5. The symmetry of this type of arrangement leads to use of the *lever arm* rule.

If $A + B = C$, then point C lies on a line *between* A and B, and *nearer* to the quantity present in the *largest* amount. If more B is added, C will lie nearer B. When two streams are subtracted, the resulting stream will lie *outside* the region of a line between the two subtracted streams, and *nearer* the one present in the larger amount. Thus if $C - B = A$, and C is larger than A, then A will lie on a straight line through C and B nearer to C than B. Look at Fig. 5.5 and imagine that B is removed from C. A represents what is left after B is taken from C.

The *inverse* lever arm principle tells us how far to go along the line through A and B to find C if $A + B = C$, or how far to go along the line through C and B if $C - B = A$. In the case of $A + B = C$, we saw that the weight (or mole) ratio of A to B was given by Eq. (5.13). From the symmetry of Fig. 5.5, you can see that if the ratio of A to B is equal to the ratio of the projections of line segments BC and AC on the H or x axes, then the amount of $A + B$ should be proportional to the sum of the line segments $(x_C - x_B) + (x_A - x_C)$, or $(H_C - H_B) + (H_A - H_C)$. Furthermore, the amount of A should be proportional to the measured distance $\overline{BC}$ and the amount of B proportional to the distance $\overline{AC}$. Also, the ratio of A to B should be

$$\frac{A}{B} = \frac{\text{distance } \overline{BC}}{\text{distance } \overline{AC}} \qquad (5.15)$$

This relation is the *inverse lever arm rule*. The distances $\overline{BC}$, $\overline{AC}$, and $\overline{AB}$ can be measured with a ruler.

As an example of the application of Eq. (5.15), if A were 100 lb and the measured distances were

$$\overline{BC} = 27.5 \text{ mm}$$

$$\overline{AC} = 49.0 \text{ mm}$$

$$\overline{AB} = 76.5 \text{ mm}$$

then

$$\frac{A}{B} = \frac{100}{B} = \frac{\overline{BC}}{\overline{AC}} = \frac{27.5 \text{ mm}}{49.0 \text{ mm}} = 0.561$$

and

$$B = \frac{100}{0.561} = 178 \text{ lb}$$

The sum of $A + B = C$ is $100 + 178 = 278$ lb. Alternatively,

$$\frac{A}{C} = \frac{BC}{AB} = \frac{27.5 \text{ mm}}{76.5 \text{ mm}} = 0.360$$

$$C = \frac{A}{0.360} = \frac{100}{0.360} = 278 \text{ lb}$$

Nonadiabatic processes. Let us now consider the circumstances under which Q is not equal to zero in Eq. (5.5). This time, for variety, let us take C as a basis so that $Q = CQ_C$. Then we have two pairs of equations to work with again,

$$AH_A + BH_B = C(H_C - Q_C) \qquad (5.16)$$
$$Ax_A + Bx_B = Cx_C \qquad (5.4)$$

Briefly, Eqs. (5.16) and (5.4) tell us that on an H–x diagram points A, B, and C lie on a straight line if the coordinates of the points are $A(H_A, x_A)$, $B(H_B, x_B)$, and

$C[(H_c - Q_c), x_c]$, respectively,[5] as illustrated in Fig. 5.6. Since the coordinates of C are really fictitious coordinates and the enthalpy projection of C is $(H_c - Q_c)$, we shall designate the point through which the line passes $[(H_c - Q_c), x_c]$ by the more appropriate notation C_q. If A or B had been chosen as the basis, we would have set up diagrams such as Figs. 5.7 and 5.8.

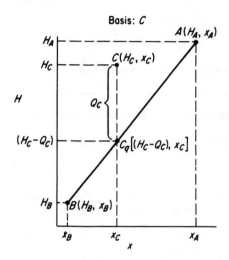

Fig. 5.6 Inverse lever arm rule applied to a nonadiabatic process.

A different series of figures could be drawn for other typical problems, as, for example, C being split up into two streams with heat being absorbed or evolved. Additional details related to more complex problems will be found in the references on enthalpy–concentration charts at the end of the chapter.

[5]Proof of these relations is analogous to the previous development.
Material balances:

$$Ax_A + Bx_B = Cx_C = (A + B)x_C = Ax_C + Bx_C \tag{a}$$

$$A(x_A - x_C) = B(x_C - x_B) \tag{b}$$

Energy balance:

$$AH_A + BH_B = C(H_C - Q_C) = (A + B)(H_C - Q_C)$$

$$= A(H_C - Q_C) + B(H_C - Q_C) \tag{c}$$

$$A[H_A - (H_C - Q_C)] = B(H_C - Q_C) - H_B \tag{d}$$

Thus

$$\frac{A}{B} = \frac{x_C - x_B}{x_A - x_C} = \frac{(H_C - Q_C) - H_B}{H_A - (H_C - Q_C)} \tag{e}$$

and, rearranging,

$$\frac{H_A - (H_C - Q_C)}{x_A - x_C} = \frac{(H_C - Q_C) - H_B}{x_C - x_B} \tag{f}$$

A similar equation can be obtained for the ratio A/C.

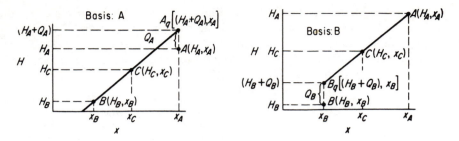

Figs. 5.7 and 5.8 Inverse lever arm principle for different bases.

EXAMPLE 5.7 Graphical Use of Enthalpy–Concentration Charts

Do Example 5.5 again by graphical means.

Solution

Plot the known points A and C on the H–x diagram for NaOH. The known points are as follows (with the given coordinates or data):

solution		x	H (Btu/lb)	T (°F)
Initial	(A)	0.73	468	350
Final	(C)	0.10		80
Added	(D)	0.00		80
	(E)	0.00		32 (ice)

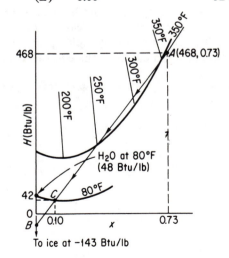

Fig. E5.7a

What we shall do is remove solution A from the final solution C to obtain solution B (i.e., subtract $C - A$ to get B as in Fig. E5.7a). This is the reverse of adding A to B to get C.

Draw a line through A and C until it intersects the $x = 0$ axis (for pure water). The measured (with a ruler) ratio of the lines $\overline{BC}$ to $\overline{AC}$ is the ratio of 73% NaOH solution to pure water.

$$\text{Basis: 100 lb of 73\% NaOH solution at 350°F } (A)$$

$$\frac{100 \text{ lb } (A)}{\text{lb } H_2O \ (B)} = \frac{\overline{BC}}{\overline{AC}} = \frac{2.85 \text{ units}}{17.95 \text{ units}} = 0.159$$

$$\frac{100}{0.159} = \text{lb } H_2O = 630 \text{ lb ice plus liquid water } (B)$$

To obtain the proportions of ice and water, we want to subtract D from B to get E as in Fig. E5.7b.

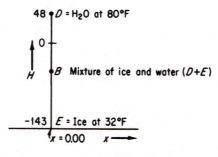

Fig. E5.7b

We measure the distances $\overline{DE}$ and $\overline{BE}$. Then

$$\frac{\text{lb } H_2O \ (D)}{\text{lb total } (B)} = \frac{\overline{BE}}{\overline{DE}} = \frac{3.65 \text{ units}}{5.95 \text{ units}} = 0.613$$

$$\text{lb } H_2O \ (D) = 630(0.613) = 386 \text{ lb}$$

$$\text{lb ice } (E) = 630 - 386 = 244 \text{ lb}$$

EXAMPLE 5.8 Graphical Use of Enthalpy–Concentration Charts

Do Example 5.6 again by graphical means.

Solution

See Fig. E5.8.

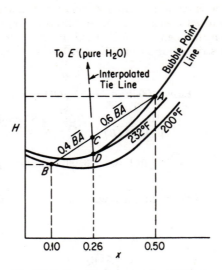

Fig. E5.8 Graphical use of enthalpy charts.

Again plot the known data on an H–x chart. Point C, the sum of solutions A and B, is found graphically by noting by the inverse lever arm rule that

$$\frac{\text{distance } \overline{BC}}{\text{distance } \overline{BA}} = \frac{\text{lb } A}{\text{lb } A + B} = \frac{400}{1000} = 0.4$$

Now measure $\overline{BA}$ on Fig. E5.8 and plot $\overline{BC}$ as $0.4(\overline{BA})$. This fixes point C.

Since C is in the two-phase region, draw a tie line through C (parallel to the nearby tie lines) and measure the distances $\overline{CD}$ and $\overline{ED}$. They will give the H_2O evaporated as follows:

$$\frac{\overline{CD}}{\overline{ED}} = \frac{\text{lb } H_2O \text{ evaporated}}{1000 \text{ lb mixture}}$$

The final concentration of the solution and the temperature of the solution can be read off the H–x chart as before as

$$T = 232°F \qquad x = 26\%$$

EXAMPLE 5.9 Graphical Use Enthalpy–Concentration Charts with Processes Involving Heat Transfer

An acetic acid–water mixture is being concentrated as shown in Fig. E5.9a. How much heat is added or removed?

Solution

First we have to prepare an enthalpy concentration chart by plotting the data for the acetic acid–water system from Appendix I.

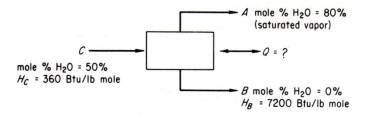

A mole % H₂O = 80%
(saturated vapor)

Q = ?

B mole % H₂O = 0%
H_B = 7200 Btu/lb mole

C

mole % H₂O = 50%
H_C = 360 Btu/lb mole

Fig. E5.9a Graphical use of enthalpy–concentration charts with processes involving heat transfer.

The coordinates of all the points, *A*, *B*, and *C*, are known. These points can be plotted on the *H–x* chart, and then some basis chosen, *A*, *B*, or *C*.

Basis: 1 lb mole of *A*

The governing material and energy balances in this case are (assuming that *Q* is added and is plus)

$$Cx_C = Ax_A + Bx_B$$

$$Q + CH_C = AH_A + BH_B = AQ_A + CH_C$$

or

$$A(H_A - Q_A) + BH_B = CH_C$$

Connect points *B* and *C* with a line, and extend the line toward the right-hand axis (see Fig. E5.9b). What we are looking for is the point A_q located at the coordinates $H = (H_A - Q_A)$ and $x = 0.80$ on the extension of the line $\overline{BC}$ through *B* and *C*. To reach A_q from *A*, we must subtract Q_A; by measuring on Fig. E5.9c, we find a plus value has to be subtracted from H_A, or

$$Q_A = +24,000 \text{ Btu/lb mole } A \qquad \text{(heat added)}$$

If point A_q had fallen above *A*, heat would have been removed from the system and Q_A would have been negative.

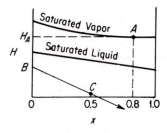

Fig. E5.9b

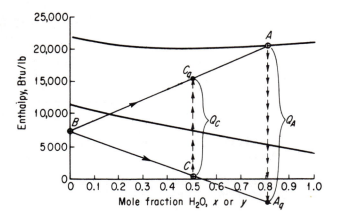

Fig. E5.9c

If we convert our answer to the basis of 1 lb mol of C, we find that

$$\frac{A}{C} = \frac{\overline{BC}}{\overline{AB}} = \frac{0.5 - 0}{0.8 - 0} = 0.625$$

or

$$C = \frac{1}{0.625} = 1.60 \text{ lb mol}$$

Basis: 1 lb mol of C

$$Q_C = \frac{24{,}000 \text{ Btu}}{\text{lb mol } A} \left| \frac{1.00 \text{ lb mol } A}{1.60 \text{ lb mol } C} \right. = 15{,}000 \text{ Btu/lb mol } C$$

As an alternative solution, we could have selected C as the basis to start with and used the equations

$$Cx_C = Ax_A + Bx_B$$
$$Q + CH_C = AH_A + BH_B = C(H_C + Q_C)$$

In this case, join A and B and find point C_q at $x = 0.50$ on the line $\overline{AB}$. Measure $\overline{C_qC}$; it is 14,900 Btu, which is close enough to the value of 15,000 Btu/lb mol C calculated above.

A similar calculation might have been made with B as the basis, by joining points A and C and extending to the $x = 0$ axis in order to get B_q. Q_B would be measured from B_q to B on the $x = 0$ axis.

Self-Assessment Test

1. Use the NH_3–H_2O enthalpy–concentration chart in Appendix I to determine for 1 atm pressure what the mass fraction ammonia is at equilibrium in the gas and liquid phases for a liquid temperature of 130°F.

2. Estimate the heat of vaporization of an ethanol–water mixture at 1 atm and an ethanol mass fraction of 0.50 from the enthalpy-concentration chart in Appendix I.

3. Prepare an enthalpy–concentration chart similar to the one in Appendix I for the acetic acid–water system.

4. From the data in the diagram, compute the ratio of vapor (V) to feed (F) in terms of specific enthalpy differences and compute the amount of V and L per 1000 kg of F.

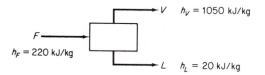

5. For the sulfuric acid–water system, what are the phase(s), composition(s), and enthalpy(ies) existing at $\hat{H} = 120$ Btu/lb and $t = 260°F$?

6. Examine Fig. 5.5 and fill in the blanks for the quantities A, B, and C, respectively: quantity _____ is proportional to the line length _____.

7. Examine Fig. 5.7 and repeat problem 6.

8. Redo Example 5.9 with B as the basis. Do you get the same result?

Section 5.3   Humidity Charts and Their Use

> *Your objectives in studying this section are to be able to:*
>
> 1. Define humidity, humid heat, humid volume, dry-bulb temperature, wet-bulb temperature, humidity chart, moist volume, and adiabatic cooling line.
> 2. Explain and show by use of equations why the slope of the wet-bulb lines are the same as the adiabatic cooling lines for water–air mixtures.
> 3. Prepare a humidity chart given relations for the heat capacities of air and water.
> 4. Use the humidity chart to determine the properties of moist air, and to calculate enthalpy changes and solve heating and cooling problems involving moist air.

In Chap. 3 we discussed humidity, condensation, and vaporization. In this section we apply simultaneous material and energy balances to solve problems involving humidification, air conditioning, water cooling, and the like. Before proceeding, you should review briefly the sections in Chap. 3 dealing with vapor pressure.

Recall that the *humidity* $\mathcal{H}$ is the mass (lb or kg) of water vapor per mass (lb or kg) of bone-dry air (some texts use moles of water vapor per mole of dry air as the humidity) or, as indicated by Eq. (3.52),

$$\mathcal{H} = \frac{18 p_{H_2O}}{29(p_T - p_{H_2O})} = \frac{18 n_{H_2O}}{29(n_T - n_{H_2O})} \qquad (5.17)$$

Table 5.1 lists some of the pertinent notation and data needed in preparing humidity charts.

TABLE 5.1   PERTINENT DATA FOR HUMIDITY CHARTS

Symbol	Meaning	SI Value	American Engineering Value
$C_{p_{air}}$	Heat capacity of air	1.00 kJ/(kg)(K)	0.24 Btu/(lb)(°F)
$C_{p_{H_2O\ vapor}}$	Heat capacity of water vapor	1.88 kJ/(kg)(K)	0.45 Btu/(lb)(°F)
$\Delta \hat{H}_{vap}$	Specific heat of vaporization of water at 0°C (32°F)	4502 kJ/kg	1076 Btu/lb
$\Delta \hat{H}_{air}$	Specific enthalpy of air		
$\Delta \hat{H}_{H_2O\ vapor}$	Specific enthalpy of water vapor		

You will also find it indispensable to learn the following definitions and relations.

(a) The *humid heat* is the heat capacity of an air–water vapor mixture expressed on the *basis of 1 lb or kg of bone-dry air*. Thus the humid heat C_S is

$$C_S = C_{p_{air}} + (C_{p_{H_2O\ vapor}})(\mathcal{H}) \qquad (5.18)$$

where the heat capacities are all per mass and not per mole. Assuming that the heat capacities of air and water vapor are constant under the narrow range of conditions experienced for air-conditioning and humidification calculations, we can write in American engineering units

$$C_S = 0.240 + 0.45(\mathcal{H}) \qquad \text{Btu/(°F)(lb dry air)} \qquad (5.19)$$

or in SI units,

$$C_S' = 1.00 + 1.88(\mathcal{H}) \qquad \text{kJ/(K)(kg dry air)} \qquad (5.19a)$$

(b) The *humid volume* is the volume of 1 lb or kg of dry air plus the water vapor in the air. In the American engineering system,

$$\hat{V} = \frac{359\ \text{ft}^3}{1\ \text{lb mol}} \left| \frac{1\ \text{lb mol air}}{29\ \text{lb air}} \right| \frac{T_{°F} + 460}{32 + 460}$$

$$+ \frac{359\ \text{ft}^3}{1\ \text{lb mol}} \left| \frac{1\ \text{lb mol H}_2\text{O}}{18\ \text{lb H}_2\text{O}} \right| \frac{T_{°F} + 460}{32 + 460} \left| \frac{\mathcal{H}\ \text{lb H}_2\text{O}}{\text{lb air}} \right. \qquad (5.20)$$

$$= (0.730 T_{°F} + 336)\left(\frac{1}{29} + \frac{\mathcal{H}}{18}\right)$$

where $\hat{V}$ is in ft³/lb dry air. In the SI system,

$$\hat{V} = \frac{22.4 \text{ m}^3}{1 \text{ kg mol}} \left| \frac{1 \text{ kg mol air}}{29 \text{ kg air}} \right| \frac{T_K}{273}$$

$$+ \frac{22.4 \text{ m}^3}{1 \text{ kg mol}} \left| \frac{1 \text{ kg mol H}_2\text{O}}{18 \text{ kg H}_2\text{O}} \right| \frac{T_K}{273} \left| \frac{\mathcal{H} \text{ kg H}_2\text{O}}{\text{kg air}} \right. \quad (5.20a)$$

$$= 2.83 \times 10^{-3} T_K + 4.56 \times 10^{-3} \mathcal{H}$$

where $\hat{V}$ is in m³/kg dry air.

(c) The *dry-bulb temperature* (T_{DB}) is the ordinary temperature you always have been using for a gas in °F or °C (or °R or K).

(d) The *wet-bulb temperature* (T_{WB}) you may guess, even though you may never have heard of this term before, has something to do with water (or other liquid, if we are concerned not with humidity but with saturation) evaporating from around an ordinary mercury thermometer bulb. Suppose that you put a wick, or porous cotton cloth, on the mercury bulb of a thermometer and wet the wick. Next you either (1) whirl the thermometer in the air as in Fig. 5.9 (this apparatus is called a sling psychrometer when the wet-bulb and dry-bulb thermometers are mounted together), or (2) set up a fan to blow rapidly on the bulb at 1000 ft³/min or more. What happens to the temperature recorded by the wet-bulb thermometer?

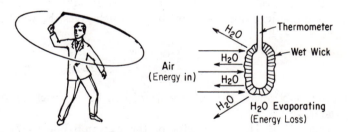

Fig. 5.9 Wet-tulb temperature obtained with a sling psychrometer.

As the water from the wick evaporates, the wick cools down and continues to cool until the rate of energy transferred to the wick by the air blowing on it equals the rate of loss of energy caused by the water evaporating from the wick. We say that the temperature of the bulb with the wet wick at equilibrium is the wet-bulb temperature. (Of course, if water continues to evaporate, it eventually will all disappear, and the wick temperature will rise.) The final temperature for the process described above will lie on the 100% relative humidity curve (saturated-air curve).

Suppose that we set up a graph on which the vertical axis is the humidity and the horizontal axis is the dry-bulb temperature. How should we plot the change of temperature as the T_{DB} of the thermometer changes to reach T_{WB}?

The equation for the wet-bulb lines is based on a number of assumptions, a detailed discussion of which is beyond the scope of this book. Nevertheless, the idea of the wet-bulb temperature is based on the equilibrium between the *rates* of energy transfer to the bulb and evaporation of water. Rates of processes are a topic that we have not discussed. The fundamental idea is that a large amount of air is brought into contact with a little bit of water and that presumably the evaporation of the water leaves the temperature and humidity of the air unchanged. Only the temperature of the water changes. The equation of the wet-bulb line is an energy balance,

$$h_c(T - T_{WB}) = k'_g \, \Delta \hat{H}_{vap}(\mathcal{H}_{WB} - \mathcal{H}) \tag{5.21}$$
$$\underset{\text{heat transfer to water}}{\phantom{h_c(T - T_{WB})}} \quad \underset{\text{heat transfer from water}}{\phantom{k'_g \, \Delta \hat{H}_{vap}(\mathcal{H}_{WB} - \mathcal{H})}}$$

where h_c = heat transfer coefficient for convection to the bulb
$\quad k'_g$ = mass transfer coefficient
$\Delta \hat{H}_{vap}$ = latent heat of vaporization
$\quad \mathcal{H}$ = humidity of moist air
$\quad T$ = temperature of moist air

Next, we can form the ratio

$$\frac{\mathcal{H}_{WB} - \mathcal{H}}{T_{WB} - T} = -\frac{h_c}{(k'_g) \, \Delta \hat{H}_{vap}} \tag{5.22}$$

to get the slope of the wet bulb line. For water only, it so happens that $h_c/k'_g \simeq C_S$ (i.e., the numerical value is about 0.25), which gives the wet-bulb lines the slope of

$$\frac{\mathcal{H}_{WB} - \mathcal{H}}{T_{WB} - T} = -\frac{C_S}{\Delta \hat{H}_{vap}} \tag{5.23}$$

For other substances, the value of h_c/k'_g can be as much as twice the value of C_s for water.

Figure 5.10 shows the plot of the wet-bulb line as $T_{DB} \longrightarrow T_{WB}$. The line is approximately straight and has a negative slope. Does this result agree with Eq. (5.23)?

Another type of process of some importance occurs when adiabatic cooling or humidification takes place between air and water that is recycled as illustrated in Fig. 5.11. In this process the air is both cooled and humidified (its water content rises) while a little bit of the recirculated water is evaporated. At *equilibrium*, in the steady state, the temperature of the air is the same as the temperature of the water, and the exit air is saturated at this temperature. By making an overall energy balance around the process ($Q = 0$), we can obtain the equation for the adiabatic cooling of the air. The equation, when plotted on the humidity chart, yields what is known as an adiabatic cooling line. We take the equilibrium temperature of the water, T_s, as a reference temperature rather than 0°C or 32°F. Do you see why? We ignore the small amount of makeup water or assume that it enters at T_s. The energy balance is

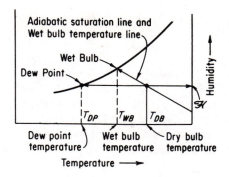

Fig. 5.10 General layout of the humidity chart showing the location of the wet-bulb and dry-bulb temperatures, the dew point and dew-point temperature, and the adiabatic saturation line and wet-bulb line.

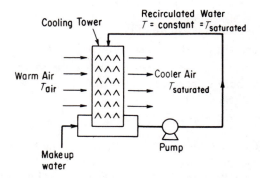

Fig. 5.11 Adiabatic humidification with recycle of water.

$$
\overbrace{C_{p_{\text{air}}}(T_{\text{air}} - T_S)}^{\substack{\text{enthalpy of air}\\\text{entering}}} + \mathcal{H}_{\text{air}}\overbrace{[\Delta\hat{H}_{\text{vap H}_2\text{O at } T_s} + C_{p_{\text{H}_2\text{O vapor}}}(T_{\text{air}} - T_S)]}^{\substack{\text{enthalpy of water vapor}\\\text{in air entering}}}
$$

$$
= \underbrace{C_{p_{\text{air}}}(T_S - T_S)}_{\substack{\text{enthalpy of air}\\\text{leaving}}} + \mathcal{H}_S\underbrace{[\Delta\hat{H}_{\text{vap H}_2\text{O at } T_s} + C_{p_{\text{H}_2\text{O vapor}}}(T_S - T_S)]}_{\substack{\text{enthalpy of water vapor}\\\text{in air leaving}}}
$$

(5.24)

This can be reduced to

$$
T_{\text{air}} = \frac{\Delta\hat{H}_{\text{vap H}_2\text{O at } T_s}(\mathcal{H}_S - \mathcal{H}_{\text{air}})}{C_{p_{\text{air}}} + C_{p_{\text{H}_2\text{O vapor}}}\mathcal{H}_{\text{air}}} + T_S
$$

(5.25)

which is the equation for adiabatic cooling.

Notice that this equation can be written as

$$\frac{\mathcal{H}_s - \mathcal{H}}{T_S - T_{air}} = -\frac{C_s}{\Delta \hat{H}_{vap \, at \, T_s}} \tag{5.26}$$

Compare Eq. (5.26) with Eq. (5.23). Can you conclude that the wet-bulb process equation, for water only, is essentially the same as the adiabatic cooling equation?

Of course! We have the nice feature that two processes can be represented by the same set of lines. For a detailed discussion of the uniqueness of this coincidence, consult any of the references at the end of the chapter. For most other substances besides water, the two equations will have different slopes.

Now that you have an idea of what the various features portrayed on the *humidity chart* (psychrometric chart) are, let us look at the chart itself [Fig. 5.12(a) and (b)]. It is nothing more than a graphical means to assist in the presentation of material and energy balances in water vapor–air mixtures and various associated parameters. Its skeleton consists of a humidity ($\mathcal{H}$)–temperature (T_{DB}) set of coordinates together with the additional parameters (lines) of

(a) Constant relative humidity
(b) Constant moist volume (humid volume)
(c) Adiabatic cooling lines which are the same (for water vapor only) as the wet-bulb or psychrometric lines
(d) The 100% relative humidity (identical to the 100% absolute humidity) curve or saturated-air curve

With any two values known, you can pinpoint the air-moisture condition on the chart and determine all the other associated values.

Off to the left of the 100% relative humidity line you will observe scales showing the enthalpy per mass of dry air of a saturated air–water vapor mixture. Enthalpy adjustments for air less than saturated are shown on the chart itself by a series of curves. The enthalpy of the wet air, in energy/mass of dry air, is

$$\Delta \hat{H} = \Delta \hat{H}_{air} + \Delta \hat{H}_{H_2O \, vapor}(\mathcal{H}) \tag{5.27}$$

We should mention at this point that the reference conditions for the humidity chart are liquid water at 32°F (0°C) and 1 atm (not the vapor pressure of H_2O) for water, and 0°F and 1 atm for air. The chart is suitable for use only at normal atmospheric conditions and must be modified[6] if the pressure is significantly different than 1 atm. If you wanted to, you could calculate the enthalpy values shown on the chart directly from tables listing the enthalpies of air and water vapor by the methods described in Chap. 4, or you could, by making use of Eq. (5.27), compute the enthalpies with reasonable accuracy from the following equation

[6]See G. E. McElroy, *U.S. Bur. Mines Rept. Invest. No. 4615*, December 1947, or other Carrier charts. See *Educational Materials*, Carrier Corp., Syracuse, N.Y., 1980.

$$\Delta \hat{H} = \underset{\text{enthalpy for air}}{C_{p_{air}}(T - T_{ref})_{air}} + \mathcal{H}[\underset{\substack{\text{heat of vaporization} \\ \text{of water at } T_{ref}}}{\Delta \hat{H}_{vap}} + \underset{\substack{\text{enthalpy for} \\ \text{water vapor}}}{C_{p_{air}}(T - T_{ref})_{water}}] \qquad (5.28)$$

In American engineering units, Eq. (5.28) is

$$H = 0.240 T_{°F} + \mathcal{H}(1061 + 0.45 T_{°F}) \qquad (5.29)$$

because $C_s = 0.240 + 0.45\mathcal{H}$ and $T_{WB} = T_S$. Thus the wet-bulb process equation, for water only, is essentially the same as the adiabatic cooling equation. For other materials these two equations have different slopes.

Only two of the quantities in Eq. (5.28) are variables, if T_S is known, because $\mathcal{H}_S$ is the humidity of saturated air at T_S and $\Delta \hat{H}_{vap\ H_2O\ at\ T_s}$ is fixed by T_S. Thus, for any value of T_S, you can make a plot of Eqs. (5.25) and/or (5.29) on the humidity chart in the form of $\mathcal{H}$ vs. T_{air}. These curves, which are essentially linear, will intersect the 100% relative humidity curve at $\mathcal{H}_S$ and T_S, as described earlier.

The adiabatic cooling lines are lines of almost constant enthalpy for the entering air–water mixture, and you can use them as such without much error (1 or 2%). However, if you want to correct a saturated enthalpy value for the deviation which exists for a less-than-saturated air–water vapor mixture, you can employ the enthalpy deviation lines which appear on the chart and which can be used as illustrated in the examples below. Any process that is not a wet-bulb process or an adiabatic process with recirculated water can be treated by the usual material and energy balances, taking the basic data for the calculation from the humidity charts. If there is any increase or decrease in the moisture content of the air in a psychrometric process, the small enthalpy effect of the moisture added to the air or lost by the air may be included in the energy balance for the process to make it more exact as illustrated in Examples 5.11 and 5.13.

You can find further details regarding the construction of humidity charts in the references at the end of the chapter. Tables are also available listing all the thermodynamic properties (p, $\hat{V}$, $\mathcal{H}$, and $\Delta \hat{H}$) in great detail.[7] Although we shall be discussing humidity charts exclusively, charts can be prepared for mixtures of any two substances in the vapor phase, such as CCl_4 and air or acetone and nitrogen, by use of Eqs. (5.17)–(5.28) if all the values of the physical constants for water and air are replaced by those of the desired gas and vapor. The equations themselves can be used for humidity problems if charts are too inaccurate or are not available.

EXAMPLE 5.10 Properties of Moist Air from the Humidity Chart

List all the properties you can find on the humidity chart in American engineering units for moist air at a dry-bulb temperature of 90°F and a wet-bulb temperature of 70°F.

[7]Byron Engelbach, *Microfilms of Psychrometric Tables*, University Microfilms, Ann Arbor, Mich., 1953.

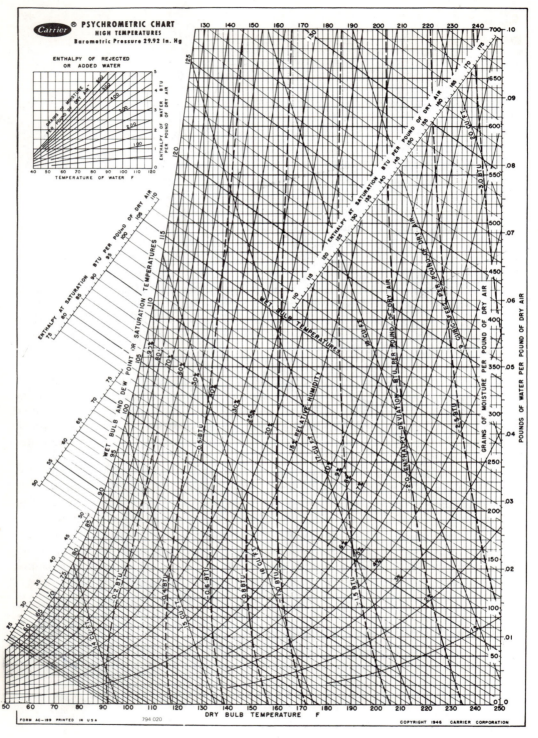

Fig. 5.12(a)   (Courtesy of Carrier Corporation.)

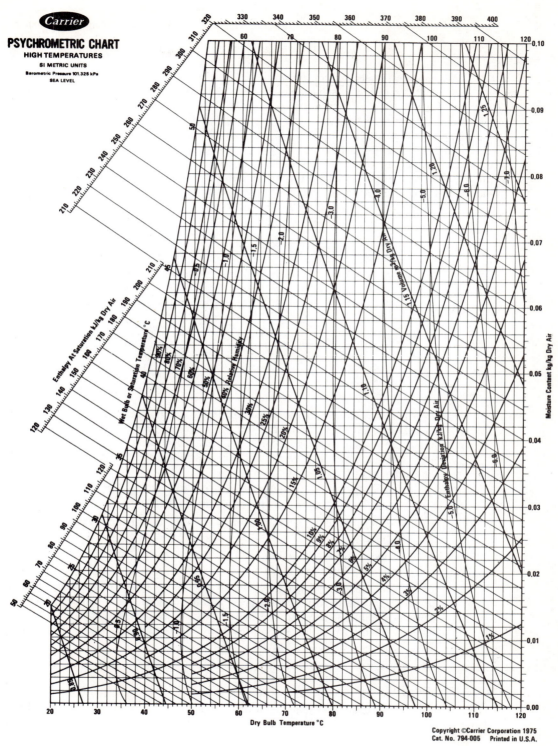

Fig. 5.12(b) (Courtesy of Carrier Corporation.)

Solution

A diagram will help explain the various properties obtained from the humidity chart. See Fig. E5.10. You can find the location of point A for 90°F DB (dry bulb) and 70°F WB (wet bulb) by following a vertical line at $T_{DB} = 90°F$ until it crosses the wet-bulb line for 70°F. This wetbulb line can be located by searching along the 100% humidity line until the saturation temperature of 70°F is reached, or, alternatively, by proceeding up a vertical line at 70°F until it intersects the 100% humidity line. From the wet-bulb temperature of 70°F, follow the adiabatic cooling line (which is the same as the wet-bulb temperature line on the humidity chart) to the right until it intersects the 90°F DB line. Now that point A has been fixed, you can read the other properties of the moist air from the chart.

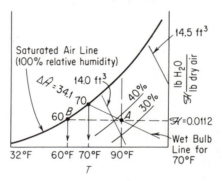

Fig. E5.10

(a) *Dew point.* When the air at A is cooled at constant pressure (and in effect at *constant humidity*), as described in Chap. 3, it eventually reaches a temperature at which the moisture begins to condense. This is represented by a horizontal line, a constant-humidity line, on the humidity chart, and the dew point is located at B, or about 60°F.

(b) *Relative humidity.* By interpolating between the 40% $\mathcal{RH}$ and 30% $\mathcal{RH}$ lines with a ruler, you can find that point A is at about 37% $\mathcal{RH}$.

(c) *Humidity* $(\mathcal{H})$. You can read the humidity from the right-hand ordinate as 0.0112 lb H_2O/lb air.

(d) *Humid volume.* By interpolation again between the 14.0- and the 14.5-ft³ lines, you can find the humid volume to be 14.097 ft³/lb dry air.

(e) *Enthalpy.* The enthalpy value for saturated air with a wet-bulb temperature of 70°F is $\Delta\hat{H} = 34.1$ Btu/lb dry air (a more accurate value can be obtained from psychrometric tables if needed). The enthalpy deviation for less-than-saturated air is about -0.2 Btu/lb of dry air; consequently, the actual enthalpy of air at 37% $\mathcal{RH}$ is $34.1 - 0.2 = 33.9$ Btu/lb of dry air.

EXAMPLE 5.11 Heating at Constant Humidity

Moist air at 38°C and 48% $\mathcal{RH}$ is heated in your furnace to 86°C. How much heat has to be added per cubic meter of initial moist air, and what is the final dew point of the air?

Solution

As shown in Fig. E5.11, the process goes from point A to point B on a horizontal line of constant humidity. The initial conditions are fixed at $T_{DB} = 38°C$ and 48% RH. Point B is fixed by the intersection of the horizontal line from A and the vertical line at 86°C. The dew point is unchanged in this process and is located at C at 24.8°C.

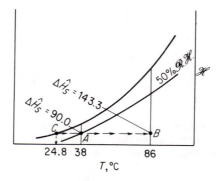

Fig. E5.11

The enthalpy values are as follows (all in kJ/kg of dry air):

point	$\Delta \hat{H}_{satd}$	δH	$\Delta \hat{H}_{actual}$
A	90.0	+0.5	90.5
B	143.3	−3.3	140.0

Also, at A the volume of the moist air is 0.91 m³/kg of dry air. Consequently, the heat added is ($Q = \Delta \hat{H}$)140.0 − 90.5 = 49.5 kJ/kg of dry air.

$$\frac{49.5 \text{ kJ}}{\text{kg dry air}} \left| \frac{1 \text{ kg dry air}}{0.91 \text{ m}^3} \right. = 54.4 \text{ kJ/m}^3 \text{ initial moist air}$$

EXAMPLE 5.12 Cooling and Humidification

One way of adding moisture to air is by passing it through water sprays or air washers. See Fig. E5.12a. Normally, the water used is recirculated rather than wasted. Then, in the steady state, the water is at the adiabatic saturation temperature, which is the same as the wet-bulb temperature. The air passing through the washer is cooled, and if the contact time between the air and the water is long enough, the air will be at the wet-bulb temperature also. However, we shall assume that the washer is small enough so that the air does not reach the wet-bulb temperature; instead, the following conditions prevail:

	T_{DB} (°C)	T_{WB} (°C)
Entering air:	40	22
Exit air:	27	

Find the moisture added per kilogram of dry air.

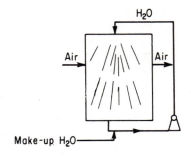

Fig. E5.12a

Solution

The whole process is assumed to be *adiabatic*, and, as shown in Fig. E5.12b, takes place between points A and B along the adiabatic cooling line. The wet-bulb temperature remains constant at 22°C. Humidity values are

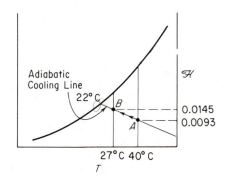

Fig. E5.12b

	$\mathcal{H} \left(\dfrac{\text{kg } H_2O}{\text{kg air}} \right)$
B	0.0145
A	0.0098
Difference:	$0.0052 \; \dfrac{\text{kg } H_2O}{\text{kg dry air}}$ added

---

EXAMPLE 5.13 Cooling and Dehumidification

A process that takes moisture out of the air by passing the air through water sprays sounds peculiar but is perfectly practical as long as the water temperature is below the dew point of the air. Equipment such as shown in Fig. E5.13a would do the trick. If the entering air has a dew point of 70°F and is at 40% $\mathcal{R}\mathcal{H}$, how much heat has to be removed by the cooler, and how much water vapor is removed, if the exit air is at 56°F with a dew point of 54°F?

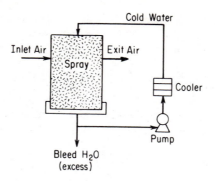

Fig. E5.13a

Solution

From Fig. 5.12 in American engineering units the initial and final values of the enthalpies and humidities are

	A	B
$\mathcal{H}\left(\dfrac{\text{grains } H_2O}{\text{lb dry air}}\right)$	111	62
$\Delta\hat{H}\left(\dfrac{\text{Btu}}{\text{lb dry air}}\right)$	$41.3 - 0.2 = 41.1$	$23.0 - 0 = 23.0$

Look at Fig. E5.13b for the relative positions of A and B. The grains of H_2O removed are

$$111 - 62 = 49 \text{ grains/lb dry air}$$

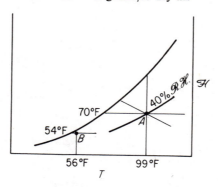

Fig. E5.13b

The cooling duty is approximately

$$41.1 - 23.0 = 18.1 \text{ Btu/lb dry air}$$

In the upper left of the humidity chart is a little insert that gives the value of the small correction factor for the water condensed from the air which leaves the system. Assuming that the water leaves at the dew point of 54°F, read for 49 grains a correction of −0.15 Btu/lb of dry air. You could calculate the same value by taking the enthalpy of liquid water from the steam tables and saying,

$$\frac{22 \text{ Btu}}{\text{lb } H_2O} \left| \frac{1 \text{ lb } H_2O}{7000 \text{ grains}} \right| \frac{49 \text{ grains rejected}}{1 \text{ lb dry air}} = 0.154 \text{ Btu/lb dry air}$$

The energy (enthalpy) balance will then give us the cooling load:

$$\underbrace{41.1}_{\substack{\Delta H \\ \text{air in}}} - \underbrace{23.0}_{\substack{\Delta H \\ \text{air out}}} - \underbrace{0.15}_{\substack{\Delta H \\ H_2O \text{ out}}} = 17.9 \text{ Btu/lb dry air}$$

EXAMPLE 5.14 Combined Material and Energy Balances for a Cooling Tower

You have been requested to redesign a water-cooling tower that has a blower with a capacity of 8.30×10^6 ft³/hr of moist air (at 80°F and a wet-bulb temperature of 65°F). The exit air leaves at 95°F and 90°F wet bulb. How much water can be cooled in pounds per hour if the water to be cooled is not recycled, enters the tower at 120°F, and leaves the tower at 90°F?

Solution

Enthalpy, humidity, and humid volume data taken from the humidity chart are as follows (see Fig. E5.14):

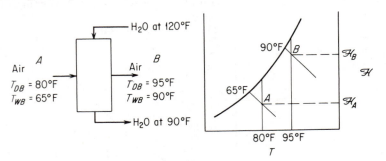

Fig. E5.14

	A	B
$\mathcal{H} \left(\dfrac{\text{lb } H_2O}{\text{lb dry air}} \right)$	0.0098	0.0297
$\mathcal{H} \left(\dfrac{\text{grains } H_2O}{\text{lb dry air}} \right)$	69	208
$\Delta \hat{H} \left(\dfrac{\text{Btu}}{\text{lb dry air}} \right)$	$30.05 - 0.12 = 29.93$	$55.93 - 0.10 = 55.83$
$\hat{V} \left(\dfrac{\text{ft}^3}{\text{lb dry air}} \right)$	13.82	14.65

The cooling-water exit rate can be obtained from an energy balance around the process.

$$\text{Basis: } 8.30 \times 10^6 \text{ ft}^3/\text{hr of moist air}$$

$$\frac{8.30 \times 10 \text{ ft}^3 \mid \text{lb dry air}}{13.82 \text{ ft}^3} = 6.00 \times 10^5 \text{ lb dry air/hr}$$

The enthalpy of the entering water stream is (reference temperature is 32°F and 1 atm)

$$\Delta \hat{H} = C_{P_{H_2O}}\Delta T = 1(120 - 32) = 88 \text{ Btu/lb H}_2\text{O}$$

and that of the exit stream is 58 Btu/lb H_2O. [The value from the steam tables at 120°F for liquid water of 87.92 Btu/lb H_2O is slightly different since it represents water at its vapor pressure (1.69 psia) based on reference conditions of 32°F and liquid water at its vapor pressure.] Any other datum could be used instead of 32°F for the liquid water. For example, if you chose 90°F, one water stream would not have to be taken into account because its enthalpy would be zero.

The loss of water to the air is

$$0.0297 - 0.0098 = 0.0199 \text{ lb H}_2\text{O/lb dry air}$$

(a) *Material balance for water stream:*

Let $W = $ lb H_2O entering the tower in the water stream per lb dry air

Then

$W - 0.0199 = $ lb H_2O leaving tower in the water stream per lb dry air

(b) *Energy balance (enthalpy balance) around the entire process:*

air and water in air entering   **water stream entering**
$$\frac{29.93 \text{ Btu} \mid 6.00 + 10^5 \text{ lb dry air}}{\text{lb dry air}} + \frac{88 \text{ Btu} \mid W \text{ lb H}_2\text{O} \mid 6.00 \times 10^5 \text{ lb dry air}}{\text{lb H}_2\text{O} \mid \text{lb dry air}}$$

air and water in air leaving
$$= \frac{55.83 \text{ Btu} \mid 6.00 \times 10^5 \text{ lb dry air}}{\text{lb dry air}}$$

water stream leaving
$$+ \frac{58 \text{ Btu} \mid (W - 0.0199) \text{ lb H}_2\text{O} \mid 6.00 \times 10^5 \text{ lb dry air}}{\text{lb H}_2\text{O} \mid \text{lb dry air}}$$

$$29.93 + 88W = 55.83 + 58(W - 0.0199)$$

$$W = 0.825 \text{ lb H}_2\text{O/lb dry air}$$

$$W - 0.0199 = 0.805 \text{ lb H}_2\text{O/lb dry air}$$

The total water leaving the tower is

$$\frac{0.805 \text{ lb H}_2\text{O} \mid 6.00 \times 10^5 \text{ lb dry air}}{\text{lb dry air} \mid \text{hr}} = 4.83 \times 10^5 \text{ lb /hr}$$

Self-Assessment Test

1. What is the difference between the wet- and dry-bulb temperatures?

2. Can the wet-bulb temperature ever be higher than the dry-bulb temperature?

3. Explain why the slope of the wet-bulb lines are essentially the same as the slope of the adiabatic cooling lines for gaseous air and water mixtures.

4. Estimate for air at 70°C dry-bulb temperature, 1 atm, and 15% relative humidity the:
    (a) kg H_2O/kg of dry air
    (b) m^3/kg of dry air
    (c) Wet-bulb temperature (in °C)
    (d) Specific enthalpy
    (e) Dew point (in °C)

5. Calculate the following properties of moist air at 1 atm and compare with values read from the humidity chart.
    (a) The humidity of saturated air at 120°F
    (b) The enthalpy of air in part (a) per pound of dry air
    (c) The volume per pound of dry air of part (a)
    (d) The humidity of air at 160°F with a wet-bulb temperature of 120°F

6. Humid air at 1 atm and 200°F, and containing 0.0645 lb of H_2O/lb of dry air, enters a cooler at the rate of 1000 lb of dry air per hour (plus accompanying water vapor). The air leaves the cooler at 100°F, saturated with water vapor (0.0434 lb of H_2O/lb of dry air). Thus 0.0211 lb H_2O is condensed per pound of dry air. How much heat is transferred to the cooler?

7. A cooling tower that uses a cold-water spray provides a method of cooling and dehumidifying a school. During the day, the average number of students in the school is 100 and the average heat-generation rate per person is 800 Btu/hr. Suppose that the ambient conditions outside the school in the summer are expected to be 100°F and 95% ℛℋ. You run this air through the cooler-dehumidifier and then mix the saturated exit air with recirculated air from the exhaust of the school building. You need to supply the mixed air to the building at 70°F and 60% ℛℋ and keep the recirculated air leaving the building at not more than 72°F. Leakage occurs from the building of the 72°F air also. Calculate:
    (a) The volumetric rate of air recirculation per hour in cubic feet at 70°F and 60% ℛℋ
    (b) The volume of fresh air required at entering conditions
    (c) The heat transferred in the cooler-dehumidifier from the inlet air per hour

Section 5.4   Complex Problems

> *Your Objective in Studying This Section Is to be Able to:*
>
> 1. Solve combined energy and mass balances for complex steady-state processes with or without chemical reactions.

In this book we have treated only small segments of material and energy balance problems. Put a large number of these segments together and you will have a complex industrial process. We do not have the space to describe the details of any specific process, but for such information you can consult the references at the end of the chapter. It is always wise to read about a process and gain as much information as you can about the stoichiometry and energy relations involved before undertaking to make any calculations. Do not be unnerved by the complexity of a large-scale detailed plant. With the techniques you have accumulated while studying this text you will find that you will be able to break up the overall scheme into smaller sections involving a manageable number of streams that can be handled in the same way you have handled the material and energy balances in this text. Perhaps you can find a tie element (or make one up) that will permit you to work from one unit of the process to the next. Example 5.15 briefly outlines, without much descriptive explanation, how you should apply your knowledge of material and energy balances to a not-too-complex alcohol plant.

Problems in design and plant operation which involve simultaneous material and energy balances are frequently solved using digital computer programs. Table 5.2 lists a number of such programs, many of which are proprietary. *Chemical Engineering* [p. 145 (June 5, 1978)] initiated a list of computer programs that encompass

TABLE 5.2 PROGRAMS THAT SOLVE COMBINED MATERIAL AND ENERGY BALANCE PROBLEMS FOR LARGE PLANTS

Name	Where Developed	Reference
ASPEN	MIT	*Annual Reports*, 1977, 1978, 1979, 1980 available from NTIS, Springfield, Va. 22161
CHESS	University of Houston	R. L. Motard, H. M. Lee, and R. W. Barkley, *CHESS User's Guide*, 3rd ed., Department of Chemical Engineering, Washington University, St. Louis, Mo., 1978
CONCEPT MARK IV	Computer Aided Design Centre, Cambridge, England	*CADC Concept Mark IV User Manual*, CADC, Cambridge CB3 OHB, England, 1979
DPS	Computer Aided Design Centre, Cambridge, England	*Dynamic Process Simulator*, CADC, Cambridge CB3 OHB, England, 1980
EQUITRAN	Mitsui Toutsu, Japan	1975
FLOWTRAN	Monsanto Co., St. Louis, Mo.	J. D. Seader, W. D. Seider, and A. C. Pauls, *FLOWTRAN Simulation—An Introduction*, 3rd ed., Ulrich's Bookstore, Ann Arbor, Mich., 1979
GENIE	ICI Agricultural Div., Billingham, England	S. R. Metcalfe and J. D. Perkins, *Trans. Inst. Chem. Eng.*, v. 56, pp. 210–214 (1978)
GPSP	McDonnell-Douglas Automation	Box 516, St. Louis, Mo. 63166
MBP II	Sood, M. K., and Reklaitis, G. V.	*Material Balance Program-II User's Manual*, School Chemical Engineering Purdue University, Lafayette, Ind., 1977
PBDS	Tokyo Engineering, Japan	1974
SWAPSCO	Stone & Webster, Boston	T. Utsumi, "Computer Package Aids Systems Engineering," *Oil Gas J.*, p. 100 (June 8, 1970)
UOS	Bonner & Moore Assoc., Houston, Tex.	

all aspects of chemical engineering. Such programs usually employ a modular structure, meaning that subroutines are prepared for pumps, compressors, reactors,
distillation columns, and so on. To be able to solve the balances for even one of these
pieces of equipment, you need to have a bank of data for the stream properties. Most
programs furnish the necessary physical data and make provision for you to provide
any missing information.

Assuming that subroutines have been prepared and are available that simulate
the behavior of individual pieces of process equipment and produce numerical
outputs for varying inputs and process parameters, your main task as the user is to
designate the interconnections between the units and the choice of the subroutines to
represent each unit. In some instances you can also select the subroutine to be used
to solve the large sets of nonlinear equations that must be solved simultaneously. As
illustrated in Fig. 5.13, executive routines call the subroutines in the proper order and
solve the material and energy balances usually by iterative methods because many of
the equations used to represent complex processes are nonlinear. In the list of references at the end of the chapter you can find surveys of the techniques of computer
solution of material and energy balances. One of the dangers of using such programs
is that inexperience leads to use of the wrong data or misinterpretation of the output
calculations. Be careful! Table 5.2 lists some of the programs available that are user-
oriented. Evans et al.[8] give a very lucid state of the art survey of heat and material balance programs, and additional references can be found at the end of the chapter.

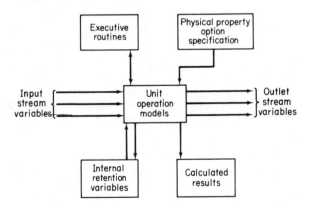

Fig. 5.13 Flow of information to and from a unit operation model.

EXAMPLE 5.15 A More Complex Problem

The Blue Ribbon Sour Mash Company plans to make commercial alcohol by a process
shown in Fig. E5.15a. Grain mash is fed through a heat exchanger where it is heated to 170°F.
The alcohol is removed as 60% by weight alcohol from the first fractionating column; the

[8]L. B. Evans et al., *Chem. Eng. Progr.*, v. 64, no. 4, p. 39 (1968).

bottoms contain no alcohol. The 60% alcohol is further fractionated to 95% alcohol and essentially pure water in the second column. Both stills operate at a 3:1 reflux ratio and heat is supplied to the bottom of the columns by steam. Condenser water is obtainable at 80°F. The operating data and physical properties of the streams have been accumulated and are listed for convenience:

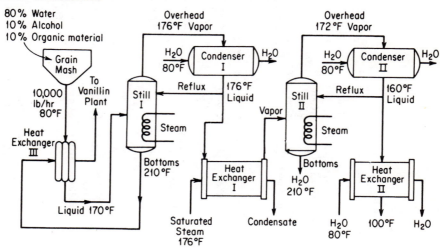

Fig. E5.15a

stream	state	b.p. (°F)	C_p [Btu/(lb)(°F)] liquid	vapor	heat of vaporization (Btu/lb)
Feed	Liquid	170	0.96	—	950
60% alcohol	Liquid or vapor	176	0.85	0.56	675
Bottoms I	Liquid	212	1.00	0.50	970
95% alcohol	Liquid or vapor	172	0.72	0.48	650
Bottoms II	Liquid	212	1.00	0.50	970

Make a complete material balance on the process and

(a) Determine the weight of the following streams per hour:
   (1) Overhead product, column I
   (2) Reflux, column I
   (3) Bottoms, column I
   (4) Overhead product, column II
   (5) Reflux, column II
   (6) Bottoms, column II

(b) Calculate the temperature of the bottoms leaving heat exchanger III.

(c) Determine the total heat input to the system in Btu/hr.

(d) Calculate the water requirements for each condenser and heat exchanger II in gal/hr if the maximum exit temperature of water from this equipment is 130°F.

Solution

Let us call the alcohol A and the organic matter M.

Phase 1: Material Balances

Basis: 10,000 lb/hr of mash

(a) *Material balance around column I and condenser:*

In

Feed $\begin{cases} A \\ H_2O \\ M \end{cases}$ $\begin{array}{ll} (0.10)(10,000) = & 1,000\ \text{lb} \\ (0.80)(10,000) = & 8,000\ \text{lb} \\ (0.10)(10,000) = & 1,000\ \text{lb} \end{array}$

$\overline{10,000\ \text{lb}}$

Out

Product $\begin{cases} A \\ H_2O \end{cases}$ $\begin{array}{ll} & = 1,000\ \text{lb} \\ \left(\dfrac{1000}{0.60}\right) - 1000 = & 667\ \text{lb} \end{array}$

Bottoms $\begin{cases} H_2O \\ M \end{cases}$ $\begin{array}{ll} (8000 - 667) = & 7,333\ \text{lb} \\ & = 1,000\ \text{lb} \end{array}$

$\overline{10,000\ \text{lb}}$

(b) *Material balance around column I only:*

In

Feed $\begin{cases} A \\ H_2O \\ M \end{cases}$ $\begin{array}{ll} = & 1,000\ \text{lb} \\ = & 8,000\ \text{lb} \\ = & 1,000\ \text{lb} \end{array}$

Reflux $\begin{cases} A \\ H_2O \end{cases}$ $\begin{array}{ll} (3)(1000) & 3,000\ \text{lb} \\ (3)(667) & 2,000\ \text{lb} \end{array}$

$\overline{15,000\ \text{lb}}$

Out

Overhead vapor $\begin{cases} A \\ H_2O \end{cases}$ $\begin{array}{ll} 4(1000) & 4,000\ \text{lb} \\ 4(667) & 2,667\ \text{lb} \end{array}$

Bottoms $\begin{cases} H_2O \\ M \end{cases}$ $\begin{array}{ll} = & 7,333\ \text{lb} \\ = & 1,000\ \text{lb} \end{array}$

$\overline{15,000\ \text{lb}}$

(c) *Material balance around column II and condenser:*

In

Feed $\begin{cases} A \\ H_2O \end{cases}$ $\begin{array}{ll} = & 1,000\ \text{lb} \\ = & 667\ \text{lb} \end{array}$

$\overline{1,667\ \text{lb}}$

Out

(d) Overhead product $\begin{cases} A \\ H_2O \end{cases}$ $\begin{array}{ll} & = 1,000\ \text{lb} \\ \left(\dfrac{1000}{0.95} - 1000\right) = & 50\ \text{lb} \end{array}$

Bottoms $\{H_2O$ $\quad (667 - 50) = \quad 617\ \text{lb}$

$\overline{1,667\ \text{lb}}$

(d) *Material balance on column II cutting reflux and overhead vapor:*

In

Feed $\begin{cases} A \\ H_2O \end{cases}$ $\begin{array}{ll} = & 1,000\ \text{lb} \\ = & 667\ \text{lb} \end{array}$

(e) Reflux $\begin{cases} A \\ H_2O \end{cases}$ $\begin{array}{ll} 3(1000) & 3,000\ \text{lb} \\ (3)(50) & 150\ \text{lb} \end{array}$

$\overline{4,817\ \text{lb}}$

Out

Vapor $\begin{cases} A \\ H_2O \end{cases}$ $\begin{array}{ll} (4)(1000) & = 4,000\ \text{lb} \\ (4)(50) & = 200\ \text{lb} \end{array}$

Bottoms $\{H_2O$ $\quad (667 - 50) = \quad 617\ \text{lb}$

$\overline{4,817\ \text{lb}}$

Phase 2: Energy Balances

(a) *Energy balance on heat exchanger III (Fig. E5.15b):*

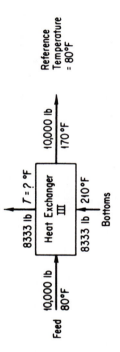

Fig. E5.15b

		In		
Feed	$(10{,}000)(0.96)(80 - 80) =$	0 Btu		
Bottoms	$(8333)(1.0)(210 - 80) =$	1,084,000 Btu		

		Out	
Feed	$(10{,}000)(0.96)(170 - 80) =$	864,000 Btu	
Bottoms	$(833)(1.0)(T - 80) =$	$= (8333T - 666{,}000)$ Btu	

$$1{,}084{,}000 = 8333T - 666{,}000 + 864{,}000$$
$$886{,}000 = 8333T$$
$$T = 106.5°F$$

(b) *Energy balance around heat exchanger III inlet, and overhead vapor and reflux line from I:*

Reference temperature $= 176°F$

In

Feed	$10{,}000(0.96)(80 - 176) =$	$-921{,}000$
Reflux	$5000(0.85)(176 - 176) =$	0
Steam	ΔH_{S_1}	
	$\overline{\Delta H_{S_1} - 921{,}000}$	

Out

Bottoms to vanillin plant	$8333(1.0)(106.5 - 176) =$	$-579{,}000$
Vapor	$6667(0.85)(176 - 176) + 6667(675) =$	$4{,}500{,}000$
		$\overline{3{,}921{,}000}$

$$\Delta H_{S_1} - 921{,}000 = 3{,}921{,}000$$
$$\Delta H_{S_1} = 4{,}842{,}000 \text{ Btu/hr}$$

505

(c) *Energy balance around heat exchanger I outlet and overhead vapor and reflux from II:*

Reference temperature = 172°F

In

Feed to II	$1667(0.85)(176 - 172) =$	5,650
	$\Delta H_{\text{vaporization} \equiv \text{steam at } 176°\text{F}}$	
	$+1667(675) =$	1,125,000
Reflux	$3150(0.72)(160 - 172) =$	−27,200
Steam		ΔH_{S_2}

$$\overline{\Delta H_{S_2} + 1,103,450}$$

Out

Vapor	$4200(0.72)(172 - 172) + 4200(650) =$	2,725,000
Bottoms	$617(1.0)(210 - 172) =$	23,400
		$\overline{2,748,400}$

$$\Delta H_{S_2} + 1,103,450 = 2,748,400$$
$$\Delta H_{S_2} = 1,645,000 \text{ Btu/hr}$$

(c)

(d) *Total energy input into system:*

$$\Delta H_{S_1} + \Delta H_{S_2} + \text{steam heater} = 4,842,000 + 1,645,000 + 1,125,000 = 7,612,000 \text{ Btu/hr}$$

(e) *Energy balance on condenser I:*

$W = $ lb H₂O/hr; Reference temperature = 176°F

In

Vapor	$6667(0.85)(176 - 176) + (6667)(675) =$	4,500,000
Cooling H₂O	$W(1.0)(80 - 176) =$	−96W
		$\overline{-96W + 4,500,000}$

Out

Condensate	$6667(0.85)(176 - 176) =$	0
Cooling H₂O	$W(1.0)(130 - 176) =$	−46W
		$\overline{-46W}$

$$50W = 4,500,000 \quad \text{or} \quad W = \frac{4,500,000}{50} = 90,000 \text{ lb H}_2\text{O/hr}$$

$$\frac{90,000 \text{ lb H}_2\text{O/hr}}{8.345 \text{ lb/gal}} = 10,800 \text{ gal H}_2\text{O/hr}$$

(d₁)

(f) *Energy balance on condenser II:*

$$W = \text{lb } H_2O/hr; \qquad \text{Reference temperature} = 172°F$$

	In	*Out*	
Vapor	$4200(0.72)(172 - 172) + (4200)(650) =$	$4200(0.72)(160 - 172) =$	$-36,000$
Cooling H_2O	$W(1.0)(80 - 172) =$	$W(1.0)(172 - 130) =$	$-42W$

$$\quad \underline{2,725,000} \qquad\qquad \underline{-42W}$$
$$\underline{-92W} \qquad\qquad\qquad -42W - 36,000$$
$$-92W + 2,725,000$$

$$50W = 2,761,000 \qquad \text{or} \qquad W = \frac{2,761,000}{50} = 55,200 \text{ lb } H_2O/hr$$

$$\frac{55,200 \text{ lb } H_2O}{8.345 \text{ lb/gal}} = 6620 \text{ gal } H_2O/hr \longleftarrow \text{(d}_2\text{)}$$

(g) *Energy balance on heat exchanger II:*

$$W = \text{lb } H_2O/hr; \qquad \text{Reference temperature} = 80°F$$

	In	*Out*	
Condensate	$1050(0.72)(160 - 80) =$	$1050(0.72)(100 - 80) =$	$15,120$
Cooling H_2O	$W(1.0)(80 - 80) =$	$W(1.0)(130 - 80) =$	$50W$

$$\underline{60,500} \qquad\qquad\qquad \underline{50W}$$
$$\underline{0} \qquad\qquad\qquad\quad 50W + 15,120$$
$$60,500$$

$$50W + 15,120 = 60,500$$

$$W = 907 \text{ lb } H_2O/hr \qquad \text{or} \qquad 109 \text{ gal } H_2O/hr \longleftarrow \text{(d}_3\text{)}$$

SUPPLEMENTARY REFERENCES

Combustion of Solid, Liquid, and Gaseous Fuels

1. Gaydon, A. G. *Flames*, 3rd ed., Chapman & Hall, London, 1970.
2. Glassman, I., *Combustion*, Academic Press, New York, 1977.
3. Griswold, John, *Fuels, Combustion and Furnaces*, McGraw-Hill, New York, 1946.
4. Lewis, W. K., A. H. Radasch, and H. C. Lewis, *Industrial Stoichiometry*, 2nd ed., McGraw-Hill, New York, 1954.

Gas Producers and Synthetic Gas

1. *Coal Gasification and Liquefaction Newsletter.*
2. Franzen, J. E., and E. K. Goeke, "Gasify Coal for Petrochemicals," *Hydrocarbon Processing*, pp. 134–138 (November 1976).
3. Gumz, Wilhelm, *Gas Producers and Furnaces*, Wiley, New York, 1958.
4. Massey, L. G., ed., *Coal Gasification*, American Chemical Society, Washington, D.C., 1974.
5. Nawacki, P., *Coal Liquefaction Processes*, Noyes Data Corp., Park Ridge, N.J. 07656, 1979.
6. Pelofsky, A. H., ed., *Coal Conversion Technology*, American Chemical Society, Washington, D.C., 1979.
7. Timmins, C., "The Future Role of Gasification Processes," *Gas Eng. Management*, v. 20, p. 89 (February 1980).

Enthalpy–Concentration Charts

1. Brown, G. G., et al., *Unit Operations*, Wiley, New York, 1950. (Concerned with H–x chart calculations.)
2. Ellis, S. R. M., *Chem. Eng. Sci.*, v. 3, p. 287 (1954). (Concerned with H–x chart calculations.)
3. Lemlich, Robert, Chad Gottschlich, and Ronald Hoke, *Chem. Eng. Data Ser.*, v. 2, p. 32 (1957). (Concerned with H–x chart construction.)
4. McCabe, W. L., and J. C. Smith, *Unit Operations of Chemical Engineering*, 3rd ed., McGraw-Hill, New York, 1976. (Concerned with H–x chart calculations.)
5. Othmer, D. F., et al., *Ind. Eng. Chem.*, v. 51, p. 89 (1959). (Concerned with H–x chart construction.)
6. Robinson, C. S., and E. R. Gilliland, *Elements of Fractional Distillation*, 3rd ed., McGraw-Hill, New York, 1939. (Concerned with H–x chart calculations.)
7. Treybal, R. E., *Mass Transfer Operations*, McGraw-Hill, New York, 1980. (Concerned with H–x chart calculations.)

Humidity Charts and Calculations

1. McCabe, W. L., and J. C. Smith, *Unit Operations of Chemical Engineering*, 3rd ed., McGraw-Hill, New York, 1976.
2. Nelson, R., "Material Properties in SI Units, Part 4," *Chem. Engr. Progr.*, pp. 83–85 (May 1980).
3. Treybal, R. E., *Mass Transfer Operations*, McGraw-Hill, New York, 1980.

Industrial Process Calculations

1. Lewis, W. K., A. H. Radasch, and H. C. Lewis, *Industrial Stoichiometry*, 2nd ed., McGraw-Hill, New York, 1954.
2. Nelson, W. L., *Petroleum Refinery Engineering*, 4th ed., McGraw-Hill, New York, 1958.
3. Shreve, R. N., and J. A. Brink, *The Chemical Process Industries*, 4th ed., McGraw-Hill, New York, 1977.
4. Williams, F. A., *Combustion Theory*, Addison-Wesley, Reading, Mass., 1965.

Solution of Material and Energy Balances via Digital Computers

1. Chen, C. C., and L. B. Evans, "Computer Programs for Chemical Engineers," *Chem. Eng.*, pp. 167–73 (May 21, 1979).
2. Evans, L. B., et al., "Aspen: An Advanced System for Process Engineering," paper presented at 12th Symposium on Computer Applications in Chemical Engineering, Montreaux, Switzerland, 1979.
3. Motard, R. L., M. Shacham, and E. M. Rosen, "Steady State Chemical Process Simulation," *Amer. Inst. Chem. Eng. J.*, v. 21, pp. 417–36 (1975).
4. Rosen, E. M., and A. C. Pauls, "Computer Aided Process Design—The FLOWTRAN System," *Computers and Chem. Eng.*, v. 1, pp. 11–21 (1977).
5. Westerberg, A. W., H. P. Hutchinson, R. L. Motard, and P. Winter, *Process Flow Sheeting*, Cambridge University Press, Cambridge, 1979.

PROBLEMS

Section 5.1

5.1. Limestone ($CaCO_3$) is converted into CaO in a continuous vertical kiln (see Fig. P5.1). Heat is supplied by combustion of natural gas (CH_4) in direct contact with the limestone using 50% excess air. Determine the kilograms of $CaCO_3$ that can be processed per kilograms of natural gas. Assume that the following mean heat capacities (relative to 25°C) apply:

$$C_{p_m} \text{ of } CaCO_3 = 234 \text{ J/(g mol)(°C)}$$

$$C_{p_m} \text{ of } CaO = 111 \text{ J/(g mol)(°C)}$$

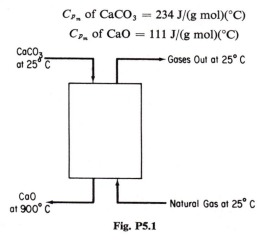

Fig. P5.1

5.2. A vertical lime kiln is charged with pure limestone ($CaCO_3$) and pure coke (carbon), both at 25°C. Dry air at 25°C is blown in at the bottom to burn the coke to CO_2, which provides the necessary heat for decomposition of the carbonate. The lime (CaO) leaves the bottom of the kiln at 950°F and contains no carbon or $CaCO_3$. The kiln gases leave at 600°F and contain no free oxygen. The molal ratio of $CaCO_3$: C in the the charge is 1.5: 1. Calculate:

(a) The analysis of the kiln gases

(b) The heat loss from the kiln per pound mole of coke fed to the kiln

You may use the following heat capacities as constants:

Component	C_p [Btu/(lb mol)(°F)]
$CaCO_3$	24.2
CaO	12.3
C	3.72

5.3. A feed stream of 16,000 lb/hr of 7% by weight NaCl solution is concentrated to a 40% by weight solution in an evaporator. The feed enters the evaporator, where it is heated to 180°F. The water vapor from the solution and the concentrated solution leave at 180°F. Steam at the rate of 15,000 lb/hr enters at 230°F and leaves as condensate at 230°F. See Fig. P5.3.

(a) What is the temperature of the feed as it enters the evaporator?

(b) What weight of 40% NaCl is produced per hour?

Assume that the following data apply:

$$\text{Mean } C_p \text{ 7\% NaCl soln:} \quad 0.92 \text{ Btu/(lb)(°F)}$$

$$\text{Mean } C_p \text{ 40\% NaCl soln:} \quad 0.85 \text{ Btu/(lb)(°F)}$$

$$\Delta \hat{H}_{vap} \text{ H}_2\text{O at 180°F} = 990 \text{ Btu/lb}$$

$$\Delta \hat{H}_{vap} \text{ H}_2\text{O at 230°F} = 959 \text{ Btu/lb}$$

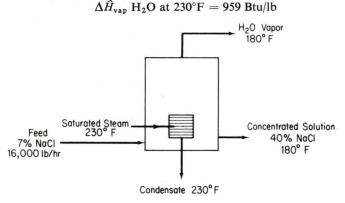

Fig. P5.3

5.4. One way to get pure N_2 is to remove the CO_2 from a combustion gas that contains no O_2. A gas composed of 20% CO_2 by volume and 80% N_2 is passed into a tower where it contacts a 20% by weight NaOH solution. The CO_2 forms Na_2CO_3 and the N_2 is sent to be dried. See Fig. P5.4.

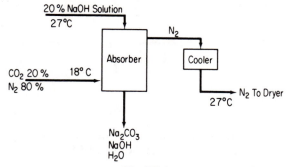

Fig. P5.4

(a) Assuming that 300 m³/hr of the gas mixture measured at 18°C, 100 kPa, is passed through the tower, what weight of 20% NaOH solution is required if 20% excess is desired in the entering liquid?

(b) If the temperature of the N_2 and the Na_2CO_3 solution leaving the tower is the same, what is this temperature if the NaOH solution enters at 27°C and the gas enters at 18°C? [C_p liquid NaOH soln = 3.8 J/(g)(°C); C_p liquid NaOH–Na_2CO_3 soln = 3.85 J/(g)(°C).]

(c) How many joules must be removed in the cooler per hour if the N_2 is to be ooled to 18°C?

Assume that the $\Delta \hat{H}_f^\circ$ of NaOH is -426.60 kJ/g mol.

5.5. C. C. Hunicke and C. L. Wagner in U.S. Patent 2,005,422 (June 18, 1935) concentrate sulfite waste liquor (s.w.l.) from the pulping of wood by blowing hot stack gas through sparger pipes immersed in the liquid. The foam that forms is knocked down by a spray of liquid s.w.l. feed. Because most of the substances in s.w.l. are colloidal, its vapor pressure and thermal properties may be considered to be the same as those of water. The flue-gas analyzes 12.0% CO_2, 7.0% O_2, and 81.0% N_2, and has a dew point of 90°F. It enters the apparatus at 800°F. The s.w.l. from the blowpits is fed into the apparatus at 120°F. The weather is partly cloudy, temperature 80°F, barometer 29.84 in. Hg, and the wind is 10 mi/hr NNE. Calculate:

(a) The boiling or equilibrium temperature of the s.w.l. in the evaporator

(b) The cubic feet of stack gas entering per pound of water evaporated

5.6. Determine the number of degrees of freedom for the condenser shown in Fig. P5.6.

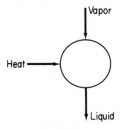

Fig. P5.6

5.7. Determine the number of degrees of freedom for the reboiler shown in Fig. P5.7. What variables should be specified to make the solution of the material and energy balances determinate?

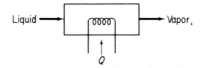

Liquid → Vapor,

Q

Fig. P5.7

5.8. If to the equilibrium stage shown in Example 5.2 you add a feed stream, determine the number of degrees of freedom. See Fig. P5.8.

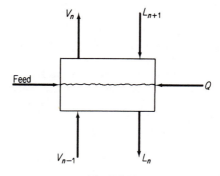

V_n L_{n+1}

Feed   Q

V_{n-1} L_n

Fig. P5.8

5.9. How many variables must be specified for the furnace shown in Fig. P5.9?

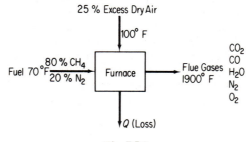

25 % Excess Dry Air

100° F

Fuel 70°F 80 % CH$_4$ Furnace Flue Gases CO$_2$
           20 % N$_2$ 1900° F   CO
                                                H$_2$O
                                                N$_2$
                                                O$_2$

Q (Loss)

Fig. P5.9

5.10. Read Problem 4.89. How many additional variables have to have their values specified to make the problem determinate?

5.11. Figure P5.11 shows a simple absorber or extraction unit. S is the absorber oil (or fresh solvent), and F is the feed from which material is to be recovered. Each stage has a Q (not shown); the total number of equilibrium stages is N. What is the number of degrees of freedom for the column? What variables should be specified?

Fig. P5.11

Section 5.2

5.12. Draw the saturated-liquid line (1 atm isobar) on an enthalpy concentration chart for caustic soda solutions. Show the 200°F, 250°F, 300°F, 350°F, and 400°F isotherms in the liquid region, and draw tie lines for these temperatures to the vapor region. Using the data from *N.B.S. Circular 500*, show where molten NaOH would be on this chart. Take the remaining data from the following table and the steam tables.

Temp. (°F)	x, Satd. Liquid Conc., Wt Fraction	$\Delta \hat{H}$, Enthalpy, Satd. Liquid (Btu/lb)
200	0	168
250	0.34	202
300	0.53	314
350	0.68	435
400	0.78	535

C_p for NaOH solutions [Btu/(lb)(°F)]

Wt % NaOH	Temperature (°F)				
	32	60	100	140	180
10	0.882	0.897	0.911	0.918	0.922
20	0.842	0.859	0.875	0.884	0.886
30		0.837	0.855	0.866	0.869
40		0.815	0.826	0.831	0.832
50			0.769	0.767	0.765

SOURCE: J. W. Bertetti and W. L. McCabe, *Ind. Eng. Chem.*, v. 28, p. 375 (1936).

NOTE: All H_2O tie lines extend to pure water vapor.

5.13. An evaporator at atmospheric pressure is designed to concentrate 10,000 lb/hr of a 10% NaOH solution at 70°F into a 40% solution. The steam pressure inside the steam chest is 40 psig. Determine the pounds of steam needed per hour if the exit strong caustic preheats the entering weak caustic in a heat exchanger, leaving the heat exchanger at 100°F.

5.14. A 50% by weight sulfuric acid solution is to be made by mixing the following:
(1) Ice at 32°F
(2) 80% H_2SO_4 at 100°F
(3) 20% H_2SO_4 at 100°F

How much of each must be added to make 1000 lb of the 50% solution with a final temperature of 100°F if the mixing is adiabatic?

5.15. Saturated steam at 300°F is blown continuously into a tank of 30% H_2SO_4 at 70°F. What is the highest concentration of liquid H_2SO_4 that can result from this process?

5.16. One thousand pounds of 10% NaOH solution at 100°F is to be fortified to 30% NaOH by adding 73% NaOH at 200°F. How much 73% solution must be used? How much cooling must be provided so that the final temperature will be 70°F?

5.17. In a battery plant pure H_2SO_4 at 80°F is mixed with pure water at 60°F make a 20% solution at 70°F. How much heat was removed per pound of final solution? If the mixing had been adiabatic, what would have been the temperature of the final solution?

5.18. For the ammonia–water system at . . . psia, calculate the unknown quantities for each of the three cases below (see Fig. P5.18).

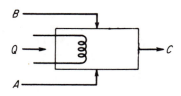

Fig. P5.18

Stream	Wt % NH₃	Enthalpy (Btu/lb)	Amount (lb)
(a) A	10	Satd. liquid	150
B	70	Satd. vapor	300
C	Unknown	Unknown	Unknown
$Q = -400,000$ Btu			
(b) A	80	Satd. vapor	Unknown
B	10	1700	Unknown
C	50	100	100
$Q =$ unknown			
(c) A	90	1200	100
B	Unknown	1500	Unknown
C	35	800	400
$Q =$ unknown			

5.19. A mixture of ammonia and water in the vapor phase, saturated at 250 psia and containing 80% by weight ammonia, is passed through a condenser at a rate of 10,000 lb/hr. Heat is removed from the mixture at the rate of 5,800,000 Btu/hr while the mixture passes through a cooler. The mixture is then expanded to a pressure of 100 psia and passes into a separator. A flow sheet of the process is given Fig. P5.19. If the heat loss from the equipment to the surroundings is neglected, determine the composition of the liquid leaving the separator, by:

(a) A material and energy balance set of equations

(b) Using the enthalpy–concentration diagram method

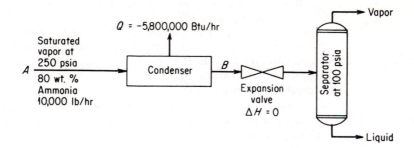

Fig. P5.19

5.20. **(a)** Plot an enthalpy–concentration diagram for the ethyl alcohol-water system at 760 mm Hg total pressure (enthalpy of the mixture in Btu/lb vs. weight fraction ethanol). Plot the diagram on 11- by 17-in. graph paper. Be sure to include the following: (1) saturated-vapor line, (2) saturated-liquid line, and (3) lines of constant temperature in the subcooled-liquid region from 0 to 200°F. Notice that even though the EtOH–water system is a nonideal system, the saturated-vapor and liquid lines are almost straight lines and could have been drawn from the enthalpies of the pure components without serious error.

(b) In each of the four cases diagrammed in Fig. P5.20 the following data[9] apply:

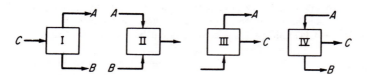

Fig. P5.20

Stream	Wt % EtOH	Condition	Amount
A	80	Unknown	Unknown
B	10	70°F	Unknown
C	60	Superheated vapor at 800 Btu/lb	100 lb

(1) For diagram I:
   (a) By a material and energy balance, find the weights of streams A and B and the enthalpy of stream A.

[9] *References for data:* L. W. Cornell and R. E. Montonna, *Ind. Eng. Chem.*, v. 25, pp. 1331–1335 (1933); J. H. Perry, *Chemical Engineers' Handbook*, 2nd ed., McGraw-Hill, New York, 1941, p. 1364; W. A. Noyes and R. R. Warfle, *J. Amer. Chem. Soc.*, v. 23, p. 463 (1901).

(b) Prove by analytical geometry that A, B, and C will lie on a straight line when plotted on an H-x diagram.

(c) Repeat part (a), but this time use the method of Ponchon instead of a system of material and energy balance equations.

(2) Repeat 1(c) for diagram II.

(3) Repeat 1(c) for diagram III.

(4) Repeat 1(c) for diagram IV.

[*Note:* For part (a) you must plot the actual data and show your points clearly. For part (b), use $8\frac{1}{2}$-by-11 graph paper with the abscissa marked from 0.1 to 0.9 weight fraction. If you use more than one graph for the solution of part (b), you may plot the saturated-vapor and liquid lines from the enthalpies of the pure components and connect them with a straight line. See Appendix I.]

Section 5.3

5.21. Construct a "saturation" chart for CO_2 and acetone at 1 atm pressure and include:

(1) 100%, 50%, and 25% saturation lines

(2) Saturated acetone vs. temperature line

(3) Latent heat of acetone vs. temperature

(4) "Humid heat" vs. saturation

(5) Adiabatic cooling lines for adiabatic saturation temperatures of 40°, 80°, and 100°F.

(6) Wet- and dry-bulb temperature lines for temperatures of 40°, 80°, and 100°F.

Take the vapor pressure data for acetone from any reference book. Acetone, $C_{p(\text{avg})}$ = 0.347 cal/(g)(°C). Acetone, latent heat of vaporization:

t (°C)	cal/g
0	134.74
20	131.87
40	128.05
60	123.51
80	118.26
100	112.76

$h_c/k_y' = 0.38$ for acetone vapor.

5.22. Moist air at 100 kPa, a dry-bulb temperature of 90°C, and a wet-bulb temperature of 46°C is enclosed in a rigid container. The container and its contents are cooled to 43°C.

(a) What is the molar humidity of the cooled moist air?

(b) What is the final total pressure in atm in the container?

(c) What is the dew point in °C of the cooled moist air?

5.23. (a) Calculate:

(1) The humidity of air saturated at 120°F

(2) The saturated volume at 120°F

(3) The adiabatic saturation temperature and wet-bulb temperature of air having a dry-bulb $t = 120°F$ and a dew point $= 60°F$

(4) The percent saturation when the air in (3) is cooled to $82°F$

(5) The pounds of water condensed/100 lb of moist air in (3) when the air is cooled to $40°F$

(b) Compute the wet-bulb temperature and percent *relative* humidity before and after for the following cases:

Dry-Bulb t (°F)	Percent Humidity	New Dry-Bulb t (°F)
80	40	100
150	60	100
100	10	60
60	95	150

5.24. A wet material is to be dried by the passage over it of hot combustion gases. Can the ordinary humidity chart for air–water vapor mixtures be employed in the design of a drier for such a purpose?

5.25. What advantage is derived from plotting humidity charts on a molar basis?

5.26. A rotary dryer operating at atmospheric pressure dries 10 tons/day of wet grain at 70°F, from a moisture content of 10% to 1% moisture. The air flow is countercurrent to the flow of grain, enters at 225°F dry-bulb and 110°F wet-bulb temperature, and leaves at 125°F dry-bulb. See Fig. P5.26. Determine:

(a) The humidity of the entering and leaving air

(b) The water removal in pounds per hour

(c) The daily product output in pounds per day

(d) The heat input to the dryer

Assume that there is no heat loss from the dryer, that the grain is discharged at 110°F, and that its specific heat is 0.18.

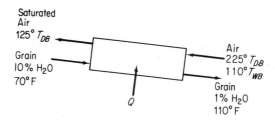

Fig. P5.26

5.27. Temperatures (°F) taken around a forced-draft cooling tower are as follows:

	In	Out
Air	85	90
Water	102	89

The wet-bulb temperature of the entering air is 77°F. Assuming the air leaving the tower to be saturated, calculate with the aid of the data given below:
(a) The humidity of the entering air
(b) The pounds of dry air through the tower per pound of water into the tower
(c) The percentage of water vaporized in passing through the tower

t (°F)	77	85	89	90	100	102
Saturated $\mathfrak{JC}$	0.0202	0.0264	0.030	0.031	0.0425	0.0455

C_p air $= 0.24$; C_p water vapor $= 0.45$; $\Delta H_v = 1040$ Btu/lb at any of the temperatures encountered in this problem.

5.28. A dryer produces 180 kg/hr of a product containing 8% water from a feed stream that contains 1.25 g of water per gram of dry material. The air enters the dryer at 100°C dry-bulb and a wet bulb temperature of 38°C; the exit air leaves at 53°C dry-bulb and 60% relative humidity. Part of the exit air is mixed with the fresh air supplied at 21°C, 52% relative humidity, as shown in Fig. P5.28. Calculate the air and heat supplied to the heater, neglecting any heat lost by radiation, used in heating the conveyor trays, and so forth. The specific heat of the product is 0.18.

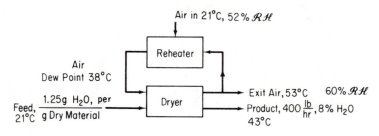

Fig. P5.28

5.29. Air, dry bulb 38°C, wet bulb 27°C, is scrubbed with water to remove dust. The water is maintained at 24°C. Assume that the time of contact is sufficient to reach complete equilibrium between air and water. The air is then heated to 93°C by passing it over steam coils. It is then used in an adiabatic rotary drier from which it issues at 49°C. It may be assumed that the material to be dried enters and leaves at 46°C. The material loses 0.05 kg H_2O per kilogram of product. The total product is 1000 kg/hr.
(a) What is the humidity:
    (1) Of the initial air?
    (2) After the water sprays?
    (3) After reheating?
    (4) Leaving the drier?

(b) What is the percent humidity at each of the above points?
(c) What is the total weight of dry air used per hour?
(d) What is the total volume of air leaving the drier?
(e) What is the total amount of heat supplied to the cycle in joules per hour?

5.30. Your boss wants to air-condition a service building 100 ft long by 60 ft wide by 16 ft average height by cooling and dehumidifying the necessary fresh air with cold water in a spray chamber. The average occupancy of the building is 100 persons/hr with a total emission of 800 Btu/(person)(hr). The severest atmospheric conditions for the city are 100°F with 95% humidity. It is believed that air at 70°F and 60% humidity will be satisfactory, provided that the total circulation is sufficient to hold the temperature rise of the air to 2°F. Neglecting radiation from the building and using the sensible heat of the recirculated air for reheating, calculate:
(a) The volume of recirculated air at the inlet conditions
(b) The volume of fresh air under the worst conditions
(c) The tons of refrigeration (1 ton = 12,000 Btu/hr) required

5.31. Sodium chloride from an evaporator contains 88% NaCl, the balance water. This material is to be dried in a countercurrent rotary drier, using hot air at a dry bulb of 60°C, 10% relative humidity. This hot air stream is made up of fresh air with a humidity of 0.0145 mole of water per mole of dry air plus a recycled stream of moist air from the drier. This recycle stream represents 15% by weight of the total air in the drier. Required drier conditions are 40 kg of dry air for every kilogram of dry stock. Compute the percent water content on a wet basis of the dried stock leaving the drier and also the humidity of the air leaving the drier. Is this process satisfactory?

5.32. Clean and air-conditioned air must be furnished to two rooms in which solid-state devices for digital computers are produced. Low-humidity air is supplied to the first room and flows from it to the second room. Additional low-humidity air is supplied to the second room, as shown in Fig. P5.32. The humidity of the low-humidity air is 0.0040 kg of water/kg of dry air, and the humidity of the air leaving the first room is 0.0055. The humidity of the air leaving the second room is 0.0065. Water is evaporated in the first room at the rate of 2.03 kg/hr and in the second room at the rate of 2.53 kg/hr. Calculate the low-humidity air flow rate to the two rooms and the total air flow rate, both in terms of kilograms of dry air per hour.

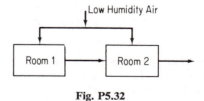

Fig. P5.32

Section 5.4

5.33. A process involving catalytic dehydrogenation in the presence of hydrogen is known as *hydroforming*. In World War II this process was of importance in helping to meet the demand for toluene for the manufacture of explosives. Toluene, benzene, and other aromatic materials can be economically produced from naptha feeds in this way. After the toluene is separated from the other components, it is condensed and cooled

in a process such as that shown in the flow sheet (Fig. P5.33). For every 100 kg of stock charged to the system, 27.5 kg of a toluene and water mixture (9.1% by weight water) is produced as overhead vapor and condensed by the charge stream. Calculate:
(a) The temperature of the charge stock after it leaves the condenser
(b) The pounds of cooling water required per hour

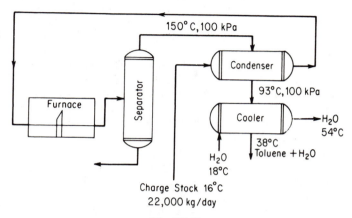

Fig. P5.33

Stream	C_p [J/(g)(°C)]	Boiling Point (K)	Latent Heat Vaporization (kJ/g)
$H_2O(l)$	4.2	373.2	2.26
$H_2O(g)$	2.1	—	—
Toluene(l)	1.7	383.8	0.364
Toluene(g)	1.3	—	—
Charge stock	2.1	—	—

5.34. Toluene, manufactured by the conversion of *n*-heptane with a Cr_2O_3-on-Al_2O_3 catalyst,

$$CH_3CH_2CH_2CH_2CH_2CH_2CH_3 \longrightarrow \underset{\text{CH}_3}{\bigcirc} + 4H_2$$

by the method of hydroforming described in Problem 5.33 is recovered by use of a solvent. See Fig. P5.34 for the process and conditions.

The yield of toluene is 15% based on the *n*-heptane charged to the reactor. Assume that 10 kg of solvent are used per kilogram of toluene in the extractors.
(a) Calculate how much heat has to be added or removed from the catalytic reactor to make it isothermal at 425°C.
(b) Find the temperature of the *n*-heptane and solvent stream leaving the mixer-settlers if both streams are at the same temperature.
(c) Find the temperature of the solvent stream after it leaves the heat exchanger.
(d) Calculate the heat duty of the fractionating column in kJ/kg of *n*-heptane feed to the process.

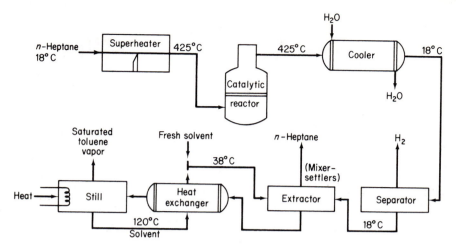

Fig. P5.34

	$-\Delta H_f^{\circ}*$ (kJ/g mol)	C_p [J/(g)(°C)]		$\Delta H_{vaporization}$ (kJ/kg)	Boiling Point (K)
		Liquid	Vapor		
Toluene†	12.00	2.22	2.30	364	383.8
n-Heptane	−224.4	2.13	1.88	318	371.6
Solvent	—	1.67	2.51	—	434

*As liquids.

†The heat of solution of toluene in the solvent is −23 J/g toluene.

5.35. The steam flows for a plant are shown in Fig. P5.35. Write the material and energy balances for the system and calculate the unknown quantities in the diagram (*A* to *F*). There are two main levels of steam flow: 600 psig and 50 psig. Use the steam tables for the enthalpies.

5.36. One hundred thousand pounds of a mixture of 50% benzene, 40% toluene, and 10% *o*-xylene is separated every day in a distillation-fractionation plant as shown on the flow sheet.

	Boiling Point (°C)	C_p Liquid [cal/(g)(°C)]	Latent Heat of Vap. (cal/g)	C_p Vapor [cal/(g)(°C)]
Benzene	80	0.44	94.2	0.28
Toluene	109	0.48	86.5	0.30
o-Xylene	143	0.48	81.0	0.32
Charge	90	0.46	88.0	0.29
Overhead T_I	80	0.45	93.2	0.285
Residue T_I	120	0.48	83.0	0.31
Residue T_{II}	143	0.48	81.5	0.32

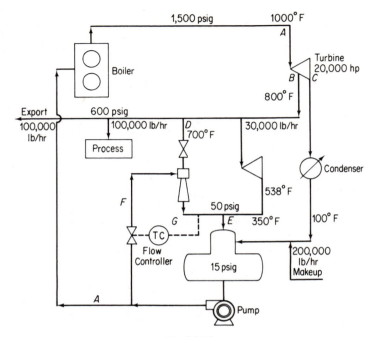

Fig. P5.35

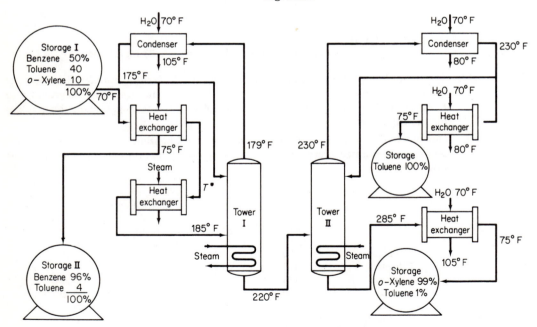

Fig. P5.36 Flow chart for problem.

The reflux ratio for tower I is 6:1; the reflux ratio for tower II is 4:1; the charge to tower I is liquid; the charge to tower II is liquid. Compute:

(a) The temperature of the mixture at the outlet of the heat exchanger (marked as T^*)

(b) The Btu supplied by the steam reboiler in each column

(c) The quantity of cooling water required in gallons per day for the whole plant

(d) The energy balance around tower I

5.37. Sulfur dioxide emission from coal-burning power plants causes serious atmospheric pollution in the eastern and midwestern portions of the United States. Unfortunately, the supply of low-sulfur coal is insufficient to meet the demand. Processes presently under consideration to alleviate the situation include coal gasification followed by desulfurization and stack-gas cleaning. One of the more promising stack-gas-cleaning processes involves reacting SO_2 and O_2 in the stack gas with a solid metal oxide sorbent to give the metal sulfate and then thermally regenerating the sorbent and absorbing the resulting SO_3 to produce sulfuric acid. Recent laboratory experiments indicate that sorption and regeneration can be carried out with several metal oxides, but no pilot or full-scale processes have yet been put into operation.

You are asked to provide a preliminary design for a process that will remove 95% of the SO_2 from the stack gas of a 1000-MW power plant. Some data are given below and in the flow diagram of the process (Fig. P5.37). The sorbent consists of fine particles of a dispersion of 30% by weight CuO in a matrix of inert porous Al_2O_3. This solid reacts in the fluidized-bed sorber at 315°C. Exit solid is sent to the regenerator, where SO_3 is evolved at 700°C, converting all the $CuSO_4$ present back to CuO. The fractional conversion of CuO to $CuSO_4$ that occurs in the sorber is called α and is an important design variable. You are asked to carry out your calculations for α = 0.2, 0.5, and 0.8. The SO_3 produced in the regenerator is swept out by recirculating air. The SO_3-laden air is sent to the acid tower, where the SO_3 is absorbed in recirculating sulfuric acid and oleum, part of which is withdrawn as salable by-products. You will notice that the sorber, regenerator, and perhaps the acid tower are adiabatic; their temperatures are adjusted by heat exchange with incoming streams. Some of the heat exchangers (nos. 1 and 3) recover heat by countercurrent exchange between the feed and exit streams. Additional heat is provided by withdrawing flue gas from the power plant at any desired high temperature up to 1100°C and then returning it at a lower temperature. Cooling is provided by water at 25°C. As a general rule, the temperature difference across heat-exchanger walls separating the two streams should average about 28°C. The nominal operating pressure of the whole process is 100 kPa. The three blowers provide 6 kPa additional head for the pressure losses in the equipment, and the acid pumps have a discharge pressure of 90 kPa gauge. You are asked to write the material and energy balances and some equipment specifications as follows:

(a) Sorber, regenerator, and acid tower. Determine the flow rate, composition, and temperature of all streams entering and leaving.

(b) Heat exchangers. Determine the heat load, and flow rate, temperature, and enthalpy of all streams.

(c) Blowers. Determine the flow rate and theoretical horsepower.

(d) Acid pump. Determine the flow rate and theoretical horsepower.

Use SI units. Also, use a basis of 100 kg of coal burned for all your calculations; then convert to the operating basis at the end of the calculations.

Power plant operation. The power plant burns 340 metric tons/hr of coal having

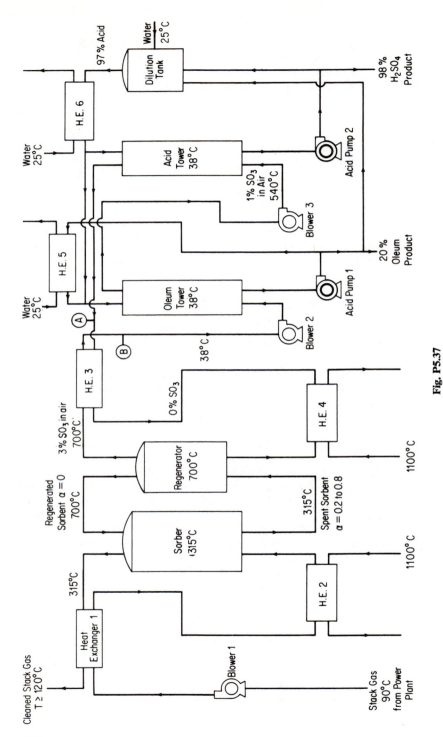

Fig. P5.37

the analysis given below. The coal is burned with 18% excess air, based on complete combustion to CO_2, H_2O, and SO_2. In the combustion only the ash and nitrogen is left unburned; all the ash has been removed from the stack gas.

Element	Wt %
C	76.6
H	5.2
O	6.2
S	2.3
N	1.6
Ash	8.1

DATA ON SOLIDS

	Al_2O_3		CuO		$CuSO_4$	
T (K)	C_p	H_T-H_{298}	C_p	H_T-H_{298}	C_p	H_T-H_{298}
298	79.04	0.00	42.13	0.00	98.9	0.00
400	96.19	9.00	47.03	4.56	114.9	10.92
500	106.10	19.16	50.04	9.41	127.2	23.05
600	112.5	30.08	52.30	14.56	136.3	36.23
700	117.0	41.59	54.31	19.87	142.9	50.25
800	120.3	53.47	56.19	25.40	147.7	64.77
900	122.8	65.65	58.03	31.13	151.0	79.71
1000	124.7	77.99	59.87	37.03	153.8	94.98

Units of C_p are J/(g mol)(K); units of H are kJ/g mol.

5.38. Sulfuric acid is a basic raw material used in a wide range of industries. It has been claimed that the state of civilization of a country can be determined by the amount of sulfuric acid consumed per capita. Sulfuric acid was one of the earliest known acids because it could be made by the absorption in H_2O of the SO_3 formed when naturally occurring sulfates or sulfides were roasted. Innumerable person-years of design effort have gone into developing methods of obtaining economic production of sulfuric acid from various starting materials. The source of sulfur in this problem is sulfur itself. From an overall viewpoint the preparation of sulfuric acid is quite simple. The sulfur is burned to SO_2,

$$S(s) + O_2(g) \longrightarrow SO_2(g)$$

which is then oxidized to SO_3 with additional air,

$$SO_2(g) + \tfrac{1}{2}O_2(g) \longrightarrow SO_3(g)$$

and the SO_3 formed is finally absorbed in water to yield H_2SO_4,

$$SO_3(g) + H_2O(l) \longrightarrow H_2SO_4(l)$$

The details of the sulfuric acid plant are shown in the process flow sheet (Fig. P5.38). Fifty percent excess air (based on the oxidation of sulfur to SO_2) is used.

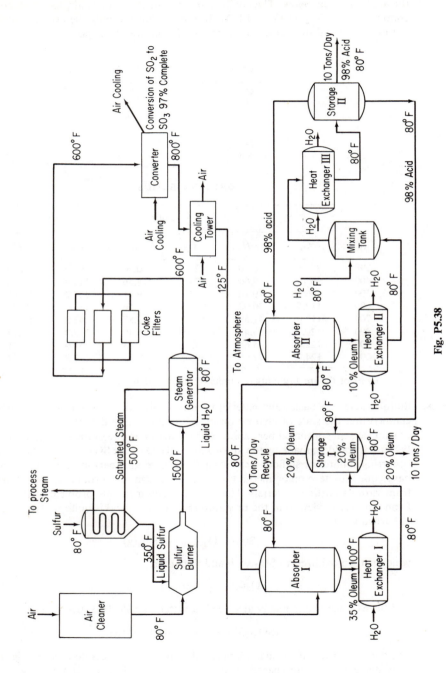

Fig. P5.38

(a) Determine the amount of heat necessary to add to the melting tank per day.

(b) Calculate the quantity of heat lost from the burner per day.

(c) Calculate the heat transferred per day from the converter by the cooling air.

(d) Calculate the amount of 98% H_2SO_4 recirculated through the 98% absorber per day.

(e) What percentage of the total amount of SO_3 produced per day is absorbed in the oleum absorber and what percentage in the 98% H_2SO_4 absorber?

Heat of solution data can be assumed to be as follows:

Solution	$\Delta H_{solution}$
SO_3 in 20% oleum	5 Btu/lb of 30% oleum formed
SO_3 in 98% H_2SO_4	50 Btu/lb of 10% oleum formed
20% oleum in 98% H_2SO_4	40 Btu/lb of 10% oleum formed
H_2O in 10% oleum	120 Btu/lb of 98% H_2SO_4 formed

Note: % oleum equals the wt % free SO_3 in the H_2SO_4–SO_3 mixture. All H_2O is considered combined with SO_3 to form H_2SO_4.

Use the following values for the heat capacities of the streams:

Stream	C_p [Btu/(lb)(°F)]
S(s)	0.163
S(l)	0.235
98% H_2SO_4	0.5
5% oleum	0.45
10% oleum	0.43
20% oleum	0.40
35% oleum	0.38

Note: C_p in Btu/(lb mol)(°F) and T in °R.

Other data can be found in Appendix E.

5.39. When coal is distilled by heating without contact with air, a wide variety of solid, liquid, and gaseous products of commercial importance is produced, as well as some significant air pollutants. The nature and amounts of the products produced depend on the temperature used in the decomposition and the type of coal. At low temperatures (400 to 750°C) the yield of synthetic gas is small relative to the yield of liquid products, whereas at high temperatures (above 900°C) the reverse is true. For the typical process flow sheet, shown in Fig. P5.39:

(a) How many tons of the various products are being produced?

(b) Make an energy balance around the primary distillation tower and benzol tower.

(c) How much (in pounds) of 40% NaOH solution is used per day for the purification of the phenol?

(d) How much 50% H_2SO_4 is used per day in the pyridine purification?

(e) What weight of Na_2SO_4 is produced per day by the plant?

(f) How many cubic feet of gas per day is produced? What percent of the gas (volume) is needed for the ovens?

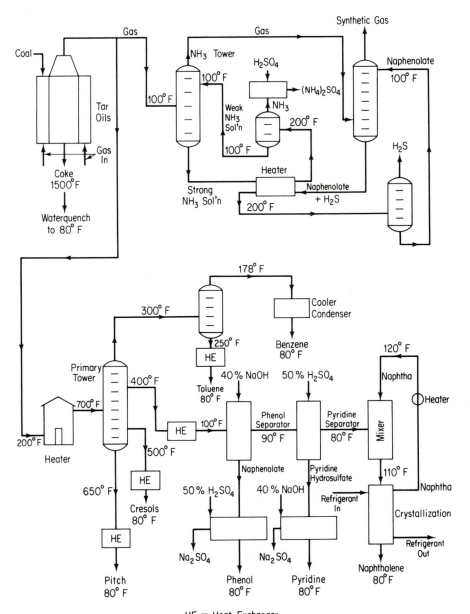

HE = Heat Exchanger

Fig. 5.39

Products Produced per Ton of Coal Charged	Mean C_p Liquid (cal/g)	Mean C_p Vapor (cal/g)	Mean C_p Solid (cal/g)	Melting Point (°C)	Boiling Point (°C)
Synthetic gas—10,000 ft³					
(555 Btu/ft³)					
$(NH_4)_2SO_4$, 22 lb					
Benzol, 15 lb	0.50	0.30	—	—	60
Toluol, 5 lb	0.53	0.35	—	—	109.6
Pyridine, 3 lb	0.41	0.28	—	—	114.1
Phenol, 5 lb	0.56	0.45	—	—	182.2
Naphthalene, 7 lb	0.40	0.35	0.281 $+0.00111T_{°F}$	80.2	218
Cresols, 20 lb	0.55	0.50	—	—	202
Pitch, 40 lb	0.65	0.60	—	—	400
Coke, 1500 lb	—	—	0.35		—

	ΔH_{vap} (cal/g)	ΔH_{fusion} (cal/g)
Benzol	97.5	—
Toluol	86.53	—
Pyridine	107.36	—
Phenol	90.0	—
Naphthalene	75.5	35.6
Cresols	100.6	—
Pitch	120	—

PROBLEM TO PROGRAM ON THE COMPUTER

5.1. Determine the values of the unknown quantities in Fig. CP5.1 by solving the following set of linear material and energy balances that represent the steam balance:

(**1**) $181.60 - x_3 - 132.57 - x_4 - x_5 = -y_1 - y_2 + y_3 + y_4 = 5.1$

(**2**) $1.17x_3 - x_6 = 0$

(**3**) $132.57 - 0.745x_7 = 61.2$

(**4**) $x_5 + x_7 - x_8 - x_9 - x_{10} + x_{15} = y_7 + y_8 - y_3 = 99.1$

(**5**) $x_8 + x_9 + x_{10} + x_{11} - x_{12} - x_{13} = -y_7 = -8.4$

(**6**) $x_6 - x_{15} = y_6 - y_5 = 24.2$

(**7**) $-1.15(181.60) + x_3 - x_6 + x_{12} + x_{16} = 1.15y_1 - y_9 + 0.4 = 19.7$

(**8**) $181.60 - 4.594x_{12} - 0.11x_{16} = -y_1 + 1.0235y_9 + 2.45 = 35.05$

(**9**) $-0.0423(181.60) + x_{11} = 0.0423y_1 = 2.88$

(**10**) $-0.016(181.60) + x_4 = 0$

(**11**) $x_8 - 0.0147x_{16} = 0$

(**12**) $x_5 - 0.07x_{14} = 0$

(**13**) $-0.0805(181.60) + x_9 = 0$

(**14**) $x_{12} - x_{14} + x_{16} = 0.4 - y_9 = -97.9$

There are four levels of steam: 680, 215, 170, and 37 psia. The 14 x_i, $i = 3, \ldots, 16$, are the unknowns and the y_i are given parameters for the system. Both x_i and y_i have the units of 10^3 lb/hr.

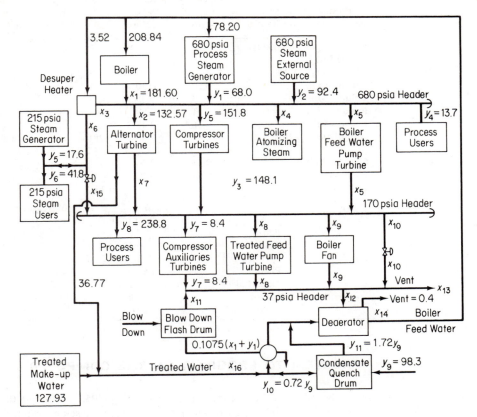

Fig. CP5.1

INDUSTRIAL CHEMICAL DATA

**ANNUAL
PRODUCTION,
PRICES
(USA 1981)**

HOW MADE	MAJOR END USES	MAJOR DERIVATIVES	
	Ethanol	CH_3CH_2OH	
Hydration of ethylene	Solvents, alone or as derivatives, 85%; food 10%	Ethyl acetate and other esters 15%, ethyl amines 5%, glycol ethers 5%	1.2×10^9 lb $1.95–$2.10/gal

Adapted from Chemical and Engineering News by permission of the American Chemical Society.

UNSTEADY-STATE MATERIAL
AND ENERGY BALANCES

In previous chapters all the material and all the energy balances you have encountered, except for the batch energy balances, were *steady-state* balances, that is, balances for processes in which the accumulation term was zero. Now it is time for us to focus our attention briefly on *unsteady-state* processes. These are processes in which quantities or operating conditions within the system *change with time*. Sometimes you will hear the word *transient state* applied to such processes. The unsteady state is somewhat more complicated than the steady state and in general problems involving unsteady-state processes are somewhat more difficult to formulate and solve than those involving steady-state processes. However, a wide variety of important industrial problems fall into this category, such as the startup of equipment, batch heating or reactions, the change from one set of operating conditions to another, and the perturbations that develop as process conditions fluctuate. In this chapter we consider only one category of unsteady-state processes, but it is the one that is the most widely used, the lumped or macroscopic balances.

Section 6.1   Unsteady-State Material and Energy Balances

> *Your objectives in studying this section are to be able to:*
>
> 1. Write down the macroscopic unsteady-state material and energy balances in words and in symbols .
> 2. Solve single linear ordinary differential material or energy balance equations given the initial conditions.
> 3. Take a word problem and translate it into a differential equation(s).

The basic expression in words, Eq. (2.1) or (4.22), for either the material or the energy balance should by now be well known to you and is repeated below as Eq. (6.1). However, you should also realize that this equation can be applied at various levels, or strata, of description. In other words, the engineer can portray the operation of a real process by writing balances on a number of physical scales. A typical illustration of this concept might be in meterology, where the following different degrees of detail can be used on a descending scale of magnitude in the real world:

Global weather pattern

Local weather pattern

Individual clouds

Convective flow in clouds

Molecular transport

The molecules themselves

Similarly, in chemical engineering we write material and energy balances from the viewpoint of various scales of information:

(a) Molecular and atomic balances

(b) Microscopic balances

(c) Dispersion balances

(d) Plug flow balances

(e) Macroscopic balances (overall balances, lumped balances)

in decreasing order of degree of detail about a process.[1] In this chapter, the type of balance to be described and applied is the simplest one, the macroscopic balance [(e) in the list above].

The macroscopic balance ignores all the detail within a system and consequently

---

[1] Additional information together with applications concerning these various types of balances can be found in D. M. Himmelblau and K. B. Bischoff, *Process Analysis and Simulation*, Swift Publishing Co., Austin, Tex., 1980.

results in a balance about the entire system. Only time remains as an independent variable in the balance. The dependent variables, such as concentration and temperature, are not functions of position but represent overall averages throughout the entire volume of the system. In effect, the system is assumed to be sufficiently well mixed so that the output concentrations and temperatures are equivalent to the concentrations and temperatures inside the system.

To assist in the translation of Eq. (6.1)

$$
\begin{Bmatrix} \text{accumulation} \\ \text{or depletion} \\ \text{within the} \\ \text{system} \end{Bmatrix} = \begin{Bmatrix} \text{transport into} \\ \text{system through} \\ \text{system} \\ \text{boundary} \end{Bmatrix} - \begin{Bmatrix} \text{transport out} \\ \text{of system} \\ \text{through system} \\ \text{boundary} \end{Bmatrix}
$$
$$
+ \begin{Bmatrix} \text{generation} \\ \text{within} \\ \text{system} \end{Bmatrix} - \begin{Bmatrix} \text{consumption} \\ \text{within} \\ \text{system} \end{Bmatrix} \qquad (6.1)
$$

into mathematical symbols, you should refer to Fig. 6.1. Equation (6.1) can be applied to the mass of a single component or to the total amount of material or energy in the system. Let us write each of the terms in Eq. (6.1) in mathematical symbols for a very small time interval Δt. Let the accumulation be positive in the direction in which time is positive, that is, as time increases from t to $t + \Delta t$. Then, using the component mass balance as an example, the accumulation will be the mass of A in the system at time $t + \Delta t$ minus the mass of A in the system at time t,

$$
\text{accumulation} = \rho_A V|_{t+\Delta t} - \rho_A V|_t
$$

where[2] ρ_A = mass of component A per unit volume
V = volume of the system

Note that the net dimensions of the accumulation term are the mass of A.

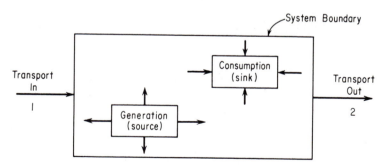

Fig. 6.1 General unsteady-state process with transport in and out and internal generation and consumption.

[2]The symbol $|_t$ means that the quantities preceding the vertical line are evaluated at time t, or time $t + \Delta t$, or at surface S_1, or at surface S_2, as the case may be.

We shall split the mass transport across the system boundary into two parts, transport through defined surfaces S_1 and S_2, whose areas are known, and transport across the system boundary through other (undefined) surfaces. The net transport of A into (through S_1) and out of (through S_2) the system through defined surfaces can be written as

$$\text{net flow across boundary via } S_1 \text{ and } S_2 = \rho_A v S \, \Delta t \, |_{S_1} - \rho_A v S \, \Delta t \, |_{S_2}$$

where v = fluid velocity in a duct of cross section S
    S = defined cross-sectional area perpendicular to material flow

Again note that the net dimensions of the transport term are the mass of A. Other types of transport across the system boundary can be represented by

$$\text{net residual flow across boundary} = \dot{w}_A \, \Delta t$$

where $\dot{w}_A$ is the rate of mass flow of component A through the system boundaries other than the defined surfaces S_1 and S_2.

Finally, the net generation–consumption term will be assumed to be due to a chemical reaction r_A:

$$\text{net generation-consumption} = \dot{r}_A \, \Delta t$$

where $\dot{r}_A$ is the net rate of generation–consumption of component A by chemical reaction.

Introduction of all these terms into Eq. (6.1) gives Eq. (6.2). Equations (6.3) and (6.4) can be developed from exactly the same type of analysis. Try to formulate Eqs. (6.3) and (6.4) yourself.

Component material balances:

$$\underbrace{\rho_A V \,|_{t+\Delta t} - \rho_A V \,|_t}_{\text{accumulation}} = \underbrace{\rho_A v S \, \Delta t \,|_{S_1} - \rho_A v S \, \Delta t \,|_{S_2}}_{\substack{\text{transport through defined}\\\text{boundaries}}} + \underbrace{\dot{w}_A \, \Delta t}_{\substack{\text{transport}\\\text{through}\\\text{other}\\\text{boundaries}}} + \underbrace{\dot{r}_A \, \Delta t}_{\substack{\text{genera-}\\\text{tion or}\\\text{consump-}\\\text{tion}}} \qquad (6.2)$$

Total material balance:

$$\underbrace{\rho V \,|_{t+\Delta t} - \rho V \,|_t}_{\text{accumulation}} = \underbrace{\rho v S \, \Delta t \,|_{S_1} + \rho v S \, \Delta t \,|_{S_2}}_{\substack{\text{transport through defined}\\\text{boundaries}}} + \underbrace{\dot{w} \, \Delta t}_{\substack{\text{transport}\\\text{through}\\\text{other}\\\text{boundaries}}} \qquad (6.3)$$

Energy balance:

$$\underbrace{E\,|_{t+\Delta t} - E\,|_t}_{\text{accumulation}} = \underbrace{\left| \left(\hat{H} + \frac{v^2}{2} + gh \right) \dot{m} \, \Delta t \right|_{S_1} - \left| \left(\hat{H} + \frac{v^2}{2} + gh \right) \dot{m} \, \Delta t \right|_{S_2}}_{\text{transport through defined boundaries}}$$

$$+ \underbrace{\dot{Q} \, \Delta t}_{\text{heat}} - \underbrace{\dot{W} \, \Delta t}_{\text{work}} + \underbrace{\dot{B} \, \Delta t}_{\substack{\text{transport}\\\text{through}\\\text{other}\\\text{boundaries}}} \qquad (6.4)$$

where $\dot{B}$ = rate of energy transfer accompanying $\dot{w}$
$\dot{m}$ = rate of mass transfer through defined surfaces
$\dot{Q}$ = rate of heat transfer to system
$\dot{W}$ = rate of work done by the system
$\dot{w}$ = rate of total mass flow through the system boundaries other than through the defined surfaces S_1 and S_2
ρ = total mass, per unit volume

The other notation for the material and energy balances is identical to that of Chaps. 2 and 4; note that the work, heat, generation, and mass transport are now all expressed as rate terms (mass or energy per unit time).

If each side of Eq. (6.2) is divided by Δt, we obtain

$$\frac{\rho_A V|_{t+\Delta t} - \rho_A V|_t}{\Delta t} = \rho_A v S|_{S_1} - \rho_A v S|_{S_2} + \dot{w}_A + \dot{r}_A \tag{6.5}$$

Similar relations can be obtained from Eqs. (6.3) and (6.4). Next, if we take the limit of each side of Eq. (6.5) as $\Delta t \longrightarrow 0$, we get

$$\frac{d(\rho_A V)}{dt} = -\Delta(\rho_A v S) + \dot{w}_A + \dot{r}_A \tag{6.6}$$

Similar treatment of the total mass balance and the energy balance yields the following two equations:

$$\frac{d(\rho V)}{dt} = -\Delta(\rho v S) + \dot{w} \tag{6.7}$$

$$\frac{d(E)}{dt} = -\Delta\left[\left(\hat{H} + \frac{v^2}{2} + gh\right)\dot{m}\right] + \dot{Q} - \dot{W} + \dot{B} \tag{6.8}$$

Can you get Eqs. (6.7) and (6.8) from (6.3) and (6.4), respectively? Try it.

The relation between the energy balance given by Eq. (6.8), which has the units of energy per unit time, and the energy balance given by Eq. (4.24), which has the units of energy, should now be fairly clear. Equation (4.24) represents the integration of Eq. (6.8) with respect to time, expressed formally as follows:

$$E_{t_2} - E_{t_1} = \int_{t_1}^{t_2} \{-\Delta[(\hat{H} + \hat{K} + \hat{P})\dot{m}] + \dot{Q} + \dot{B} - \dot{W}\}\,dt \tag{6.9}$$

The quantities designated in Eqs. (4.24) or (4.24a) without the superscript dot are the respective integrated values from Eq. (6.9).

To solve one or a combination of the very general equations (6.6), (6.7), or (6.8) analytically may be quite difficult, and in the following examples we shall have to restrict our analyses to simple cases. If we make enough (reasonable) assumptions and work with simple problems, we can consolidate or eliminate enough terms of the equations to be able to integrate them and develop some analytical answers. Numerical solutions are also possible for nonlinear equations.

In the solution of unsteady-state problems you have to apply the usual procedures of problem solving discussed in Chaps 1, 2, and 4. Two major tasks exist:

(a) To set up the unsteady-state equation(s)

(b) To solve it (them) once the equation(s) is (are) established

After you draw a diagram of the system and set down all the information available, you should try to recognize the important variables and represent them by letters. Next, decide which variable is the independent one, and label it. The independent variable is the one you select to have one or a series of values, and then the other variables are all dependent ones—they are determined by the first variable(s) and the initial state of the process you selected. Which is chosen as an independent and which as a dependent variable is usually fixed by the problem but may be arbitrary. Although there are no general rules applicable to all cases, the quantities that appear prominently in the statement of the problem are usually the best choices. You will need as many equations as you have dependent variables.

As mentioned before, in the macroscopic balance the independent variable is time. In mathematics, when the quantity x varies with time, we consider dx to be the change in x that occurs during time dt if the process continues through the interval beginning at time t. Let us say that t is the independent variable and x is the dependent one. In solving problems you can use one or a combination of Eqs. (6.6), (6.7), and (6.8) directly, or alternatively you can proceed, as shown in some of the examples, to set up the differential equations from scratch exactly in the same fashion as Eqs. (6.2)–(6.4) were formulated. For either approach the objective is to translate the problem statement in words into one or more simultaneous differential equations having the form

$$\frac{dx}{dt} = f(x, t) \tag{6.10}$$

Then, assuming this differential equation can be solved, x can be found as a function of t. Of course, we have to know some initial condition(s) or at one (or more) given time(s) know the value(s) of x.

If dt and dx are always considered positive when increasing, you can use Eq. (6.1) without having any difficulties with signs. However, if you for some reason transfer an output to the left-hand side of the equation, or the equation is written in some other form, you should take great care in the use of signs (see Example 6.2 below). We are now going to examine some very simple unsteady-state problems which are susceptible to reasonably elementary mathematical analysis. You can (and will) find more complicated examples in texts dealing with all phases of mass transfer, heat transfer, and fluid dynamics.

EXAMPLE 6.1 Unsteady-State Material Balance without Generation

A tank holds 100 gal of a water–salt solution in which 4.0 lb of salt is dissolved. Water runs into the tank at the rate of 5 gal/min and salt solution overflows at the same rate. If the mixing in the tank is adequate to keep the concentration of salt in the tank uniform at all times, how much salt is in the tank at the end of 50 min? Assume that the density of the salt solution is essentially the same as that of water.

Solution

We shall set up from scratch the differential equations that describe the process.
Step 1: Draw a picture, and put down the known data . See Fig. E6.1.

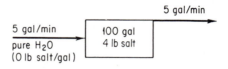

Fig. E6.1

Step 2: Choose the independent and dependent variables. Time, of course, is the independent variable, and either the salt quantity or concentration in the tank can be the dependent variable. Suppose that we make the mass (quantity) of salt the dependent variable. Let $x = $ lb of salt in the tank at time t.
Step 3: Write the known value of x at a given value of t. This is the initial condition:

$$\text{at } t = 0 \qquad x = 4.0 \text{ lb}$$

Step 4: It is easiest to make a total material balance and a component material balance on the salt. (No energy balance is needed because the system can be assumed to be isothermal.)
    Total balance:

$$[m_{\text{tot}} \text{ lb}]_{t+\Delta t} - [m_{\text{tot}} \text{ lb}]_t = \frac{5 \text{ gal}}{\text{min}} \left| \frac{1 \text{ ft}^3}{7.48 \text{ gal}} \right| \frac{\rho_{\text{H}_2\text{O}} \text{ lb}}{\text{ft}^3} \left| \Delta t \text{ min} \right.$$

$$- \frac{5 \text{ gal}}{\text{min}} \left| \frac{1 \text{ ft}^3}{7.48 \text{ gal}} \right| \frac{\rho_{\text{soln}} \text{ lb}}{\text{ft}^3} \left| \Delta t \text{ min} \right. = 0$$

This equation tells us that the flow of water into the tank equals the flow of solution out of the tank if $\rho_{\text{H}_2\text{O}} = \rho_{\text{soln}}$ as assumed. Otherwise, there is an accumulation term.
    Salt balance:

$$[x \text{ lb}]_{t+\Delta t} - [x \text{ lb}]_t = 0 - \frac{5 \text{ gal}}{\text{min}} \left| \frac{x \text{ lb}}{100 \text{ gal}} \right| \Delta t \text{ min}$$

Dividing by Δt and taking the limit as Δt approaches zero,

$$\lim_{\Delta t \to 0} \frac{[x]_{t+\Delta t} - [x]_t}{\Delta t} = -0.05x$$

or

$$\frac{dx}{dt} = -0.05x \tag{a}$$

Notice how we have kept track of the units in the normal fashion in setting up these equations. Because of our assumption of uniform concentration of salt in the tank, the concentration of salt leaving the tank is the same as that in the tank, or x lb/100 gal of solution.

Step 5: Solve the unsteady-state material balance on the salt. By separating the independent and dependent variables, we get

$$\frac{dx}{x} = -0.05 \, dt$$

This equation is easily integrated between the definite limits of

$$t = 0 \qquad x = 4.0$$
$$t = 50 \qquad x = \text{the unknown value of } X \text{ lb}$$

$$\int_{4.0}^{x} \frac{dx}{x} = -0.05 \int_{0}^{50} dt$$

$$\ln \frac{X}{4.0} = -2.5 \qquad \ln \frac{4.0}{X} = 2.5$$

$$\frac{4.0}{X} = 12.2 \qquad\qquad X = \frac{4.0}{12.2} = 0.328 \text{ lb salt}$$

An equivalent differential equation to Eq. (a) can be obtained directly from the component mass balance in the form of Eq. (6.6) if we let ρ_A = concentration of salt in the tank at any time t in terms of lb/gal:

$$\frac{d(\rho_A V)}{dt} = -\left(\frac{5 \text{ gal}}{\text{min}} \middle| \frac{\rho_A \text{ lb}}{\text{gal}}\right) - 0$$

If the tank holds 100 gal of solution at all times, V is a constant and equal to 100, so that

$$\frac{d\rho_A}{dt} = -\frac{5\rho_A}{100} \tag{b}$$

The initial conditions are

$$\text{at } t = 0 \qquad \rho_A = 0.04$$

The solution of Eq. (b) would be carried out exactly as the solution of Eq. (a).

EXAMPLE 6.2 Unsteady-State Material Balance without Generation

A square tank 4 m on a side and 10 m high is filled to the brim with water. Find the time required for it to empty through a hole in the bottom 5 cm² in area.

Solution

Step 1: Draw a diagram of the process, and put down the data. See Fig. E6.2a.

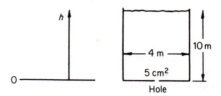

Fig. E6.2a

Step 2: Select the independent and dependent variables. Again, time will be the independent variable. We could select the quantity of water in the tank as the dependent variable, but since the cross section of the tank is constant, let us choose h, the height of the water in the tank, as the dependent variable .

Step 3: Write the known value of h at a given value of t:

$$\text{at } t = 0 \qquad h = 10 \text{ m}$$

Step 4: Develop the unsteady-state balance(s) for the process. In an elemental time Δt, the height of the water in the tank drops Δh. The mass of water leaving the tank is in the form of a cylinder 5 cm² in area, and we can calculate the quantity as

$$\frac{5 \text{ cm}^2}{} \left| \left(\frac{1 \text{ m}}{100 \text{ cm}}\right)^2 \right| \frac{v^* \text{m}}{\text{s}} \left| \frac{\rho \text{ kg}}{\text{m}^3} \right| \frac{\Delta t \text{ s}}{} = 5 \times 10^{-4} v^* \rho \, \Delta t \qquad \text{kg}$$

where ρ = density of water
v^* = average velocity of the water leaving the tank

The depletion of water inside the tank in terms of the variable h, expressed in pounds, is

$$\frac{16 \text{ m}^2}{} \left| \frac{\rho \text{ kg}}{\text{m}^3} \right| h \text{ m} \Big|_{t+\Delta t} - \frac{16 \text{ m}^2}{} \left| \frac{\rho \text{ kg}}{\text{m}^3} \right| h \text{ m} \Big|_{t} = 16 \, \rho \Delta h \qquad \text{kg}$$

An overall material balance indicates that

$$\textbf{accumulation = in } - \qquad \textbf{out}$$
$$16\rho \, \Delta h = 0 - 5 \times 10^{-4} \rho v^* \, \Delta t \qquad \text{(a)}$$

Although Δh is a negative value, our equation takes account of this automatically. The accumulation is positive if the right-hand side is positive, and the accumulation is negative if the right-hand side is negative. Here the accumulation is really a depletion. You can see that the term ρ, the density of water, cancels out, and we could just as well have made our material balance on a volume of water.

Equation (a) becomes

$$\frac{\Delta h}{\Delta t} = -\frac{5 \times 10^{-4} \, v^*}{16}$$

Taking the limit as Δh and Δt approach zero, we get

$$\frac{dh}{dt} = -\frac{5 \times 10^{-4} \, v^*}{16} \tag{b}$$

Step 5: Unfortunately, this is an equation with one independent variable, t, and two dependent variables, h and v^*. We must find another equation to eliminate either h or v^* if we want to obtain a solution. Since we want our final equation to be expressed in terms of h, the next step is to find some function that relates v^* to h and t, and then we can substitute for v^* in the unsteady-state equation.

We shall employ the steady-state mechanical energy balance for an incompressible fluid, discussed in Chap. 4, to relate v^* and h. See Fig. E6.2b. Recall from Eq. (4.30) with $W = 0$ and $E_v = 0$ that

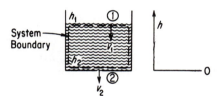

Fig. E6.2b

$$\Delta\left(\frac{v^2}{2} + gh\right) = 0 \tag{c}$$

We assume that the pressures are the same at section ①—the water surface—and section ②—the hole—for the system consisting of the water in the tank. Equation (c) reduces to

$$\frac{v_2^2 - v_1^2}{2} + g(h_2 - h_1) = 0 \tag{d}$$

where v_2 = exit velocity through the 5 cm² hole at boundary ②
v_1 = velocity of the water in the tank at boundary ①

If $v_1 \simeq 0$, a reasonable assumption for the water in the large tank at any time, at least compared to v_2, we have

$$v_2^2 = -2g(0 - h_1) = 2gh$$
$$v_2 = \sqrt{2gh} \tag{e}$$

Because the exit-stream flow is not frictionless and because of turbulence and orifice effects in the exit hole, we must correct the value of v given by Eq. (e) for frictionless flow by an empirical adjustment factor as follows:

$$v_2 = c\sqrt{2gh} = v^* \tag{f}$$

where c is an orifice correction which we could find (from a text discussing fluid dynamics) has a value of 0.62 for this case. Thus

$$v^* = 0.62\sqrt{2(9.80)h} = 2.74\sqrt{h} \qquad \text{m/s}$$

Let us substitute this relation into Eq. (b) in place of v^*. Then we obtain

$$\frac{dh}{dt} = \frac{(5.4 \times 10^{-4})(2.74)(h)^{1/2}}{16}$$

or

$$-1.08 \times 10^4 \, \frac{dh}{h^{1/2}} = dt \tag{g}$$

Equation (g) can be integrated between

$$h = 10 \text{ m} \qquad \text{at } t = 0$$

and

$$h = 0 \text{ m} \qquad \text{at } t = \theta, \quad \text{the unknown time}$$

$$-1.08 \times 10^4 \int_{10}^{0} \frac{dh}{h^{1/2}} = \int_0^\theta dt$$

to yield θ,

$$\theta = 1.08 \times 10^4 \int_0^{10} \frac{dh}{h^{1/2}} = 1.08 \times 10^4 \left[2\sqrt{h} \right]_0^{10} = 6.83 \times 10^4 \text{ s}$$

EXAMPLE 6.3 Material Balance in Batch Distillation

A small still is separating propane and butane at 135°C and initially contains 10 kg moles of a mixture whose composition is $x = 0.30$ ($x =$ mole fraction butane). Additional mixture ($x_F = 0.30$) is fed at the rate of 5 kg mol/hr. If the total volume of the liquid in the still is constant, and the concentration of the vapor from the still (x_D) is related to x_S as follows:

$$x_D = \frac{x_S}{1 + x_S}$$

how long will it take for the value of x_S to change from 0.30 to 0.40? What is the steady-state ("equilibrium") value of x_S in the still (i.e., when x_S becomes constant)? See Fig. E6.3.

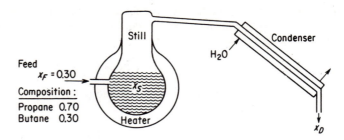

Fig. E6.3

Solution

Since butane and propane form ideal solutions, we do not have to worry about volume changes on mixing or separation. Only the material balance is needed to answer the questions posed. If t is the independent variable and x_S the dependent variable, we can say that

Butane Balance (C_4): The input to the still is

$$\frac{5 \text{ mol feed}}{\text{hr}} \left| \frac{0.30 \text{ mol } C_4}{\text{mol feed}} \right| \Delta t \text{ hr}$$

The output from the still is equal to the amount condensed,

$$\frac{5 \text{ mol cond.}}{\text{hr}} \left| \frac{x_D \text{ mol } C_4}{\text{mol cond.}} \right| \Delta t \text{ hr}$$

The accumulation is

$$\frac{10 \text{ mol in still}}{} \left| \frac{x_S \text{ mol } C_4}{\text{mol in still}} \right|_{t+\Delta t} - \frac{10 \text{ mol in still}}{} \left| \frac{x_S \text{ mol } CH_4}{\text{mol in still}} \right|_{t} = 10 \, \Delta x_S$$

Our unsteady-state material balance is then

$$\textbf{accumulation} = \textbf{in} \quad - \quad \textbf{out}$$
$$10 \, \Delta x_S = 1.5 \, \Delta t - 5 x_D \, \Delta t$$

or, dividing by Δt and taking the limit as Δt approaches zero,

$$\frac{dx_S}{dt} = 0.15 - 0.5 x_D$$

As in Example 6.2, it is necessary to reduce the equation to one dependent variable by substituting for x_D:

$$x_D = \frac{x_S}{1 + x_S}$$

Then

$$\frac{dx_S}{dt} = 0.15 - \frac{x_S}{1 + x_S}(0.5)$$

$$\frac{dx_S}{0.15 - \dfrac{0.5x_S}{1 + x_S}} = dt$$

The integration limits are

$$\text{at } t = 0 \qquad x_S = 0.30$$
$$t = \theta \qquad x_S = 0.40$$

$$\int_{0.30}^{0.40} \frac{dx_S}{0.15 - [0.5x_S/(1 + x_S)]} = \int_0^\theta dt = \theta$$

$$\int_{0.30}^{0.04} \frac{(1 + x_S)\, dx_S}{0.15 - 0.35x_S} = \theta = \left[-\frac{x_S}{0.35} - \frac{1}{(0.35)^2} \ln(0.15 - 0.35x_S) \right]_{0.30}^{0.40}$$

$$\theta = 5.85 \text{ hr}$$

If you did not know how to integrate the equation analytically, or if you only had experimental data for x_D as a function of x_S instead of the given equation, you could always integrate the equation graphically as shown in Example 6.6 or integrate numerically on a digital computer.

The steady-state value of x_S is established at infinite time or, alternatively, when the accumulation is zero. At that time,

$$0.15 = \frac{0.5x_S}{1 + x_S} \qquad \text{or} \qquad x_S = 0.428$$

The value of x_S could never be greater than 0.428 for the given conditions.

EXAMPLE 6.4 Unsteady-State Chemical Reaction

A compound dissolves in water at a rate proportional to the product of the amount undissolved and the difference between the concentration in a saturated solution and the concentration in the actual solution at any time. A saturated solution of compound contains 40 g/ 100 g H_2O. In a test run starting with 20 kg of undissolved compound in 100 kg of pure compound is found that 5 kg is dissolved in 3 hr. If the test continues, how many kilograms of compound will remain undissolved after 7 hr? Assume that the system is isothermal.

Solution

Step 1: See Fig. E6.4.
Step 2: Let the dependent variable x = the kilograms of *undissolved* compound at time t (the independent variable).

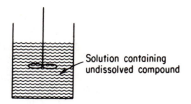

Solution containing
undissolved compound

Fig. E6.4

Step 3: At any time t the concentration of dissolved compound is

$$\frac{20 - x}{100} \left(\frac{\text{kg compound}}{\text{lb } H_2O}\right)$$

The rate of dissolution of compound according to the problem statement is

$$\text{rate } \frac{\text{kg}}{\text{hr}} = k(x \text{ kg}) \left[\frac{40 \text{ kg compound}}{100 \text{ kg } H_2O} - \frac{(20 - x) \text{ kg compound}}{100 \text{ kg } H_2O}\right]$$

where k is the proportionality constant. We could have utilized a concentration measured on a volume basis rather than one measured on a weight basis, but this would just change the value and units of the constant k.

Next we make an unsteady-state balance for the compound in the solution (the system):

accumulation = in − out +   generation

$$\frac{dx}{dt} = 0 - 0 + kx\left(\frac{40}{100} - \frac{20 - x}{100}\right)$$

or

$$\frac{dx}{dt} = \frac{kx}{100}(20 + x)$$

Step 4: Solution of the unsteady-state equation:

$$\frac{dx}{x(20 + x)} = \frac{k}{100} dt$$

This equation can be split into two parts and integrated,

$$\frac{dx}{x} - \frac{dx}{x + 20} = \frac{k}{5} dt$$

or integrated directly for the conditions

$$\begin{aligned}
t_0 &= 0 & x_0 &= 20 \\
t_1 &= 3 & x_1 &= 15 \\
t_2 &= 7 & x_2 &= ?
\end{aligned}$$

Since k is an unknown, one extra pair of conditions is given in order to evaluate k.

To find k, we integrate between the limits stated in the test run:

$$\int_{20}^{15} \frac{dx}{x} - \int_{20}^{15} \frac{dx}{x+20} = \frac{k}{5} \int_{0}^{3} dt$$

$$\left[\ln \frac{x}{x+20} \right]_{20}^{15} = \frac{k}{5}(3-0)$$

$$k = -0257$$

(Note that the negative value of k corresponds to the actual physical situation in which undissolved material is consumed.) To find the unknown amount of undissolved compound at the end of 7 hr, we have to integrate again:

$$\int_{20}^{x_2} \frac{dx}{x} - \int_{20}^{x_2} \frac{dx}{x+20} = \frac{-0.257}{5} \int_{0}^{7} dt$$

$$\left[\ln \frac{x}{x+20} \right]_{20}^{x_2} = \frac{-0.0257}{5}(7-0)$$

$$x_2 = 10.7 \text{ kg}$$

EXAMPLE 6.5 Overall Unsteady-State Process

A 15% Na_2SO_4 solution is fed at the rate of 12 lb/min into a mixer that initially holds 100 lb of a 50–50 mixture of Na_2SO_4 and water. The exit solution leaves at the rate of 10 lb/min. Assuming uniform mixing, what is the concentration of Na_2SO_4 in the mixer at the end of 10 min? Ignore any volume changes on mixing.

Solution

In contrast to Example 6.1, both the water and the salt concentrations, and the total material in the vessel, change with time in this problem.

Step 1: See Fig. E6.5.

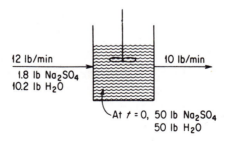

12 lb/min
1.8 lb Na_2SO_4
10.2 lb H_2O

10 lb/min

At $t = 0$, 50 lb Na_2SO_4
50 lb H_2O

Fig. E6.5

Step 2:

Let x = fraction Na_2SO_4 in the tank at time t, $\left(\dfrac{\text{lb } Na_2SO_4}{\text{lb total}} \right)$

y = total pounds of material in the tank at time t

Step 3: At time $t = 0$, $x = 0.50$, $y = 100$.

Step 4: Material balances between time t and $t + \Delta t$:

(a) Total balance:

$$\text{accumulation} \quad = \quad \text{in} \quad - \quad \text{out}$$

$$[y]_{t+\Delta t} - [y]_t = \Delta y = \frac{12 \text{ lb}}{\text{min}} \bigg| \frac{\Delta t \text{ min}}{} - \frac{10 \text{ lb}}{\text{min}} \bigg| \frac{\Delta t \text{ min}}{}$$

or

$$\frac{dy}{dt} = 2 \tag{a}$$

(b) Na_2SO_4 balance:

$$\text{accumulation}$$

$$\left[\frac{x \text{ lb Na}_2\text{SO}_4}{\text{lb total}} \bigg| y \text{ lb total} \right]_{t+\Delta t} - \left[\frac{x \text{ lb Na}_2\text{SO}_4}{\text{lb total}} \bigg| y \text{ lb total} \right]_t$$

$$= \qquad\qquad \text{in} \qquad\qquad - \qquad\qquad \text{out}$$

$$= \frac{1.8 \text{ lb Na}_2\text{SO}_4}{\text{min}} \bigg| \frac{\Delta t \text{ min}}{} - \frac{10 \text{ lb total}}{\text{min}} \bigg| \frac{x \text{ lb Na}_2\text{SO}_4}{\text{lb total}} \bigg| \frac{\Delta t \text{ min}}{}$$

or

$$\frac{d(xy)}{dt} = 1.8 - 10x \tag{b}$$

Step 5: There are a number of ways to handle the solution of these two differential equations with two unknowns. Perhaps the easiest is to differentiate the xy product in Eq. (b) and substitute Eq. (a) into the appropriate term containing dy/dt:

$$\frac{d(xy)}{dt} = x \frac{dy}{dt} + y \frac{dx}{dt}$$

To get y, we have to integrate Eq. (a):

$$\int_{100}^{y} dy = \int_{0}^{t} 2 \, dt$$

$$y - 100 = 2t$$

$$y = 100 + 2t$$

Then

$$\frac{d(xy)}{dt} = x(2) + (100 + 2t) \frac{dx}{dt} = 1.8 - 10x$$

$$\frac{dx}{dt} (100 + 2t) = 1.8 - 12x$$

$$\int_{0.50}^{x} \frac{dx}{1.8 - 12x} = \int_{0}^{10} \frac{dt}{100 + 2t}$$

$$-\frac{1}{12} \ln \frac{1.8 - 12x}{1.8 - 6.0} = \frac{1}{2} \ln \frac{100 + 20}{100}$$

$$x = 0.267 \frac{\text{lb Na}_2\text{SO}_4}{\text{lb total solution}}$$

An alternative method of setting up the dependent variable would have been to let $x = $ lb of Na_2SO_4 in the tank at the time t. Then the Na_2SO_4 balance would be

$$\frac{dx}{dt} \frac{\text{lb } Na_2SO_4}{\text{min}} = \frac{1.8 \text{ lb } Na_2SO_4}{\text{min}} - \frac{10 \text{ lb total}}{\text{min}} \left| \frac{x \text{ lb } Na_2SO_4}{y \text{ lb total}} \right.$$

$$\frac{dx}{dt} = 1.8 - \frac{10x}{100 + 2t}$$

or

$$\frac{dx}{dt} + \frac{10}{100 + 2t} x = 1.8$$

The solution to this linear differential equation gives the pounds of Na_2SO_4 after 10 min; the concentration then can be calculated as x/y. To integrate the linear differential equation, we introduce the integrating factor $e^{\int (dt/10 + 0.2t)}$:

$$xe^{\int (dt/10 + 0.2t)} = \int (e^{\int (dt/10 + 0.2t)})(1.8) \, dt + C$$

$$x(10 + 0.2t)^5 = 1.8 \int (10 + 0.2t)^5 \, dt + C$$

at $x = 50$, $t = 0$, and $C = 35 + 10^5$;

$$x = 1.5(10 + 0.2t) + \frac{35 \times 10^5}{(10 + 0.2t)^5}$$

at $t = 10$ min;

$$x = 1.5(12) + \frac{35 \times 10^5}{(12)^5} = 32.1$$

The concentration is

$$\frac{x}{y} = \frac{32.1}{100 + 20} = 0.267 \frac{\text{lb } Na_2SO_4}{\text{lb total solution}}$$

EXAMPLE 6.6 Unsteady-State Energy Balance with Graphical Integration

Five thousand pounds of oil initially at 60°F is being heated in a stirred (perfectly mixed) tank by saturated steam which is condensing in the steam coils at 40 psia. If the rate of heat transfer is given by Newton's heating law,

$$\dot{Q} = \frac{dQ}{dt} = h(T_{\text{steam}} - T_{\text{oil}})$$

where Q is the heat transferred in Btu and h is the heat transfer coefficient in the proper units, how long does it take for the discharge from the tank to rise from 60°F to 90°F? What is the maximum temperature that can be achieved in the tank?

Additional Data:

$$\text{motor horsepower} = 1 \text{ hp; efficiency is } 0.75$$

$$\text{entering oil flow rate} = 1018 \text{ lb/hr at a temperature of } 60°F$$

$$\text{discharge oil flow rate} = 1018 \text{ lb/hr at a temperature of } T$$

$$h = 291 \text{ Btu/(hr)(°F)}$$

$$C_{p_{oil}} = 0.5 \text{ Btu/(lb)(°F)}$$

Solution

The process is shown in Fig. E6.6a. The independent variable will be t, the time; the dependent variable will be the temperature of the oil in the tank, which is the same as the temperature of the oil discharged. The material balance is not needed because the process, insofar as the quantity of oil is concerned, is assumed to be in the steady state.

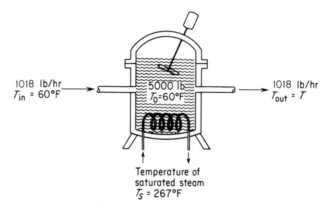

Fig. E6.6a

Our next step is to set up the unsteady-state energy balance for the interval Δt. Let T_S = the steam temperature and T = the oil temperature:

$$\text{accumulation} = \text{input} - \text{output}$$

A good choice for a reference temperature for the enthalpies is 60°F because this choice makes the input enthalpy zero.

enthalpy of the
input steam

$$\frac{1018 \text{ lb}}{\text{hr}} \left| \frac{0.5 \text{ Btu}}{\text{(lb)(°F)}} \right| \overbrace{(T_{\text{in}} - T_{\text{ref}})}^{60°F - 60°F} \left| \Delta t \text{ hr} \right. = 0$$

heat transfer

$$h(T_S - T) = \frac{291 \text{ Btu}}{\text{(hr)(°F)}} \left| (267 - T)°F \right| \Delta t \text{ hr}$$

$\left.\vphantom{\begin{array}{c} a \\ b \\ c \end{array}}\right\}$ **input (Btu)**

enthalpy of the
output stream

$$\frac{1018 \text{ lb}}{\text{hr}} \left| \frac{0.5 \text{ Btu}}{\text{(lb)(°F)}} \right| (T - 60)°F \left| \Delta t \text{ hr} \right.$$

output (Btu)

enthalpy
change
inside
tank

$$\left\{ \begin{array}{l} \left[\dfrac{5000\ lb}{}\middle|\dfrac{0.5\ Btu}{(lb)(°F)}\middle|\dfrac{(T-60)°F}{}\right]_{t+\Delta t} \\[2em] -\left[\dfrac{5000\ lb}{}\middle|\dfrac{0.5\ Btu}{(lb)(°F)}\middle|\dfrac{(T-60)°F}{}\right]_{t} \end{array} \right. \qquad \text{accumulation (Btu)}$$

The rate of change of energy inside the tank is nothing more than the rate of change of internal energy, which, in turn, is essentially the same as the rate of change of enthalpy (i.e., $dE/dt = dU/dt = dH/dt$) because $d(pV)/dt \simeq 0$.

We use Eq. (6.8) with $\dot{B} = 0$. The energy introduced by the motor enters the tank as $\dot{W}$ and is:

$$\dot{W} = \frac{3\ hp}{4} \middle| \frac{0.7608\ Btu}{(sec)(hp)} \middle| \frac{3600\ sec}{1\ hr} = \frac{1910\ Btu}{hr}$$

Then

$$2500\frac{dT}{dt} = 291(267 - T) - 509(T - 60) + 1910$$

$$\frac{dT}{dt} = 44.1 - 0.32T$$

$$\int_{60}^{90} \frac{dT}{44.1 - 0.32T} = \int_{0}^{\theta} dt = \theta$$

An analytical solution to this equation gives $\theta = 1.52$ hr. For illustrative purposes, so you can see how equations too complex to integrate directly can be handled, let us graphically integrate the left-hand side of the equation. We set up a table and choose values of T. We want to calculate $1/(44.1 - 0.32T)$ and plot this quantity vs. T. The area under the curve from $T = 60°$ to $T = 90°F$ should be 1.52 hr.

T	$0.32T$	$44.1 - 0.32T$	$\dfrac{1}{44.1 - 0.32T} = \phi$
60	19.2	24.9	0.0402
70	22.4	21.7	0.0462
80	25.6	18.5	0.0540
90	28.8	15.3	0.0655

These data are plotted in Fig. E6.6b.

The area below $\phi = 0.030$ is

$$(90 - 60)(0.030 - 0) = 0.90\ hr$$

The area above $\phi = 0.030$ but under the curve is

$$(311)(0.002) = 0.622\ hr$$

The total area $= 0.90 + 0.622 = 1.52$ hr.

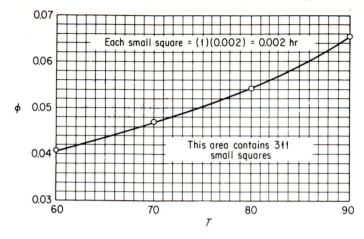

Fig. E6.6b

Self-Assessment Test

1. In a batch type of chemical reactor, do we have to have the starting conditions to predict the yield?

2. Is time the independent or dependent variable in macroscopic unsteady-state equations?

3. How can you obtain the steady-state balances from Eqs. (6.6), (6.7), and (6.8)?

4. Which of the following plots of the response vs. time are not transient state processes?

5. Group the following words in sets of synonyms: (1) response, (2) input variable, (3) parameter, (4) state variable, (5) system parameter, (6) initial condition, (7) output, (8) independent variable, (9) dependent variable, (10) coefficient, (11) output variable, (12) constant.

6. A chemical inhibitor must be added to the water in a boiler to avoid corrosion and scale. The inhibitor concentration must be maintained between 4 and 30 ppm. The boiler system always contains 100,000 kg of water and the blowdown (purge) rate is 15,000 kg/hr. The makeup water contains no inhibitor. An initial 2.8 kg of inhibitor is added to the water and thereafter 2.1 kg is added periodically. What is the maximum time interval until the first addition of inhibitor?

7. In a reactor a small amount of undesirable by-product is continuously formed at the rate of 0.5 lb/hr when the reactor contains a steady inventory of 10,000 lb of material. The reactor inlet stream contains traces of the by-product (40 ppm), and the effluent stream is 1400 lb/hr. If on startup, the reactor contains 5000 pm of the by-product, what will will be the concentration in ppm in 10 hr?

SUPPLEMENTARY REFERENCES

1. Himmelblau, D. M., and K. B. Bischoff, *Process Analysis and Simulation,* Swift Publishing Co., Austin, Tex., 1980.

2. Jenson, V. G., and G. V. Jeffreys, *Mathematical Methods in Chemical Engineering,* Academic Press, New York, 1963.

3. Marshall, W. R., and R. L. Pigford, *The Application of Differential Equations to Chemical Engineering Problems,* University of Delaware, Newark, 1942.

4. Sherwood, T. K., and C. E. Reid, *Applied Mathematics in Chemical Engineering,* 2nd ed., McGraw-Hill, New York, 1957.

PROBLEMS

Chapter 6

6.1. A tank containing 100 kg of a 60% brine (60% salt) is filled with a 10% salt solution at the rate of 10 kg/min. Solution is removed from the tank at the rate of 15 kg/min. Assuming complete mixing, find the kilograms of salt in the tank after 10 min.

6.2. A defective tank of 1500 ft³ volume containing 100% propane gas (at 1 atm) is to be flushed with air (at 1 atm) until the propane concentration is reduced to less than 1%. At that concentration of propane the defect can be repaired by welding. If the flow rate of air into the tank is 30 ft³/min, for how many minutes must the tank be flushed out? Assume that the flushing operation is conducted so that the gas in the tank is well mixed.

6.3. A 2% uranium oxide slurry (2 lb UO_2/100 lb H_2O) flows into a 100-gal tank at the rate of 2 gal/min. The tank initially contains 500 lb of H_2O and no UO_2. The slurry is well mixed and flows out at the same rate at which it enters. Find the concentration of slurry in the tank at the end of 1 hr.

6.4. The catalyst in a fluidized-bed reactor of 200-m³ volume is to be regenerated by contact with a hydrogen stream. Before the hydrogen can be introduced into the reactor, the O_2 content of the air in the reactor must be reduced to 0.1%. If pure N_2 can be fed into the reactor at the rate of 20 m³/min, for how long should the reactor be purged with N_2? Assume that the catalyst solids occupy 6% of the reactor volume and that the gases are well mixed.

6.5. An advertising firm wants to get a spherical inflated sign out of a warehouse. The sign is 20 ft in diameter and is filled with H_2 at 15 psig. Unfortunately, the door to the warehouse is only 19 ft high by 20 ft wide. The maximum rate of H_2 that can be safely vented from the balloon is 5 ft³/min (measured at room conditions). How long will it take to get the sign small enough to just pass through the door?
   (a) First assume that the pressure inside the balloon is constant so that the flow rate is constant.

(b) Then assume the amount of H_2 escaping is proportional to the volume of the balloon, and initially is 5 ft³/min.

(c) Could a solution to this problem be obtained if the amount of escaping H_2 were proportional to the pressure difference inside and outside the balloon?

6.6. A plant at Canso, Nova Scotia, makes fish-protein concentrate (FPC). It takes 6.6 kg of whole fish to make 1 kg of FPC, and therein is the problem—to make money, the plant must operate most of the year. One of the operating problems is the drying of the FPC. It dries in the fluidized dryer at a rate approximately proportional to its moisture content. If a given batch of FPV loses one-half of its initial moisture in the first 15 min, how long will it take to remove 90% of the water in the batch of FPC?

6.7. Water flows from a conical tank at the rate of $0.020(2 + h^2)$ m³/min, as shown in Fig. P6.7. If the tank is initially full, how long will it take for 75% of the water to flow out of the tank? What is the flow rate at that time?

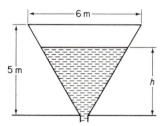

Fig. P6.7

6.8. A sewage disposal plant a big concrete holding tank of 100,000 gal capacity. It is three-fourths full of liquid to start with and contains 60,000 lb of organic material in suspension. Water runs into the holding tank at the rate of 20,000 gal/hr and the solution leaves at the rate of 15,000 gal/hr. How much organic material is in the tank at the end of 3 hr?

6.9. Suppose that in Problem 6.8 the bottom of the tank is covered with sludge (precipitated organic material) and that the stirring of the tank causes the sludge to go into suspension at a rate proportional to the difference between the concentration of sludge in the tank at any time and 10 lb of sludge/gal. If no organic material were present, the sludge would go into suspension at the rate of 0.05 lb/(min) (gal solution) when 75,000 gal of solution are in the tank. How much organic material is in the tank at the end of 3 hr?

6.10. In a chemical reaction the products X and Y are formed according to the equation

$$C \longrightarrow X + Y$$

The rate at which each of these products is being formed is proportional to the amount of C present. Initially: $C = 1$, $X = 0$, $Y = 0$. Find the time for the amount of X to equal the amount of C.

6.11. A tank is filled with water. At a given instant two orifices in the side of the tank are opened to discharge the water. The water at the start is 3 m deep and one orifice is 2 m below the top while the other one is 2.5 m below the top. The coefficient of discharge of each orifice is known to be 0.61. The tank is a vertical right circular cylinder 2 m in diameter. The upper and lower orifices are 5 and 10 cm in diameter, respectively. How long will be required for the tank to be drained so that the water level is at a depth of 1.5 m?

6.12. Suppose that you have two tanks in series, as diagrammed in Fig. P6.12. The volume of liquid in each tank remains constant because of the design of the overflow lines. Assume that each tank is filled with a solution containing 10 lb of A, and that the tanks contain 100 gal of aqueous solution each. If fresh water enters at the rate of 10 gal/hr, what is the concentration of A in each tank at the end of 3 hr? Assume complete mixing in each tank and ignore any change of volume with concentration.

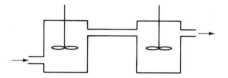

Fig. P6.12

6.13. A well-mixed tank has a maximum capacity of 100 gal and it is initially half full. The discharge pipe at the bottom is very long and thus it offers resistance to the flow of water through it. The force that causes the water to flow is the height of the water in the tank, and in fact that flow is just proportional to this height. Since the height is proportional to the total volume of water in the tank, the volumetric flow rate of water out, q_o, is

$$q_o = kV$$

The flow rate of water into the tank, q_i, is constant. Use the information given in Fig. P6.13 to decide whether the amount of water in the tank increases, decreases, or remains the same. If it changes, how much time is required to completely empty or fill the tank, as the case may be?

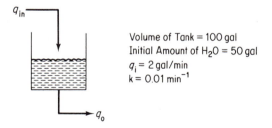

Volume of Tank = 100 gal
Initial Amount of H_2O = 50 gal
q_i = 2 gal/min
k = 0.01 min^{-1}

Fig. P6.13

6.14. A stream containing a radioactive fission product with a decay constant of 0.01 hr^{-1} (i.e., $dn/dt = 0.01\,n$), is run into a holding tank at the rate of 100 gal/hr for 24 hr. Then the stream is shut off for 24 hr. If the initial concentration of the fission product was 10 mg/liter and the tank volume is constant at 10,000 gal of solution (owing to an overflow line), what is the concentration of fission product:

(a) At the end of the first 24-hr period?

(b) At the end of the second 24-hr period?

What is the maximum concentration of fission product? Assume complete mixing in the tank.

6.15. A radioactive waste that contains 1500 ppm of ^{92}Sr is pumped into a holding tank that contains 0.40 m^3 at the rate of $1.5 \times 10^{-3} \text{ m}^3/\text{min}$. ^{92}Sr decays as follows:

$$^{92}Sr \longrightarrow {}^{92}Y \longrightarrow {}^{92}Zr$$

half-life: 2.7 hr   3.5 hr

If the tank contains clear water initially and the solution runs out at the rate of $1.5 \times 10^{-3} \text{ m}^3/\text{min}$, assuming perfect mixing:

(a) What is the concentration of Sr, Y, and Zr after 1 day?

(b) What is the equilibrium concentration of Sr and Y in the tank?

The rate of decay of such isotopes is $dN/dt = -\lambda N$, where $\lambda = 0.693/t_{1/2}$ and the half-life is $t_{1/2}$. $N = $ moles.

6.16. A cylinder contains 3 m^3 of pure oxygen at atmospheric pressure. Air slowly pumped into the tank and mixes uniformly with the contents, an equal volume of which is forced out of the tank. What is the concentration of oxygen in the tank after 9 m^3 of air has been admitted?

6.17. Suppose that an organic compound decomposes as follows:

$$C_6H_{12} \longrightarrow C_4H_8 + C_2H_4$$

If 1 mole of C_6H_{12} exists at $t = 0$ but no C_4H_8 and C_2H_4, set up equations showing the moles of C_4H_8 and C_2H_4 as a function of time. The rates of formation of C_4H_8 and C_4H_8 are each proportional to the number of moles of C_6H_{12} present.

6.18. A large tank is connected to a smaller tank by means of a valve. The large tank contains N_2 at 690 kPa while the small tank is evacuated. If the valve leaks between the two tanks and the rate of leakage of gas is proportional to the pressure difference between the two tanks $(p_1 - p_2)$, how long does it take for the pressure in the small tank to be one-half its final value? The instantaneous initial flow rate with the small tank evacuated is 0.091 kg mol/hr.

	Tank 1	Tank 2
Initial pressure (kPa)	700	0
Volume (m³)	30	15

Assume that the temperature in both tanks is constant and is 20°C.

6.19. The following chain reactions take place in a constant-volume batch tank:

$$A \xrightarrow{k_1} B \xrightarrow{k_2} C$$

Each reaction is first order and irreversible. If the initial concentration of A is C_{A_0}

and if only A is present initially, find an expression for C_B as a function of time. Under what conditions will the concentration of B be dependent primarily on the rate of reaction of A?

6.20. Consider the following chemical reaction in a constant-volume batch tank:

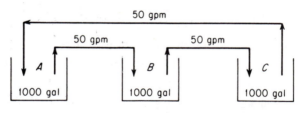

All the indicated reactions are first order. The initial concentration of A is C_{A_0}, and nothing else is present at that time. Determine the concentrations of A, B, and C as functions of time.

6.21. Tanks A, B, and C are each filled with 1000 gal of water. See Fig. P6.21. Workers have instructions to dissolve 2000 lb of salt in each tank. By mistake, 3000 lb is dissolved in each of tanks A and C and none in B. You wish to bring all the compositions to within 5% of the specified 2 lb/gal. If the units are connected A–B–C–A by three 50-gpm pumps,

(a) Express concentrations C_A, C_B, and C_C in terms of t (time).

(b) Find the shortest time at which all concentrations are within the specified range. Assume the tanks are all well mixed.

Fig. P6.21

6.22. Determine the time required to heat a 10,000-lb batch of liquid from 60°F to 120°F using an external, counterflow heat exchanger having an area of 300 ft². Water at 180°F is used as the heating medium and flows at a rate of 6000 lb/hr. An overall heat transfer coefficient of 40 Btu/(hr)(ft²)(°F) may be assumed; use Newton's law of heating. The liquid is circulated at a rate of 6000 lb/hr, and the specific heat of the liquid is the same as that of water (1.0). Assume that the residence time of the liquid in the external heat exchanger is very small and that there is essentially no holdup of liquid in this circuit.

6.23. A ground material is to be suspended in water and heated in preparation for a chemical reaction. It is desired to carry out the mixing and heating simultaneously in a tank equipped with an agitator and a steam coil. The cold liquid and solid are to be added continuously and the heated suspension will be withdrawn at the same rate. One method of operation for starting up is to (1) fill the tank initially with water and solid in the proper proportions, (2) start the agitator, (3) introduce fresh water and solid in the proper proportions and simultaneously begin to withdraw the suspension for reaction, and (4) turn on the steam. An estimate is needed of the time required, after the steam is turned on, for the temperature of the effluent suspension to reach a certain elevated temperature.

(a) Using the nomenclature given below, formulate a differential equation for this process. Integrate the equation to obtain n as a function of B and ϕ (see nomenclature).

(b) Calculate the time required for the effluent temperature to reach 180°F if the initial contents of the tank and the inflow are both at 120°F and the steam temperature is 220°F. The surface area for heat transfer is 23.9 ft², and the heat transfer coefficient is 100 Btu/(hr)(ft²)(°F). The tank contains 6000 lb, and the rate of flow of both streams is 1200 lb/hr. In the proportions used, the specific heat of the suspension may be assumed to be 1.00.

If the area available heat transfer is doubled, how will the time required be affected? Why is the time with the larger area less than half that obtained previously? The heat transferred is $Q = UA(T_{\text{tank}} - T_{\text{steam}})$.

Nomenclature

W = weight of tank contents, lb

G = rate of flow of suspension, lb/hr

T_s = temperature of steam, °F

T = temperature in tank at any instant, perfect mixing assumed, °F

T_0 = temperature of suspension introduced into tank; also initial temperature of tank contents, °F

U = heat-transfer coefficient, Btu/(hr)(ft²)(°F)

A = area of heat-transfer surface, ft²

C_p = specific heat of suspension, Btu/(lb)(°F)

t = time elapsed from the instant the steam is turned on, hr

n = dimensionless time, Gt/W

B = dimensionless ratio, UA/GC_p

ϕ = dimensionless temperature (relative approach to the steam temperature) $(T - T_0)/(T_s - T_0)$

6.24. Consider a well-agitated cylindrical tank in which the heat transfer surface is in the form of a coil that is distributed uniformly from the bottom of the tank to the top of the tank. The tank itself is completely insulated. Liquid is introduced into the tank at a uniform rate, starting with no liquid in the tank, and the steam is turned on at the instant that liquid flows into the tank.

(a) Using the nomenclature of Problem 6.23, formulate a differential equation for this process. Integrate this expression to obtain an equation for ϕ as a function of B and f, where f = fraction filled = W/W_{filled}.

(b) If the heat transfer surface consists of a coil of 10 turns of 1-in.-OD tubing 4 ft in diameter, the feed rate is 1200 lb/hr, the heat capacity of the liquid is 1.0 Btu/(lb)(°F), the heat transfer coefficient is 100 Btu/(hr)(°F)(ft²) of covered area, the steam temperature is 200°F, and the temperature of the liquid introduced into the tank is 70°F, what is the temperature in the tank when it is completely full? What is the temperature when the tank is half full? The heat transfer is given by $Q = UA(T_{tank} - T_{steam})$.

6.25. A cylindrical tank 5 ft in diameter and 5 ft high is full of water at 70°F. The water is to be heated by means of a steam jacket around the sides only. The steam temperature is 230°F, and the overall coefficient of heat transfer is constant at 40 Btu/(hr)(ft²)(°F). Use Newton's law of cooling (heating) to estimate the heat transfer. Neglecting the heat losses from the top and the bottom, calculate the time necessary to raise the temperature of the tank contents to 170°F. Repeat, taking the heat losses from the top and the bottom into account. The air temperature around the tank is 70°F, and the overall coefficient of heat transfer for both the top and the bottom is constant at 10 Btu/(hr)(ft²)(°F).

INDUSTRIAL CHEMICAL DATA

HOW MADE	MAJOR END USES	MAJOR DERIVATIVES	ANNUAL PRODUCTION, PRICES (USA 1981)
	Methanol	C_3HOH	
Reaction of carbon monoxide and hydrogen made by reforming natural gas or heavier hydrocarbon fractions	Polymers used as adhesives, fibers, and engineering plastics 60%	Formaldehyde 45%, acetic acid 10%, methyl tert-butyl ether 5%	1.2×10^9 gal $.75/gal
	Formaldehyde	O‖H—C—H	
Catalytic oxidation of methanol	Adhesives 60%, plastics 15%	Urea-formaldehyde resins 30%, phenol-formaldehyde resins 25%, polyacetal resins 5%, butanediol 5%, pentaerythritol 5%	6×10^9 lb $0.07–$0.08/lb
	Potash (as K_2O)	KCl, K_2SO_4, $K_2SO_4 \cdot 2MgSO_4$	
Shaft and solution mining of potassium salts, mainly the chloride, and subsequent purification	Fertilizer 95%		2.9×10^6 tons $115–$130/ton

Adapted from *Chemical and Engineering News* by permission of the American Chemical Society.

ANSWERS
TO SELF-ASSESSMENT TESTS

Section 1.1

1. 2.10×10^{-5} m³/s
2. (**a**) 252 lb$_m$; (**b**) 29.6 lb$_f$
3. Examine the conversion factors inside the cover.
4. c is dimensionless.
5. 32.174 (ft)(lb$_m$)/(s²)(lb$_f$)
6. (a)
7. A has the same units as k; B has the units of T.
8. (**a**) 28.349, 0.454; (**b**) 1.609; (**c**) 37.85

Section 1.2

1. 60.05
2. A kilogram mole contains 6.02×10^{26} molecules, whereas a pound mole contains $(6.02 \times 10^{26})(0.454)$ molecules.
3. 0.123 kg mol NaCl/kg mol H_2O
4. 1.177 lb mol
5. 0.121 kg/s

Section 1.3

1. (a) F; (b) T; (c) F; (d) T
2. 13.6 g/cm³
3. 62.4 lb$_m$/ft³ (10³ kg/m³)
4. This means that the density at 10°C of liquid HCN is 1.2675 times the density of water at 4°C.
5. 0.79314 g/cm³ (Note that you need the density of water at 60°F.)
6. 9
7. (a) 63%; (b) 54.3; (c) 13.8
8. 8.11 ft³
9. (a) 0.33; (b) 18.7
10. (a) $C_4 - 0.50$, $C_5 - 0.30$, $C_6 - 0.20$; (b) $C_4 - 0.57$, $C_5 - 0.28$, $C_6 - 0.15$; (c) $C_4 - 57$, $C_5 - 28$, $C_6 - 15$; (d) 62.2 kg/kg mol

Section 1.4

1. See p. 28.
2. (a) 100 kg or 100 lb
    (b) 1 ton of water
    (c) The stated mass for each case
    (d) 100 cm³ mixture

Section 1.5

1. (a) 0°C and 100°C; (b) 32°F and 212°F
2. $\Delta°F(1.8) = \Delta°C$.
3. Yes. Yes.
4. $92.76 + 0.198T_{°F}$.

5.

°C	°F	K	°R
−40.0	−40.0	233	420
25.0	77.0	298	537
425	796	698	1256
−234	−390	38.8	69.8

6. Immerse in ice-water bath and mark 0°C. Immerse in boiling water at 1 atm pressure and mark 100°C. Interpolate between 0°C and 100°C in desired intervals.

Section 1.6

1. Gauge pressure + barometric pressure = absolute pressure.
2. See p. 44.
3. Barometric pressure − vacuum pressure = absolute pressure.
4. (a) 15.5; (b) 106.6 kPa; (c) 1.052; (d) 35.6
5. (a) Gauge pressure; (b) barometric pressure, absolute pressure; (c) 50 in. Hg
6. In the absence of the barometric pressure, assume 101.3 kPa; then absolute pressure is 61.3 kPa.

Section 1.7

1. See p. 49.
2. (a) *Chemical Engineers' Handbook* and/or *Properties of Gases and Liquids*
   (b) *Technical Data Book—Petroleum Refining*
   (c) *Handbook of Physics and Chemistry* and others
   (d) *Chemical Engineers' Handbook*

Section 1.8

1. A serial sequence is adequate; some steps can be in parallel. Feedback may be inserted at various stages.
3. (a) Find a reason to remember; talk to others who are interested to ascertain what motivated them. Write down the reason for periodic pep talks with yourself.
   (b) Get a "bird's-eye view" of the whole section. Use the figure numbered 0 to see the relation to other sections. Look at a similar textbook. Ask for help; do not wait.
   (c) After scanning the serious reading count.
   (d) Take a few minutes as you read to mentally test yourself on what you have just read to reinforce learning. Continue to test at periodic intervals.
   (e) Practice using the information for an assumed quiz; predict quiz questions; use text examples or problems as samples.

Section 1.9

1. (a) $C_9H_{18} + \frac{27}{2}O_2 \longrightarrow 9CO_2 + 9H_2O$
   (b) $FeS_2 + 11O_2 \longrightarrow 2Fe_2O_3 + 8SO_2$
2. 3.08
3. 323
4. (a) $CaCO_3 - 43.4\%$; $CaO - 56.6\%$; (b) 0.308
5. (a) H_2O; (b) $NaCl$; (c) $NaCl$, H_2O, $NaOH$ (assuming that the gas escapes)

Section 2.1

1, 2. The systems are somewhat arbitrary, as are the time intervals selected, but (a) and (c) can be closed systems (ignoring evaporation) and (b) open.
3. See Eqs. (2.1), (2.2), (2.3), and (2.3a).
4. (a) batch; (b) flow; (c) flow; (d) flow.
5. Accumulation is zero.
6. Fig. 2.4—No, because many streams have no values and some streams, such as gases, are not included in the diagram. Fig. 2.5—Yes.
7. Equation (2.1) or simplifications thereof.
8. (a) T; (b) T is C is regarded as an element; F if the exit compound is CO_2 and the entering reactant is C; (c) F
9. No chemical reaction takes place, or with reaction, the moles of reactants may equal the moles of product by accident.

Section 2.2

1. A set of unique values for the variables in the equations representing the problem.
2. (a) one; (b) three; (c) three
3. (a) two; (b) two of these three: acetic acid, water, total; (c) two; (d) feed of the

10% solution and mass fraction ω of the acetic acid in P; **(e)** 14% acetic acid, 86% water
4. Not for a unique solution. Only two are independent.
5. Substitution; Gauss–Jordan elimination; use of determinants.
6. Select the most accurate equations that will provide a unique solution. (Or use least squares if all must be used.)
7. The sum of the mass or mole fractions in one stream is unity.
8. Collect additional data so as to specify or evaluate the values of the excess unknown variables.
9. $F, D, P, \omega_{D2}, \omega_{P1}$.
10. Only two independent material balances can be written. The sum of the mass fractions for streams D and P yields two additional independent equations. One value of F, D, or P must be specified to obtain a solution. Note that specifying W_{D2} or W_{P1} will not help.

Section 2.3

1. Orsat (dry basis) does not include the water vapor in the analysis.
2. SO_2 is not included in the gas analysis.
3. See Eq. (2.7).
4. Yes. See Eq. (2.6).
5. More
6. 4.5%
7. 11.0% CO_2, 7.0% O_2, 82.0% N_2
8. 178 kg/hr

Section 2.4

1. Three components = three independent equations. Any three of the four (1, 2, 3, total) will do.
2. $A = 40$ kg/hr; $B = 60$ kg/hr
3. 3.7321, 0.2680
4. $x = 0.2268$; $y = 0.3696$
5. **(a)** 252; **(b)** 1.063; **(c)** 2.31; **(d)** 33.8%
6. 33.3 kg
7. 186 kg
8. **(a)** 23.34; **(b)** 66.5%

Section 2.5

1. 0.6%
2. 0.82%

Section 2.6

1. Recycle is feedback from downstream to upstream in the materials flow bypassing one or more units. Bypassing is feedforward bypassing one or more units.
2. Purge is a side stream usually small in quantity relative to main stream that is bled from the main stream to remove impurities from the system.
3. The stoichiometric ratio
4. **(a)** $R = 3000$ kg/hr; **(b)** air $= 85,100$ kg/hr
5. **(a)** 908 recycled and 3.2 purged; **(b)** 9.05% conversion
6. **(a)** 1570 kg/hr; **(b)** 242 kg/hr

Section 3.1-1

1. $pV = nRT$
2. T: absolute temperature in degrees; p: absolute pressure in mass/(length)(time)^2; V: (length)^3/mole; n: mole; R: (mass)(length)^2/(time)^2(mole)(degree)

3. See p. 204.
4. 1883 ft^3
5. 2.98 kg
6. 1.32

Section 3.1-2

1. 0.0493 kg/m^3
2. 1.02 (lb CH$_4$/ft^3 CH$_4$ at 70°F and 2 atm)/(lb air/ft^3 at S.C.)

Section 3.1-3

1. See Eqs. (3.3) and (3.6).
2. (a) N$_2$, 0.28 psia; CH$_4$, 10.9 psia; C$_2$H$_6$, 2.62 psia all when $V = 2$ ft^3 at 120°F
   (b) N$_2$, 0.04 ft^3; CH$_4$, 1.58 ft^3; C$_2$H$_6$, 0.38 ft^3 all at 13.8 psia and 120°F
   (c) Same as the mole fractions
3. (a) 11.8 psia at 2 ft^3 and 120°F; (b) 0.28 psia at 2 ft^3 and 120°F

Section 3.2-1

1. See the last paragraph in Sec. 3.2-1. Also, they can be used for interpolation and extrapolation.
2. (a) Least squares and theory; (b) least squares would give the least error.
3. Comparison with experimental data; reduction in limit of ideal gas to $pV = nRT$; go through the critical point; holding one variable constant, the residual expression should be correct; derivatives give proper functions.
4. The volume occupied by the moles of the gas—by using $(V - nb)$, and the attractive forces existing between the molecules of the gas—by adding $[p + (n^2a/V^2)]$.
5. b is m^3; a is $(K)^{1/2}(m)^6(kPa)$.
6. $V = 0.60$ ft^3
7. 314 K
8. (a) 50.7 atm; (b) 34.0 atm
9. 0.316 m^3/kg mol

Section 3.2-2

1. $\hat{V}_{c_i} = RT_c/P_c$. It can be used to calculate $\hat{V}_{r_i}$, which is a parameter on the Nelson and Obert charts.
2. (a) No; (b) 5.08 ft^3; (c) 1.35 ft^3
3. 1.65 kg
4. 14.9 atm

Section 3.2-3

1. (a) 203 atm
   (b) 289 atm
   (c) 300 atm
   (d) 262 atm

Section 3.3

1. See Fig. 3.12.
2. Ice at its vapor pressure changes to liquid water at 32°F (at 0.0886 psia) and the

pressure increases as shown in Fig. 3.12 with vapor and liquid in equilibrium. At 250°F, the pressure is 29.82 psia.

  3. (a) 75 psia (5.112 atm); (b) it sublimes.
  4. Experimental $p^* = 219.9$ mm Hg; predicted 220.9 mm Hg
  5. 74.4°C

Section 3.4

  1. The partial pressure of the vapor equals the vapor pressure of the gas. Liquid and vapor are in equilibrium.
  2. Yes; yes
  3. 21°C; benzene
  4. 0.0373
  5. 4.00 lb
  6. (a) Both gas; (b) some liquid water, residual is gas; (c) both gas; (d) some liquid water, residual is gas.

Section 3.5

  1. 0.063
  2. 57.3%

  3. 86% $\mathcal{RJC}$

Section 3.6

  1. 53°C (126°F)
  2. 71.2% H_2O

  3. 94.0%; 100%

Section 3.7

  1. No, gas and liquid in equilibrium. The triple point in the p–T projection is actually a line on the pVT surface. The pressure and temperature are fixed but the volume is not fixed.
  2. Nothing. The points are not realizable equilibrium values.
  3. Yes for a liquid–vapor if the process goes above the critical point. Probably not for a solid–vapor unless a solid–liquid region can be formed with a smooth property transition.
  4. See Fig. 3.23.
  5. (a) $C = 3$, $\mathcal{P} = 1$, $F = 4$; (b) $C = 2$, $\mathcal{P} = 2$, $F = 2$
  6. No. $C = 2$, $\mathcal{P} = 6$, F is negative, which is not possible.
  7. Two

Section 4.1

  1. Intensive—independent of quantity of material; extensive—dependent of quantity of material. Measurable—temperature, pressure; unmeasurable—internal energy, enthalpy. State variable—difference in value between two states depends only on the states; path variable—difference in two states depends on trajectory reaching in the final state.
  2. Heat: energy transfer across a system boundary due to a temperature difference. Work: energy transfer across a system boundary by means of a vector force acting through a vector displacement on the boundary.

3. (a) Open; (b) open; (c) open; (d) closed; (e) closed
4. (a)
6. Decrease
7. (a) $K_0 = 0, P_0 = 4900$ J; (b) $K_f = 0$; (c) $P_f = 0$; (c) $\Delta K = 0, \Delta P = -4900$ J (a decrease); (d) 1171 cal, 4.64 Btu
8. 1.26×10^5 J

Section 4.2

1. $C_p = 4.25 + 0.002T$
2. $\frac{7}{2}R$
3. (a) cal/(g mol)(ΔK); (b) cal/(g mol)(ΔK)(K); (c) cal/(g mol)(ΔK)(K^2); (d) cal/(g mol)(ΔK)(K^3)
4. 7.11 cal/(g mol)(K)
5. (a) 0.58 cal/(g)(°C); (b) 0.49 cal/(g)(°C); experimental $= 0.535$ cal/(g)(°C)
6. $27,115 + 6.55T - 9.98 \times 10^{-4}T^2$

Section 4.3

1. 32,970 J (7880 cal)
2. 0.058
3. (a) Liquid; (b) two phase; (c) two phase; (d) vapor; (e) vapor
4. -5635 Btu
5. 0.424 Btu/(lb mol)(°F)
6. 13,900 Btu
7. 178.7 J/kg

Section 4.4

1. -334 Btu/lb
2. 6567 Btu
3. 1991 kJ/kg; experimental $= 2025.9$ kJ/kg

Section 4.5-1

1. (a) $T = 90.19$ K; (b) $m = 1.48 \times 10^7$ g; (c) $x = 3.02 \times 10^{-4}$; (d) 49.6 hr; (e) from a handbook, $T = 100$ K approx (for process at constant volume)
2. (a) Down; (b) probably up, depending on room temperature; (c) down
3. (a) $T_1 = 166.2°$C, $T_2 = 99.63°$C; (b) $Q = -6.40 \times 10^4$ J
4. 596 Btu
5. (a) 0; (b) 0; (c) 0; (d) if ideal gas $\Delta T = 0$, $T_2 =$ room temperature; (e) 0.26 atm
6. 7 m

Section 4.5-2

1. 12,200 lb/hr
2. 99°C

3. 1847 watts (2.48 hp)
4. -19.7 Btu/s

Section 4.6

1. (2)
2. Thermal—presumably internal energy or heat—not freely convertible to mechanical energy. The latter types of energy are freely convertible one to the other.

3. Process in which total of freely convertible forms of energy are not reduced.
4. $W = 27.6$ kJ; $Q = -27.6$ kJ; $\Delta E = \Delta H = 0$
5. 38.8 hp

Section 4.7-1

1. -74.83 kJ/g mol CH_4 (-17.88 kcal)
2. 597.32 kJ/g mol C_6H_6 (142.76 kcal)

Section 4.7-2

1. -890.36 kJ/g mol CH_4
2. -148.53 k cal/g mol
3. No

4. -24.82 k cal/g mol C_3H_8
5. 230 Btu/ft³ at 60°F and 30.0 in. Hg

Section 4.7-3

1. No. The result depends on the values in Eq. (4.38) and may be positive or negative.
2. (a) $p \simeq 2$ atm; (b) $Q_v = -3.55$ kcal/g mol CO (evolved)

Section 4.7-4

1. 125 Btu/ft³ at SC.

Section 4.7-5

1. $-248,100$ J ($-59,299$ kcal)
2. 6020 Btu/lb methane

3. 0.52 lb

Section 4.7-6

1. 825°C

Section 4.8

1. Almost always
2. (a) HNO_3, HCl, or H_2SO_4 in water; (b) NaCl, KCl, NH_4NO_3 in water
3. (a) 25°C and 1 atm; (b) 0
4. $-4,655$ cal/g mol soln (heat transfer is from solution to surroundings)
5. $Q = -1.48 \times 10^6$ Btu

Section 5.1

1. Yes. See Appendix L.
2. For one phase, $N_{sp} - 1 + 2$ intensive variable (from the phase rule) plus one extensive variable are needed to completely specify a stream.
3. 9
4. Any number from 7 to 11 may be acceptable, depending on assumptions concerning the relations among the pressures in the reactor and mixer streams.
5. (a) 11,310,000 Btu/hr; (b) 2,950 lb/hr

Section 5.2

1. $\omega_V = 0.85$; $\omega_L = 0.15$
2. 500 Btu/lb

4. $V = 190$ kg, $L = 810$ kg
5. Two phase; for the liquid $\omega_{H_2O} = 0.50$ and $\hat{H} = 8$ Btu/lb; for the vapor $\omega_{H_2O} = 1.00$ and $\hat{H} = 1174$ Btu/lb
6. A and $\overline{BC}$; B and $\overline{AC}$; C and $\overline{AB}$
7. A and $\overline{BC}$; B and $\overline{AC}$; C and $\overline{AB}$

Section 5.3

1. t_{DB} = gas temperature; t_{WB} = temperature registered at equilibrium between evaporating water around temperature sensor and surroundings at t_{DB}
2. No
3. See Eqs. (5.21)–(5.26).
4. (a) 0.03 kg/kg dry air; (b) 1.02 m³/kg dry air; (c) 38°C$_{in}$; (d) 151 kJ/kg dry air; (e) 31.5°C
5. (a) $\mathcal{H} = 0.0808$ lb H_2O/lb dry air; (b) $H = 118.9$ Btu/lb dry air; (c) $V = 16.7$ ft³/lb dry air; (d) $\mathcal{H} = 0.0710$ lb H_2O/lb dry air
6. 49,700 Btu
7. (a) 1.94×10^6 ft³/hr; (b) 3.00×10^4 ft³/hr; (c) 2.44×10^5 Btu/hr

Chapter 6

1. The solution of the differential equation requires the initial conditions to be able to predict the response at other times.
2. Independent
3. Let the derivatives vanish (delete them from the equations).
4. (a)
5. (1), (4), (7), (9), (11); (2); (8); (3), (5), (10), (12); (6)
6. 13 hr
7. 0.212 lb

ATOMIC WEIGHTS AND NUMBERS

Appendix B

TABLE B.1 RELATIVE ATOMIC WEIGHTS, 1965 (BASED ON THE ATOMIC MASS OF $^{12}C = 12$)

The values for atomic weights given in the table apply to elements as they exist in nature, without artificial alteration of their isotopic composition, and, further, to natural mixtures that do not include isotopes of radiogenic origin.

Name	Symbol	Atomic Number	Atomic Weight	Name	Symbol	Atomic Number	Atomic Weight
Actinium	Ac	89	—	Mercury	Hg	80	200.59
Aluminium	Al	13	26.9815	Molybdenum	Mo	42	95.94
Americium	Am	95	—	Neodymium	Nd	60	144.24
Antimony	Sb	51	121.75	Neon	Ne	10	20.183
Argon	Ar	18	39.948	Neptunium	Np	93	—
Arsenic	As	33	74.9216	Nickel	Ni	28	58.71
Astatine	At	85	—	Niobium	Nb	41	92.906
Barium	Ba	56	137.34	Nitrogen	N	7	14.0067
Berkelium	Bk	97	—	Nobelium	No	102	—
Beryllium	Be	4	9.0122	Osmium	Os	75	190.2
Bismuth	Bi	83	208.980	Oxygen	O	8	15.9994
Boron	B	5	10.811	Palladium	Pd	46	106.4
Bromine	Br	35	79.904	Phosphorus	P	15	30.9738
Cadmium	Cd	48	112.40	Platinum	Pt	78	195.09
Caesium	Cs	55	132.905	Plutonium	Pu	94	—
Calcium	Ca	20	40.88	Polonium	Po	84	—
Californium	Cf	98	—	Potassium	K	19	39.102
Carbon	C	6	12.01115	Praseodym	Pr	59	140.907
Cerium	Ce	58	140.12	Promethium	Pm	61	—
Chlorine	Cl	17	35.453^b	Protactinium	Pa	91	—
Chromium	Cr	24	51.996^b	Radium	Ra	88	—
Cobalt	Co	27	58.9332	Radon	Rn	86	—
Copper	Cu	29	63.546^b	Rhenium	Re	75	186.2
Curium	Cm	96	—	Rhodium	Rh	45	102.905
Dysprosium	Dy	66	162.50	Rubidium	Rb	37	84.57
Einsteinium	Es	99	—	Ruthenium	Ru	44	101.07
Erbium	Er	68	167.26	Samarium	Sm	62	150.35
Europium	Eu	63	151.96	Scandium	Sc	21	44.956
Fermium	Fm	100	—	Selenium	Se	34	78.96
Fluorine	F	9	18.9984	Silicon	Si	14	28.086
Francium	Fr	87	—	Silver	Ag	47	107.868
Gadolinium	Gd	64	157.25	Sodium	Na	11	22.9898
Gallium	Ga	31	69.72	Strontium	Sr	38	87.62
Germanium	Ge	32	72.59	Sulfur	S	16	32.064
Gold	Au	79	196.967	Tantalum	Ta	73	180.948
Hafnium	Hf	72	178.49	Technetium	Tc	43	—
Helium	He	2	4.0026	Tellurium	Te	52	127.60
Holmium	Ho	67	164.930	Terbium	Tb	65	158.924
Hydrogen	H	1	1.00797	Thallium	Tl	81	204.37
Indium	In	49	114.82	Thorium	Th	90	232.038
Iodine	I	53	126.9044	Thulium	Tm	59	168.934
Iridium	Ir	77	192.2	Tin	Sn	50	118.69
Iron	Fe	26	55.847	Titanium	Ti	22	47.90
Krypton	Kr	36	83.80	Tungsten	W	74	183.85
Lanthanum	La	57	138.91	Uranium	U	92	238.03
Lawrencium	Lr	103	—	Vanadium	V	23	50.942
Lead	Pb	82	207.19	Xenon	Xe	54	131.30
Lithium	Li	3	6.939	Ytterbium	Yb	70	173.04
Lutetium	Lu	71	174.97	Yttrium	Y	39	88.905
Magnesium	Mg	12	24.312	Zinc	Zn	30	65.37
Manganese	Mn	25	54.9380	Zirconium	Zr	40	91.22
Mendelevium	Md	101	—				

SOURCE: *Comptes Rendus*, 23rd IUPAC Conference, 1965, Butterworth's, London, 1965, pp. 177–178.

Appendix C

STEAM TABLES

Note: **Tables in SI and American Engineering units for superheated steam are on the chart in the pocket in the back of the book.**

Absolute pressure = atmospheric pressure − vacuum.

Barometer and vacuum columns may be corrected to mercury at 32°F by subtracting 0.00009 × (t − 32) × column height, where t is the column temperature in °F.

One inch of mercury at 32°F = 0.4912 lb/in.^{2}

Example:

Barometer reads 30.17 in. at 70°F. Vacuum column reads 28.26 in. at 80°F. Pounds pressure = (30.17 − 0.00009 × 38 × 30.17) − 28.26 − 0.00009 × 48 × 28.26) = 1.93 in. of mercury at 32°F.

Saturation temperature (from table) = 100°F.

TABLE C.1 SATURATED STEAM: TEMPERATURE TABLE

Temp. Fahr. t	Absolute pressure Lb/in.² p	Absolute pressure In. Hg 32°F	Sat. Liquid v_f	Specific volume Evap. v_{fg}	Sat. Vapor v_g	Sat. Liquid h_f	Enthalpy Evap. h_{fg}	Sat. Vapor h_g
32	0.0886	0.1806	0.01602	3305.7	3305.7	0	1075.1	1075.1
34	0.0961	0.1957	0.01602	3060.4	3060.4	2.01	1074.9	1076.0
36	0.1041	0.2120	0.01602	2836.6	2836.6	4.03	1072.9	1076.9
38	0.1126	0.2292	0.01602	2632.2	2632.2	6.04	1071.7	1077.7
40	0.1217	0.2478	0.01602	2445.1	2445.1	8.05	1070.5	1078.6
42	0.1315	0.2677	0.01602	2271.8	2271.8	10.06	1069.3	1079.4
44	0.1420	0.2891	0.01602	2112.2	2112.2	12.06	1068.2	1080.3
46	0.1532	0.3119	0.01602	1965.5	1965.5	14.07	1067.1	1081.2
48	0.1652	0.3364	0.01602	1829.9	1829.9	16.07	1065.9	1082.0
50	0.1780	0.3624	0.01602	1704.9	1704.9	18.07	1064.8	1082.9
52	0.1918	0.3905	0.01603	1588.4	1588.4	20.07	1063.6	1083.7
54	0.2063	0.4200	0.01603	1482.4	1482.4	22.07	1062.5	1084.6
56	0.2219	0.4518	0.01603	1383.5	1383.5	24.07	1061.4	1085.5
58	0.2384	0.4854	0.01603	1292.7	1292.7	26.07	1060.2	1086.3
60	0.2561	0.5214	0.01603	1208.1	1208.1	28.07	1059.1	1087.2
62	0.2749	0.5597	0.01604	1129.7	1129.7	30.06	1057.9	1088.0
64	0.2949	0.6004	0.01604	1057.1	1057.1	32.06	1056.8	1088.9
66	0.3162	0.6438	0.01604	989.6	989.6	34.06	1055.7	1089.8
68	0.3388	0.6898	0.01605	927.0	927.0	36.05	1054.5	1090.6
70	0.3628	0.7387	0.01605	868.9	868.9	38.05	1053.4	1091.5
72	0.3883	0.7906	0.01606	814.9	814.9	40.04	1052.3	1092.3
74	0.4153	0.8456	0.01606	764.7	764.7	42.04	1051.2	1093.2
76	0.4440	0.9040	0.01607	718.0	718.0	44.03	1050.1	1094.1
78	0.4744	0.9659	0.01607	674.4	674.4	46.03	1048.9	1094.9
80	0.5067	1.032	0.01607	633.7	633.7	48.02	1047.8	1095.8
82	0.5409	1.101	0.01608	595.8	595.8	50.02	1046.6	1096.6
84	0.5772	1.175	0.01608	560.4	560.4	52.01	1045.5	1097.5
86	0.6153	1.253	0.01609	527.6	527.6	54.01	1044.4	1098.4
88	0.6555	1.335	0.01609	497.0	497.0	56.00	1043.2	1099.2
90	0.6980	1.421	0.01610	468.4	468.4	58.00	1042.1	1100.1
92	0.7429	1.513	0.01611	441.7	441.7	59.99	1040.9	1100.9
94	0.7902	1.609	0.01611	416.7	416.7	61.98	1039.8	1101.8
96	0.8403	1.711	0.01612	393.2	393.2	63.98	1038.7	1102.7
98	0.8930	1.818	0.01613	371.3	371.3	65.98	1037.5	1103.5
100	0.9487	1.932	0.01613	350.8	350.8	67.97	1036.4	1104.4
102	1.0072	2.051	0.01614	331.5	331.5	69.96	1035.2	1105.2
104	1.0689	2.176	0.01614	313.5	313.5	71.96	1034.1	1106.1
106	1.1338	2.308	0.01615	296.5	296.5	73.95	1033.0	1107.0
108	1.2020	2.447	0.01616	280.7	280.7	75.94	1032.0	1107.9
110	1.274	2.594	0.01617	265.7	265.7	77.94	1030.9	1108.8
112	1.350	2.749	0.01617	251.6	251.6	79.93	1029.7	1109.6
114	1.429	2.909	0.01618	238.5	238.5	81.93	1028.6	1110.5
116	1.512	3.078	0.01619	226.2	226.2	83.92	1027.5	1111.4
118	1.600	3.258	0.01620	214.5	214.5	85.92	1026.4	1112.3
120	1.692	3.445	0.01620	203.45	203.47	87.91	1025.3	1113.2
122	1.788	3.640	0.01621	193.16	193.18	89.91	1024.1	1114.0
124	1.889	3.846	0.01622	183.44	183.46	91.90	1023.0	1114.9
126	1.995	4.062	0.01623	174.26	174.28	93.90	1021.8	1115.7
128	2.105	4.286	0.01624	165.70	165.72	95.90	1020.7	1116.6

v = specific volume, ft³/lb. h = enthalpy, Btu/lb.

TABLE C.1 (*cont.*)

Temp. Fahr. t	Absolute pressure		Specific volume			Enthalpy		
	Lb/in.² p	In. Hg 32°F	Sat. Liquid v_f	Evap. v_{fg}	Sat. Vapor v_g	Sat. Liquid h_f	Evap. h_{fg}	Sat. Vapor h_g
130	2.221	4.522	0.01625	157.55	157.57	97.89	1019.5	1117.4
132	2.343	4.770	0.01626	149.83	149.85	99.89	1018.3	1118.2
134	2.470	5.029	0.01626	142.59	142.61	101.89	1017.2	1119.1
136	2.603	5.300	0.01627	135.73	135.75	103.88	1016.0	1119.9
138	2.742	5.583	0.01628	129.26	129.28	105.88	1014.9	1120.8
140	2.887	5.878	0.01629	123.16	123.18	107.88	1013.7	1121.6
142	3.039	6.187	0.01630	117.37	117.39	109.88	1012.5	1122.4
144	3.198	6.511	0.01631	111.88	111.90	111.88	1011.3	1123.2
146	3.363	6.847	0.01632	106.72	106.74	113.88	1010.2	1124.1
148	3.536	7.199	0.01633	101.82	101.84	115.87	1009.0	1124.9
150	3.716	7.566	0.01634	97.18	97.20	117.87	1007.8	1125.7
152	3.904	7.948	0.01635	92.79	92.81	119.87	1006.7	1126.6
154	4.100	8.348	0.01636	88.62	88.64	121.87	1005.5	1127.4
156	4.305	8.765	0.01637	84.66	84.68	123.87	1004.4	1128.3
158	4.518	9.199	0.01638	80.90	80.92	125.87	1003.2	1129.1
160	4.739	9.649	0.01639	77.37	77.39	127.87	1002.0	1129.9
162	4.970	10.12	0.01640	74.00	74.02	129.88	1000.8	1130.7
164	5.210	10.61	0.01642	70.79	70.81	131.88	999.7	1131.6
166	5.460	11.12	0.01643	67.76	67.78	133.88	998.5	1132.4
168	5.720	11.65	0.01644	64.87	64.89	135.88	997.3	1133.2
170	5.990	12.20	0.01645	62.12	62.14	137.89	996.1	1134.0
172	6.272	12.77	0.01646	59.50	59.52	139.89	995.0	1134.9
174	6.565	13.37	0.01647	57.01	57.03	141.89	993.8	1135.7
176	6.869	13.99	0.01648	56.64	54.66	143.90	992.6	1136.5
178	7.184	14.63	0.01650	52.39	52.41	145.90	991.4	1137.3
180	7.510	15.29	0.01651	50.26	50.28	147.91	990.2	1138.1
182	7.849	15.98	0.01652	48.22	48.24	149.92	989.0	1138.9
184	8.201	16.70	0.01653	46.28	46.30	151.92	987.8	1139.7
186	8.566	17.44	0.01654	44.43	44.45	153.93	986.6	1140.5
188	8.944	18.21	0.01656	42.67	42.69	155.94	985.3	1141.3
190	9.336	19.01	0.01657	40.99	41.01	157.95	984.1	1142.1
192	9.744	19.84	0.01658	39.38	39.40	159.95	982.8	1142.8
194	10.168	20.70	0.01659	37.84	37.86	161.96	981.5	1143.5
196	10.605	21.59	0.01661	36.38	36.40	163.97	980.3	1144.3
198	11.057	22.51	0.01662	34.98	35.00	165.98	979.0	1145.0
200	11.525	23.46	0.01663	33.65	33.67	167.99	977.8	1145.8
202	12.010	24.45	0.01665	32.37	32.39	170.01	976.6	1146.6
204	12.512	25.47	0.01666	31.15	31.17	172.02	975.3	1147.3
206	13.031	26.53	0.01667	29.99	30.01	174.03	974.1	1148.1
208	13.568	27.62	0.01669	28.88	28.90	176.04	972.8	1148.8
210	14.123	28.75	0.01670	27.81	27.83	178.06	971.5	1149.6
212	14.696	29.92	0.01672	26.81	26.83	180.07	970.3	1150.4
215	15.591		0.01674	25.35	25.37	186.10	968.3	1151.4
220	17.188		0.01677	23.14	23.16	188.14	965.2	1153.3
225	18.915		0.01681	21.15	21.17	193.18	961.9	1155.1
230	20.78		0.01684	19.371	19.388	198.22	958.7	1156.9
235	22.80		0.01688	17.761	17.778	203.28	955.3	1158.6
240	24.97		0.01692	16.307	16.324	208.34	952.1	1160.4
245	27.31		0.01696	15.010	15.027	213.41	948.7	1162.1
250	29.82		0.01700	13.824	13.841	218.48	945.3	1163.8
255	32.53		0.01704	12.735	12.752	223.56	942.0	1165.6
260	35.43		0.01708	11.754	11.771	228.65	938.6	1167.3
265	38.54		0.01713	10.861	10.878	233.74	935.3	1169.0
270	41.85		0.01717	10.053	10.070	238.84	931.8	1170.6
275	45.40		0.01721	9.313	9.330	243.94	928.2	1172.1

v = specific volume, ft³/lb. h = enthalpy, Btu/lb.

TABLE C.1 (*cont.*)

Temp. Fahr. t	Absolute pressure Lb/in.² p	In. Hg 32°F	Specific volume Sat. Liquid v_f	Specific volume Evap. v_{fg}	Specific volume Sat. Vapor v_g	Enthalpy Sat. Liquid h_f	Enthalpy Evap. h_{fg}	Enthalpy Sat. Vapor h_g
280	49.20		0.01726	8.634	8.651	249.06	924.6	1173.7
285	53.25		0.01731	8.015	8.032	254.18	921.0	1175.2
290	57.55		0.01735	7.448	7.465	259.31	917.4	1176.7
295	62.13		0.01740	6.931	6.948	264.45	913.7	1178.2
300	67.01		0.01745	6.454	6.471	269.60	910.1	1179.7
305	72.18		0.01750	6.014	6.032	274.76	906.3	1181.1
310	77.68		0.01755	5.610	5.628	279.92	902.6	1182.5
315	83.50		0.01760	5.239	5.257	285.10	898.8	1183.9
320	89.65		0.01765	4.897	4.915	290.29	895.0	1185.3
325	96.16		0.01771	4.583	4.601	295.49	891.1	1186.6
330	103.03		0.01776	4.292	4.310	300.69	887.1	1187.8
335	110.31		0.01782	4.021	4.039	305.91	883.2	1189.1
340	117.99		0.01788	3.771	3.789	311.14	879.2	1190.3
345	126.10		0.01793	3.539	3.557	316.38	875.1	1191.5
350	134.62		0.01799	3.324	3.342	321.64	871.0	1192.6
355	143.58		0.01805	3.126	3.144	326.91	866.8	1193.7
360	153.01		0.01811	2.940	2.958	332.19	862.5	1194.7
365	162.93		0.01817	2.768	2.786	337.48	858.2	1195.7
370	173.33		0.01823	2.607	2.625	342.79	853.8	1196.6
375	184.23		0.01830	2.458	2.476	348.11	849.4	1197.5
380	195.70		0.01836	2.318	2.336	353.45	844.9	1198.4
385	207.71		0.01843	2.189	2.207	358.80	840.4	1199.2
390	220.29		0.01850	2.064	2.083	364.17	835.7	1199.9
395	233.47		0.01857	1.9512	1.9698	369.56	831.0	1200.6
400	247.25		0.01864	1.8446	1.8632	374.97	826.2	1201.2
405	261.67		0.01871	1.7445	1.7632	380.40	821.4	1201.8
410	276.72		0.01878	1.6508	1.6696	385.83	816.6	1202.4
415	292.44		0.01886	1.5630	1.5819	391.30	811.7	1203.0
420	308.82		0.01894	1.4806	1.4995	396.78	806.7	1203.5
425	325.91		0.01902	1.4031	1.4221	402.28	801.6	1203.9
430	343.71		0.01910	1.3303	1.3494	407.80	796.5	1204.3
435	362.27		0.01918	1.2617	1.2809	413.35	791.2	1204.6
440	381.59		0.01926	1.1973	1.2166	418.91	785.9	1204.8
445	401.70		0.01934	1.1367	1.1560	424.49	780.4	1204.9
450	422.61		0.01943	1.0796	1.0990	430.1ſ	774.9	1205.0
455	444.35		0.0195	1.0256	1.0451	435.74	769.3	1205.0
460	466.97		0.0196	0.9745	0.9941	441.42	763.6	1205.0
465	490.43		0.0197	0.9262	0.9459	447.10	757.8	1204.9
470	514.70		0.0198	0.8808	0.9006	452.84	751.9	1204.7
475	539.90		0.0199	0.8379	0.8578	458.59	745.9	1204.5
480	566.12		0.0200	0.7972	0.8172	464.37	739.8	1204.2
485	593.28		0.0201	0.7585	0.7786	470.18	733.6	1203.8
490	621.44		0.0202	0.7219	0.7421	476.01	727.3	1203.3
495	650.59		0.0203	0.6872	0.7075	481.90	720.8	1202.7
500	680.80		0.0204	0.6544	0.6748	487.80	714.2	1202.0
505	712.19		0.0206	0.6230	0.6436	493.8	707.5	1201.3
510	744.55		0.0207	0.5932	0.6139	499.8	700.6	1200.4
515	777.96		0.0208	0.5651	0.5859	505.8	693.6	1199.4
520	812.68		0.0209	0.5382	0.5591	511.9	686.5	1198.4
525	848.37		0.0210	0.5128	0.5338	518.0	679.2	1197.2
530	885.20		0.0212	0.4885	0.5097	524.2	671.9	1196.1
535	923.45		0.0213	0.4654	0.4867	530.4	664.4	1194.8
540	962.80		0.0214	0.4433	0.4647	536.6	656.7	1193.3

v = specific volume, ft³/lb. h = enthalpy, Btu/lb.

TABLE C.1 *(cont.)*

Temp. Fahr. t	Absolute pressure		Specific volume			Enthalpy		
	Lb/in.^{2} p	In. Hg 32°F	Sat. Liquid v_f	Evap. v_{fg}	Sat. Vapor v_g	Sat. Liquid h_f	Evap. h_{fg}	Sat. Vapor h_g
545	1003.6		0.0216	0.4222	0.4438	542.9	648.9	1191.8
550	1045.6		0.0218	0.4021	0.4239	549.3	640.9	1190.2
555	1088.8		0.0219	0.3830	0.4049	555.7	632.6	1188.3
560	1133.4		0.0221	0.3648	0.3869	562.2	624.1	1186.3
565	1179.3		0.0222	0.3472	0.3694	568.8	615.4	1184.2
570	1226.7		0.0224	0.3304	0.3528	575.4	606.5	1181.9
575	1275.7		0.0226	0.3143	0.3369	582.1	597.4	1179.5
580	1326.1		0.0228	0.2989	0.3217	588.9	588.1	1177.0
585	1378.1		0.0230	0.2840	0.3070	595.7	578.6	1174.3
590	1431.5		0.0232	0.2699	0.2931	602.6	568.8	1171.4
595	1486.5		0.0234	0.2563	0.2797	609.7	558.7	1168.4
600	1543.2		0.0236	0.2432	0.2668	616.8	548.4	1165.2
605	1601.5		0.0239	0.2306	0.2545	624.1	537.7	1161.8
610	1661.6		0.0241	0.2185	0.2426	631.5	526.6	1158.1
615	1723.4		0.0244	0.2068	0.2312	638.9	515.3	1154.2
620	1787.0		0.0247	0.1955	0.2202	646.5	503.7	1150.2
625	1852.4		0.0250	0.1845	0.2095	654.3	491.5	1145.8
630	1919.8		0.0253	0.1740	0.1993	662.2	478.8	1141.0
635	1989.0		0.0256	0.1638	0.1894	670.4	465.5	1135.9
640	2060.3		0.0260	0.1539	0.1799	678.7	452.0	1130.7
645	2133.5		0.0264	0.1441	0.1705	687.3	437.6	1124.9
650	2208.8		0.0268	0.1348	0.1616	696.0	422.7	1118.7
655	2286.4		0.0273	0.1256	0.1529	705.2	407.0	1112.2
660	2366.2		0.0278	0.1167	0.1445	714.4	390.5	1104.9
665	2448.0		0.0283	0.1079	0.1362	724.5	372.1	1096.6
670	2532.4		0.0290	0.0991	0.1281	734.6	353.3	1087.9
675	2619.2		0.0297	0.0904	0.1201	745.5	332.8	1078.3
680	2708.4		0.0305	0.0810	0.1115	757.2	310.0	1067.2
685	2800.4		0.0316	0.0716	0.1032	770.1	284.5	1054.6
690	2895.0		0.0328	0.0617	0.0945	784.2	254.9	1039.1
695	2992.7		0.0345	0.0511	0.0856	801.3	219.1	1020.4
700	3094.1		0.0369	0.0389	0.0758	823.9	171.7	995.6
705	3199.1		0.0440	0.0157	0.0597	870.2	77.6	947.8
705.34*	3206.2		0.0541	0	0.0541	910.3	0	910.3

*Critical temperature. v = specific volume, ft^3/lb. h = enthalpy, Btu/lb.

SOURCE: Combustion Engineering, Inc.

PHYSICAL PROPERTIES
OF VARIOUS ORGANIC
AND INORGANIC SUBSTANCES

General Sources of Data for Tables on the Physical Properties, Heat Capacities, and Thermodynamic Properties in Appendices D, E, and F

1. Kobe, Kenneth A., and Associates, "Thermochemistry of Petrochemicals," Reprint from *Petroleum Refiner*, Gulf Publishing Co., Houston, Tex., Jan. 1949–July 1958. (Enthalpy tables D.2–D. and heat capacities of several gases in Table E.1, Appendix E.)
2. Lange, N. A., *Handbook of Chemistry*, 12th ed., McGraw-Hill, New York, 1979.
3. Maxwell, J. B., *Data Book on Hydrocarbons*, Van Nostrand Reinhold, New York, 1950.
4. Perry, J. H., and C. H. Chilton, eds., *Chemical Engineers' Handbook*, 5th ed., McGraw-Hill, New York, 1973.
5. Rossini, Frederick D., et al., "Selected Values of Chemical Thermodynamic Properties," from *National Bureau of Standards Circular 500*, U.S. Government Printing Office, Washington, D.C., 1952.
6. Rossini, Frederick D., et al., "Selected Values of Physical and Thermodynamic Properties of Hydrocarbons and Related Compounds," American Petroleum Institute Research Project 44, 1953 and subsequent years.
7. Weast, Robert C., *Handbook of Chemistry and Physics*, 59th ed., CRC Press, West Palm Beach, Florida, 1979.

TABLE D.1 PHYSICAL PROPERTIES OF VARIOUS ORGANIC AND INORGANIC SUBSTANCES*

To convert to kcal/g mol multiply by 0.2390; to Btu/lb mol multiply by 430.2.

Compound	Formula	Formula Wt	Sp Gr	Melting Temp. (K)	$\Delta\hat{H}$ Fusion (kJ/g mol)	Normal b.p. (K)	$\Delta\hat{H}$ Vap. at b.p. (kJ/g mol)	T_c (°K)	p_c (atm)	$\hat{V}_c$ (cm³/g mol)	z_c
Acetaldehyde	$C_2H_4O_2$	44.05	$0.7831^{8/4°}$	149.5		293.2		461.0	57.1	171	0.200
Acetic acid	CH_3CHO	60.05	1.049	289.9	12.09	390.4	24.4	594.8	47.0		
Acetone	C_3H_6O	58.08	0.791	178.2	5.69	329.2	30.2	508.0	61.6	213	0.238
Acetylene	C_2H_2	26.04	0.9061(A)	191.7	3.7	191.7	17.5	309.5	37.2	113	0.274
Air			1.000					132.5			
Ammonia	NH_3	17.03	$0.817^{-79°}$ / 0.597(A)	195.40	5.653	239.73	23.35	405.5	111.3	72.5	0.243
Ammonium carbonate	$(NH_4)_2CO_3\cdot H_2O$	114.11				(decomposes at 331 K)					
Ammonium chloride	NH_4Cl	53.50	$2.53^{17°}$			(decomposes at 623 K)					
Ammonium nitrate	NH_4NO_3	80.05	$1.725^{25°}$	442.8	5.4	decomposes at 483.2 K					
Ammonium sulfate	$(NH_4)_2SO_4$	132.14	1.769	786		(decomposes at 786 K after melting)					
Aniline	C_6H_7N	93.12	1.022	266.9		457.4		699	52.4		
Benzaldehyde	C_6H_5CHO	106.12	1.046	247.16		452.16	38.40				
Benzene	C_6H_6	78.11	0.879	278.693	9.837	353.26	30.76	562.6	48.6	260	0.274
Benzoic acid	$C_7H_6O_2$	122.12	$1.316^{28/4°}$	395.4		523.0					
Benzyl alcohol	C_7H_8O	108.13	1.045	257.8		478.4					
Boron oxide	B_2O_3	69.64	1.85	723	22.0						
Bromine	Br_2	159.83	$3.119^{20°}$ / 5.87(A)	265.8	10.8	331.78	31.0	584	102	144	0.306
1, 2-Butadiene	C_4H_6	54.09	$0.6522^{20°}$	136.7		283.3		446			
1, 3-Butadiene	C_4H_6	54.09	0.621	164.1		268.6		425	42.7	221	0.271
Butane	$n\text{-}C_4H_{10}$	58.12	0.579	134.83	4.661	272.66	22.31	425.17	37.47	255	0.374
iso-Butane	$iso\text{-}C_4H_{10}$	58.12	0.557	113.56	4.540	261.43	21.29	408.1	36.0	263	0.283

1-Butene	C_4H_8	56.10	0.60	87.81	3.848	266.91	21.92	419.6	39.7	240	0.277
Butyl phthalate	*see Dibutyl phthalate*										
n-Butyric acid	$n\text{-}C_4H_8O_2$	88.10	0.958	267		437.1		628	52.0	290	0.293
iso-Butyric acid	$iso\text{-}C_4H_8O_2$	88.10	0.949	226		427.7		609			
Calcium arsenate	$Ca_3(AsO_4)_2$	398.06		1723							
Calcium carbide	Ca_2C_2	64.10	$2.22^{18°}$	2573							
Calcium carbonate	$CaCO_3$	100.09	2.93	(decomposes at 1098 K)							
Calcium chloride	$CaCl_2$	110.99	$2.152^{15°}$	1055	28.4						
	$CaCl_2 \cdot H_2O$	129.01									
	$CaCl_2 \cdot 2H_2O$	147.03									
	$CaCl_2 \cdot 6H_2O$	219.09	$1.78^{17°}$	303.4	37.3	(—6H$_2$O at 473 K)					
Calcium cyanamide	$CaCN_2$	80.11	2.29								
Calcium cyanide	$Ca(CN)_2$	92.12									
Calcium hydroxide	$Ca(OH)_2$	74.10	2.24	(—H$_2$O at 853 K)							
Calcium oxide	CaO	56.08	2.62	2873		3123					
Calcium phosphate	$Ca_3(PO_4)_2$	310.19	3.14	1943	50						
Calcium silicate	$CaSiO_3$	117.17	2.915	1803							
Calcium sulfate (gypsum)	$CaSO_4 \cdot 2H_2O$	172.18	2.32	(—1½H$_2$O at 301°K)	48.62						
Carbon	C	12.010	2.26	3873	46.0	4473					
Carbon dioxide	CO_2	44.01	1.53(A)	$217.0^{5.2\text{ atm}}$	8.32	(sublimes at 195 K)		304.2	72.9	94	0.275
Carbon disulfide	CS_2	76.14	$1.261^{22°/20°}$ 2.63(A)	161.1	4.39	319.41	26.8	552.0	78.0	170	0.293
Carbon monoxide	CO	28.01	0.968(A)	68.10	0.837	81.66	6.042	133.0	34.5	93	0.294
Carbon tetrachloride	CCl_4	153.84	1.595	250.3	2.5	349.9		556.4	45.0	276	0.272
Chlorine	Cl_2	70.91	2.49(A)	172.16	6.406	239.10	20.41	417.0	76.1	124	0.276

*Sources of data are listed at the beginning of Appendix D.

SP gr = 20°C/4°C unless specified. Sp gr for gas referred to air (A).

TABLE D.1 (*cont.*)

Compound	Formula	Formula Wt	Sp Gr	Melting Temp. (K)	$\Delta \hat{H}$ Fusion (kJ/g mol)	Normal b.p. (K)	$\Delta \hat{H}$ Vap. at b.p. (kJ/g mol)	T_c (°K)	p_c (atm)	$\hat{V}_c$ (cm³/g mol)	z_c
Chlorobenzene	C_6H_5Cl	112.56	1.107	228		405.26	36.5	632.4	44.6	308	0.265
Chloroform	$CHCl_3$	119.39	$1.4892^{0°}$	209.5		334.2		536.0	54.0	240	0.294
Chromium	Cr	52.01	7.1								
Copper	Cu	63.54	8.92	1356.2	13.0	2855	305				
Cumene	C_9H_{12}	120.19	0.862	177.125	7.1	425.56	37.5	636	31.0	440	0.260
Cupric sulfate	$CuSO_4$	159.61	$3.6051^{5°}$	(decomposes at 873 K)							
Cyclohexane	C_6H_{12}	84.16	0.779	279.83	2.677	353.90	30.1	553.7	40.4	308	0.274
Cyclopentane	C_5H_{10}	70.13	0.745	179.71	0.6088	322.42	27.30	511.8	44.55	260	0.27
Decane	$C_{10}H_{22}$	142.28	$0.730^{20°}$	243.3		447.0		619.0	20.8	602	0.2476
Dibutyl phthalate	$C_8H_{22}O_4$	278.34	$1.045^{21°}$			613					
Diethyl ether	$(C_2H_5)_2O$	74.12	$0.7082^{25°}$	156.86	7.301	307.76	26.05	467	35.6	281	0.261
Ethane	C_2H_6	30.07	1.049(A)	89.89	2.860	184.53	14.72	305.4	48.2	148	0.285
Ethanol	C_2H_6O	46.07	0.789	158.6	5.021	351.7	38.6	516.3	63.0	167	0.248
Ethyl acetate	$C_4H_8O_2$	88.10	0.901	189.4		350.2		523.1	37.8	286	0.252
Ethyl benzene	C_8H_{10}	106.16	0.867	178.185	9.163	409.35	36.0	619.7	37.0	360	0.260
Ethyl bromide	C_2H_5Br	108.98	1.460	154.1		311.4		504	61.5	215	0.320
Ethyl chloride	CH_3CH_2Cl	64.52	$0.9031^{0°}$	134.83	4.452	285.43	25	460.4	52.0	199	0.274
3-Ethyl hexane	C_8H_{18}	114.22	0.7169			391.69	34.3	567.0	26.4	466	0.264
Ethylene	C_2H_4	28.05	0.975(A)	103.97	3.351	169.45	13.54	283.1	50.5	124	0.270
Ethylene glycol	$C_2H_6O_2$	62.07	$1.1131^{9°}$	260	11.23	470.4	56.9				
Ferric oxide	Fe_2O_3	159.70	5.12	1833		(decomposes at 1833 K)					
Ferric sulfide	Fe_2S_3	207.90	4.3	(decomposes)		(decomposes)					
Ferrous sulfide	FeS	87.92	4.84	1466							
Formaldehyde	H_2CO	30.03	$0.815^{-20°}$	154.9		253.9	24.5				
Formic acid	CH_2O_2	46.03	1.220	281.46	12.7	373.7	22.3				
Glycerol	$C_3H_8O_3$	92.09	$1.260^{50°}$	291.36	18.30	563.2					
Helium	He	4.00	0.1368(A)	3.5	0.02	4.216	0.084	5.26	2.26	58	0.304
Heptane	C_7H_{16}	100.20	0.684	182.57	14.03	371.59	31.69	540.2	27.0	426	0.260

Compound	Formula	Mol. wt.	Sp. gr.								
Hexane	C_6H_{14}	86.17	0.659	177.84	13.03	341.90	28.85	507.9	29.9	368	0.264
Hydrogen	H_2	2.016	0.06948(A)	13.96	0.12	20.39	0.904	33.3	12.8	65	0.304
Hydrogen chloride	HCl	36.47	1.268(A)	158.94	1.99	188.11	16.15	324.6	81.5	87	0.266
Hydrogen fluoride	HF	20.01	1.15	238		293		503.2			
Hydrogen sulfide	H_2S	34.08	1.1895(A)	187.63	2.38	212.82	18.67	373.6	88.9	98	0.284
Iodine	I_2	253.8	$4.93^{20°}$	386.5	15	457.4		826.0			
Iron	Fe	55.85	7.7	1808		3073	353				
Iron oxide	Fe_3O_4	231.55	5.2	1867	138	(decomposes at 1867 K after melting)					
Lead	Pb	207.21	$11.337^{20°}$	600.6	5.10	2023	180				
Lead oxide	PbO	223.21	9.5	1159	11.7	1745	213				
Magnesium	Mg	24.32	1.74	923	9.2	1393	132				
Magnesium chloride	$MgCl_2$	95.23	$2.325^{25°}$	987	43.1	1691	137				
Magnesium hydroxide	$Mg(OH)_2$	58.34	2.4	(decomposes at 623 K)							
Magnesium oxide	MgO	40.32	3.65	3173	77.4	3873					
Mercury	Hg	200.61	$13.546^{20°}$								
Methane	CH_4	16.04	0.554(A)	90.68	0.941	111.67	8.180	190.7	45.8	99	0.290
Methanol	CH_3OH	32.04	0.792	175.26	3.17	337.9	35.3	513.2	78.5	118	0.222
Methyl acetate	$C_3H_6O_2$	74.08	0.933	174.3		330.3		506.7	46.3	228	0.254
Methyl amine	CH_5N	31.06	$0.699^{-11°}$	180.5		266.37^{58mm}		429.9	73.6		
Methyl chloride	CH_3Cl	50.49	1.785(A)	175.3		249		416.1	65.8	143	0.276
Methyl ethyl ketone	C_4H_8O	72.10	0.805	186.1		352.6					
Methyl cyclohexane	C_7H_{14}	98.18	0.769	146.58	6.751	374.10	31.7	572.2	34.32	344	0.251
Molybdenum	Mo	95.95	10.2								
Napthalene	$C_{10}H_8$	128.16	1.145	353.2		491.0					
Nickel	Ni	58.69	$8.90^{20°}$	1725		3173					
Nitric acid	HNO_3	63.02	1.502	231.56	10.47	359	30.30				
Nitrobenzene	$C_6H_5O_2N$	123.11	1.203	278.7		483.9					
Nitrogen	N_2	28.02	12.5(D)	63.15	0.720	77.34	5.577	126.2	33.5	90	0.291
Nitrogen dioxide	NO_2	46.01	1.448	263.86	7.334	294.46	14.73	431.0	100.0	82	0.232

*Sources of data are listed at the beginning of Appendix D.

SP gr = 20°C/4°C unless specified. Sp gr for gas referred to air (A).

TABLE D.1 (*cont.*)

Compound	Formula	Formula Wt	Sp Gr	Melting Temp. (K)	$\Delta\hat{H}$ Fusion (kJ/g mol)	Normal b.p. (K)	$\Delta\hat{H}$ Vap. at b.p. (kJ/g mol)	T_c (°K)	p_c (atm)	$\hat{V}_c$ (cm³/g mol)	z_c
Nitrogen oxide	NO	30.01	1.0367(A)	109.51	2.301	121.39	13.78	19.2	65.0	58	0.256
Nitrogen pentoxide	N_2O_5	108.02	1.63[18°]	303		320					
Nitrogen tetraoxide	N_2O_4	92	1.448[20°]	263.7		294.3		431.0	99.0		
Nitrogen trioxide	N_2O_3	76.02	1.447[2°]	171		276.5					
Nitrous oxide	N_2O	44.02	1.226[-89°] / 1.530(A)	182.1		184.4		309.5	71.7	96.3	0.272
n-Nonane	C_9H_{20}	128.25	0.718	219.4		423.8		595	23		
n-Octane	C_8H_{18}	114.22	0.703	216.2		398.7		595.0	22.5	543	0.250
Oxalic acid	$C_2H_2O_4$	90.04	1.90	(decomposes at 459 K)							
Oxygen	O_2	32.00	1.1053(A)	54.40	0.443	90.19	6.820	154.4	49.7	74	0.290
n-Pentane	C_5H_{12}	72.15	0.6301[8°]	143.49	8.393	309.23	25.77	469.8	33.3	311	0.269
iso-Pentane	iso-C_5H_{12}	72.15	0.6219[9°]	113.1		300.9		461.0	32.9	308	0.268
1-Pentane	C_5H_{10}	70.13	0.641	107.96	4.937	303.13		474	39.9		
Phenol	C_6H_5OH	94.11	1.071[25°]	315.66	11.43	454.56		692.1	60.5		
Phenyl hydrazine	$C_6H_8N_2$	108.14	1.0972[3°]	292.76	16.43	51.66					
Phosphoric acid	H_3PO_4	98.00	1.834[18°]	315.51	10.5	($-\frac{1}{2}H_2O$ at 486 K)					
Phosphorus (red)	P_4	123.90	2.20	863	81.17	863	41.84				
Phosphorus (white)	P_4	123.90	1.82	317.4	2.5	553	49.71				
Phosphorus pentoxide	P_2O_5	141.95	2.387	(sublimes at 523 K)							
Propane	C_3H_8	44.09	1.562(A)	85.47	3.524	231.09	18.77	369.9	42.0	200	0.277
Propene	C_3H_6	42.08	1.498(A)	87.91	3.002	225.46	18.42	365.1	45.4	181	0.274
Propionic acid	$C_3H_6O_2$	74.08	0.993	252.2		414.4		612.5	53.0		

Compound	Formula	Mol. wt.	Sp gr	Melting point (K)	Heat of fusion	Boiling point (K)	Heat of vaporization	T_c	P_c	V_c	Z_c
n-Propyl alcohol	C_3H_8O	60.09	0.804	146		370.2		536.7	49.95	220	0.251
iso-Propyl alcohol	C_3H_8O	60.09	0.785	183.5		355.4		508.8	53.0	219	0.278
n-Propyl benzene	C_9H_{12}	120.19	0.862	173.660	8.54	432.38	38.2	638.7	31.3	429	0.257
Silicon dioxide	SiO_2	60.09	2.25	1883	8.54	2503					
Sodium bisulfate	$NaHSO_4$	120.07	2.742	455							
Sodium carbonate·10H$_2$O (sal soda)	$Na_2CO_3 \cdot 10H_2O$	286.15	1.46	306.5		($-H_2O$ at 306.5 K)					
Sodium carbonate (soda ash)	Na_2CO_3	105.99	2.533	1127	33.4	(decomposes)					
Sodium chloride	$NaCl$	58.45	2.163	1081	28.5	1738	171				
Sodium cyanide	$NaCN$	49.01		835	16.7	1770	155				
Sodium hydroxide	$NaOH$	40.00	2.130	592	8.4	1663					
Sodium nitrate	$NaNO_3$	85.00	2.257	583	15.9	(decomposes at 653 K)					
Sodium nitrite	$NaNO_2$	69.00	$2.168^{0°}$	544		(decomposes at 593 K)					
Sodium sulfate	Na_2SO_4	142.05	2.698	1163	24.3						
Sodium sulfide	Na_2S	78.05	1.856	1223	6.7						
Sodium sulfite	Na_2SO_3	126.05	$2.633^{15°}$	(decomposes)							
Sodium thiosulfate	$Na_2S_2O_3$	158.11	1.667								
Sulfur (rhombic)	S_8	256.53	2.07	386	10.0	717.76	84				
Sulfur (monoclinic)	S_8	256.53	1.96	392	14.17	717.76	84				
Sulfur chloride (mono)	S_2Cl_2	135.05	1.687	193.0		411.2	36.0				
Sulfur dioxide	SO_2	64.07	2.264(A)	197.68	7.402	263.14	24.92	430.7	77.8	122	0.269
Sulfur trioxide	SO_3	80.07	2.75(A)	290.0	24.5	316.5	41.8	491.4	83.8	126	0.262
Sulfuric acid	H_2SO_4	98.08	$1.834^{18°}$	283.51	9.87	(decomposes at 613 K)					
Toluene	$C_6H_5CH_3$	92.13	0.866	178.169	6.619	383.78	33.5	593.9	40.3	318	0.263
Water	H_2O	18.016	$1.004^{4°}$	273.16	6.009	373.16	40.65	647.4	218.3	56	0.230
m-Xylene	C_8H_{10}	106.16	0.864	225.288	11.57	412.26	34.4	619	34.6	390	0.27
o-Xylene	C_8H_{10}	106.16	0.880	247.978	13.60	417.58	36.8	631.5	35.7	380	0.26
p-Xylene	C_8H_{10}	106.16	0.861	286.423	17.11	411.51	36.1	618	33.9	370	0.25
Zinc	Zn	65.38	7.140	692.7	6.673	1180	114.8				
Zinc sulfate	$ZnSO_4$	161.44	$3.74^{15°}$	(decomposes at 1013 K)							

*Sources of data are listed at the beginning of Appendix D.

SP gr = 20°C/4°C unless specified. Sp gr for gas referred to air (A).

TABLE D.2 ENTHALPIES OF PARAFFINIC HYDROCARBONS, C_1–C_6 (J/G MOL)

To convert to Btu/lb mol, multiply by 0.4306.

K	C_1	C_2	C_3	n-C_4	i-C_4	n-C_5	n-C_6
273	0						
291	630	912	1,264	1,709	1,658	2,125	2,545
298	879	1,277	1,771	2,394	2,328	2,976	3,563
300	950	1,383	1,919	2,592	2,522	3,222	3,858
400	4,740	7,305	10,292	13,773	13,623	17,108	20,463
500	9,100	14,476	20,685	27,442	27,325	34,020	40,622
600	14,054	22,869	32,777	43,362	43,312	53,638	64,011
700	19,585	32,342	46,417	61,186	61,220	75,604	90,123
800	25,652	42,718	61,337	80,600	80,767	99,495	118,532
900	32,204	53,931	77,404	101,378	101,754	125,101	148,866
1,000	39,204	65,814	94,432	123,428	123,971	152,213	181,041
1,100	46,567	78,408	112,340	146,607	147,234	180,665	214,764
1,200	54,308	91,504	131,042	170,707	171,418	210,246	249,868
1,300	62,383	105,143	150,331	195,727	196,480	240,872	286,143
1,400	70,709	119,202	170,205	221,375	222,212	272,378	323,465
1,500	79,244	133,678	190,581	247,650	248,571	304,595	361,539
1,600	88,031						
1,800	106,064						
2,000	124,725						
2,200	143,804						
2,500	173,050						

TABLE D.3 ENTHALPIES OF MONOOLEFINIC HYDROCARBONS, C_2–C_4 (J/G MOL)

To convert to Btu/lb mol, multiply by 0.4306.

K	Ethylene	Propylene	1-Butene	iso-Butene	cis-2-Butene	trans-2-Butene
273	0	0	0	0	0	0
291	753	1,104	1,536	1,538	1,354	1,520
298	1,054	1,548	2,154	2,154	1,895	2,125
300	1,125	1,665	2,313	2,322	2,020	2,263
400	6,008	8,882	12,455	12,367	11,070	12,112
500	11,890	17,572	24,765	24,468	22,346	23,995
600	18,648	27,719	38,911	38,425	35,614	37,668
700	26,158	39,049	54,643	53,889	50,542	53,011
800	34,329	51,379	71,755	70,793	66,985	69,705
900	43,053	64,642	89,997	88,826	84,684	87,654
1,000	52,258	78,742	109,286	107,947	103,470	106,692
1,100	61,923	93,470	129,494	123,846	123,260	126,649
1,200	71,964	108,825	150,456	148,866	143,845	147,402
1,300	82,341	124,683	172,087	170,414	165,184	168,824
1,400	92,968	141,000	194,304	188,363	187,150	190,874
1,500	103,888	157,736	217,065	215,183	209,702	213,467

TABLE D.4 ENTHALPIES OF NITROGEN AND SOME OF ITS OXIDES (J/G MOL)

To convert to Btu/lb mol, multiply by 0.4306.

K	N_2	NO	N_2O	NO_2	N_2O_4
273	0	0	0	0	0
291	524	537	681	658	1,384
298	728	746	951	917	1,937
300	786	801	9,660	985	2,083
400	3,695	3,785	13,740	4,865	10,543
500	6,644	6,811	18,179	9,070	19,915
600	9,627	9,895	22,919	13,564	30,124
700	12,652	13,054	27,924	18,305	
800	15,756	16,292	33,154	23,242	
900	18,961	19,597	38,601	28,334	
1,000	22,171	22,970	44,258	33,551	
1,100	25,472	26,392	50,115	38,869	
1,200	28,819	29,861	56,170	44,266	
1,300	32,216	33,371	62,425	49,731	
1,400	35,639	36,915	68,868	55,258	
1,500	39,145	40,488	75,504	60,826	
1,750	47,940	49,505			
2,000	56,902	58,634			
2,250	65,981	67,856			
2,500	75,060	77,127			

TABLE D.5 ENTHALPIES OF SULFUR COMPOUNDS (J/G MOL)

To convert to Btu/lb mol, multiply by 0.4306.

K	S_2	SO_2	SO_3	H_2S	CS_2
273	0	0	0	0	0
291	579	706	899	607	807
298	805	984	1,255	845	1,125
300	869	1,064	1,338	909	1,217
400	4,196	5,234	6,861	4,372	5,995
500	7,652	9,744	13,033	7,978	11,108
600	11,192	14,514	19,832	11,752	16,455
700	14,790	19,501	27,154	15,706	21,974
800	18,426	24,647	34,748	19,840	27,631
900	22,087	29,915	42,676	24,145	33,388
1,000	25,769	35,275	50,835	28,610	39,220
1,100	29,463	40,706	59,203	33,216	45,103
1,200	33,174	46,191	67,738	37,953	51,044
1,300	36,898	51,714	76,399	42,802	57,027
1,400	40,630	57,320		47,739	63,052
1,500	44,371	62,927		52,802	69,119
1,600	48,116	68,533		57,906	75,186
1,700	51,881	74,182		63,094	81,295
1,800	55,605	79,872		68,324	87,361
1,900	59,370	85,520			
2,000	63,136	91,253			
2,500	82,006	119,871			
3,000	100,959	148,657			

HEAT
CAPACITY EQUATIONS

TABLE E.1 HEAT CAPACITY EQUATIONS FOR ORGANIC AND INORGANIC COMPOUNDS*

Forms: (1) $C_p^\circ = a + b(T) + c(T)^2 + d(T)^3$;
(2) $C_p^\circ = a + b(T) + c(T)^{-2}$.
Units of C_p° are J/(g mol)(K or °C).
To convert to cal/(g mol)(K or °C) = Btu/(lb mol)(°R or °F), multiply by 0.2390.

Compound	Formula	Mol. Wt.	State	Form	T	a	$b \cdot 10^2$	$c \cdot 10^5$	$d \cdot 10^9$	Temp. Range (in T)
Acetone	CH_3COCH_3	58.08	g	1	°C	71.96	20.10	−12.78	34.76	0–1200
Acetylene	C_2H_2	26.04	g	1	°C	42.43	6.053	−5.033	18.20	0–1200
Air		29.0	g	1	°C	28.94	0.4147	0.3191	−1.965	0–1500
			g	1	K	28.09	0.1965	0.4799	−1.965	273–1800
Ammonia	NH_3	17.03	g	1	°C	35.15	2.954	0.4421	−6.686	0–1200
Ammonium sulfate	$(NH_4)_2SO_4$	132.15	c	1	K	215.9				275–328
Benzene	C_6H_6	78.11	l	1	K	62.55	23.4			279–350
			g	1	°C	74.06	32.95	−25.20	77.57	0–1200
Isobutane	C_4H_{10}	58.12	g	1	°C	89.46	30.13	−18.91	49.87	0–1200
n-Butane	C_4H_{10}	58.12	g	1	°C	92.30	27.88	−15.47	34.98	0–1200
Isobutene	C_4H_8	56.10	g	1	°C	82.88	25.64	−17.27	50.50	0–1200
Calcium carbide	CaC_2	64.10	c	2	K	68.62	1.19	-8.66×10^{10}	—	298–720
Calcium carbonate	$CaCO_3$	100.09	c	2	K	82.34	4.975	-12.87×10^{10}	—	273–1033
Calcium hydroxide	$Ca(OH)_2$	74.10	c	1	K	89.5				276–373
Calcium oxide	CaO	56.08	c	2	K	41.84	2.03	-4.52×10^{10}		273–1173
Carbon	C	12.01	c†	2	K	11.18	1.095	-4.891×10^{10}		273–1373
Carbon dioxide	CO_2	44.01	g	1	°C	36.11	4.233	−2.887	7.464	0–1500
Carbon monoxide	CO	28.01	g	1	°C	28.95	0.4110	0.3548	−2.220	0–1500
Carbon tetrachloride	CCl_4	153.84	l	1	K	93.39	12.98			273–343
Chlorine	Cl_2	70.91	g	1	°C	33.60	1.367	−1.607	6.473	0–1200
Copper	Cu	63.54	c	1	K	22.76	0.06117			273–1357
Cumene (Isopropyl benzene)	C_9H_{12}	120.19	g	1	°C	139.2	53.76	−39.79	120.5	0–1200
Cylohexane	C_6H_{12}	84.16	g	1	°C	94.140	49.62	−31.90	80.63	0–1200

†Graphite.

TABLE E.1 (*cont.*)

Compound	Formula	Mol. Wt.	State	Form	T	a	b · 10²	c · 10⁵	d · 10⁹	Temp. Range (in T)
Cylopentane	C_5H_{10}	70.13	g	1	°C	73.39	39.28	−25.54	68.66	0–1200
Ethane	C_2H_6	30.07	g	1	°C	49.37	13.92	−5.816	7.280	0–1200
Ethyl alcohol	C_2H_6O	46.07	l	1	°C	103.1				0
			l	1	°C	158.8				100
Ethylene	C_2H_4	28.05	g	1	°C	61.34	15.72	−8.749	19.83	0–1200
Ferric oxide	Fe_2O_3	159.70	g	1	°C	+40.75	11.47	−6.891	17.66	0–1200
			c	2	K	103.4	6.711	-17.72×10^{10}	—	273–1097
Formaldehyde	CH_2O	30.03	g	1	°C	34.28	4.268	0.0000	−8.694	0–1200
Helium	He	4.00	g	1	°C	20.8				All
n-Hexane	C_6H_{14}	86.17	l	1	°C	216.3				20–100
			g	1	°C	137.44	40.85	−23.92	57.66	0–1200
Hydrogen	H_2	2.016	g	1	°C	28.84	0.00765	0.3288	−0.8698	0–1500
Hydrogen bromide	HBr	80.92	g	1	°C	29.10	−0.0227	0.9887	−4.858	0–1200
Hydrogen chloride	HCl	36.47	g	1	°C	29.13	−0.1341	0.9715	−4.335	0–1200
Hydrogen cyanide	HCN	27.03	g	1	°C	35.3	2.908	1.092		0–1200
Hydrogen sulfide	H_2S	34.08	g	1	°C	33.51	1.547	0.3012	−3.292	0–1500
Magnesium chloride	$MgCl_2$	95.23	c	1	K	72.4	1.58			273–991
Magnesium oxide	MgO	40.32	c	2	K	45.44	0.5008	-8.732×10^{10}		273–2073
Methane	CH_4	16.04	g	1	°C	34.31	5.469	0.3661	−11.00	0–1200
			g	1	K	19.87	5.021	1.268	−11.00	273–1500
Methyl alcohol	CH_3OH	32.04	l	1	°C	75.86				0
						82.59				40
			g	1	°C	42.93	8.301	−1.87	−8.03	0–700

Name	Formula	Mol. wt.	State		Units					Range
Methyl cyclohexane	C_7H_{14}	98.18	g	1	°C	121.3	56.53	−37.72	100.8	0–1200
Methyl cyclopentane	C_6H_{12}	84.16	g	1	°C	98.83	45.857	−30.44	83.81	0–1200
Nitric acid	HNO_3	63.02	l	1	°C	110.0				25
Nitric oxide	NO	30.01	g	1	°C	29.50	0.8188	−0.2925	0.3652	0–3500
Nitrogen	N_2	28.02	g	1	°C	29.00	0.2199	0.5723	−2.871	0–1500
Nitrogen dioxide	NO_2	46.01	g	1	°C	36.07	3.97	−2.88	7.87	0–1200
Nitrogen tetraoxide	N_2O_4	92.02	g	1	°C	75.7	12.5	−11.3		0–300
Nitrous oxide	N_2O	44.02	g	1	°C	37.66	4.151	−2.694	10.57	0–1200
Oxygen	O_2	32.00	g	1	°C	29.10	1.158	−0.6076	1.311	0–1500
n-Pentane	C_5H_{12}	72.15	l	1	°C	155.4	43.68			0–36
			g	1	°C	114.8	34.09	−18.99	42.26	0–1200
Propane	C_3H_8	44.09	g	1	°C	68.032	22.59	−13.11	31.71	0–1200
Propylene	C_3H_6	42.08	g	1	°C	59.580	17.71	−10.17	24.60	0–1200
Sodium carbonate	Na_2CO_3	105.99	c	1	K	121				288–371
Sodium carbonate ·10H₂O	Na_2CO_3 $\cdot 10H_2O$	286.15	c	1	K	535.6				298
Sulfur	S	32.07	c‡	1	K	15.2	2.68			273–368
			c§	1	K	18.5	1.84			368–392
Sulfuric acid	H_2SO_4	98.08	l	1	°C	139.1	15.59			10–45
Sulfur dioxide	SO_2	64.07	g	1	°C	38.91	3.904	−3.104	8.606	0–1500
Sulfur trioxide	SO_3	80.07	g	1	°C	48.50	9.188	−8.540	32.40	0–1000
Toluene	C_7H_8	92.13	l	1	°C	148.8				0
			l	1	°C	181.2				100
Water	H_2O	18.016	g	1	°C	94.18	38.00	−27.86	80.33	0–1200
			l	1	°C	75.4				0–100
			g	1	°C	33.46	0.6880	0.7604	−3.593	0–1500

‡Rhombic.   §Monoclinic.

TABLE E.1 (cont.)

The following relations have the units of cal/(g mol)(°C or K) = Btu/(lb mol)(°F or °R), and T is in °F or °R.

Compound	Formula	Wt.	Mol. State	Form	T	a	$b \cdot 10^2$	$c \cdot 10^5$	$d \cdot 10^9$	Temp. Range (in T)
Acetylene	C_2H_2	26.04	g	1	°F	9.89	0.8273	−0.3783	0.7457	32–2200
Air			g	1	°F	6.900	0.02884	0.02429	−0.08052	32–2700
			g	1	°R	6.713	0.02609	0.03540	−0.08052	492–3200
Ammonia	NH_3	17.03	g	1	°F	8.2765	0.39006	0.035245	−0.2740	32–2200
Nitrogen	N_2	28.02	g	1	°F	6.895	0.07624	−0.007009		
Oxygen	O_2	32.00	g	1	°F	7.104	0.07851	−0.005528		
Carbon dioxide	CO_2	44.01	g	1	°F	8.448	0.5757	−0.2159	0.3059	0–3500
Carbon monoxide	CO	28.01	g	1	°F	6.865	0.08024	−0.007367		
Cyclohexane	C_6H_{12}	84.16	l	1	°F	37.05				50
					°F	41.26				150
			g	1	°F	30.84				50
Cyclopentane	C_5H_{10}	70.13	l	1	°F	32.95				100
Toluene	$C_6H_5 \cdot CH_3$		l	1	°F	20.869	5.293	−2.086	3.929	32–2200

*Sources of data are listed at the beginning of Appendix D.

Appendix F

HEATS
OF FORMATION
AND COMBUSTION

**TABLE F.1 HEATS OF FORMATION AND HEATS OF COMBUSTION
OF COMPOUNDS AT 25 C*†**

Standard states of products for $\Delta \hat{H}_c^\circ$ are $CO_2(g)$, $H_2O(l)$, $N_2(g)$, $SO_2(g)$, and $HCl(aq)$. To convert to Btu/lb mol, multiply by 430.6.

Compound	Formula	State	$-\Delta \hat{H}_f^\circ$ (kJ/g mol)	$-\Delta \hat{H}_c^\circ$ (kJ/g mol)
Acetic acid	CH_3COOH	l	409.19	871.69
		g		919.73
Acetaldethyde	CH_3CHO	g	166.4	1192.36
Acetone	C_3H_6O	aq, 200	410.03	
		g	216.69	1821.38
Acetylene	C_2H_2	g	−226.75	1299.61
Ammonia	NH_3	l	67.20	
		g	46.191	382.58
Ammonium carbonate	$(NH_4)_2CO_3$	c		
		aq	941.86	
Ammonium chloride	NH_4Cl	c	315.4	

Compound	Formula	State	$-\Delta \hat{H}_f^{\circ}$ (kJ/g mol)	$-\Delta \hat{H}_c^{\circ}$ (kJ/g mol)
Ammonium hydroxide	NH_4OH	aq	366.5	
Ammonium nitrate	NH_4NO_3	c	366.1	
		aq	339.4	
Ammonium sulfate	$(NH_4)_2SO_4$	c	1179.3	
		aq	1173.1	
Benzaldehyde	C_6H_5CHO	l	88.83	
		g	40.0	
Benzene	C_6H_6	l	−48.66	3267.6
		g	−82.927	3301.5
Boron oxide	B_2O_3	c	1263	
		l	1245.2	
Bromine	Br_2	l	0	
		g	−30.7	
n-Butane	C_4H_{10}	l	147.6	2855.6
		g	124.73	2878.52
Isobutane	C_4H_{10}	l	158.5	2849.0
		g	134.5	2868.8
1-Butene	C_4H_8	g	−1.172	2718.58
Calcium arsenate	$Ca_3(AsO_4)_2$	c	3330.5	
Calcium carbide	CaC_2	c	62.7	
Calcium carbonate	$CaCO_3$	c	1206.9	
Calcium chloride	$CaCl_2$	c	794.9	
Calcium cyanamide	$CaCN_2$	c	352	
Calcium hydroxide	$Ca(OH)_2$	c	986.59	
Calcium oxide	CaO	c	635.6	
Calcium phosphate	$Ca_3(PO_4)_2$	c	4137.6	
Calcium silicate	$CaSiO_3$	c	1584	
Calcium sulfate	$CaSO_4$	c	1432.7	
		aq	1450.5	
Calcium sulfate (gypsum)	$CaSO_4 \cdot 2H_2O$	c	2021.1	
Carbon	C	c	0	393.51
		Graphite (β)		
Carbon dioxide	CO_2	g	393.51	
		l	412.92	
Carbon disulfide	CS_2	l	−87.86	1075.2
		g	−115.3	1102.6
Carbon monoxide	CO	g	110.52	282.99
Carbon tetrachloride	CCl_4	l	139.5	352.2
		g	106.69	384.9
Chloroethane	C_2H_5Cl	g	105.0	1421.1
Cumene (isopropylbenzene)	$C_6H_5CH(CH_3)_2$	l	41.20	5215.44
		g	−3.93	5260.59
Cupric sulfate	$CuSO_4$	c	769.86	
		aq	843.12	
Cyclohexane	C_6H_{12}	l	156.2	3919.9
		g	123.1	3953.0

Compound	Formula	Mol. wt	State	$-\Delta \hat{H}_f^\circ$ (kJ/g mol)	$-\Delta \hat{H}_c^\circ$ (kJ/g mol)
Cyclopentane	C_5H_{10}		l	105.8	3290.9
			g	77.23	3319.5
Ethane	C_2H_6		g	84.667	1559.9
Ethyl alcohol	C_2H_5OH		l	277.63	1366.91
			g	235.31	1409.25
Ethyl benzene	$C_6H_5 \cdot C_2H_5$		l	12.46	4564.87
			g	−29.79	4607.13
Ethyl chloride	C_2H_5Cl		g	105	
Ethylene	C_2H_4		g	−52.283	1410.99
Ethylene chloride	C_2H_3Cl		g	−31.38	1271.5
3-Ethyl hexane	C_8H_{18}		l	250.5	5470.12
			g	210.9	5509.78
Ferric chloride	$FeCl_3$		c	403.34	
Ferric oxide	Fe_2O_3		c	822.156	
Ferric sulfide	FeS_2	See iron sulfide			
Ferrosoferric oxide	Fe_3O_4		c	1116.7	
Ferrous chloride	$FeCl_2$		c	342.67	303.76
Ferrous oxide	FeO		c	267	
Ferrous sulfide	FeS		c	95.06	
Formaldehyde	H_2CO		g	115.89	563.46
n-Heptane	C_7H_{16}		l	224.4	4816.91
			g	187.8	4853.48
n-Hexane	C_6H_{14}		l	198.8	4163.1
			g	167.2	4194.753
Hydrogen	H_2		g	0	285.84
Hydrogen bromide	HBr		g	36.23	
Hydrogen chloride	HCl		g	92.312	
Hydrogen cyanide	HCN		g	−130.54	
Hydrogen sulfide	H_2S		g	20.15	562.589
Iron sulfide	FeS_2		c	177.9	
Lead oxide	PbO		c	219.2	
Magnesium chloride	$MgCl_2$		c	641.83	
Magnesium hydroxide	$Mg(OH)_2$		c	924.66	
Magnesium oxide	MgO	40.32	c	601.83	
Methane	CH_4	16.041	g	74.84	890.4
Methyl alcohol	CH_3OH	32.042	l	238.64	726.55
			g	201.25	763.96
Methyl chloride	CH_3Cl	50.49	g	81.923	766.63†
Methyl cyclohexane	C_7H_{14}	98.182	l	190.2	4565.29
			g	154.8	4600.68
Methyl cyclopentane	C_6H_{12}	84.156	l	138.4	3937.7
			g	106.7	3969.4
Nitric acid	HNO_3	63.02	l	173.23	
			aq	206.57	
Nitric oxide	NO	30.01	g	−90.374	
Nitrogen dioxide	NO_2	46.01	g	−33.85	

Compound	Formula	Mol. wt	State	$-\Delta \hat{H}_f^\circ$ (kJ/g mol)	$-\Delta \hat{H}_c^\circ$ (kJ/g mol)
Nitrous oxide	N_2O	44.02	g	−81.55	
n-Pentane	C_5H_{12}	72.15	l	173.1	3509.5
			g	146.4	3536.15
Phosphoric acid	H_3PO_4	98.00	c	1281	
			aq (1H$_2$O)	1278	
Phosphorus	P_4	123.90	c	0	
Phosphorus pentoxide	P_2O_5	141.95	c	1506	
Propane	C_3H_8	44.09	l	119.84	2204.0
			g	103.85	2220.0
Propene	C_3H_6	42.078	g	−20.41	2058.47
n-Propyl alcohol	C_3H_8O	60.09	g	255	2068.6
n-Propylbenzene	$C_6H_5 \cdot CH_2 \cdot C_2H_5$	120.19	l	38.40	5218.2
			g	−7.824	5264.5
Silicon dioxide	SiO_2	60.09	c	851.0	
Sodium bicarbonate	$NaHCO_3$	84.01	c	945.6	
Sodium bisulfate	$NaHSO_4$	120.07	c	1126	
Sodium carbonate	Na_2CO_3	105.99	c	1130	
Sodium chloride	$NaCl$	58.45	c	411.00	
Sodium cyanide	$NaCN$	49.01	c	89.79	
Sodium nitrate	$NaNO_3$	85.00	c	466.68	
Sodium nitrite	$NaNO_2$	69.00	c	359	
Sodium sulfate	Na_2SO_4	142.05	c	1384.5	
Sodium sulfide	Na_2S	78.05	c	373	
Sodium sulfite	Na_2SO_3	126.05	c	1090	
Sodium thiosulfate	$Na_2S_2O_3$	158.11	c	1117	
Sulfur	S	32.07	c (rhombic)	0	
			c (monoclinic)	−0.297	
Sulfur chloride	S_2Cl_2	135.05	l	60.3	
Sulfur dioxide	SO_2	64.066	g	296.90	
Sulfur trioxide	SO_3	80.066	g	395.18	
Sulfuric acid	H_2SO_4	98.08	l	811.32	
			aq	907.51	
Toluene	$C_6H_5 \cdot CH_3$	92.13	l	−11.99	3909.9
			g	−50.000	3947.9
Water	H_2O	18.016	l	285.840	
			g	241.826	
m-Xylene	$C_6H_4(CH_3)_2$	106.16	l	25.42	4551.86
			g	−17.24	4594.53
o-Xylene	$C_6H_4(CH_3)_2$	106.16	l	24.44	4552.86
			g	−19.00	4596.29
p-Xylene	$C_6H_4(CH_3)_2$	106.16	l	24.43	4552.86
			g	−17.95	4595.25
Zinc sulfate	$ZnSO_4$	161.45	c	978.55	
			aq	1059.93	

*Sources of data are given at the beginning of Appendix D, References 1, 4, and 5.
†Standard state HCl(g).

VAPOR PRESSURES

TABLE G.1 VAPOR PRESSURES OF VARIOUS SUBSTANCES

Antoine equation:

$$\ln (p^*) = A - \frac{B}{C + T}$$

where p^* = vapor pressure, mm Hg
T = temperature, K
A, B, C = constants

Name	Formula	Range (K)	A	B	C
Acetic acid	$C_2H_4O_2$	290–430	16.8080	3405.57	−56.34
Acetone	C_3H_6O	241–350	16.6513	2940.46	−35.93
Ammonia	NH_3	179–261	16.9481	2132.50	−32.98
Benzene	C_6H_6	280–377	15.9008	2788.51	−52.36
Carbon disulfide	CS_2	288–342	15.9844	2690.85	−31.62
Carbon tetrachloride	CCl_4	253–374	15.8742	2808.19	−45.99
Chloroform	$CHCl_3$	260–370	15.9732	2696.79	−46.16
Cyclohexane	C_6H_{12}	280–380	15.7527	2766.63	−50.50
Ethyl acetate	$C_4H_8O_2$	260–385	16.1516	2790.50	−57.15
Ethyl alcohol	C_2H_6O	270–369	16.9119	3803.98	−41.68
Ethyl bromide	$C_2H_5B_r$	226–333	15.9338	2511.68	−41.44
n-Heptane	C_7H_{16}	270–400	15.8737	2911.32	−56.51
n-Hexane	C_6H_{14}	245–370	15.8366	2697.55	−48.78
Methyl alcohol	CH_4O	257–364	18.5875	3626.55	−34.29
n-Pentane	C_5H_{12}	220–330	15.8333	2477.07	−39.94
Sulfur dioxide	SO_2	195–280	16.7680	2302.35	−35.97
Toluene	$C_6H_5CH_3$	280–410	16.0137	3096.52	−53.67
Water	H_2O	284–441	18.3036	3816.44	−46.13

SOURCE: R. C. Reid, J. M. Prausnitz, and T. K. Sherwood, *The Properties of Gases and Liquids*, 3rd ed., McGraw-Hill, New York, 1977, Appendix A.

Appendix H

HEATS
OF SOLUTION
AND DILUTION

TABLE H.1 INTEGRAL HEATS OF SOLUTION AND DILUTION AT 25°C

Formula	Description	State	$-\Delta\hat{H}_f^\circ$ (kJ/g mol)	$-\Delta\hat{H}_{soln}^\circ$ (kJ/g mol)	$-\Delta\hat{H}_{dil}^\circ$ (kJ/g mol)
HCl		g	92.311		
	in 1 H_2O	aq	118.536	26.225	26.225
	2 H_2O	aq	141.130	48.818	22.593
	3 H_2O	aq	149.163	56.852	8.033
	4 H_2O	aq	153.515	61.203	4.351
	5 H_2O	aq	156.360	64.048	2.845
	10 H_2O	aq	161.799	69.487	5.439
	20 H_2O	aq	164.088	71.776	2.288
	30 H_2O	aq	164.903	72.592	.815
	40 H_2O	aq	165.334	73.002	.410
	50 H_2O	aq	165.590	73.257	.255
	100 H_2O	aq	166.159	73.847	.589
	200 H_2O	aq	166.514	74.203	.355
	300 H_2O	aq	166.678	74.366	.163
	400 H_2O	aq	166.770	74.458	.092
	500 H_2O	aq	166.832	74.521	.062

Formula	Description	State	$-\Delta\hat{H}_f^\circ$ (kJ/g mol)	$-\Delta\hat{H}_{soln}^\circ$ (kJ/g mol)	$-\Delta\hat{H}_{dil}^\circ$ (kJ/g mol)
	700 H_2O	aq	166.920	74.609	.087
	1,000 H_2O	aq	166.995	74.684	.075
	2,000 H_2O	aq	167.134	74.822	.138
	3,000 H_2O	aq	167.192	74.881	.058
	4,000 H_2O	aq	167.226	74.914	.451
	5,000 H_2O	aq	167.242	74.931	.016
	7,000 H_2O	aq	167.284	74.973	.041
	10,000 H_2O	aq	167.305	74.994	.020
	20,000 H_2O	aq	167.351	75.040	.046
	50,000 H_2O	aq	167.389	75.077	.037
	100,000 H_2O	aq	167.410	75.098	.020
	∞ H_2O	aq	167.456	75.144	.046
NaOH	crystalline II		426.726		
	in 3 H_2O	aq	455.612	28.869	28.869
	4 H_2O	aq	461.156	34.434	5.564
	5 H_2O	aq	464.486	37.739	3.305
	10 H_2O	aq	469.227	42.509	4.769
	20 H_2O	aq	469.591	42.844	.334
	30 H_2O	aq	469.457	42.718	.125
	40 H_2O	aq	469.340	42.593	.125
	50 H_2O	aq	469.252	42.509	.083
	100 H_2O	aq	469.059	42.342	.167
	200 H_2O	aq	469.026	42.258	.083
	300 H_2O	aq	469.047	42.300	.041
	500 H_2O	aq	469.097	42.383	.083
	1,000 H_2O	aq	469.189	42.467	.083
	2,000 H_2O	aq	469.285	42.551	.083
	5,000 H_2O	aq	469.386	42.676	.125
	10,000 H_2O	aq	469.448	42.718	.041
	50,000 H_2O	aq	469.528	42.802	.083
	∞ H_2O	aq	469.595	42.886	.083
H_2SO_4		liq	811.319		
	in 0.5 H_2O	aq	827.051	15.731	15.731
	1.0 H_2O	aq	839.394	28.074	12.343
	1.5 H_2O	aq	848.222	36.902	8.823
	2 H_2O	aq	853.243	41.923	5.021
	3 H_2O	aq	860.314	48.994	7.071
	4 H_2O	aq	865.376	54.057	5.063
	5 H_2O	aq	869.351	58.032	3.975
	10 H_2O	aq	878.347	67.027	8.995
	25 H_2O	aq	883.618	72.299	5.272
	50 H_2O	aq	884.664	73.345	1.046
	100 H_2O	aq	885.292	73.973	0.628
	500 H_2O	aq	888.054	76.734	2.761
	1,000 H_2O	aq	889.894	78.575	1.841
	5,000 H_2O	aq	895.752	84.433	5.858
	10,000 H_2O	aq	898.388	87.069	2.636
	100,000 H_2O	aq	904.957	93.637	6.568
	500,000 H_2O	aq	906.630	95.311	1.674
	∞ H_2O	aq	907.509	96.190	0.879

SOURCE: F. D. Rossini et al., "Selected Values of Chem. Thermo. Properties," *Natl. Bur. Std. Circ. 500*, U.S. Government Printing Office, Washington, D.C., 1952.

ENTHALPY–
CONCENTRATION DATA

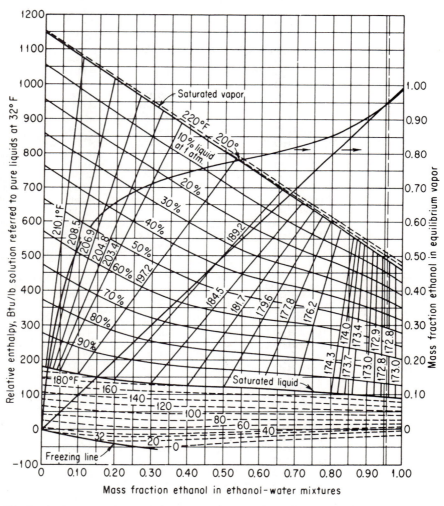

Fig. I.1 Enthalpy–composition diagram for the ethanol–water system, showing liquid and vapor phases in equilibrium at 1 atm.

TABLE I.1 ENTHALPY–CONCENTRATION DATA FOR THE SINGLE-PHASE LIQUID REGION AND ALSO THE SATURATED VAPOR OF THE ACETIC ACID–WATER SYSTEM AT 1 ATMOSPHERE

Reference state: Liquid water at 32°F and 1 atm; solid acid at 32°F and 1 atm.

| Liquid or Vapor | | Enthalpy—Btu/lb Liquid Solution | | | | | | Enthalpy |
Mole Fraction Water	Weight Fraction Water	20°C 68°F	40°C 104°F	60°C 140°F	80°C 176°F	100°C 212°F	Satu-rated Liquid	Saturated Vapor Btu/lb
0.00	0.00	93.54	111.4	129.9	149.1	169.0	187.5	361 8
0.05	0.01555	93.96	112.2	130.9	150.4	170.6	186.9	
0.10	0.03225	93.82	112.3	131.5	151.3	172.0	186.5	374.6
0.20	0.0698	92.61	111.9	131.7	152.2	173.8	185.2	395.3
0.30	0.1140	90.60	110.7	131.3	152.6	175.0	183.8	423.7
0.40	0.1667	87.84	108.9	130.6	152.9	176.1	182.9	461.4
0.50	0.231	83.96	106.3	129.1	152.7	177.1	182.4	510.5
0.55	0.268	81.48	104.5	128.1	152.5	177.5	182.3	
0.60	0.3105	78.53	102.5	126.9	152.1	178.0	182.0	573.4
0.65	0.358	75.36	100.2	125.5	151.6	178.3	181.9	
0.70	0.412	71.72	97.71	123.9	151.1	178.8	181.7	656.0
0.75	0.474	67.59	94.73	122.2	149.9	179.3	182.1	
0.80	0.545	62.88	91.43	120.2	149.7	179.9	181.6	767.3
0.85	0.630	57.44	87.56	117.8	148.9	180.5	181.6	
0.90	0.730	51.03	83.02	115.1	147.8	180.8	181.7	921.6
0.95	0.851	43.74	77.65	111.7	146.4	181.2	181.5	
1.00	1.00	36.06	71.91	107.7	143.9	180.1	180.1	1150.4

SOURCE: Data calculated from miscellaneous literature sources and smoothed.

TABLE I.2 VAPOR–LIQUID EQUILIBRIUM DATA FOR THE ACETIC ACID–WATER SYSTEM; PRESSURE = 1 ATMOSPHERE

x Mole Fraction Water in the Liquid	y Mole Fraction Water in the Vapor
0.020	0.035
0.040	0.069
0.060	0.103
0.080	0.135
0.100	0.165
0.200	0.303
0.300	0.425
0.400	0.531
0.500	0.627
0.600	0.715
0.700	0.796
0.800	0.865
0.900	0.929
0.940	0.957
0.980	0.985

SOURCE: Data from L. W. Cornell and R. E. Montonna, *Ind. Eng. Chem.*, v. 25, pp. 1331–1335 (1933).

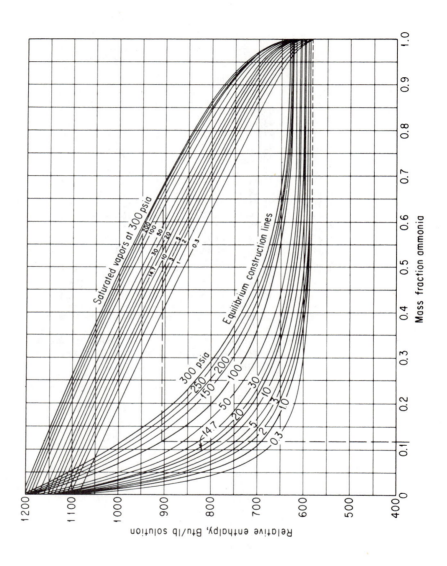

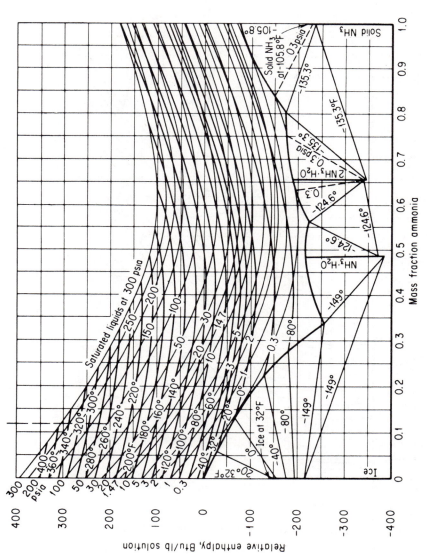

Reference states: water at 32°F and liquid ammonia at −40°F. To determine equilibrium compositions, erect a vertical line from any liquid composition at its saturation or boiling point, and locate its intersection with the appropriate equilibrium construction line. A horizontal line from this intersection will intersect the appropriate saturated vapor line at the desired equilibrium vapor composition.

Fig. I.2 Enthalpy–concentration chart for NH_3–H_2O.

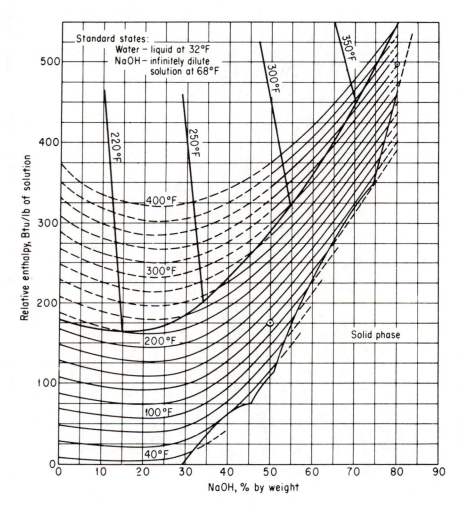

Fig. I.3 Enthalpy–concentration chart for sodium hydroxide–water.

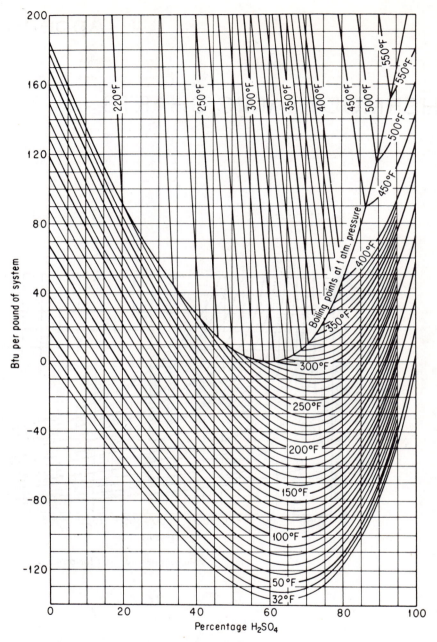

Fig. I.4 Enthalpy-concentration of sulfuric acid–water system relative to pure components. (Water and H_2SO_4 at 32°F and own vapor pressure.) (Data from International Critical Tables, © 1943 O. A. Hougen and K. M. Watson.)

Appendix J

THERMODYNAMIC CHARTS

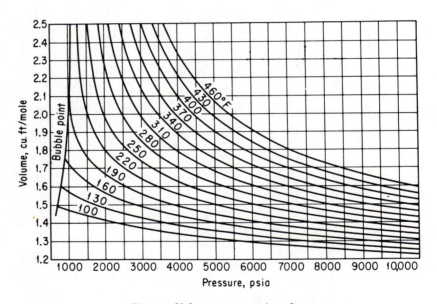

Fig. J.1 Volume–pressure chart for propane.

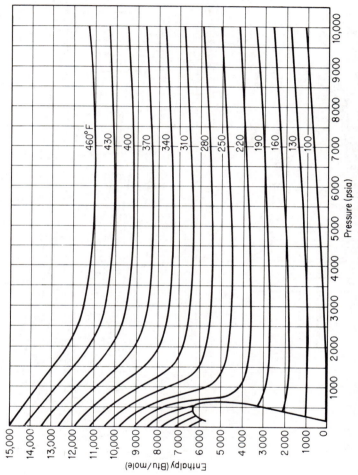

Fig. J.2 Enthalpy–pressure chart for propane.

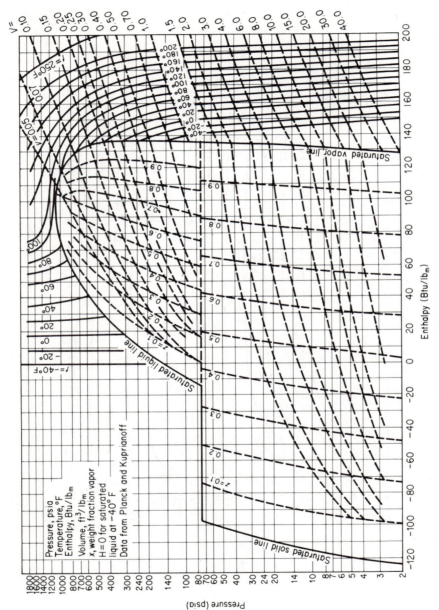

Fig. J.3 Pressure–enthalpy chart for carbon dioxide.

PHYSICAL PROPERTIES
OF PETROLEUM FRACTIONS

In the early 1930s, tests were developed which characterized petroleum oils and petroleum fractions, so that various physical characteristics of petroleum products could be related to these tests. Details of the tests can be found in *Petroleum Products and Lubricants*, an annual publication of the Committee D-2 of the American Society for Testing Materials.[1] These tests are not scientifically exact, and hence the procedure used in the tests must be followed faithfully if reliable results are to be obtained. However, the tests have been adopted because they are quite easy to perform in the ordinary laboratory and because the properties of petroleum fractions can be predicted from the results. The specifications for fuels, oils, and so on, are set out in terms of these tests plus many other properties, such as the flashpoint, the percent sulfur, and the viscosity.

Over the years various phases of the initial work have been extended and development of a new characterization scheme using the pseudocompound approach

[1]Report of Committee D-2, ASTM, Philadelphia, annually.

is evolving. Daubert[2] summarizes the traditional and new methods insofar as predicting molecular weights, pseudocritical temperature and pressure, acentric factor, and characterization factors.

In this appendix we present the results of the work of Smith and Watson and associates,[3-5] who related petroleum properties to a factor known as the *characterization factor* (sometimes called the *UOP characterization factor*). It is defined as

$$K = \frac{(T_B)^{1/3}}{S}$$

where $K = UOP$ characterization factor
  $T =$ cubic average boiling point, °R
  $S =$ specific gravity at 60°F

This factor has been related to many of the other simple tests and properties of petroleum fractions, such as viscosity, molecular weight, critical temperature, and percentage of hydrogen, so that it is quite easy to estimate the factor for any particular sample. Furthermore, tables of the UOP characterization factor are available for a wide variety of common types of petroleum fractions as shown in Table K.1 for typical liquids.

TABLE K.1   TYPICAL UOP CHARACTERIZATION FACTORS

Type of Stock	K	Type of Stock	K
Pennsylvania crude	12.2–12.5	Propane	14.7
Mid-Continent crude	11.8–12.0	Hexane	12.8
Gulf Coast crude	11.0–11.8	Octane	12.7
East Texas crude	11.9	Natural gasoline	12.7–12.8
California crude	10.8–11.9	Light gas oil	10.5
Benzene	9.5	Kerosene	10.5–11.5

Van Winkle[6] discusses the relationships among the volumetric average boiling point, the molal average boiling point, the cubic average boiling point, the weight average boiling point, and the mean average boiling point, and illustrates how the K and other properties of petroleum fractions can be evaluated from experimental data. In Table K.2 are shown the source, boiling-point basis, and any special limitations of the various charts in this appendix.

[2]T. E. Daubert, "Property Predictions," *Hydrocarbon Proc.*, pp. 107–110 (March 1980).

[3]R. L. Smith and K. M. Watson, *Ind. Eng. Chem.*, v. 29, p. 1408 (1937).

[4]K. M. Watson and E. F. Nelson, *Ind. Eng. Chem.*, v. 25, p. 880 (1933).

[5]K. M. Watson, E. F. Nelson, and G. B. Murphy, *Ind. Eng. Chem.*, v. 27, p. 1460 (1935).

[6]M. Van Winkle, *Petrol. Refiner*, v. 34, pp. 136–138 (June 1955).

TABLE K.2 INFORMATION CONCERNING CHARTS IN APPENDIX K

1. *Specific heats of hydrocarbon liquids*
   Source: J. B. Maxwell, *Data Book on Hydrocarbons*, Van Nostrand Reinhold, New York, 1950, p. 93 (original from M. W. Kellogg Co.).
   Description: A chart of C_p (0.4 to 0.8) vs. t (0 to 1000°F) for petroleum fractions from 0 to 120°API.
   Boiling-point basis: Volumetric average boiling point, which is equal to graphical integration of the differential ASTM distillation curve (Van Winkle's "exact method").
   Limitations: This chart is not valid at temperatures within 50°F of the pseudocritical temperatures.
2. *Vapor pressure of hydrocarbons*
   Source: Maxwell, *Data Book on Hydrocarbons*, p. 42.
   Description: Vapor pressure (0.002 to 100 atm) vs. temperature (50 to 1200°F) for hydrocarbons with normal boiling point of 100 to 1200°F (C_4H_{10} and C_5H_{12} lines shown).
   Boiling-point basis: Normal boiling points (pure hydrocarbons).
   Limitations: These charts apply well to all hydrocarbon series except the lowest-boiling members of each series.
3. *Heat of combustion of fuel oils and petroleum fractions*
   Source: Maxwell, *Data Book on Hydrocarbons*, p. 180.
   Description: Heats of combustion above 60°F (17,000 to 25,000 Btu/lb) vs. gravity (0 to 60°API) with correction for sulfur and inerts included (as shown on chart).
4. *Properties of petroleum fractions*
   Source: O. A. Hougen and K. M. Watson, *Chemical Process Principles Charts*, Wiley, New York, 1946, Chart 3.
   Description: °API (−10 to 90°API) vs. boiling point (100 to 1000°F) with molecular weight, critical temperature, and K factors as parameters.
   Boiling-point basis: Use cubic average boiling point when using the K values; use mean average boiling point when using the molecular weights.
5. *Heats of vaporization of hydrocarbons and petroleum fractions at 1.0 atm pressure*
   Source: Hougen and Watson, *Chemical Process Principles Chart*, Chart 68.
   Description: Heats of vaporization (60 to 180 Btu/lb) vs. mean average boiling point (100 to 1000°F) with molecular weight and API gravity as parameters.
   Boiling-point basis: Mean average boiling point.

Riazi[7] has proposed an equation to predict the basic properties of crude and products based on the equation

$$\text{value of property} = aT_B^b S^c$$

where T_B is the normal boiling point in °R and S is the specific gravity at 60°F; a, b, and c are empirical constants.

[7]M. R. Riazi, Ph.D. dissertation, Pennsylvania State University, University Park, Pa., 1980.

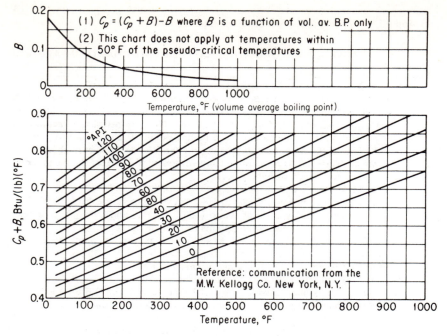

Fig. K.1 Specific heats of hydrocarbon liquids.

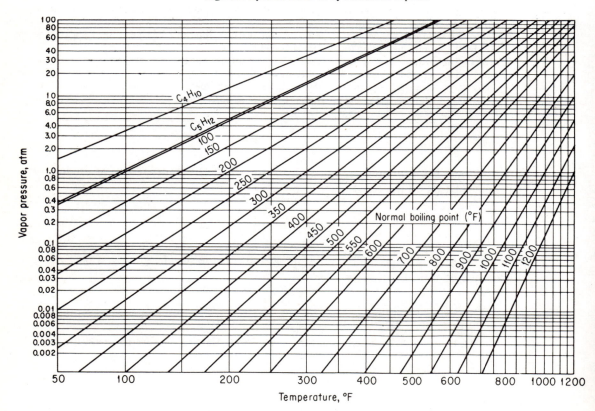

Fig. K.2 Vapor pressure of hydrocarbons.

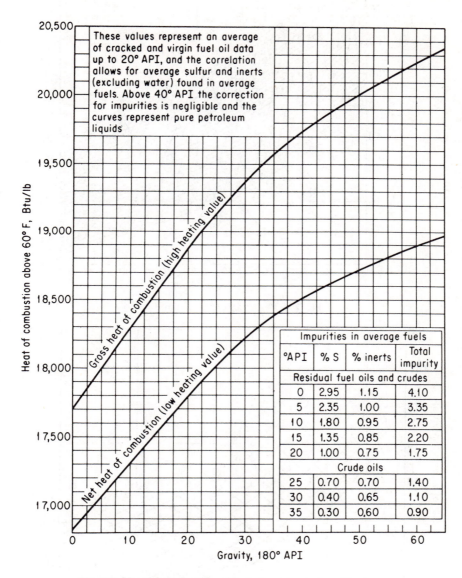

These values represent an average of cracked and virgin fuel oil data up to 20° API, and the correlation allows for average sulfur and inerts (excluding water) found in average fuels. Above 40° API the correction for impurities is negligible and the curves represent pure petroleum liquids

Impurities in average fuels			
°API	% S	% inerts	Total impurity
Residual fuel oils and crudes			
0	2.95	1.15	4.10
5	2.35	1.00	3.35
10	1.80	0.95	2.75
15	1.35	0.85	2.20
20	1.00	0.75	1.75
Crude oils			
25	0.70	0.70	1.40
30	0.40	0.65	1.10
35	0.30	0.60	0.90

Gross heat of combustion (high heating value)

Net heat of combustion (low heating value)

Fig. K.3 Heat of combustion of fuel oils and petroleum fractions.

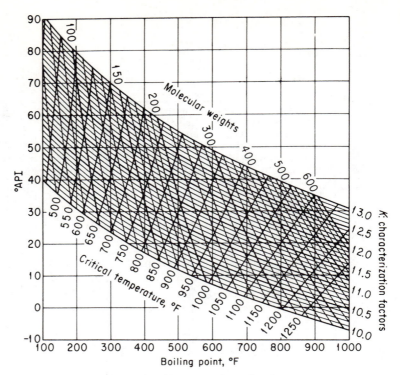

Fig. K.4 Properties of petroleum fractions.

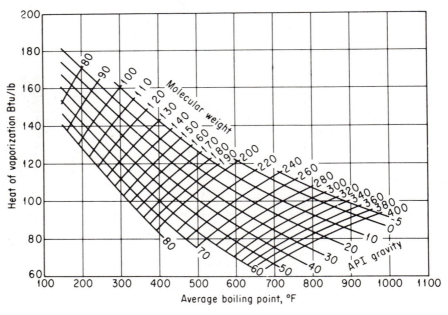

Fig. K.5 Heats of vaporization of hydrocarbons and petroleum fractions at 1.0 atm pressure.

Appendix L

SOLUTION OF SETS
OF EQUATIONS

L.1 Independent Linear Equations

If you write several *linear* material balances, say m in number, they will take the form

$$a_{11}x_1 + a_{12}x_2 + \cdots + a_{1n}x_n = b_1$$
$$a_{21}x_1 + a_{22}x_2 + \cdots + a_{2n}x_n = b_2 \qquad \text{(L.1)}$$
$$\cdots$$
$$a_{m1}x_1 + a_{m2}x_2 + \cdots + a_{mn}x_n = b_m$$

or in compact matrix notation

$$\mathbf{ax} = \mathbf{b} \qquad \text{(L.1a)}$$

where $x_1, x_2, \ldots, x_n$ represent the unknown variables, and the a_{ij} and the b_i represent the constants and known variables. As an example of Eq. (L.1), we can write the three-component mass balances corresponding to Fig. 2.6(c):

$$0.50(100) = 0.80(P) + 0.005(W)$$

$$0.40(100) = 0.05(P) + 0.925(W)$$

$$0.10(100) = 0.15(P) + 0.025(W)$$

With m equations in n unknown variables, three cases can be distinguished:

(a) There is no set of x's that satisfies Eq. (L.1).

(b) There is a unique set of x's that satisfies Eq. (L.1).

(c) There is an infinite number of sets of x's that satisfy Eq. (L.1).

Figure L.1 represents each of the three cases geometrically in two dimensions. Case 1 is usually termed inconsistent, whereas cases 2 and 3 are consistent; but to the engineer who is interested in the solution of practical problems, case 3 is as unsatisfying as case 1. Hence case 2 will be termed *determinate*, and case 3 will be termed *indeterminate*.

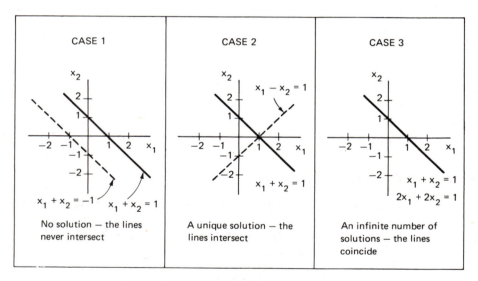

Fig. L.1 Types of solutions of linear equations.

To ensure that a system of equations represented by (L.1) has a unique solution, it is necessary to first show that (L.1) is consistent, that is, that the coefficient matrix **a** and the augmented matrix [**a, b**] must have the same rank r. Then, if $n = r$, the system (L.1) is determinate, whereas if $r < n$, as may be the case, then the number $(n - r)$ variables must be specified in some manner or determined by optimization procedures. If the equations are independent, $m = r$.

As an illustration of these ideas, consider the case of the three equations corresponding to Fig. 2.6(c), where we have more equations than unknowns ($m > n$).

The matrix [**a**, **b**] is

$$[\mathbf{a}, \mathbf{b}] = \begin{bmatrix} 0.80 & 0.005 & 50 \\ 0.50 & 0.925 & 40 \\ 0.15 & 0.025 & 10 \end{bmatrix}$$

Note how the rank of **a** (the rank of a matrix is given by the size of the largest non-zero determinant that can be formed from the matrix), the matrix composed of the first two columns, can at the most be 2, whereas the rank of [**a**, **b**] is 3. To obtain a consistent set of equations, one of the three material balances must be eliminated, leaving two equations in two unknowns, P and W, that have a unique solution ($m = r = 2$ and $n = 2$). It would probably be best to pick the two equations in which the coefficients were known with the greatest precision.

As another example, consider a set of 10 equations involving 16 unknowns ($m < n$) so that ($n - m$) = 6. Consequently, six more variables must be specified in some manner before the system of equations becomes determinate.

Next, suppose that you are interested in solving n linear independent equations in n unknown variables:

$$\left. \begin{array}{l} a_{11}x_1 + a_{12}x_2 + \cdots + a_{1n}x_n = b_1 \\ a_{21}x_1 + a_{22}x_2 + \cdots + a_{2n}x_n = b_2 \\ \qquad\qquad \cdot \ \ \cdot \ \ \cdot \\ a_{n1}x_1 + a_{n2}x_2 + \cdots + a_{nn}x_n = b_n \end{array} \right\} \ \mathbf{ax} = \mathbf{b} \qquad\qquad (\text{L.1b})$$

In general there are two ways to solve Eq. (L.1b) for $x_1, \ldots, x_n$: elimination techniques and iterative techniques. Both are easily executed by computer programs. We shall illustrate the Gauss–Jordan elimination method. The other techniques can be found in texts on matrices, linear algebra, and numerical analysis.

The essence of the Gauss–Jordan method is to transform Eq. (L.1b) into Eq. (L.2) by elementary operations on Eq. (L.1b):

$$\begin{array}{l} x_1 + 0 + \cdots + 0 = b_1' \\ 0 + x_2 + \cdots + 0 = b_2' \\ \qquad \cdot \ \ \cdot \ \ \cdot \\ 0 + 0 + \cdots + x_n = b_n' \end{array} \qquad\qquad (\text{L.2})$$

Equation (L.2) has a solution for $x_1, \ldots, x_n$ that can be obtained by inspection.

To illustrate the elementary operations that are required, consider the following set of three equations in three unknowns:

(1) $4x_1 + 2x_2 + x_3 = 15$

(2) $20x_1 + 5x_2 - 7x_3 = 0$

(3) $8x_1 - 3x_2 + 5x_3 = 24$

The augmented matrix is

$$\begin{bmatrix} 4 & 2 & 1 & 15 \\ 20 & 5 & -7 & 0 \\ 8 & -3 & 5 & 24 \end{bmatrix}$$

Take the a_{11} element as a pivot. To make it 1 and the other elements in the first column zero, carry out the following elementary operations shown in order for each row:

(a) Subtract $(\frac{20}{4})$ [Eq. (1)] from Eq. (2)

(b) Subtract $(\frac{8}{4})$ [Eq. (1)] from Eq. (3)

(c) Multiply Eq. (1) by $\frac{1}{4}$

to get

$$\begin{bmatrix} 1 & \frac{1}{2} & \frac{1}{4} & \frac{15}{4} \\ 0 & -5 & -12 & -75 \\ 0 & -7 & 3 & -6 \end{bmatrix}$$

new eq. no.

(1a)
(2a)
(3a)

Carry out the following elementary operations to make the pivot element a_{22} equal to 1 and the other elements in the second column equal to zero:

(d) Subtract $[(\frac{1}{2})/-5][$Eq. (2a)$]$ from Eq. (1a)

(e) Subtract $(-7/-5)$ [Eq. (2a)] from Eq. (3a)

(f) Multiply Eq. (2a) by $(1/-5)$

to obtain

$$\begin{bmatrix} 1 & 0 & -\frac{19}{20} & -\frac{15}{4} \\ 0 & 1 & \frac{12}{5} & 15 \\ 0 & 0 & \frac{99}{5} & 99 \end{bmatrix}$$

new eq. no.

(1b)
(2b)
(3b)

Another series of elementary operations (left for you to propose) leads to a 1 for the element a_{33} and zeros for the other two elements in the third column:

$$\begin{bmatrix} 1 & 0 & 0 & 1 \\ 0 & 1 & 0 & 3 \\ 0 & 0 & 1 & 5 \end{bmatrix}$$

The solution to the original set of equations is

$$x_1 = 1$$
$$x_2 = 3$$
$$x_3 = 5$$

as can be observed from the augmentation column.

To obtain good accuracy and avoid numerical errors, the choice of the pivot should be made by scanning all the eligible coefficients and chosing the one with the greatest magnitude for the next pivot. For example, you might choose a_{21} for the first pivot, and then find that a_{31} would be the next pivot and finally a_{22} the last pivot to give

$$\begin{bmatrix} 0 & 1 & 0 & | & 3 \\ 1 & 0 & 0 & | & 1 \\ 0 & 0 & 1 & | & 5 \end{bmatrix}$$

EXAMPLE L.1

Determine the number of components (which is the same as the number of independent material balances) for a process involving the following two competing reactions:

$$CO + 2H_2 \longrightarrow CH_3OH$$
$$CO + 3H_2 \longrightarrow CH_4 + H_2O$$

Solution

Prepare a matrix in which the rows are the atomic species for which the balances are to be made and the columns are chemical compounds entering and leaving the process. Each element in the matrix is the number of atoms in the chemical compound.

	H_2	H_2O	CO	CH_4	CH_3OH
H	2	2	0	4	4
O	0	1	1	0	1
C	0	0	1	1	1

Transform the matrix by elementary operations so that there are 1's on the main diagonal starting at the (1, 1) position, and only zeros below the main diagonal. The sum of the diagonal elements [starting at the (1, 1) position] is the *rank* of the matrix that is equivalent to the number of components. Do you get three for the matrix above?

The number of components is not always equal to the number of atomic species, as shown in the next example.

EXAMPLE L.2

Determine the number of components for a process involving the following reaction:

$$SO_3 + H_2O \longrightarrow H_2SO_4$$

Solution

Form the matrix and determine its rank.

$$
\begin{array}{c}
 \\
H \\
S \\
O
\end{array}
\begin{array}{ccc}
H_2O & SO_3 & H_2SO_4 \\
\left[\begin{array}{ccc}
2 & 0 & 2 \\
0 & 1 & 1 \\
1 & 3 & 4
\end{array}\right]
\end{array}
$$

Are you able to make the transformation to

$$
\begin{array}{c}
 \\
H \\
S \\
O
\end{array}
\begin{array}{ccc}
H_2O & SO_3 & H_2SO_4 \\
\left[\begin{array}{ccc}
1 & 0 & 1 \\
0 & 1 & 1 \\
0 & 0 & 0
\end{array}\right]
\end{array}
$$

Note that the rank of the matrix is 2, not 3; hence only two components exist for independent material balances.

To summarize, you will find the information flow diagram in Fig. L.2 helpful.

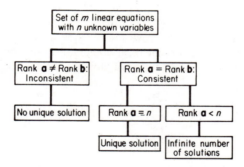

Fig. L.2 **a** is the m × n matrix of the coefficients in the equations, and **b** is the m × (n + 1) augmented matrix.

L.2 Nonlinear Independent Equations

The precise criteria used to ascertain if a linear system of equations is determinate cannot be neatly extended to nonlinear systems of equations. Furthermore, the solution of sets of nonlinear equations requires the use of computer codes that may fail to solve your problem for one or more of a variety of reasons. The problem to be solved can be written as

$$\left. \begin{array}{l} f_1(x_1, \ldots, x_n) = 0 \\ f_2(x_1, \ldots, x_n) = 0 \\ \quad \cdot \quad \cdot \quad \cdot \\ f_n(x_1, \ldots, x_n) = 0 \end{array} \right\} f(\mathbf{x}) = 0 \qquad\qquad (L.3)$$

Generalized linear methods first linearize Eqs. (L.3) at $\mathbf{x}^{(k)}$ and then iteratively solve the linear system for $\mathbf{x}^{(k+1)}$:

$$\mathbf{a}(\mathbf{x}^{(k)})(\mathbf{x}^{(k+1)} - \mathbf{x}^{(k)}) + f(\mathbf{x}^{(k)}) = 0$$

Typical methods are:

(a) Newton–Gauss–Seidel
(b) Newton-successive over relaxation
(c) Secant-successive over relaxation
(d) Secant-alternating directions

A second route of approach is to minimize the sum of the squares of the individual equations by an unconstrained minimization technique. If you choose either method, it is wise to use a prepared computer code that has been verified by many test problems so that you can have some reasonable confidence in the results provided by the computer code.

Appendix M

ANSWERS
TO SELECTED PROBLEMS

Chapter 1

1.2 (1) Five: International nautical mile (6076.1 feet); British nautical mile 6080 feet); U.S. survey nautical mile (6076.103 feet); statute mile (5280 feet); U.S. survey statute mile (5279.98 feet)

(2) U.S. oil barrel (42 gal); U.S. liquid barrel (31.5 gal); Imperial barrel (36 Imperial gal); U.S. dry barrel (26.25 gal)

(3) 55 gal (usually).

(4) 31.5 U.S. gal,

(5) The gold: 11 oz troy = 342.73 grams; 12 oz avdp = 340.18 grams

(6) Better ask

(7) The Canadian rye: British proof is 57% alcohol by volume; U.S. proof is 50%

1.6 (g): 1; (h): 2

1.7 None; 20 gal extra are required

1.11 30,000 bbl $\approx$ 1,684 $\times$ 10^5 ft^3; the slick is 2.32 $\times$ 10^6 ft^3 $\approx$ several days

1.17 1.4 $\times$ 10^3 (ft)(lb$_f$)

1.18 (b) 3.26 $\times$ 10^5 J; (c) 438 (hp)(s)

1.23 No

1.25 (1) 1.24 $\times$ 10^6; (4) 2.38 $\times$ 10^7

1.29 (a) 0.1408 lb; (b) 196.5 lb; (c) 1.8032 $\times$ 10^3 lb; (d) 3.93 $\times$ 10^3 lb

1.32 (e) 3.18 $\times$ 10^6 g; (f) 15.4 lb

1.33 (a) (2): 0.0352 lb; (b) (1): 129.8 kg

1.34 5.88 lb/ft^3

1.36   152 ft³
1.41   (a) 14,600; (b) 10,220
1.44   Weight percent: gasoline 24.7, kerosene 26.8, gas oil 28.3, isopentane 20.2
1.52   0.085 CH_4, 0.40 C_2H_6, 0.857 CO_2
1.51   28.14 lb/lb mol
1.55   (a) A 32.2%, B 27.2%, C 40.6%; (e) 0.556 kg mol
1.59   (a) 600°R, 333K, 60°C; (h) 1832°F, 2292°R, 1273K
1.62   (b) 73.3°B
1.67   1.104 × 10⁴ kPa

1.70   (a) 5.26 × 10³ lb_f/ft³; (b) 74.3 in. Hg; (c) 2.5 × 10⁵ N/m²
1.75   2.06 psia
1.79   No
1.81   3.04
1.91   (e) 2.04 g NaCl; (f) 3.44 lb NaCl; (g) 4.22 lb AgCl
1.95   5210 lb CaO
1.97   3.90 kg mol
1.103   67.1%
1.109   (a) 4.5%; (b) 3.09 kg CO_2/kg C burned
1.113   (a) 31%; (b) 77%

Chapter 2

2.4   Adequate balance
2.5   (a) unsteady state, open
2.8   2
2.10   (a) 9; (b) 7
2.13   (a) 3: F, P, W; (b) the 2 unknown compositions can be calculated; (c) No, unless a basis is selected
2.17   (a) 21% CO_2, 79% N_2; (b) 14% CO_2, 7% O_2, 79% N_2; (c) 12.5% CO_2, 1.4% CO, 7.65% O_2, 78.45% N_2
2.20 CO_2 17.8%, O_2 1.2%, N_2 81%
2.27   686%
2.29   47.8 kg
2.33   (a) 33.3%; (b) 8.0%

2.36   161 lb
2.38   (a) 3920 lb H_2O; (c) 4356 lb H_2O
2.43   (a) NH_3 55.0%, CH_4 45.0%; (b) HCN 46%, CH_4 54%
2.45   44.5%
2.53   (a) 220.5 lb; (b) 31.2%
2.57   530 kg
2.63 $53.18/ton
2.67   (a) B = 586 lb/hr; (b) D = 414 lb/hr; (c) fraction = 0.551
2.69   NaCl 62.5%, H_2O 37.5%
2.70   (a) 6.42 × 10⁵ lb CH_4/day; (b) 1.60 × 10⁵ lb; (c) 4.65%
2.75   (a) recycle = 890 kg mol, purge = 3.18 kg mol

Chapter 3

3.1   1.026 liters
3.8   390 mm Hg
3.11   27.1 kg
3.13   0.214
3.18   (a) 2790 ft³ at 60°F, 30 in. Hg gives 7.7% difference (increased cost)
3.23   3.56 ft³ air/ft³ gas
3.27   248 gal/hr
3.33   5.16 m³ gas/min
3.36 C_3H_6
3.41   1.99 × 10⁻⁴ g/cm³
3.45 p_{CO_2} = 160 kPa; p_{N_2} = 40 kPa; V_{N_2} = 0.4 m³
3.48   1520 mm Hg

3.54   524°F by van der Waals' equation; 510°F by Nelson-Obert Chart
3.58   102.8 kg by compressibility factor; 103.6 by van der Waals' eq.
3.60   (a) $13.66
3.66   (b) 71% decrease
3.68   (2): natural gas —1.23 × 10⁵ ft³ at 100°F and 175 psia
3.74   27 atm
3.77   m = −1175.5 K, p_0^* = 5.815 atm
3.79   21°C
3.85 V_{O_2} = 296 ft³ at 745 mm Hg, 25°C; $V_{C_2H_2}$ = 54 ft³ at 745 mm Hg, 25°C
3.89   3.98 kg total

3.93   (a) 51.4%; (e) 21.0%
3.97   (a) None; (b) 0.05% increase
3.100   18.75°C
3.103   (a) 750 lb H_2O; (b) 14,090 lb dry air
3.107   (a) 0.95 m³; (b) 9.06 g/m³
3.110   4.88 lb/hr

3.115   2.41 as ideal gases; 2.51 as real gases
3.118   207 lb H_2O
3.122   (a) entering water is all liquid; (b) exit water is all vapor; (c) 0.24 lb/ft³; (d) 2

Chapter 4

4.3   geothermal flux is 4.70×10^8 kW
4.7   (a) intensive; (b) extensive; (c) intensive
4.10   (a) 6.09×10^5 (ft)(lb$_f$) or 8.26×10^5 J
4.15   (a) 3.429×10^4 J/(kg mol)(°C); (c) 2.598×10^4 J/(kg mol)(°C)
4.19   (a) 2550 Btu/lb mol; (b) 8480 Btu
4.23 $C_p = 8.78 + 0.00963T - 5.29 \times 10^{-6}T^2$
4.25   (b) $C_p = 0.226 + 3.118 + 10^{-5}T°_F -2.765 \times 10^{-9}T^2_°_F$
4.31 $\Delta H = 19,580$ J; $\Delta U = 14,600$ J
4.34   Steam tables: 3.709×10^7 J; combustion gas tables: 3.695×10^7 J; C_p: 3.77×10^7 J
4.40   10.1 cal/(g mol)(°C)
4.43 -7.90×10^5 J/kg
4.48   (a) good as line is straight; $\Delta H_v = 8640$ Btu/lb mol; (b) 9110 Btu/lb mol
4.54   (a$_1$) 183 J/g mol; (a$_2$) 660 J/g mol; (a$_3$)477 J/g mol
4.58   (a) 30,000 (ft)(lb$_f$); (b) $\Delta P = 15,000$ (ft)(lb$_f$); $\Delta K = 10,000$ (ft)(lb$_f$)
4.60   (b) 500 kg/s, 1.84×10^5 J/kg; (c) 24.4 J/kg, 2.2 kg/s
4.63   22.1 lb

4.68   (d) for CO_2 $W = 3260$ Btu/lb mol; work would be less for isothermal compression
4.72   loss = 8.75×10^8 Btu/hr
4.74   (a) steady state ($\Delta E = 0$), Q may or may not be 0, $W = 0$, $\Delta H \neq 0$, $\Delta K = 0$, $\Delta P = 0$, hence $Q = \Delta H$
4.80 2.63×10^9 J/hr lost
4.85   No (1.31×10^6 J/s)
4.86   (a) 0.37 gas; (b) 98 Btu/lb
4.89   324 hp
4.97   (a) $-64,300$ cal/g mol FeO; (b) $-24,617$ cal/g mol FeO
4.99   (a) 49.271 kcal; (b) -9.83 kcal; (c) 17.889 kcal
4.101   (a) -0.681 kcal; (b) -24.4252 kcal; (c) 43,88 k cal; (i) -26.76 k cal
4.105 -3.5166×10^6 J/g mol
4.112   141 Btu/ft³
4.114   16,763 Btu/lb mol
4.120   242.61 kJ/g mol H_2
4.122   No (1253°F)
4.130   (a) $C_6H_5CH_3$: 9.5%, O_2: 19.1%, N_2: 71.4%; (b) $C_2H_5CH_3$: 8.20%, CO_2: 0.33%, H_2O: 1.42%, C_2H_5CHO: 1.23%, O_2: 17.30%, N_2 residual; (c) 335 gal/hr
4.132   276%
4.137   (b) 168°F

Chapter 5

5.1   20.2 kg $CaCO_3$/kg CH_4
5.3   (b) 2800 lb/hr
5.7   C + 4
5.9   5
5.13   9140 lb/hr
5.18   (a) C = 450 lb, $H_c = -271$ Btu/lb, $x_c = 50$ wt. % NH_3

5.20   (a) (1): A + B = C = 100; 0.8A + 0.1B = 0.6C = 60; AH$_a$ + 27B = (100)(800); B = 100 − A; A = 71.5 lb; B = 28.5 lb
5.22   (a) 0.079, (b) 87.1 kPa
5.27   (a) $\mathcal{H} = 0.01813$ lb H_2O/lb air; (b) 0.915 lb air/lb H_2O

5.30   (a) 1.96×10^6 ft³ at 70°F, 60% H per hour; (b) 3.04×10^4 ft³ at 100°F, 95% H per hour

5.34   (a) 3.83×10^8 J removed; (b) 29°C

5.36   (a) 49°C; (b) Tower 1: $6.29 + 10^7$ Btu

Chapter 6

6.2   230.5 min
6.8   34,700 lb
6.11   121.5 s

6.17   moles $C_6H_{12} = e^{-kt}$
6.21   28 min
6.24   (b) 129°F for half-full

NOMENCLATURE*

a, b, c = constants in heat capacity equation

a = constant in general

a = constant in van der Waals' equation, Eq. (3.11)

a = acceleration

$\mathbf{a}$ = coefficient matrix

A = constant in Eq. (3.40)

A = area

$°API$ = specific gravity of oil defined in Eq. (1.9)

aS = absolute saturation

b = constant in van der Waals' equation, Eq. (3.11)

B = constant in Eq. (3.40)

$\dot{B}$ = rate of energy transfer ccompanying $\dot{w}$

(c) = crystalline

C = constant in Eq. (1.1)

C = number of chemical components in the phase rule

C_p = heat capacity at constant pressure

C_{p_m} = mean heat capacity

C_v = heat capacity at constant volume

C_S = humid heat defined by Eq. (5.18)

D = distillate product

E = total energy in system = $U + K + P$

E_v = irreversible conversion of mechanical energy to internal energy

F = force

F = number of degrees of freedom in the phase rule

F = feed stream

(g) = gas

g = acceleration due to gravity

g_c = conversion factor of $\dfrac{32.174 \text{ (ft)}(\text{lb}_m)}{(\text{sec}^2)(\text{lb}_f)}$

h = distance above reference plane

h_c = heat transfer coefficient in Eq. (5.24)

$\hat{H}$ = enthalpy per unit mass or mole

H = enthalpy, with appropriate subscripts, relative to a reference enthalpy

ΔH = enthalpy change, with appropriate subscripts

$\Delta \hat{H}$ = enthalpy change per unit mass or mole

ΔH_c° = standard heat of combustion

ΔH_f° = standard heat of formation

ΔH_{rxn} = heat of reaction

ΔH_{soln} = heat of solution

ΔH_0 = constant defined in Eq. (4.50)

*Units are discussed in the text.

$\mathcal{3C}$ = humidity, lb water vapor/lb dry air
k_g' = mass transfer coefficient in Eq. (5.24)
K = kinetic energy
K = degree kelvin
(l) = liquid
l = distance
m = mass of material
m = number of equations
$\dot{m}$ = rate of mass transport through defined surfaces
mol. wt. = molecular weight
n = number of moles
n = number of unknown variables
N_d = degrees of freedom
N_p = number of equipment parameters
N_Q = number of heat transfer variables
N_r = number of independent constraints
N_s = number of streams
N_{sp} = number of chemical species
N_v = total number of variables
N_w = number of work variables
p = pressure
p = partial pressure (with a suitable subscript for the p)
p^* = vapor pressure
$\mathcal{P}$ = number of phases in the phase rule
p_c = critical pressure
p_c' = pseudocritical pressure
p_r = reduced pressure = p/p_c
p_r' = pseudoreduced pressure
p_t = total pressure in a system
P = product
P = potential energy
Q = heat transferred
Q_p = heat evolved in a constant pressure process
Q_v = heat evolved in a constant volume process
$\dot{Q}$ = rate of heat transferred (per unit time)
r = rank of a matrix in Chap. 2
$\dot{r}_A$ = rate of generation or consumption of component A (by chemical reaction)
R = universal gas constant
R = recycle stream
$\mathcal{R3C}$ = relative humidity
$\mathcal{RS}$ = relative saturation
(s) = solid
sp gr = specific gravity
S = cross-sectional area perpendicular to material flow

$$t = \text{time}$$
$$t = \text{temperature in C}° \text{ or F}°$$
$$T_{DB} = \text{dry-bulb temperature}$$
$$T_{WB} = \text{wet-bulb temperature}$$
$$T = \text{absolute temperature or temperature in general}$$
$$T_b = \text{normal boiling point (in K or }°\text{R)}$$
$$T_c = \text{critical temperature (absolute)}$$
$$T_c' = \text{pseudocritical temperature}$$
$$T_f = \text{melting point (in K)}$$
$$T_r = \text{reduced temperature} = T/T_c$$
$$T_r' = \text{pseudoreduced temperature}$$
$$U = \text{internal energy}$$
$$v = \text{velocity}$$
$$V = \text{system volume or fluid volume in general}$$
$$\hat{V} = \text{specific volume (volume per unit mass or mole)}$$
$$\hat{V} = \text{humid volume defined in Eq. (5.20)}$$
$$V_c = \text{critical volume}$$
$$\hat{V}_{ci} = \text{ideal critical volume} = RT_c/p_c$$
$$\hat{V}_{ci} = \text{pseudocritical ideal volume}$$
$$V_g = \text{volume of gas}$$
$$V_l = \text{volume of liquid}$$
$$V_r = \text{reduced volume} = V/V_c$$
$$V_{ri} = \text{ideal reduced volume} = \hat{V}/\hat{V}_{ci}$$
$$\dot{w}_A, \dot{w} = \text{rate of mass flow of component } A \text{ and total mass flow, respectively, through system boundary other than a defined surface}$$
$$W = \text{work done by the system}$$
$$W = \text{waste stream}$$
$$\dot{W} = \text{rate of work done by system (per unit time)}$$
$$x = \text{unknown variable}$$
$$x = \text{mass or mole fraction in general}$$
$$x = \text{mass or mole fraction in the liquid phase for two-phase systems}$$
$$y = \text{mass or mole fraction in the vapor phase for two-phase systems}$$
$$z = \text{compressibility factor}$$
$$z_c = \text{critical compressibility factor}$$
$$z_c' = \text{pseudocritical compressibility factor}$$
$$z_m = \text{mean compressibility factor}$$

Greek letters:

$$\alpha, \beta, \gamma = \text{constants in heat capacity equation}$$
$$\gamma = \text{constant in Eq. (3.18)}$$
$$\Delta = \text{difference between exit and entering stream; also used for final minus initial times or small time increments}$$
$$\lambda = \text{molal heat of vaporization}$$
$$\lambda_f = \text{molal latent heat of fusion at the melting point}$$

ρ = density

$\rho_A,\ \rho$ = mass of component, A, or total mass, respectively, per unit volume

ρ_L = liquid density

ρ_V = vapor density

ω = mass fraction

Subscripts:

A, B = components in a mixture

c = critical

i = any component

i = ideal state

t = at constant temperature

t = total

r = reduced state

$1, 2$ = system boundaries

Superscript:

$\wedge$ = per unit mass or per mole

$\cdot$ = per unit time (a rate)

INDEX